Springer Climate

Series editor

John Dodson, Menai, Australia

Springer Climate is an interdisciplinary book series dedicated to all climate research. This includes climatology, climate change impacts, climate change management, climate change policy, regional climate, climate monitoring and modelling, palaeoclimatology etc. The series hosts high quality research monographs and edited volumes on Climate, and is crucial reading material for researchers and students in the field, but also policy makers, and industries dealing with climatic issues. Springer Climate books are all peer-reviewed by specialists (see Editorial Advisory board). If you wish to submit a book project to this series, please contact your publisher (elodie.tronche@springer.com).

More information about this series at
http://www.springer.com/series/11741

Karl W. Steininger • Martin König •
Birgit Bednar-Friedl • Lukas Kranzl •
Wolfgang Loibl • Franz Prettenthaler
Editors

Economic Evaluation of Climate Change Impacts

Development of a Cross-Sectoral Framework and Results for Austria

Springer

Editors
Karl W. Steininger
University of Graz
Graz, Austria

Birgit Bednar-Friedl
University of Graz
Graz, Austria

Wolfgang Loibl
Austrian Institute of Technology
Vienna, Austria

Martin König
Environment Agency Austria
Vienna, Austria

Lukas Kranzl
Vienna University of Technology
Vienna, Austria

Franz Prettenthaler
Institute for Economic and Innovation Research
Graz, Austria

Additional material to this book can be downloaded from http://extras.springer.com

ISSN 2352-0698 ISSN 2352-0701 (electronic)
ISBN 978-3-319-12456-8 ISBN 978-3-319-12457-5 (eBook)
DOI 10.1007/978-3-319-12457-5
Springer Cham Heidelberg New York Dordrecht London

Library of Congress Control Number: 2014959279

The results reported in this book have been acquired in a project financially supported by the Austrian Climate and Energy Fund (COIN project, carried out within the Austrian Climate Research Programme, ACRP).

Printed on acid-free paper

Springer International Publishing AG Switzerland is part of Springer Science+Business Media (www.springer.com)

Preface

Our current actions determine both our own well-being and that of future generations. Beyond climate change, there are very few areas where the lag between action and potential impact is that long and where the risk of delaying an appropriate response may entail impacts of such enormous magnitude. There is thus a clear need for adequate information for society on climate change and its impacts.

Climate change impacts are multifaceted, interdependent and characterised by a high degree of uncertainty. Their analysis thus necessitates collaboration across a broad set of disciplines and expertise, and entails devising appropriate scenarios.

In this volume, we show how, at the national scale, relevant societal information on climate change impacts can be generated. Here, particular emphasis is placed on the generation of information related to economic evaluation and economic implications of climate change. A tool box enabling consistent analysis across the many fields of climate impact is developed and then applied to one particular country, Austria.

Climate scenario analysis to date indicates that the expected mean values associated with climate change damage are increasing. However, as such mean values are surrounded by a considerable amount of uncertainty, it is also crucial to consider the potential range of damage that might occur (e.g. potential higher and lower damage values). Identification of such ranges is also useful in that it helps clarify that several different types of response can and will be relevant. The paths of socio-economic development taken by our societies not only determine the extent of greenhouse gas emission mitigation, but of at least equal importance, they also determine how resilient society and its individuals will be to a changing climate (and whether they will be in a position to actively implement robust countermeasures in response). For example, questions such as whether we continue to construct infrastructure in flood prone areas, whether sealing and urban sprawl continue to increase the urban heat island effect, whether institutions such as hospitals or old people's homes are equipped to deal with heat waves (particularly in the light of expected future ageing), whether energy services of all types (heating and cooling, transport, production processes) will remain affordable—all of these questions are crucial in determining a society's level of resilience in a changing climate. Given the potential impact of 'tipping elements' at various levels, and the potential damage that can be caused by extreme events, it becomes clear that adequate risk management needs to consider also substantial mitigation policies.

This exercise was undertaken by a team comprising 18 research groups and scientific institutions. With a time span of less than one and a half years between kick-off and submission to print, including three review cycles, a substantial degree of discipline and collaboration was necessary, particularly given the high level of interdependency across the modelling approaches. It was a pleasure to be part of an endeavour where each and every scientist did his or her best to ensure the achievement of a collaborative common result.

The project was fostered greatly by its Scientific Advisory Board, comprising Paul Watkiss, Roger Street and Reimund Schwarze, who reacted immediately to all our manifold requests and reviewed the full manuscripts thoroughly. Their advice was extremely helpful. In addition, via a process of continuous consultation throughout, they wisely directed the project to its successful completion. It is our wish that every project may have such a supportive advisory board.

Günther Liebel, Helmut Hojesky and Barbara Kronberger-Kießwetter of the Austrian Ministry of Agriculture, Forestry, Environment and Water Management, as well as Jose Delgado and Tobias Orischnig of the Ministry of Finance, supplied crucial feedbacks on practical applicability of results and information demands throughout the project in a very constructive and supportive way, for which the team thanks indeed.

Matthias Themessl, at the Service Centre of the Climate Change Centre Austria (CCCA), exercised much care and thought in organising and directing the international review process. We thank all 39 reviewers for their valuable contributions. These were a great help in improving the respective chapters. Two anonymous reviewers from the publisher, Springer, then reviewed the entire volume resulting in significant improvements.

Administration of the project was substantially supported by Karin Eisner at Wegener Center, the native speaker checks were carried out by a multitude of experts, but for a substantial number of chapters we would like to express particular thanks to Laurie Conway. The uniform and attractive layout of the chapters and of the supplementary materials is owed to the careful work of Michael Kriechbaum. Matthias Themessl, Angelika Wolf and Michael Kriechbaum produced further dissemination material.

Finally, it was a pleasure to work in the production process with Barbara Feß, Johannes Glaeser and their team at Springer.

We wish to express our thanks to all of them. May the impact of this volume be seen as a reward to all.

Graz, Vienna
August 2014

Karl W. Steininger
Martin König
Birgit Bednar-Friedl
Lukas Kranzl
Wolfgang Loibl
Franz Prettenthaler

Executive Summary

The infrared absorption capacity of greenhouse gases is inducing a warming of the earth's atmosphere. Already in 1979 the World Meteorological Organization found "that it is now urgently necessary for the nations of the world: [. . .] to foresee and to prevent potential man-made changes in climate that might be adverse to the well-being of humanity," and that "it is possible that some effects on a regional and global scale may be detectable before the end of this century and become significant before the middle of the next century" (WMO 1979).

In various assessment reports published since 1990, and most recently in 2013/2014, the Intergovernmental Panel on Climate Change (IPCC) has confirmed the findings presented in the scientific literature that climate change has led to a global mean temperature increase of almost 1 °C since 1880 and that it is predominantly caused by human activities (IPCC 2013, 2014). The IPCC also reports that, left unabated, future emissions will lead to a temperature increase by the end of the twenty-first century of 3.2–5.4 °C. Even the most ambitious mitigation scenarios could potentially lead to dangerous climate change; i.e. even if global average warming is limited to 2 °C relative to pre-industrial levels (the current international goal agreed, noting that this is unlikely to be met). For most regions, particularly land-locked mountainous and continental climate zones, this implies a more substantial increase, e.g. a 4.5–6.6 °C increase by 2100 is projected for the Alpine region and thus for a country such as Austria (Jacob et al. 2013).

Due to the inertia of the climate system, societies are thus confronted with the need to adapt to climate change and—in order to avoid a further increase that gets increasingly unmanageable in the future—the need to engage in attempts to agree on and implement greenhouse gas emission mitigation policies. For both types of decisions, adaptation and mitigation, well-informed decision making requires knowledge on the type and magnitude of climate change impacts expected and on the type of information available and deducible.

During the last two decades a rich body of literature has thus developed on climate change impacts, with results put into perspective most recently in IPCC (2014). In this literature two strands can be distinguished. One is employing aggregated impact functions, within so-called Integrated Assessment Models, which have been applied mainly at the global level in order to quantify the social costs of carbon (the additional damage of an extra ton of greenhouse gas (GHG) emitted). A second strand builds upon physical impact assessments, often extended by related economic valuation.

All of these studies indicate the high demand for evaluations at the national and sub-national level, as this is where climate change materialises and where administration and governance of adaptation takes place. This also lends force to the IPCC's demands for disaggregated studies and scenarios capable of allowing for more appropriate impact assessment at the national to local level. To date, however, studies at the national and sub-national level have tended to focus solely on a few selected fields of impact (i.e. on those considered the most important).

This clear gap in the literature provides the motivation for the present volume. The objective here is to cover *as broad a* field of impacts as possible at the national level within a single comprehensive cost evaluation. To create such information at the national level, the present volume presents (a) a toolbox for deriving future climate impacts and arriving at related monetary quantification at the sectoral level, (b) the means for doing so consistently across all fields of impact, (c) a framework for impact integration in terms of a consistent macroeconomic framework in order to quantify economic feedback effects, (d) an approach for dealing with non-market impacts, e.g. impacts related to human health and biodiversity and (e) appropriate methods for considering extreme events and their 'fat tail' distribution.

Methodologically speaking, the approach presented combines a scenario-based impact assessment across all fields of impact, a computable general equilibrium (CGE) analysis so as to capture cross-sectoral linkages and economy-wide effects, and a qualitative analysis to capture additional non-market effects where monetisation is not considered appropriate.

The volume first gives an overview of climate costs at the European level. Impacts are found to amount to several percentage points of GDP by the end of the century, and are characterised by large differences in the patterns of impacts across Europe. For example, due to a combination of enhanced climate signal and higher local vulnerability, there are more negative impacts in South-Eastern Europe and the Mediterranean area.

In general, available national assessments of climate change risks and adaptation planning follow one of two approaches, i.e. either the use of top-down global Integrated Assessment Models (IAMs) which are then downscaled to reflect the national or regional scale or the use of bottom-up sectoral impact assessments which are scaled up to capture the regional or national level. On comparing the national evaluations undertaken in the UK, France, Germany and Switzerland, it becomes clear that the approach presented in this volume can indeed generate complementary information. Specifically, the new approach is helpful in the following three important areas: (a) it explicitly considers uncertainties through high impact case narratives (i.e. damage-enhancing socio-economic developments and high-damage climate change scenarios), (b) it applies consistent socio-economic scenarios and shared policy assumptions across various sectors and (c) it advances the state of the art with respect to the assessment of cross-sectoral, indirect and macroeconomic effects.

The national scale evaluation approach is designed specifically to deal with the following issues:

- Provision of a consistent overall framework
- Derivation of local indicators from climate model ensembles
- Development of shared socio-economic pathways necessary to ensure consistency across sectoral evaluations
- Creation of a toolbox for economic impact evaluation ensuring consistent evaluation
- Development of the macroeconomic modelling framework
- Macroeconomic integration of sectoral impacts while taking sufficient account of feedback effects.

We consider the methodological approach as comprehensive regarding the fields of impact and the relevant aspects of climate change costs. However, we are aware that the quantification of costs has to leave many open questions and relevant impacts which could not be quantified in this work.

In order to exemplify its use, the set of tools is applied to a single country, i.e. Austria. The following results were derived:

With respect to observed welfare damage of climate- and weather-induced extreme events in Austria, insurance data reveal annual average sums of 97 million euros (M€) in the 1980s, 129 million euros in the 1990s, and 705 million euros in the last decade. However, these figures are covering large events (catastrophes of class 5 and 6) only, and are of incomplete coverage even for this subcategory pre-2002. In the past, the most significant damage at the national scale in Austria was related to riverine flooding, valued at 3.5 billion euros in 2002 and 2.3 billion euros in 2013 (which amounted to 1.4 % and 0.7 % of GDP, respectively; all monetary values given in this summary are at prices of 2010). Non-market impacts of premature heat-related deaths can be evaluated at a current annual average 150–390 million euros. Thus, the current welfare damage of climate and weather induced extreme events in Austria is an annual average of about 1 billion euros (large events only).

We find that this has the potential to rise to 4–5 billion euros by mid-century (annual average, known knowns of impact chains only, undiscounted), with an uncertainty range of 4–9 billion euros. When extreme events and the tails of their distribution are included, even for a partial analysis focused on extremes, damages are seen to rise significantly, e.g. with an estimated increase to 40 billion euros due to riverine flooding events alone by the end of the century. These highlight the need to consider the distribution of impacts, as well as the central values.

In contrast, traditional economic measurements, such as those assessing climate change impacts on GDP, provide, at best, only a partial picture. For example, GDP losses do not account for losses in stocks (e.g. buildings) due to climate change events.

For the case of Austria, the following climate impacts were identified in detail by impact field:

Agriculture Potential average yields increase at least until the middle of the century (mostly due to a lengthening of the vegetation period as a result of higher temperatures, rather stable annual precipitation sums, and the CO_2 fertilisation effect). However, several factors are at work which tend to offset (partially or fully) such an increase, e.g. disruption caused by extreme weather events or periods, higher investment costs or / and disruptions in the functioning of ecological systems (e.g. in the effectiveness of insect pollination and biological pest control). Furthermore, the agricultural sector itself is less likely to benefit from the (uncertain) potential increase in yield than are the food and retail sectors.

Forestry In mountain forests longer vegetation periods result in increased productivity, while at low elevations in the east and in the south of Austria, drought will negatively impact on forest growth. Assuming no suitable adaptation measures are taken, increases in bark beetle infestations and possibly also storms are likely to result in yield reductions. In addition, the investment needed to maintain protection functionality against gravitational hazards in spite of losses of protective forest cover is higher than that needed to compensate for productivity loss alone.

Ecosystem Services Climate plays a major role here. Researchers have only just begun to derive the specific threshold values at which ecosystem services start to decline. In economic terms, the pollutant buffer capacities of soil and vegetation, erosion protection and the provision of drinking water are all extremely significant ecosystem services. In our assessment reported here, the only agricultural services that were investigated were insect pollination and biological pest control, and the results were taken into consideration as explained earlier for agriculture.

Human Health More intensive and frequent heat waves raise the number of deaths in the growing share of the elderly (leading, under the mid-range assumptions, to an additional 1,000 annual deaths in the period 2036–2065). In more extreme years, where the group of those vulnerable is extended to include the chronically ill, health impacts may be as much as six times higher than those found under the mid-range assumptions (more than two times higher than under the high range assumptions).

Water Supply and Sanitation By mid-century, the already high level of investment required for dealing with socio-economic development will be at least 10 % higher due to climate change implications. However, as is the case for all impact fields, but of particular importance here, only a subset of impact chains was quantified.

Catastrophe Management Already today, riverine flooding is one of the economically most important weather and climate risks in Austria. There is thus a clear need for catastrophe management, especially in terms of reducing vulnerability. However, as extreme weather events are, by their very nature, outliers, the uncertainties with respect to forecasting the flood risk for the future climate remain quite high. While for the period 1981–2010 the average annual figure for flood cost damage was about 200 million euros; forecasts for the period 2036–2065 arrive at a corresponding average annual cost figure of between 400 and 1,800 million euros. Estimates for flood events with a recurrence time of 100 years show that, as a result of climate change and increases in wealth, the cost of flood damage in the period 2036–2065 is likely to be twice as high as in 1977–2006. Such flood events would result in damage of between 5 and 7 billion euros.

Transport Even today, damage to transport infrastructure, primarily resulting from landslides or from road and rail undercutting (or washouts) caused by heavy precipitation, is already considerable (amounting to 18 million euros p.a. for road infrastructure). The extent of future damage depends directly on how traffic networks develop. Network exposure depends on the nature of network extensions. Local aspects need to be considered (e.g. geological conditions determining landslide potential, slope gradients, the risk of damage through undercutting (washouts) or wind). Depending on the duration of the disruption and on the availability of alternative routes, the indirect impact of traffic disruptions (losses in production and time) may easily exceed the direct costs of repair.

Buildings and Energy With respect to the energy needs of buildings, it was found that by the middle of the century, the savings in fossil fuel energy in the winter period more than offset the additional energy demand for cooling needed in the hot season. One potentially critical aspect, however, is the growing peak load for cooling and the discrepancy between electricity production capacity (which in Austria is based to a large extent on hydroelectric generation) and the increasing demand for cooling energy in the summer period. Higher peak demand occurs at the same time as summer drought imposes limits on traditional production, with excess demand for electricity needing to be met either by increased imports or by extending plant capacity (with quite a potential for photovoltaic electricity). Increased importing of electricity (particularly from southern European countries) not only places a higher burden on the grid network, there may be also an increase in the risk of widespread power failure and blackouts.

Manufacturing and Trade The impacts of climate change in this sector are diverse and branch-specific, and range from the need for adjustments in cooling and cooling chains, on to the impact of extreme weather events on transport networks and their related essential services. A uniform assessment of the losses in labour productivity arising from more frequent heat waves was undertaken for all branches in manufacturing and trade. By the middle of the century, the annual cost of such losses, alone in manufacturing and trade, amounts to up to 140 million euros.

Urban Green Climate change will result in even more pronounced urban heat islands. The normal cooling effect caused by vegetation is lost in the presence of sealed surfaces and buildings and will be further accelerated by additional city growth. As a result, urban areas are a few degrees warmer than their surroundings. The impact of future climate change could be limited by additional investments in green and blue infrastructure to maintain their thermal comfort service.

Tourism While rising temperatures and lower precipitation benefit summer tourism, they are detrimental to winter tourism (in its present form). In the mid-range climate scenario, by the middle of the century, the loss in winter overnight stays is expected to exceed the gain in summer overnight stays by 1.5 million. This net loss alone results in average annual costs of 300 million euros. Related macroeconomic effects lead to further costs (and magnify cost by 60 % over direct sector cost), as do changes in the sector's cost structure (e.g. increased costs for artificial snow, air conditioning, water supply, etc.) and the impact of extreme weather events.

References

IPCC (2013) Climate Change 2013: The Physical Science Basis, Working Group I Contribution to the Fifth Assessment Report of the Intergovernmental Panel on Climate Change. Oxford University Press, Oxford

IPCC (2014) Climate Change 2014: Impacts, Adaptation, and Vulnerability, Working Group II Contribution to the Fifth Assessment Report of the Intergovernmental Panel on Climate Change. Oxford University Press, Oxford

Jacob D et al (2013) EURO-CORDEX: new high-resolution climate change projections for European impact research. Reg Environ Change. doi:10.1007/s10113-013-0499-2

WMO (1979) Declaration of the World Climate Conference. World Meteorological Organization, Geneva

Testimonials

"This study is a landmark, setting a new standard for the assessment of the impacts of climate change. It stands out for the comprehensiveness of its coverage of potential impacts across different sectors of the economy. Beyond that, it innovates in three important ways. First, it clearly delineates the current vulnerability to climate (the current "stock" of climate and weather induced damages) before going on to identify the additional impacts expected to occur with future global warming. Second, it makes a serious effort to consider the "fat tail" of climate impacts, which is central to the debate on climate policy when this is viewed—as it should be—as an exercise in risk management. Third, unlike the recent US national climate assessment, it characterises the effects of climate change not just in physical, biological and social terms but also in terms of economic endpoints. This is a model for how a national assessment should be conducted!"

Michael Hanemann, Professor of Economics, Arizona State University and Professor of the Graduate School, University of California, Berkeley

"Climate change is a defining issue of our time. It triggers a broad set of impacts with significant interactions within the economy and broader society. Economic impact evaluation is of crucial importance to plan society's response. This volume develops a consistent, bottom-up approach for such an evaluation across the whole range of impact fields, acknowledging their macroeconomic feedbacks and budgetary implications. The applications are exemplified with data for Austria but this book provides core insights that could and should be applied to other countries to support appropriate societal decisions."

Thomas Sterner, Professor of Economics, University of Gothenburg

"This volume provides an essential methodological element for climate impact evaluation and the application and sharing of lessons learnt adds to the potential for transferability to other settings—both critical to stimulating action. It provides credible evidence and demonstrates the scale of the problem. The lasting value of this book will come from the methodology with its frameworks, consistent toolbox and comprehensive integration, as well as the lessons learnt and shared, exemplified through application in Austria. For this Alpine country unmitigated weather and climate induced net damages are shown to increase by mid-century at least four to eight-fold, with tail events raising damages even an order of magnitude higher."

Roger Street, Director of UK Climate Impacts Programme, University of Oxford

Contents

Part III Fields of Impact

Chapter 1
Introduction

Karl W. Steininger

Research on human-induced climate change has a long history. The Swedish scientist Svante Arrhenius quantified the impact of the infrared absorption capacity of the greenhouse gas CO_2 as early as 1896. He pointed out that cutting its concentration in the earth's atmosphere by half would produce an ice age, while doubling the concentration would result in a warming of 5–6 °C (Arrhenius 1896). After almost a century of further scientific analysis, the US National Academy of Sciences was asked by the US government administration to assess the scientific basis concerning the projection of possible future climate change resulting from anthropogenic carbon dioxide emissions. The respective report (Charney et al. 1979) found that a doubling of the earth's atmospheric CO_2 concentration was associated with a temperature increase of 1.5–4.5 °C, an assessment that has been repeatedly reconfirmed since. The report also concluded, that "it appears that the warming will eventually occur, and the associated regional climatic changes so important to the assessment of socioeconomic consequences may well be significant, but unfortunately the latter cannot yet be adequately projected" (Charney et al. 1979, p. 3).

In the same year, 1979, the World Climate Conference of the World Meteorological Organization found "that it is now urgently necessary for the nations of the world: [. . .] to foresee and to prevent potential man-made changes in climate that might be adverse to the well-being of humanity". It also concluded that "it is possible that some effects on a regional and global scale may be detectable before the end of this century and become significant before the middle of the next century" (both: WMO 1979).

A vast body of scientific literature, rigorously compiled by the Intergovernmental Panel on Climate Change (IPCC) in its Assessment Reports since 1990, has confirmed that climate change is taking place with global mean temperature increase of almost 1 °C since 1880, and that it is predominantly caused by human activities (IPCC 2013, 2014). The IPCC also reports that if left unabated, future emissions will lead to a temperature increase by the end of the twenty-first century of 3.2–5.4 °C. Even the most ambitious mitigation scenarios could potentially lead to dangerous climate change; i.e. even if global average warming is limited to 2 °C

K.W. Steininger (✉)
Wegener Center for Climate and Global Change, University of Graz, Graz, Austria

Institute of Economics, University of Graz, Graz, Austria
e-mail: karl.steininger@uni-graz.at

K.W. Steininger et al. (eds.), *Economic Evaluation of Climate Change Impacts*,
Springer Climate, DOI 10.1007/978-3-319-12457-5_1

relative to pre-industrial levels [the current international goal agreed (UNFCCC 2010)], noting that this is unlikely to be met). Given that surface air temperature above oceans will warm by less than the global average, many regions, particular land-bound mountainous and continental climate zones, will face more substantial increases; e.g. a 4.5–6.6 °C increase by 2100 is projected for the Alpine region and thus for a country such as Austria (Jacob et al. 2013, APCC 2014)[1]. Societies are thus confronted with the need to adapt to climate change—both to that already triggered by past emissions, as well as to that expected as a result of future emissions—and to weigh the need for adaptation against the need to avoid such risks in the first place, i.e. to agree on and implement greenhouse gas emission mitigation policies. While adaptation policy is mainly addressed at the national and regional level [see, for example, the EU white paper "Adapting to climate change: Towards a European framework for action" (EU Commission 2009)], mitigation obviously entails an additional, stronger, global harmonisation component, as has been addressed to date within the United Nations Framework Convention on Climate Change and its Conferences of Parties. For both types of decisions, adaptation and mitigation, well-informed decision making requires knowledge on the type and magnitude of climate change impacts expected, and on the type of information (potentially and actually) available.

During the last two decades a rich body of literature has thus developed on climate change impacts. Most recently, IPCC (2014) puts these results into perspective, and for the first time in its assessment reports devoted a separate sub-volume to the detailed assessment of impacts on the continental and regional-to-local scale. As the translation of such impacts into a uniform scale of monetary values is often considered helpful for decision making, attempts at evaluation of climate impact cost (or 'damage') have also gained momentum. In one strand of the literature, these have been put forward using aggregated impact functions, within so-called Integrated Assessment Models, which have been applied mainly at the global level in order to quantify the social costs of carbon (the additional damage of an extra ton of greenhouse gas emitted). This approach has recently been questioned on several grounds, including the use of highly simplified and thus somewhat arbitrary economic damage functions (Pindyck 2013).[2]

---

[1] This range for the Alpine region refers to the "likely" range, i.e. the 17–83 ‰. To be fully comparable with the global temperature range given by IPCC, which refers to the 5–95 ‰, the range for the Alpine region would be larger.

[2] The three most often applied Integrated Assessment Models (IAMs) to date are DICE (Dynamic Integrated Climate and Economy), PAGE (Policy Analysis of the Greenhouse Effect), and FUND (Climate Framework for Uncertainty, Negotiation, and Distribution), with model descriptions given by Nordhaus (1991, 2011) and Hope (2006)—on which the Stern review is based (Stern 2007)—and Tol 2002a, b, respectively. They are used to provide total net present values for future damage over time and to estimate the marginal social costs of carbon (the damage cost of an extra tonne of GHG emissions). Their use to this end has been questioned, most importantly for arbitrary parameter choice in social welfare functions, ill-founded climate sensitivity (the temperature increase a GHG doubling implies), arbitrary and non-empirical based climate damage functions (usually a functional relationship between temperature increase and (regional) GDP loss, for

A second strand of the literature builds upon physical impact assessments and related economic valuation. Metroeconomica (2004) for the UK is an example at the national level, as are the three large European evaluations at the continental scale: Aaheim et al. (2012), Bosello et al. (2011) and Ciscar et al. (2011, 2012). International institutions have built on this second strand in their evaluations, e.g. the World Bank (2013) for its focus at developing regions, and the EEA (2008, 2012) for Europe. An alternative, but related, third approach is to map impacts and vulnerability, but refrain from monetisation (e.g. ESPON Climate 2013).

All of these studies indicate the high demand for evaluations at the national and sub-national level, as this is where climate change materialises and where administration and governance of adaptation takes place. This also lends force to the IPCC's demands for disaggregated studies and scenarios capable of allowing for more appropriate impact assessment at the national to local level.

To date, however, studies at the national and sub-national level have tended to focus solely on a few selected fields of impact (i.e. on those considered the most important).

This clear gap in the literature provides the motivation for the present volume. The objective here is to cover *as broad a* field of impacts as possible at the national level within a single comprehensive cost evaluation. The intention here is first, to provide a toolbox such that any effort made in this direction may be applied consistently across the fields of impacts and thus to result in meaningful results at the aggregated level. Second, and by way of example, to apply the framework and methods developed to a single country, in our case to Austria. Third, to draw conclusions concerning the nature of results arrived at when undertaking such an endeavour.

Methodologically our approach draws from and combines the following:

- Scenario-Based Impact-Field-Assessment: to capture impacts at the most detailed level available
- Computable General Equilibrium (CGE) analysis: to capture cross-sectoral linkages and economy-wide effects
- Qualitative analysis: to capture additional non-market effects where monetisation is not considered appropriate or possible.

The opening section in the present volume, Part I, offers an overview of climate costs at the continental scale. In Chap. 2, Paul Watkiss gives a condensed report on the results of a regional assessment for Europe—the EU FP7 ClimateCost project, Watkiss 2011—which has combined sectoral assessments and wider economic analysis. The results show large differences in the patterns of impacts across Europe, with more negative impacts in South Eastern Europe and the Mediterranean due to a combination of the enhanced climate signal and the higher

FUND also distinguishing individual sectors), and neglect of consideration of possible catastrophic outcomes. For a detailed discussion see Pindyck (2013).

vulnerability in these regions. While this European-wide view is important, the chapter also shows there is a need for country level analysis—as is presented in the remainder of this book—in order to capture national context and insights, to allow for analysis of country specific risks, and to provide the national-level information needed to start planning for adaptation.

An overview of the available national assessments of climate change risks needed in adaptation planning is given by Reimund Schwarze in Chap. 3. This chapter covers both of the standard methodologies used for risk assessment at the aggregate national or regional level, i.e. top-down global integrated assessment models (IAMs), which are downscaled to the national or regional scale, and bottom-up sectoral impact assessments, which are up-scaled to the national or regional level. The chapter gives a comprehensive overview of approaches applied in the UK, France, Germany and Switzerland, places them in context and indicates their respective merits and shortcomings. It not only evaluates in which respects the approach presented in the present volume complements the earlier approaches, it also points out its shortcomings. The merits of the approach developed in the present volume are found to lie in the possibility of advancing cost-benefit analysis by explicitly considering uncertainties through worse case narratives; the ability to apply consistent socio-economic scenarios and shared policy assumptions across sectors; the combination of observations and projections, which can then be more easily communicated in national dialogues than in top-down models; advancing the state of the art of the assessment of cross-sectoral, indirect and macroeconomic effects; and, finally, the greater ease with which the consequences for public budget may be indicated. In terms of shortcomings, the chapter reveals that owing to important gaps in data and methods, quantitative results tend to be somewhat "conservative" and, thus need to be augmented by qualitative research.

The methodological approach needed to achieve consistent application across sectors and macroeconomic evaluation is developed in Part II. Here, in Chap. 4, Steininger et al. provide the overall framework, while Formayer et al. in Chap. 5, set forth how climate change scenarios can be used to derive local indicators from climate model ensembles. König et al. in Chap. 6, define the shared socioeconomic pathways necessary to ensure consistency across sectoral evaluations, and, finally, Bachner et al. in Chap. 7, develop the macroeconomic modelling framework and present the means by which economic impact evaluation methods may be employed consistently across sectors and what the implications are in terms of macroeconomic aggregates and feedback-effects.

Part III looks at each impact field in detail, and by way of example, explores the case of one country, Austria. Impact evaluation is provided for the following fields: Mitter et al. analyse impacts on agriculture (Chap. 8), Lexer et al. on forestry (Chap. 9), Zulka and Götzl on ecosystem services (Chap. 10), Haas et al. on human health (Chap. 11), Neunteufel et al. on water supply and sanitation (Chap. 12), Kranzl et al. on buildings, i.e. heating and cooling (Chap. 13), Kranzl et al. on electricity (Chap. 14), Bednar-Friedl et al. on transport (Chap. 15), Urban and Steininger on manufacturing and trade (Chap. 16), Loibl et al. on cities and urban green (Chap. 17), Prettenthaler et al. on riverine flooding (Chap. 18), and Köberl et al. on tourism (Chap. 19).

Aggregate evaluation is covered in Part IV of the present volume. While each of the above chapters focused on a mid-century time horizon, and only some extended the analysis even further into the future, Kettner et al. (Chap. 20) derive a more comprehensive cost assessment up to 2100 based on a Delphi-approach. These authors also identify the most relevant barriers to adaptation. Bachner et al. (Chap. 21) assess climate change impacts across all the ten sectors with quantified impacts simultaneously and draw conclusions concerning overall macroeconomic impact. Finally, Steininger et al. (Chap. 22) place the results within a broader perspective in order to give an overall evaluation of climate impacts at the national level and then consider what we may (or may not) conclude from such an endeavour.

Acknowledgements Financial support for this research by the Austrian Climate and Energy Fund for the project COIN, carried out within the Austrian Climate Research Programme (ACRP) is thankfully acknowledged.

References

APCC (Austrian Panel on Climate Change) (2014) Austrian assessment report 2014. Summary for policy makers, Österreichischer Sachstandsbericht Klimawandel. Austrian Academy of Sciences, Vienna

Aaheim A, Amundsen H, Dokken T, Wei T (2012) Impacts and adaptation to climate change in European economies. Global Environ Change 22(4):959–968

Arrhenius S (1896) On the influence of carbonic acid in the air upon the temperature of the ground. Philos Mag J Sci Ser 5(41):237–276

Bosello F, Eboli F, Parrado R, Nunes PALD, Ding H, Rosa R (2011) The economic assessment of changes in ecosystem services: an application of the CGE methodology. Economía Agraria y Recursos Naturales 11(1):161–190

Charney JG, Arakawa A, Baker DJ, Bolin B, Dickinson RE, Goody RM, Leith CE, Stommel HM, Wunsch CI (1979) Carbon dioxide and climate: a scientific assessment. Report of an ad hoc study group on carbon dioxide and climate. National Academy of Sciences, Washington, DC

Ciscar J-C, Iglesias A, Feyen L, Szabó L, Van Regemorter D, Amelung B, Nicholls R, Watkiss P, Christensen OB, Dankers R, Garrote L, Goodess CM, Hunt A, Moreno A, Richards J, Soria A (2011) Physical and economic consequences of climate change in Europe. Proc Natl Acad Sci USA 108(7):2678–2683

Ciscar J-C, Szabó L, van Regemorter D, Soria A (2012) The integration of PESETA sectoral economic impacts into the GEM-E3 Europe model: methodology and results. Clim Change 112(1):127–142

EU Commission (2009) Adapting to climate change: towards a European framework for action, White paper of the Commission of the European Communities, COM(2009) 147 final. EU Commission, Brussels

EEA (European Environment Agency) (2008) Impacts of Europe's changing climate—2008 indicator-based assessment, Joint EEA-JRC-WHO report, EEA report 4/2008. EEA, Copenhagen

EEA (European Environment Agency) (2012) Climate change, impacts and vulnerability in Europe, EEA report 12/2012. EEA, Copenhagen

ESPON Climate (2013) Climate change and territorial effects on regions and local economies, Final report, ESPON and IRPUD. TU Dortmund, Dortmund
Hope C (2006) The marginal impact of CO_2 from PAGE2002: an integrated assessment model incorporating IPCC's five reasons for concern. Integr Assess 6(1):19–56
IPCC (2013) Climate change 2013: the physical science basis. Working Group I contribution to the fifth assessment report of the intergovernmental panel on climate change. Oxford University Press, Oxford
IPCC (2014) Climate change 2014: impacts, adaptation, and vulnerability, working group II contribution to the fifth assessment report of the intergovernmental panel on climate change. Oxford University Press, Oxford
Jacob D, Peterson J, Eggert B et al (2013) EURO-CORDEX: new high-resolution climate change projections for European impact research. Reg Environ Change. doi:10.1007/s10113-013-0499-2
Metroeconomica (2004) Costing the impacts of climate change in the UK: overview of guidelines, UKCIP technical report. UKCIP, Oxford
Nordhaus WD (1991) To slow or not to slow: the economics of the greenhouse effect. Econ J 101 (407):920–937
Nordhaus WD (2011) Estimates of the social cost of carbon: background and results from the RICE-2011 model. National bureau of economic research working paper 17540
Pindyck RS (2013) Climate change policy: what do the models tell us? J Econ Lit 51(3):860–872
Stern N (2007) The economics of climate change: the Stern review. Cambridge University Press, Cambridge
Tol R (2002a) Estimates of the damage costs of climate change, Part I: Benchmark estimates. Environ Resour Econ 21(1):47–73
Tol R (2002b) Estimates of the damage costs of climate change, Part II: Dynamic estimates. Environ Resour Econ 21(2):135–160
UNFCCC (2010) The Cancun agreements, United Nations framework convention on climate change. http://unfccc.int/meetings/cancun_nov_2010/meeting/6266.php
Watkiss P (ed) (2011) The ClimateCost project. Final report. Volume 1: Europe. Stockholm Environment Institute, Sweden
WMO (1979) Declaration of the world climate conference. World Meteorological Organization, Geneva
World Bank (2013) Turn down the heat: climate extremes, regional impacts, and the case for resilience. A report for the World Bank by the Potsdam Institute for Climate Impact Research and Climate Analytics. World Bank, Washington, DC

Part I
Cost and Opportunities of Climate Change at the European Level

Chapter 2
The Cost of Climate Change in Europe

Paul Watkiss

Abstract Climate change has the potential to lead to major impacts and economic costs in Europe. This chapter reports on a recent regional assessment—the ClimateCost project—which has combined sectoral assessments and wider economic analysis to derive such estimates.

The results reveal potentially high economic costs from climate change in Europe, though these vary with the emission scenario and time period. While many of these impacts are projected to be adverse and lead to economic costs, there are also economic benefits. The results also show large differences in the patterns of impacts across Europe, with more negative impacts in South-Eastern Europe and the Mediterranean, due to a combination of the enhanced climate signal and the higher vulnerability in these regions. The analysis of different scenarios shows that mitigation (towards a 2 °C stabilisation scenario) would reduce these costs significantly, but only in the medium-long term (after 2040). There will therefore be a need for adaptation as well as mitigation, but given the high future uncertainty, this is likely to be best advanced through a framework of adaptive management.

While this European-wide view is important, the chapter also shows there is a need for country level analysis—as presented in this book—to capture national context and insights, to allow analysis of country specific risks, and to provide national-level information to start planning for adaptation.

The research leading to the results reported in this paper received funding from the European Community's Seventh Framework Programme, as part of the ClimateCost Project (Full Costs of Climate Change, Grant Agreement 212774) www.climatecost.eu. Additional support for the paper was provided by funding from the European Union Seventh Framework Programme (FP7/2007–2013) IMPACT2C Project: Quantifying projected impacts under 2 °C warming, grant agreement no. 282746.

P. Watkiss (✉)
Paul Watkiss Associates, Oxford, UK
e-mail: paul_watkiss@btinternet.com

K.W. Steininger et al. (eds.), *Economic Evaluation of Climate Change Impacts*, Springer Climate, DOI 10.1007/978-3-319-12457-5_2

2.1 Introduction

There are a wide range of potential impacts from climate change in Europe. These include impacts on the built and the natural environment, which affect many sectors. These impacts will lead to economic costs, which are often referred to as the 'costs of inaction' (the economic costs of climate change if no mitigation or adaptation takes place). These costs include all effects on society, i.e. both market and non-market impacts, and environmental, economic and social costs, rather than direct financial costs or losses alone. Many of these costs are projected to be adverse, though there will also be benefits.

The estimation of these costs is increasingly being used to provide policy input. By reporting future impacts in monetary terms, these assessments provide a common metric to compare impacts over time and across sectors. They also help to inform the debate on the costs and benefits of mitigation (i.e. the reduction of greenhouse gases emissions) and increasingly, the major risks and the prioritisation of adaptation. This information is potentially relevant at a number of different aggregation levels, addressing different objectives. It can provide input at the European level, where information on the economic costs of climate change can raise awareness on the scale of the challenge, and provide context and justification for European mitigation policy, as in the European Road Map for a low carbon economy (CEC 2011). It can also provide the economic case for adaptation, as in the EU Strategy on Adaptation (CEC 2013), with the analysis of the costs of inaction and economic benefits of adaptation.

These European estimates are the focus of this chapter. They provide important contextual information and insights, but similar analysis is also needed at the national level, as shown in Chap. 3 (Risk and Opportunity). This is because a national level assessment—as presented in this book—can analyse national risks in more detail. It can capture important local impacts that may be excluded in a European-wide assessment (e.g. impacts on Alpine regions). Finally, it can better align to national context and policies, and inform adaptation strategies, the development of which is primarily governed at the national level.

2.2 Methodological Approaches and Frameworks

Over the last few years, a wide range of methodologies have emerged for assessing the costs of climate change. These are well documented (e.g. UNFCCC 2009; Chambwera et al. 2014) and include 'bottom-up' assessments at local to sector level, as well as 'top-down' macro-economic or global assessments. The main methods are (Watkiss and Hunt 2010):

- Scenario-Based Impact-Assessment. This approach combines climate model outputs with sector impact models (or functional relationships) to estimate physical impacts, which are then valued to estimate welfare costs. These can be applied to market and non-market (e.g. health) sectors, at the European scale (e.g. Ciscar et al. 2011a, Watkiss 2012) or country level (e.g. UK CCRA 2012). However, these assessments are not able to capture cross-sectoral, economy-wide effects. There are a number of variations, including risk assessment, which focuses on extreme (probabilistic) events such as flood (using historical analogues or damage-loss relationships), and econometric based assessments, which use historical relationships between economic production and climate and then apply these to future climate scenarios.

- Computable General Equilibrium models (CGE). These provide multi-sectoral and macro-economic analysis of the economic costs of climate change. Examples include European analysis (e.g. Ciscar et al. 2011a) and national level analysis (e.g. SCCV 2007). These have the advantage of capturing cross-sectoral linkages and economy wide effects (and metrics), and they can also look at price and trade effects. However, they use aggregated representations of impacts and omit non-market impacts.
- Global economic integrated assessment models. These assess the economic costs of climate change using an integrated framework. They can be used to provide total net present values for future damages over time and to estimate the marginal social costs (the damage cost of an extra tonne of GHG emissions). These models provide valuable headline estimates, but they use highly aggregated functions, see Watkiss (2011).

These three approaches use different metrics, modelling approaches and assumptions. No one method is right or wrong—their use depends on objectives. More recently, some studies combine all approaches in a single framework, to produce complementary information. An example of such an analysis is presented in this chapter, summarising results on the economic costs of Europe from the European Commission FP7 Funded ClimateCost Project.

The study started with scenario-based sectoral impact assessment modelling. The results of this analysis were then fed into a number of CGE models to assess wider economic effects. Complementing this, the study ran a number of IAMs, assessing the effects on Europe as part of a global integrated assessment. The overall approach follows the stylised Fig. 2.1.

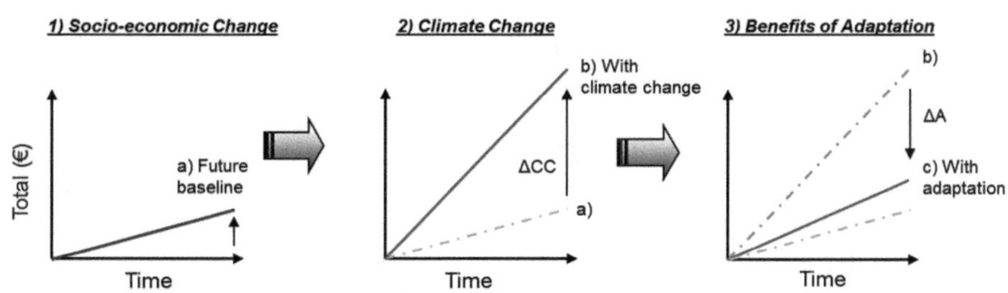

Fig. 2.1 Outline and steps of a stylised framework. *Source*: UNFCCC (2009)

1. The economic costs are first estimated for the future baseline, shown in (a). This is needed because future impacts are strongly influenced by socio-economic change, e.g. population growth, increased wealth, and these will occur even in the absence of climate change. Previous studies show that socio-economic change can be as important as climate change in determining economic costs.
2. The additional impact of climate change is added (ΔCC) to give the total effects of socio-economic change and climate change together, shown in (b). Strictly speaking, only the marginal (or net) increase above the baseline in (1) is due to climate change. Note that in some cases, socio-economic and/or climate change may lead to economic benefits, as well as costs.
3. Adaptation reduces the impacts downwards, shown in (c) as the residual costs. The reduction (ΔA) provides the economic benefits of adaptation and this can be compared against the costs of adaptation and the residual impacts after adaptation.

The aim is to express the impacts in terms of the effects on social welfare, as measured by individuals' preferences using a monetary metric. The basic approach to the costing analysis in such a framework is to multiply relevant unit values (market prices or non-market prices) by the physical impacts identified. While most studies primarily used market and non-market estimates of Willingness to Pay (WTP), in some cases cost-based estimates have been used as a proxy.

It is highlighted that the analysis below presents the results of one study only. There is a growing literature on European, regional, country and sector assessments, as reported in IPCC AR5 Europe Chapter (Kovats et al. 2014), though costing remains primarily focused on flood defences, water, energy, and agriculture sectors.

2.3 The Costs of Climate Change in Europe: Results for a European Assessment

2.3.1 Climate Model Projections

Analysis of the future impacts of climate change requires climate models. These require inputs of future GHGs based on modelled global socio-economic scenarios, in order to make projections of future changes in temperature, precipitation and other variables. The ClimateCost project considered two emissions scenarios: a medium-high baseline scenario (A1B SRES, Nakicenovic et al. 2000) and a mitigation scenario (E1, from the ENSEMBLES project, Van der Linden and Mitchell 2009), which stabilises global temperature change at about 2 °C above pre-industrial levels, using multi-model ensemble data from the ENSEMBLES project. Under a medium-high emission baseline (A1B), with no mitigation, the climate models projected that global average temperatures could rise by between 1.6 and 2.3 °C by 2041–2070, and 2.4 and 3.4 °C by 2071–2100, relative to the

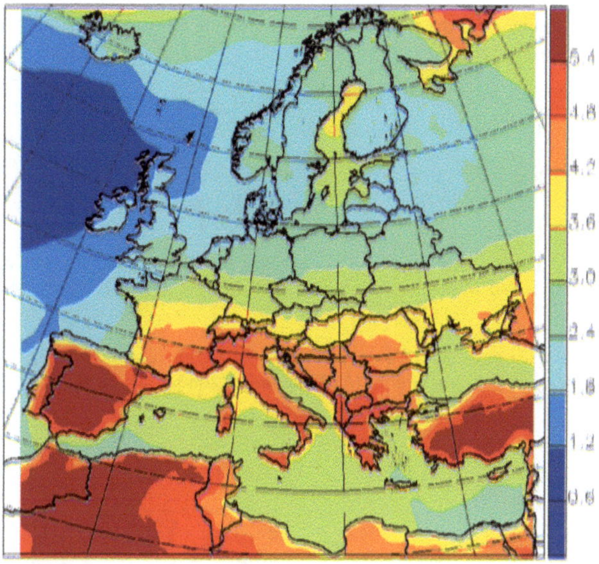

Fig. 2.2 Summer temperature change for Europe from a Regional Climate Model (1960–1999 to 2070–2099, A1B), showing the higher warming in Southern Europe. *Source*: Christensen et al. (2011)

modelled baseline period of 1961–1990. However, the models project larger temperature increases for Europe in summer. They also show a highly differentiated pattern, as shown in Fig. 2.2. Southern Europe and the Iberian Peninsula are projected to experience much higher levels of warming than the global average, with a mean increase for the latter up to 5 °C by 2071–2100. This differentiated signal is important in impacts across Europe. Under the E1 stabilisation (mitigation) scenario, future warming is significantly reduced, though only after 2040.

The projections of future precipitation change show much greater differences across scenarios, models and regions of Europe. These can be seen in Fig. 2.3. This shows the change in summer precipitation across different time periods (top), different scenarios (middle) and different climate models (bottom). There are some robust patterns of change, e.g. wetter winters are projected for Western and Northern areas but drier conditions projected all year for the South. However, in other areas (notably a band from the UK in the west across to Eastern Europe) the changes are uncertain, and the driest (left, bottom) and wettest (right, bottom) results even differ in sign (i.e. decreases vs. increases). The consideration of this uncertainty is important in analysing and reporting on future impacts, and in the subsequent analysis of adaptation.

2.3.2 *Sector Results*

The climate projections were input into sector impact assessment models. The results are summarised below, with economic costs in future periods reported in current prices to facilitate direct comparison over time.

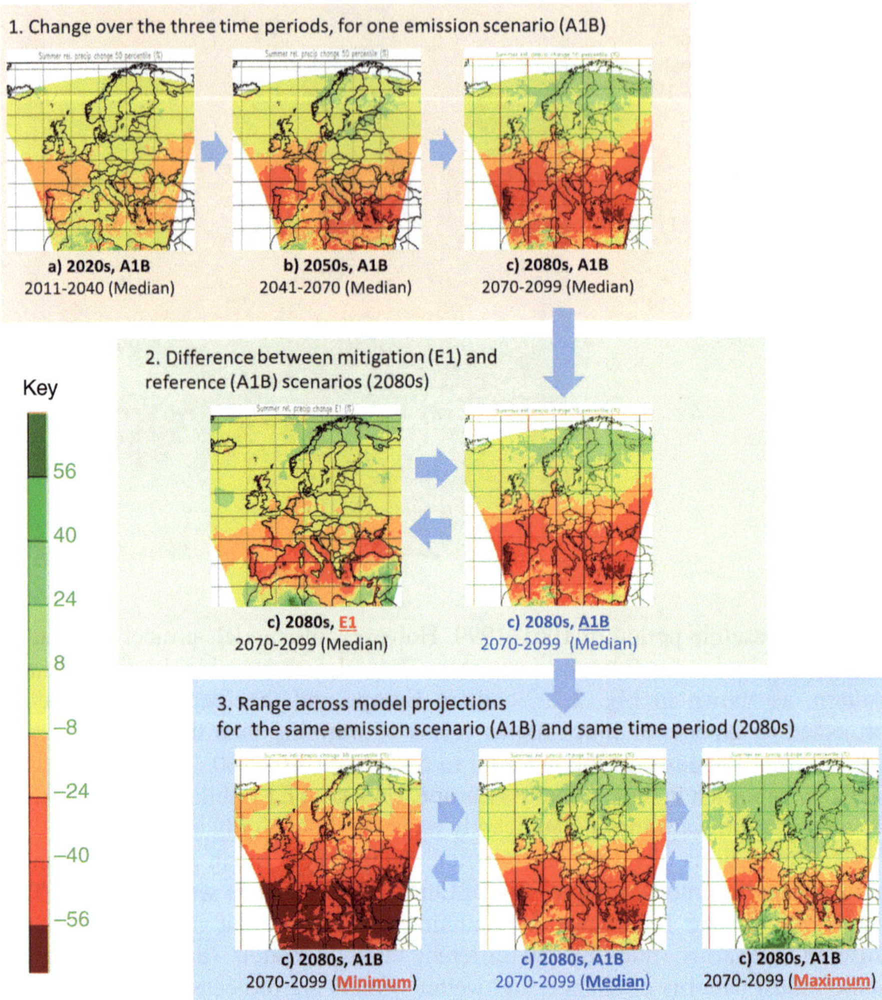

Fig. 2.3 Relative change in summer precipitation (%) from 11 RCM simulations (ENSEMBLES archive) showing trends (1) over time for the median A1B from 1961–1990 to 2011–2040, 2041–2070 and 2070–2099, (2) A1B and E1 median scenarios for 2070–2099 and (3) different model projections for the same time period and same emissions scenario (min, med, max), for 2070–2099, A1B. *Source*: Christensen et al. (2011)

2.3.2.1 Coastal Zones

Coastal zones contain high population densities, significant economic activities and important ecosystem services in Europe. These are already subject to coastal flooding and climate change has the potential to increase future risks. The project (Brown et al. 2011) considered the future economic costs to coastal zones using the DIVA model (Hinkel and Klein 2009), assessing future impacts and damage costs for the A1B and the E1 scenarios. Assuming that defences are not upgraded, it was estimated that 55,000 people/year (mid estimate) in the EU could be flooded by the 2050s (2041–2070) rising to over 250,000 people/year by the 2080s (2071–2100) (A1B scenario). The economic cost of sea level rise (including direct impacts on people, salinisation, costs of moving and land loss) is significant, estimated at around 11 billion euros/year for the 2050s, rising to 25 billion euros/year by the 2080s (mid-estimate of the combined effects of climate and socio-economic change, based on current prices, with no discounting). Additional unquantified costs will also occur due to ecosystem losses and indirect effects. The results show major differences between different Member States, with some countries projected to face much higher relative increases in coastal-flood damages. This can be seen in Fig. 2.4. The Netherlands, Belgium, Denmark, UK and Portugal are ranked in the top five for damage costs relative to GDP.

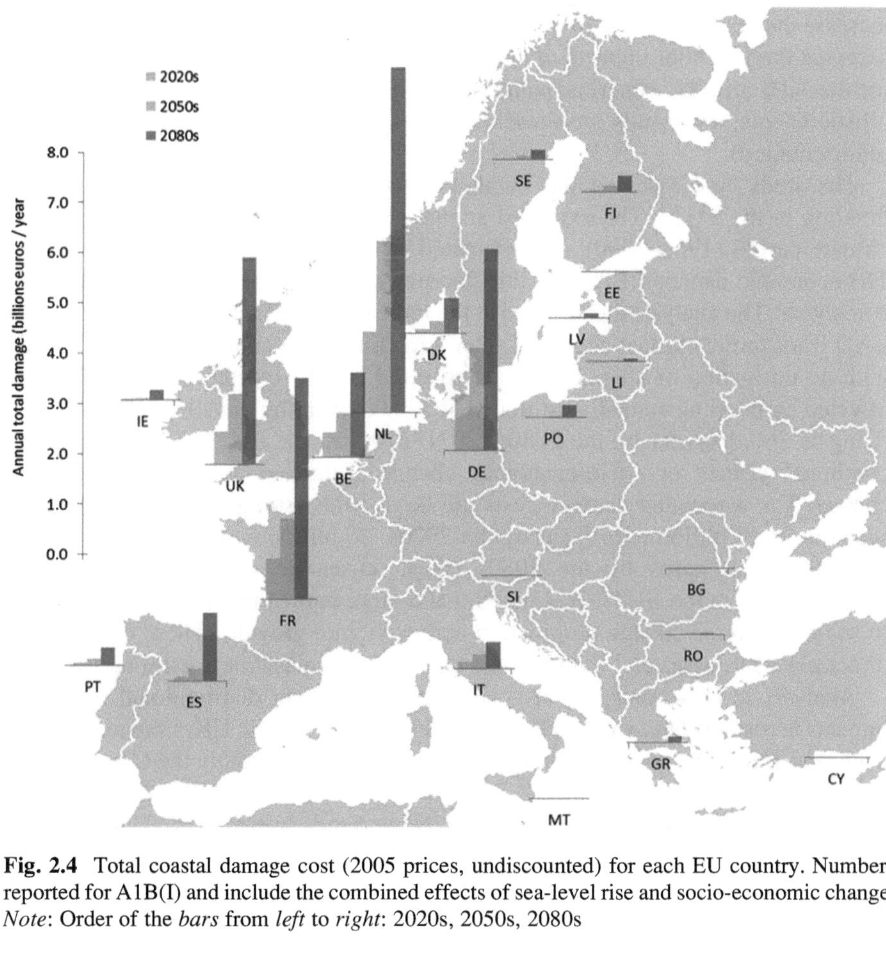

Fig. 2.4 Total coastal damage cost (2005 prices, undiscounted) for each EU country. Numbers reported for A1B(I) and include the combined effects of sea-level rise and socio-economic change. *Note*: Order of the *bars* from *left* to *right*: 2020s, 2050s, 2080s

The study also considered uncertainty, finding a wide range around these mid-point values, due to the uncertainty in emission scenarios, projected temperature and the sea level rise response. Under the E1 (mitigation) scenario, the estimated annual number of people flooded falls to 180,000 and the annual damage costs fall to 17 billion euros (mid estimates) by the 2080s. The analysis also considered an extreme sea level rise, considering a rise of more than 1.2 m by 2100. This scenario significantly increased the estimated damage costs for the EU to 156 billion euros/year (undiscounted) by the 2080s—six times higher than the mid-A1B scenario. This is an important finding, as it highlights the need for both mitigation as well as adaptation, as the chances of these extreme scenarios are significantly reduced with mitigation.

2.3.2.2 River Flooding

River floods already cause major economic costs in Europe and climate change may increase the magnitude and frequency of these events. The study (Rojas et al. 2013) assessed the potential impacts of climate change on river flood damage in Europe for the A1B and E1 scenarios, using the LISFLOOD model. As floods are probabilistic events, the results are presented as expected annual damage (EAD) costs (undiscounted).

The study first assessed the number of people potentially affected by river flooding in the EU27. The expected annual people (EAP) flooded in the baseline climate period (1961–1990) was estimated at around 167,000/year in the EU27. The economic damage (EAD) of this flooding was estimated at around 5.5 billion euros/year. The analysis then looked at the increase in the number of people and the EAD from future climate change. Under a medium-high emission baseline (A1B), with no mitigation or adaptation, the projected mean expected number of people affected by flooding annually estimated at 300,000/year by the 2050s (2041–2070), rising to 360,000/year by the 2080s (2071–2100) in the EU27. This includes the combined effects of socio-economic change (future population) and climate change. The associated damage costs are large, with estimated EAD for the A1B scenario of 20 billion euros/year by the 2020s, 46 billion euros/year by the 2050s and 98 billion euros by the 2080s (mean ensemble results, current values, undiscounted) in the EU27. It is noted that a large part of this (around half) is due to socio-economic change (population and economic growth). These only include direct physical losses and indirect impacts would increase these estimates further.

Analysis at the country level again showed a strong distributional pattern of impacts across Europe, with high climate-related costs in the UK, France, Italy and in central-European countries along major river systems (notably the Czech Republic and Hungary), as shown in Fig. 2.5. When normalised for GDP, impacts in the UK, Italy, Slovenia, Belgium and the Netherlands were high. Under an E1 stabilisation scenario, broadly equivalent to the EU 2° target, the EAD was estimated to fall to 15 billion euros by the 2020s, 42 billion euros by the 2050s and 68 billion euros by the 2080s in the EU27 (current values, undiscounted).

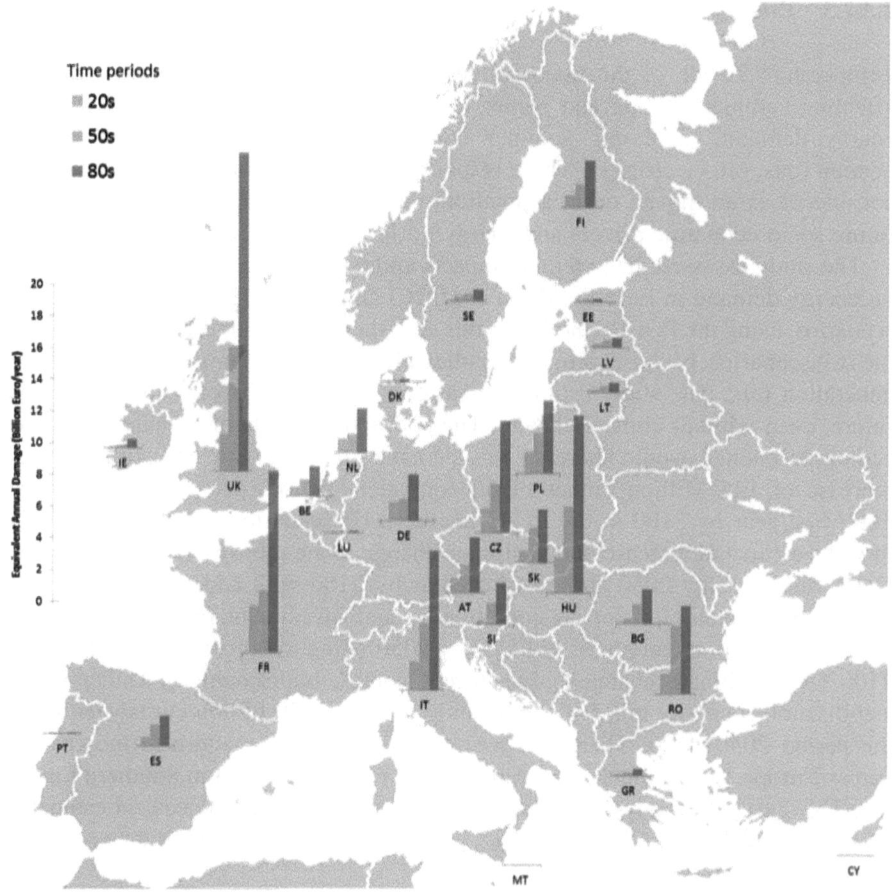

Fig. 2.5 EU27 LISFLOOD EAD from floods for A1B (ensemble mean, 12 regional climate models) (constant 2006 prices, undiscounted), with no adaptation. Values are combined climate and socio-economic change. The map shows billions of euros for the 2020s (2011–2040), 2050s (2041–2070) and 2080s (2071–2100). *Note*: Order of the *bars* from *left* to *right*: 2020s, 2050s, 2080s

The consideration of uncertainty found a very wide range around these central estimates, representing the range of results from different climate models. At the European scale, damage costs were found to vary by a factor of two (higher or lower). These differences were even more significant at the country level, with some models reporting differences in the sign of change (i.e. some models project relative reductions in future flood risk from climate change, while others project increases for the same areas). These differentiated patterns highlight the importance of more detailed national analysis, as presented in Chap. 18 (Catastrophe), and it also highlights the need to consider uncertainty in formulating adaptation strategies.

2.3.2.3 Energy

Temperature is one of the major drivers of current energy demand in Europe, affecting summer cooling and winter heating. Climate change will affect future energy demand, increasing summer cooling (electricity) but reducing winter heating (gas, oil, electricity). These responses are largely autonomous, and can be considered as an impact or an adaptation, noting they are strongly influenced by future socio-economic drivers and energy/mitigation policy.

The study assessed the potential impacts and economic costs of climate change on energy demand in Europe (Mima et al. 2012) for the A1B and E1 (mitigation) scenarios, using the POLES model. In this case there are major differences between the two scenarios, because mitigation policy affects energy demand and the energy/generation mix. The study estimated that cooling demand would increase in the future, even without climate change. The additional effects of climate and socio-economic change would increase this demand sharply, with an estimated total increase of 145 Million tonnes of oil equivalent (Mtoe) per year by 2050 and 269 Mtoe/year by 2100 in the EU27 for the A1B scenario (ensemble mean). Of this, the estimated increase due to climate change alone (above the future baseline) is 16 Mtoe/year by 2050 and 53 Mtoe/year by 2100 with additional costs (from electricity consumption for air conditioning and investment for new units) estimated at 30 billion euros/year in EU27 by 2050 rising to 109 billion euros/year by 2100 (A1B, climate change only, current values, undiscounted). Under an E1 stabilisation scenario, this fell significantly to around 20 billion euros/year across the period 2050–2100. There was a strong distributional pattern to the changes across Europe, with a much higher increase in cooling demand in Southern Europe (see Fig. 2.6). The analysis also found a wide range around the central estimates, representing different climate models, which found that the potential costs varied by ±25 % (A1B, by 2100).

The study also assessed the decrease in heating demand from climate change in Europe (a benefit). The reduction from climate change alone (over future baseline levels) was estimated at 28 Mtoe/year by 2050, rising to 65 Mtoe/year by 2100. This is approximately a 10 and 20 % fall. Under the E1 scenario, the reduction in heating demand was lower, estimated at −11 Mtoe/year by 2050 and −13 Mtoe/year by 2100. Again, there were large variations across the suite of climate models considered and large differences across regions of Europe, with the largest reductions in Western Europe. While the reduction in winter heating was larger in energy terms than the increase in cooling—from climate change alone—the relative costs of the two were similar, as cooling is more expensive than heating. The reduction in total heating demand (from climate change alone) was estimated at 34 billion euros/year in 2050 rising to 121 billion euros/year in 2100 for the EU27 under the A1B scenario (current prices, undiscounted). There are also important differences at the national level even within regions, as shown in Chap. 14 (Electricity).

Climate change will also have effects on energy supply, affecting hydro-electric generation, thermal power plant cooling and renewables (wind, biomass, solar).

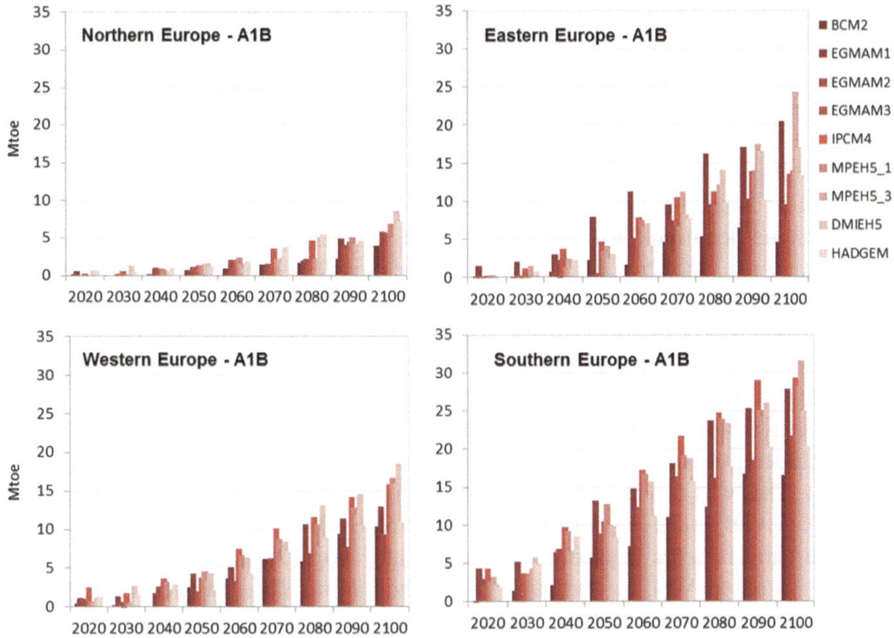

Fig. 2.6 Impact of climate change on EU27 energy consumption for space cooling (Mtoe/year) in residential and service sector by region (A1B) (individual columns show alternative climate model projections). *Source*: Mima et al. 2012. *Note*: Order of the *bars* from *left* to *right*: BCM2, EGMAM1, EGMAM2, EGMAM3, IPCM4, MPEH5_1, MPEH5_3, DMIEH5, HADGEM

The scale of these impacts was also considered using the POLES model. The impacts of climate change on hydro-electric generation was estimated to decrease European hydro-electric generation by around −3 % in 2050 and −8 % in 2100 (A1B), compared to the future baseline, though there was considerable uncertainty around the central estimates. There was also a strongly differentiated pattern of change across Europe. Finally, the effect of higher temperatures on thermal power plant cooling (reduced efficiency of nuclear and fossil) indicated power generation in Europe could be reduced by up to 2–4 % per year (A1B, 2100), though advances in plant design and anticipatory action could reduce these significantly.

2.3.2.4 Health

There are a large number of potential health impacts that could arise from climate change, directly or indirectly, including heat-related impacts, food and vector-borne disease, impacts from flooding, etc. though also some potential benefits, notably the reduction in cold-related mortality. There are also risks to health infrastructure. The study (Kovats et al. 2011) assessed potential health impacts, focusing on heat related mortality, food borne disease, coastal flooding and labour productivity, for the A1B and E1 scenarios.

The study estimated climate change (alone) would lead to additional 26,000 heat-related deaths/year by the 2020s (2011–2040), rising to 89,000/year by the 2050s (2041–2070) and 127,000/year by the 2080s (2071–2100) (A1B scenario, mid estimate). There were relatively higher levels of climate change-attributable heat deaths found in Southern Europe. The cost of these impacts varies on the valuation method used for monetising the risk of a fatality, and whether a Value of a Life Year Lost (VOLY) or a Value of a Statistical Life (VSL) is used. Using the latter, the estimated welfare costs were estimated at 31 billion euros/year by the 2020s, 103 billion euros/year by the 2050s and 147 billion euros/year by the 2080s. These values fell by over an order of magnitude when using the VOLY approach. Under the E1 scenario, these impacts are reduced significantly after 2040, falling to 69,000 deaths/year and 80 billion euros/year by the 2050s (VSL). A sensitivity was also included that took account of natural (autonomous) acclimatisation (at 0.16 °C/ decade). This was found to reduce impacts significantly, falling to 40,000/year in the 2080s (A1B) and even more in the mitigation (E1) scenario. However, there will be significant benefits from climate change in Europe from the reduction in cold related mortality (Watkiss and Hunt 2012), though these benefits occur in different regions of Europe (primarily in Western Europe). A more specific health analysis for Austria is presented in Chap. 11 (Health).

The study also considered food borne illness in Europe, focusing on Salmonellosis, which is sensitive to ambient temperature. The study estimated that climate change (alone) could lead to an additional 7,000 cases/year of salmonellosis in EU27 by 2020s, rising to 13,000 by the 2050s and 17,000 by the 2080s (A1B, mid estimate), if incidence remains at current levels (noting that a fall in baseline incidence associated with planned policy would reduce these). The economic costs were estimated at 36 million euros/year in the 2020s (A1B, current baseline), rising to 68 and 89 million euros/year in the 2050s and 2080s respectively (current price, undiscounted).

The analysis also considered additional mortality due to climate induced coastal flooding. The combined effects of climate and socio-economic change was estimated to lead to 130 deaths/year in the EU by the 2050s and 650 deaths/year in the EU by the 2080s (A1B) predominantly in Western Europe, with associated welfare costs of 151 euros/year in the 2050s and 750 million euros/year by the 2080s. The impacts on well-being (from disruption, mental health impacts) would increase these further. These fell significantly under the E1 mitigation scenario to 185 (2080s) fatalities/year and welfare costs of 214 million euros/year (2080s).

Finally, the study estimated the impact on labour productivity. Under the A1B scenario, Southern Europe was found to incur a mean loss of productivity—measured as days lost—of 0.4–0.9 % by the 2080s (with the range reflecting different future labour structures). Total productivity losses for Europe were estimated at 300–740 million euros/year in the 2080s (A1B), with most impacts in the South. These were significantly reduced under the E1 scenario to 60–150 million euros/year in the 2080s.

It is stressed that there was found a wide range of uncertainty around all these estimates—reflecting the underlying uncertainty in emissions scenario and climate-health relationships.

2.3.2.5 Agriculture

Agriculture is a highly climate sensitive sector and climate change has the potential to lead to major effects in Europe. These involve many potential climate variables, which can impact directly and indirectly on crop production, agricultural supply and value chains. They involve potentially negative effects (e.g. from lower rainfall and variability) but also positive effects (e.g. from CO_2 fertilisation and from extended growing seasons), as well as complex changes from changes in extreme events, the range and prevalence of pests and disease, etc. At the European level, these effects are also influenced from the impacts (positive and negative) at the global scale, and demand, production, trade, etc. These are also potential impacts on horticulture, viniculture, industrial crops and livestock, and on the multi-functionality role of agriculture (e.g. in landscape).

The study (Iglesias et al. 2012a, b) estimated the changes in productivity for major crops in Europe and globally, using an integrated modelling system built around the DSSAT crop simulation model. The results found that agro-climatic regions will change significantly in Europe, as a result of climate change. It also found large differences between regions, with strong distributional differences (positive and negative). In general, there were yield improvements projected for Northern Europe due to a longer growing season (and frost-free period), while crop productivity decreased in Southern Europe. The results were found to vary significantly with climate model. This was particularly evident for Central Europe, which depended strongly on the particular climate scenario and model output. This highlights the need for national level analysis: the results for Austria are presented in Chap. 8 (Agriculture).

At the aggregated level, the net changes in the EU under the A1B scenario were modest, even by the 2080s. However, at the international level, there was a more marked decrease in crop productivity, with many of the major changes (up to −50 % decreases) occurring in food-limited areas. Under the E1 scenario, Europe (overall) was not found to experience negative impacts on crop yield, though there were exceptions at the country and local level. The global impacts were also significantly reduced under this E1 scenario. These results were subsequently fed into trade and CGE models.

The analysis also considered farm-level responses (autonomous adaptation, such as additional fertiliser use and irrigation). These were found to significantly reduce negative impacts in Europe. However, these autonomous responses were found to be limited when other constraints were included e.g. water availability, pollution standards.

2.3.2.6 Co-benefits

Mitigation policy has a beneficial effect in reducing greenhouse gas (GHG) emissions and also reduces emissions of air pollutants, which leads to air quality benefits. Importantly these benefits arise immediately and are experienced locally (i.e. in Europe). These co-benefits are important in comparing scenarios. As part of the study, Holland et al. (2011) used the GAINS and ALPHA models to estimate the health and environmental benefits of achieving the EU's 2050 low carbon path, i.e. in moving from the A1B to the E1 scenario. Large benefits were found, with increased life expectancy and lower pollution related impacts estimated at 48–99 billion euros/year in 2050 for the EU27 (current prices, undiscounted). Additional benefits (energy security and diversity) were also noted, but were not valued.

2.3.2.7 The Coverage of Impacts

A key issue in assessing the economic costs of climate change is to consider (and report transparently) on the coverage of impacts. Previous studies have highlighted that the number and extent of impacts considered makes a very large difference to results (e.g. Watkiss 2011) and all current studies can only be considered partial. The estimates for Europe reported above are therefore a sub-total of the full impacts of climate change. They only include five sectors, and even within these sectors, they represent a partial coverage.

One of the important omissions in the analysis is the impact of climate change on biodiversity and ecosystem services (terrestrial, aquatic and marine). However, the analysis of these impacts—and their subsequent valuation—is very challenging. There have been some early assessments of the economic benefits that ecosystem services (forests) provide in terms of carbon sequestration (regulating services) and the impacts of climate change on these services under future scenarios (e.g. Ding et al. 2011), which indicate large potential costs. What is clear is that this sector remains a major priority for future analysis.

The impacts on tourism are also important, noting that there are strongly differentiated effects across Europe. There are projected effects on summer tourism, which is likely to affect current flows in the Mediterranean (Amelung and Moreno 2012) as conditions become less favourable, but increase them positively in northern and western countries of Europe. There will also be impacts on winter tourism (e.g. Agrawala 2007) due to the decrease in snow reliability in the mountainous regions, particularly the Alps.

In addition, there will be important impacts on many other sectors, such as transport, manufacturing and industry etc., for which European-wide estimates are only starting to emerge. The consideration of these other sectors is important to allow a comprehensive picture of the impacts of climate change on Europe, and thus to fully inform the policy debate.

These additional categories may be particularly important for individual countries, reinforcing the need for national level studies, such as COIN, and also providing the opportunity for addressing the costing gaps identified by IPCC (Kovats et al. 2014).

2.3.3 Discussion of Sector Results

The overall results reveal potentially large costs from climate change in Europe, which total several 100 billion euros/year in later years. It is stressed that these only cover a number of impacts in a number of sectors, but also that there are some potentially large economic benefits. The results also show that significant reductions in these costs can be achieved by mitigation policy consistent with the 2° goal, noting these benefits only arise after 2040.

The results also show a strong distributional pattern of impacts in different regions of Europe. The impacts of coastal zones are most important for Western Europe, while the impacts of energy for cooling are most important for the South. Overall, there is a trend of more (net) negative impacts for South-Eastern Europe and the Mediterranean (e.g. in relation to energy demand, agricultural productivity, water availability, health effects, summer tourism, ecosystems) as compared to Northern and Western Europe.

2.3.4 Computable General Equilibrium and Integrated Assessment Model Results

The sector assessments above provide key information, but they do not address the wider economic effects. The study (Ciscar et al. 2011b) addressed this by using the GEM-E3 computable general equilibrium model, using the sectoral outputs (for coastal zones, river floods, energy and agriculture) to assess the effects on overall GDP. Figure 2.7 presents the EU and regional breakdown of the overall change from climate change (A1B) on GDP, compared to the reference scenario. The study estimated a GDP change of −0.44 % by the 2050s, and −0.83 % by the 2080s. Importantly there is a strong distributional impact, with the largest GDP losses in the Southern Europe region, estimated at −2.3 % by the 2080s.

These impacts were significantly reduced under the E1 scenario (reducing down European impacts to around −0.3 % in the 2050s and the 2080s—noting that benefits occur in later years). The study also estimated the overall welfare loss using the CGE model, which was estimated to be 1.5 % for Europe by the 2080s under the A1B scenario, with again, the most negatively affected region being Southern Europe. The strong regional differences also cascade down to the national level, where economic structure and key impacts can be very different. This

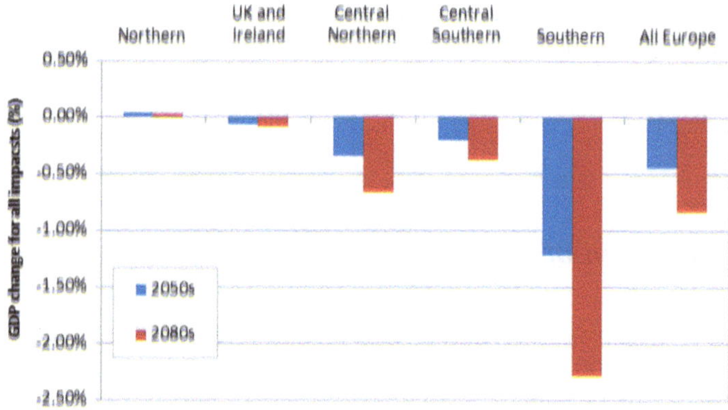

Fig. 2.7 Overall impact on GDP for four sectors (% difference with respect to the baseline) for the A1B scenario. *Source*: Ciscar et al. 2011b. *Note*: Order of the *bars* from *left* to *right*: 2050s, 2080s

highlights the importance of more detailed national analysis, as presented in Chap. 21 (Macroeconomic Evaluation).

Finally, the study included a number of the Global Economic Integrated Assessment Models. These combine the scientific and economic aspects of climate change within a single, iteractive analytical framework. The models also have mitigation modules and can therefore look at the costs and benefits of emission reductions and in some models the optimal policy trajectory. As part of the ClimateCost project, an existing IAM, the PAGE Integrated Assessment Model was updated, to PAGE09 (Hope 2011). The model was then run to look at the global and regional damage costs for the two scenarios. The results estimated damages equivalent to almost 4 % of GDP for Europe by 2100 under the A1B scenario, with a risk of extremely large costs at the tails of the distribution (in excess of 10 % of GDP equivalent). Under the E1 scenario (equivalent to the 2° target) these fell to under 1 % of GDP equivalent, and more importantly, removed the tail of extreme values. It is stressed, however, that a number of different IAM models used in the study estimated much lower values.

2.4 Towards Adaptation

While these estimates provide useful context for the consideration of mitigation and adaptation in Europe, the findings also indicate a change in focus is needed for practical adaptation. First, the framework presented in Fig. 2.1 is evolving, with an increasing focus on the current costs of climate variability (the 'adaptation deficit'). Second, the findings of the project above highlight the considerable uncertainty in the future climate projections and impacts. This necessitates a new approach, which looks to advance flexibility and robustness using iterative adaptation plans and adaptation pathways (UNFCCC 2009; Downing 2012; IPCC SREX 2012) rather than optimisation. These approaches require a greater focus on multi-model uncertainty and decision making under uncertainty. Finally, there is a greater recognition of the need to include autonomous adaptation (noting this may sometimes lead to mal-adaptation) and include emerging policy responses as these affect baseline risks, e.g. such as the introduction of heat-alert systems introduced following the 2003 heat-wave.

2.5 Future Research

These regional studies continue to evolve. The most recent projects—notably the EC FP7 funded IMPACT2C project—are now looking towards the use of the new RCP and SSP scenarios (Representative Concentration Pathways and Shared Socio-economic Pathways), ensuring a core theme of uncertainty is considered, and looking at cross-sectoral perspectives.

2.6 Conclusions

This chapter reports the results of the ClimateCost project—which has combined sectoral assessments and wider economic analysis. The results reveal potentially high economic costs from climate change in Europe. The analysis of different scenarios shows that mitigation (towards a 2 °C stabilisation scenario) would reduce these costs significantly, but only in the medium-long term (after 2040). There will therefore be a need for adaptation as well as mitigation, but given the high uncertainty, this is likely to be best advanced through a framework of adaptive management.

It also finds strong differences in the geographical patterns of impacts and economic costs across Europe, with more negative impacts in the South-East and the Mediterranean, due to a combination of the enhanced climate signal and the higher vulnerability in these regions. Therefore, while a European-wide analysis provides important insights, these regional differences highlight the need for more comprehensive and disaggregated level analysis, i.e. at the national level. Such an analysis allows the consideration of country context and policies, and can capture important local risks that may be missing from broader assessments (such as the threats to Alpine areas). The subsequent chapters of this book focus on one such study, looking in detail at Austria.

References

Agrawala S (2007) Climate change in the European Alps: adapting winter tourism and natural hazards management. Organisation for Economic Co-operation and Development (OECD), Paris

Amelung B, Moreno A (2012) Costing the impact of climate change on tourism in Europe: results of the PESETA project. Clim Change 112:83–100. doi:10.1007/s10584-011-0341-0

Brown S, Nicholl's RJ, Vafeidis A, Hinkel J, Watkiss P (2011) The impacts and economic costs of sea-level rise in Europe and the costs and benefits of adaptation. In: Watkiss P (ed) The ClimateCost project Final report Volume 1: Europe. Stockholm Environment Institute, Sweden. ISBN 978-91-86125-35-6

CCRA (2012) The UK climate change risk assessment 2012 evidence report. Presented to Parliament pursuant to Section 56 of the Climate Change Act 2008. http://randd.defra.gov.uk/Default.aspx?Menu=Menu&Module=More&Location=None&Completed=0&ProjectID=15747#RelatedDocuments. Accessed Dec 2013

CEC (2011) A roadmap for moving to a competitive low carbon economy in 2050. Communication from the Commission to the European Parliament, the Council, the European Economic and Social Committee and the Committee of the Regions. COM(2007)112. Commission of the European Communities, Brussels

CEC (2013) EU Strategy on adaptation to climate change. Communication from the Commission to the European Parliament, the Council, the European Economic and Social Committee and the Committee of the Regions. COM(2013) 216 final. Commission of the European Communities, Brussels

Chambwera M, Heal G, Dubeux C, Hallegatte S, Leclerc L, Markandya A, McCarl B, Mechler R, Neumann J (2014) Economics of adaptation, Chap. 17. IPCC WGII AR5. http://ipcc-wg2.gov/AR5/images/uploads/WGIIAR5-Chap17_FGDall.pdf

Christensen OB, Goodess CM, Harris I, Watkiss P (2011) European and global climate change projections: discussion of climate change model outputs, scenarios and uncertainty. In: Watkiss P (ed) The ClimateCost project. Final report. Volume 1: Europe. Stockholm Environment Institute, Sweden. ISBN 978-91-86125-35-6

Ciscar J-C, Iglesias A, Feyen L, Szabó L, Van Regemorter D, Amelung B, Nicholls R, Watkiss P, Christensen OB, Dankers R, Garrote L, Goodess CM, Hunt A, Moreno A, Richards J, Soria A (2011a) Physical and economic consequences of climate change in Europe. PNAS 108:2678–2683

Ciscar JC, Saveyn B, Van Regemorter D (2011b) ClimateCost deliverable 7B 1. Report – policy analysis. The ClimateCost Project

Ding H, Nunes PALD, Teelucksingh S (2011) European forests and carbon sequestration services: an economic assessment of climate change impacts. The United Nations environment programme division of environmental policy. Ecosystem Services Economics (ESE) working paper series (Environment for Development). Implementation paper N° 9, January 2011. UNEP, Nairobi, Kenya

Downing TE (2012) Views of the frontiers in climate change adaptation economics. WIREs Clim Change. doi:10.1002/wcc.157

Hinkel J, Klein RJT (2009) The DINAS-COAST project: developing a tool for the dynamic and interactive assessment of coastal vulnerability. Global Environ Change 19(3):384–395

Holland M, Amann M, Heyes C, Rafaj P, Schöpp W, Hunt A, Watkiss P (2011) The reduction in air quality impacts and associated economic benefits of mitigation policy. In: Watkiss P (ed) The ClimateCost project. Final report. Volume 1: Europe. Stockholm Environment Institute, Sweden. ISBN 978-91-86125-35-6

Hope C (2011) The PAGE09 integrated assessment model: a technical description. Cambridge University Judge Business School. Working paper series 4/2011. http://www.jbs.cam.ac.uk/research/working_papers/2011/wp1104.pdf

Iglesias A, Garrote L, Quiroga S, Moneo M (2012a) A regional comparison of the effects of climate change on agricultural crops in Europe. Clim Change 112(1):29–46

Iglesias A et al (2012b) Agriculture. In: Watkiss P (ed) The ClimateCost project. Final report. Volume 1: Europe. Stockholm Environment Institute, Sweden. ISBN 978-91-86125-35-6

IPCC (2012) In: Field CB, Barros V, Stocker TF, Qin D, Dokken DJ, Ebi KL, Mastrandrea MD, Mach KJ, Plattner G-K, Allen SK, Tignor M, Midgley PM (eds) Managing the risks of extreme events and disasters to advance climate change adaptation. A special report of working groups I and II of the intergovernmental panel on climate change. Cambridge University Press, Cambridge, p 582

Kovats S, Lloyd S, Hunt A, Watkiss P (2011) Technical policy briefing note 5: the impacts and economic costs on health in Europe and the costs and benefits of adaptation, results of the EC RTD ClimateCost project

Kovats S, Valentini R, Bouwer L, Georgopoulou E, Jacob D, Martin R, Rounsevell M, Soussana J-F (2014) Europe. IPCC WGII AR5, Chap. 23. http://ipcc-wg2.gov/AR5/images/uploads/WGIIAR5-Chap23_FGDall.pdf

Mima S, Criqui P, Watkiss P (2012) The impacts and economic costs of climate change on energy in Europe. In: Watkiss P (ed) The ClimateCost project. Final report. Volume 1: Europe. Stockholm Environment Institute, Sweden. ISBN 978-91-86125-35-6

Nakicenovic N, Alcamo J, Davis G, de Vries B, Fenhann J, Gaffin S, Gregory K, Grübler A et al (2000) Special report on emissions scenarios, working group III, intergovernmental panel on climate change (IPCC). Cambridge University Press, Cambridge, 595pp. ISBN 0 521 80493 0

Rojas R, Feyen L, Watkiss P (2013) Climate change and river floods in the European Union: socio-economic consequences and the costs and benefits of adaptation. Global Environ Change 23(6):1737–1751

Swedish Commission on Climate and Vulnerability Stockholm (2007) Sweden facing climate change – threats and opportunities, final report. http://www.regeringen.se/sb/d/574/a/96002

UNFCCC (2009) Potential costs and benefits of adaptation options: a review of existing literature. UNFCCC technical paper. F CDCeCce/mTPb/e2r0 20090/29. http://unfccc.int/resource/docs/2009/tp/02.pdf

Watkiss P, Hunt A (2010) Review of adaptation costs and benefits estimates in Europe for the European environment state and outlook report 2010. Technical report prepared for the European Environmental Agency for the European environment state and outlook report, Copenhagen, Denmark, March 2010. Published online by the European Environment Agency, Copenhagen

Watkiss P (2011) Aggregate economic measures of climate change damages: explaining the differences and implications. Clim Change 2(3):356. Published online 2 May 2011 (Wiley Interdisciplinary Reviews)

Watkiss P (ed) (2012) The ClimateCost project. Final report. Volume 1: Europe. Stockholm Environment Institute, Sweden. ISBN 978-91-86125-35-6

Watkiss P, Hunt A (2012) Projection of economic impacts of climate change in sectors of Europe based on bottom up analysis: human health. Clim Change 112(1):101–126

Van der Linden P, Mitchell JB (eds) (2009) ENSEMBLES: climate change and its impacts: summary of research and results from the ENSEMBLES project. Met Office Hadley Centre, FitzRoy Road, Exeter EX1 3PB, UK, 160pp

Chapter 3
On the State of Assessing the Risks and Opportunities of Climate Change in Europe and the Added Value of COIN

Reimund Schwarze

Abstract This paper provides an overview on how climate change impact assessment is conducted in some EU countries, strengths and weaknesses of the current approaches, and the added values of the Austrian study (COIN). It focuses on bottom-up approaches for the assessment of climate risks and opportunities (CRA) as well as costs and benefits (CBA) of climate change. Main findings are: Despite different decision making contexts all methodologies acknowledge the inevitability of "unquantifiable impacts". Uncertainties are pervasive but confidence rankings are not universally applied. Risk scorings and CBA coexist in almost all countries but are differently established in adaptation planning. An important gap in many methodologies of bottom-up is the assessment of cross-sectoral, indirect and macroeconomic effects. The COIN project advances CBA methods in several respects: It carefully defines concepts and impact chains, applies consistent socio-economic scenarios and shared policy assumptions across sectors. It covers cross-sectoral, indirect and macroeconomic effects. And it combines observations and projections, which can be more easily communicated in national dialogues than top down models. A logical next step is parallel national CRA effort—much in similarity to other European countries.

3.1 Introduction

In the face of stagnating international climate negotiations many countries in Europe have put climate change adaptation top on their political agenda, not least in order to find the 'optimal mix' of mitigation and adaptation. Many have also developed national adaptation plans in response to EU commission's white paper on adaptation to climate change (EU Commission 2009). The merits of national assessments of

R. Schwarze (✉)
European University Viadrina, Frankfurt (Oder), Germany

UFZ—Helmoltz Centre for Environmental Research, Leipzig, Germany
e-mail: schwarze@europa-uni.de; reimund.schwarze@ufz.de

K.W. Steininger et al. (eds.), *Economic Evaluation of Climate Change Impacts*, Springer Climate, DOI 10.1007/978-3-319-12457-5_3

climate change risks for adaptation planning are obvious: As decisions on adaptation are to be taken at the national/regional level and the 'costs of inaction' will be felt in national or regional contexts, adaptation is a national or regional good. At the aggregate national level some regional or even local impacts may compensate, e.g. shrinking winter tourism in some regions of a country may be compensated by larger summer tourism in others. At the same time regional disparities and other distributional concerns are increasingly coming to the fore.

Scientific analysis is widely used as a basis for national adaptation planning, however the methodologies applied as well as their scope and levels of consistency differ significantly. Vulnerability and risk assessments are frequently used in Europe. The concepts of vulnerability and risk, however, are often very differently defined. Also the coverage of sectors as well as the consideration of adaptive capacity and the consideration of the costs of adaptation is often very limited—mainly due to a lack of data and adequate scientific resources to study these impacts and costs (European Environment Agency 2007; Watkiss and Hunt 2010).

Essentially there are two methodologies for risk assessment at an aggregate national or regional level:

(a) **top-down** global integrated assessment models (IAMs), which are downscaled to the national or regional scale,[1]
(b) **bottom-up** sectoral impacts assessments, which are up-scaled to the regional or national level.

In principal also, risks can be evaluated either on non-monetary scales (**risks and opportunities**) or in monetary metrics (**costs and benefits**). In practice both could be converted by standardised monetisation factors, however.

The following case studies of national and regional assessments of climate change risks focus on bottom-up approaches evaluated in terms of risks and opportunities as well as costs and benefits.

3.2 United Kingdom

The UK Climate Impact Programme (UKCIP) is probably the earliest in Europe in its use of monetary metrics for climate change impact assessment. It started in the early 2000s with an emphasis on technical assistance and local scale sectoral case study applications in partnership with stakeholder groups. It advanced to become one of the first regional and national assessments for key sectoral impacts based

[1] For an overview of top-down-approaches see Döll (2010) and Schenker (2012). Prominent examples of global IAMs are MERGE, PAGE, FUND, D/RICE or WIAGEM, some of them having been developed further to explicitly model adaptation, e.g. AD-D/RICE and AD-WITCH. It is widely acknowledged, however, that top-down approaches are weaker in their performance for national adaptation planning than for mitigation policies because of their lack of connectedness to empirical studies. See Fisher-Vanden et al. (2013) for a survey of the issues.

Table 3.1 UK national assessment—methodologies, Metroeconomica (2004) as compiled by Hunt (2006)

	What quantified	Proxy for welfare change
Health		
Mortality	Premature deaths; years of life lost	WTP
Morbidity	Respiratory hospital admissions	WTP
Agriculture		
Crops	Δ in crop yield	Gross margin
Flooding	Δ in crop yield	
Biodiversity		
Selected species and habitats	Δ in species space	Restoration cost
Tourism		
Visitor spending	Δ in visitor number	Tourist spending
Water resources		
Drought—domestic use		
Transport		
Infrastructure subsidence	Rail buckling; road subs. Time loss	Restoration cost; WTP
Flooding and coastal inundation	Time loss	WTP
Winter disruption and maintenance	Δ in maintenance req.	Preventative/Restoration cost
Energy		
Heating	Δ in space heating req.	Δ in consumer surplus
Cooling	Δ in space cooling req.	Δ in consumer surplus
Built environment and cultural heritage		
Flooding	Flood damage to buildings	Partial WTP
Subsidence	Subsidence damage to buildings	Restoration cost

Note: WTP is an abbreviation of willingness-to-pay, a questionnaire-based approach to non-market valuation

on a comprehensive guideline for costing of impacts of climate change (Metroeconomica 2004; see Table 3.1).

The subset of sectors included was selective and it excluded the national impacts of climate change elsewhere in the world, specifically the costs of climate-induced migration. Often so, for example in the case of costs to biodiversity, it also lacked quantifiable impacts, e.g. estimates on species space impacts and monetary estimates of the restoration costs (see Table 3.2). Cost impacts are given for four classes of sensitivities (low, medium-low, medium-high, high), separating differente seasonal impacts (e.g. winter and summer mortality effects).

In further developing this methodology, UKCIP acknowledged the inevitability of "unvalued impacts" (Street 2013) and the need for a systematic method to deal with this, e.g. checklists of impacts which can be valued. It also since provides simplified **costing spreadsheets**, geared at different regional and even local

Table 3.2 UK national assessment—provisional results, Metroeconomica (2004) as compiled by Hunt (2006)

Sector/Impact	Annualised impact costs (£ million, 2004 prices) (minus sign denotes benefit)			
	2080s			
	Low (L)	Mid-L	Mid-H	High (H)
Health				
Mortality—summer	3	3	4	8
Mortality—winter	−34	−39	−44	−67
Agriculture				
Crops—mean precpn. (England only)	49	–	–	294
Flooding (England and Wales)	−1	18	2	−4
Biodiversity				
Selected species and habitats	–	–	–	–
Transport				
Infrastructure subsidence	35	49	62	101
Flooding and coastal inundation	13	19	19	26
Winter disruption and maintenance	−102	–	–	−340
Built environment and cultural heritage				
Flooding—fluv. and coastal (England and Wales)	−272	−470	419	353
Flooding—intra-urban	−131	−100	368	32
Subsidence (England only)	162	114	213	316
	Changes in consumer expenditure (£ billion, 2004 prices); minus sign denotes reduction in consumer spending			
Tourism				
Visitor spending	14.8	11.3	12.6	28.9
Energy				
Heating	−1.2	−1.3	−2.1	−2.8
Cooling	0.3	0.1	0.3	1.2

geographical scales, to enable adaptation planners and stakeholders to estimate the costs for a small number of climate impacts in a over setting over relevant time periods. The purpose of these simplified costing tools is to move towards 'orders of magnitude' estimates, in order to provide the kind of flexible decision support desired by users based on the data and budget available to them.

The Department of Environment, Food and Rural Affairs (DEFRA) finally ceased using monetary metrics in its first national **'Climate Change Risk Assessment'** (CCRA 2012). CCRA is based on a risk scoring approach. It provides a detailed analysis of more than 100 potential impacts of climate change in 11 sectors grouped under five themes: agriculture and forestry, business, health and wellbeing, buildings and infrastructure and natural environment. The risk assessment is applied in five steps: (1) Stakeholder experts are asked to assess priority risks, (2) the sensitivity of each risk to climate change is assessed, (3) scenarios of future climate and population are applied to each risk and, on this basis, each risk is

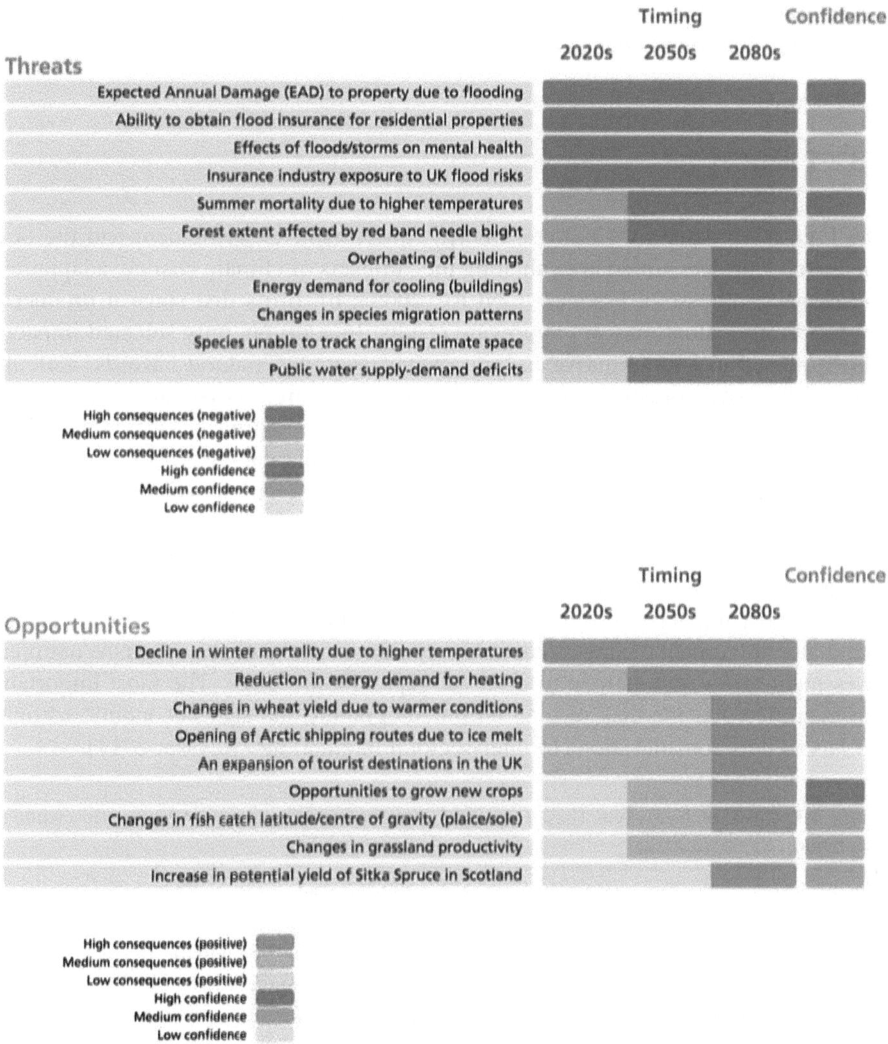

Fig. 3.1 (**a**) Selection of national threats, CCRA 2012. (**b**) Selection of national opportunities, CCRA 2012

assessed, (4) experts score each risk by magnitude and confidence,[2] (5) the scores of all risks are compared.[3] Figure 3.1a, b demonstrates a selection of negative impacts (threats) and positive impacts (opportunities) over different time scales and with

[2] Confidence scores allow for a transparent consideration of the current level of scientific understanding.

[3] Further details are available at: https://www.gov.uk/government/publications/uk-climate-change-risk-assessment-government-report

different levels of confidence. The CCRA is mandated and will be repeated every 5 years, the next one being due in 2017.

3.3 France

The Frensh Ministry for Ecology, Energy, Sustainable Development and the Sea conducted a project aimed at evaluating the "Impacts of climate change, adaptation and associated costs in France" (ONERC 2009a, b). In its first stage it produced only sectoral evaluations, not aggregate results. In addition, only selected impacts were assessed in a quantitative fashion (water resources, natural hazards, agriculture, biodiversity, forests, energy, transport infrastructure, tourism and health). The assessments in these nine sectors followed a pragmatic approach. Different costing approaches were applied in combination with available data, e.g. extrapolation of data from the 2003 heat wave into future climate scenarios (A2, B2), and GIS models of future permanent marine submersion of coastal infrastructure (only main roads) linked to a sea level rise of one metre with associated property losses.[4] A few indicative results are given in Table 3.3.

These cost impacts were intended as 'easy to understand' examples for a first round of involvement of stakeholders in preparation for the next steps in the national assessment. They also helped to discover gaps in knowledge. The most important gap identified was the lack of long term demographic and socio-economic scenarios,[5] as well as considerations of inter-sectoral interactions (Reysset 2012).

Table 3.3 Costs of heatwaves (based on 2003 observations[a]) under an A2 scenario (without adaptation), Reysset (2012)

Cereals losses	−500 M€/year
Clay soils shrinking losses	−1.3 M€/year[b]
Loss of electric power during heatwaves	−0.4 Mtoe/year (negligible)
Health	Cost depends on the pricing of human life loss

[a]Heatwaves in 2030, 2050 and 2090 were—in an approach that was deemed 'easy to perform and understand by stakeholders'—modelled as a replication of '2003-like temperatures'. Using this methods the study finds few significant effects up until 2050 (doubling to quadrupling of effects observed in 2003), but anticipates steeply rising costs of a factor 32–51 compared to the reference period (Reysset 2012, p. 3)
[b]ONERC (2009b) cites "annual damage cost going from approximately 220 M€ (reference period) to 1,300 M€ (under an A2 scenario) in 2100", ibid. p. 29

[4] Detailed impact assessments are available online (French only) at: www.developpement-durable.gouv.fr/Rapport-du-groupe-detravail,10875.html.
A summary of impacts and cost assessments in English can be accessed at: http://www.developpement-durable.gouv.fr/IMG/pdf/rapport_onerc_3_ENG_vf_2.pdf

[5] The study used the current French socio-economic situation (a scenario defined as "constant economy").

3.4 Germany

The German Federal Environment Agency (Umweltbundesamt) is the national platform for the German Adaptation Strategy (DAS). Jointly with the "Bundesländer" (federal state agencies) it is developing a common framework for Vulnerability Assessments (V-assessments) at the regional and national level. The V-assessments in Germany will be based on the IPCC definition of vulnerability which includes exposure, sensitivity and adaptive capacity (IPCC 2013). Issues relating to adaptive capacity are, in contrast to other European approaches, addressed explicitly. The assessment looks, for example, at economic adaptation capacity indicators such as GDP. It also estimates indicators for so-called 'potential/space for adaptation' in different sectors. For example, winter tourism has a small potential/space for adaptation compared, say, to agriculture. The first round of vulnerability assessments were conducted in 2010 in order to prepare the German Adaptation Strategy. The second was started at the end of 2013 (it was delayed by the century-floods experienced this summer in Germany) with a focus on implementing the DAS and a very first indicator-based monitoring of efforts undertaken so far (cf. Umweltbundesamt 2011).

Lacking a consistent general framework for V-assessments, Germany has chosen to build on two different approaches:

1. **FAVAIA** is a scientific project run by the Potsdam Institute for Climate Impact Research that seeks to develop new, science-based transdisciplinary approaches for V-assessments at different levels. It applies a top-down-approach in impact analysis and considers climatic as well as non-climatic drivers. The project develops scenarios and evaluates potential adaptation measures while taking several examples of cross-sectoral interactions into consideration.
2. "**Vulnerability Network**" links relevant actors working at the higher level federal authorities with the aim of establishing an overall picture of Germany's vulnerability to climate change. The 'vulnerability picture' of Germany draws on existing regional and sectoral vulnerability assessments in a bottom-up approach.

Bringing together the results of these two different approaches (science- and policy-driven, top-down and bottom-up) will be a demanding task—and it will take time. There are severe methodological challenges involved, not least including the extent to which different kinds of information from case studies, expert judgements, statistical analyses, etc. can be rendered comparable and commensurable. It is also unclear how such information can be 'normalised' and weighted in order to derive aggregate results, given the difficult normative decisions involved. Further, it is not yet clear how issues of uncertainty can best be integrated into V-assessments. Because of these ongoing methodological gaps, which are difficult to communicate to stakeholders and political agencies, there is currently a debate in Germany over whether or not V-assessments are needed for the DAS at all.[6]

[6] Further information see: www.umweltbundesamt.de/en/topics/climate-energy/climate-change-adaptation

3.5 Switzerland

The Federal Office for the Environment of Switzerland (BAFU) has developed a methodological framework for **'Climate related risk and opportunity assessment'** (**CRA-CH**). This framework was applied in a 2013 pilot study assessment for the canton Aargau, while the first national assessment is due to be completed in 2016 based on this pilot study. The aim of the national assessment is to provide a scientific basis for setting priorities for the implementation of the Swiss adaptation strategy. The starting point comprises a matrix that combines climate-related hazards and effects with impact areas in order to assess the relevance of impacts (see Fig. 3.2).

Selected impacts are studied under two climate scenarios (2060-moderate and 2060-severe[7]), comparing today's socio-economic situation with one socio-economic-scenario for 2060 given different demographic and socio-economic assumptions. Risks are quantified where possible, and standardised monetisation factors are provided (see Table 3.4).

Figure 3.3 summarises the results for quantifiable impacts on six sectors (health, agriculture, forestry, energy, housing and infrastructure, water management) in the pilot study focusing on the canton of Aargau. Positive values indicate opportunities, while negative values indicate costs of climate change under different climate

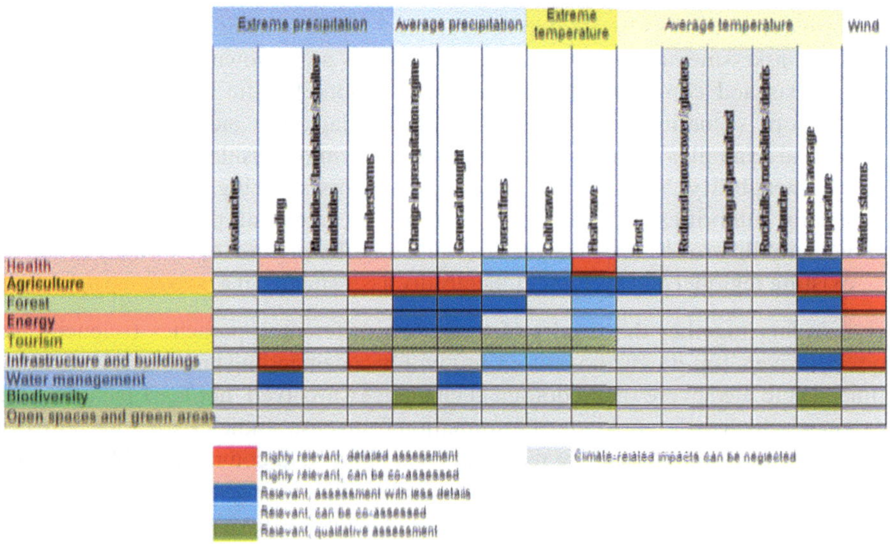

Fig. 3.2 Matrix of climate-related hazards and effects in canton Aargau, Ernst Basler + Partner (2013b), p. 15

[7] 2060-moderate assumes a global reduction in GHG emissions of 90 % in 2100, 2060-severe assumes business as usual.

Table 3.4 Monetisation factors, Basler + Partner (2013a), p. 19

Indicator	Unit	Monetisation
Economy		
Greater yields	CHF	1:1
Lower yields	CHF	1:1
Property loss or costs	CHF	1:1
Society		
Persons affected by heat	Nb. of persons—v. hot days	50 CHF
Persons affected by cold	Nb. of persons—v. cold days	10 CHF
Persons evacuated	Number	10,000 CHF
Persons affected by natural disaster	Number	10,000 CHF
Persons affected by loss of residential areas	Number	10,000 CHF
Jobs lost (social dimension)	Number	100,000 CHF
Jobs created (social dimension)	Number	100,000 CHF
Persons affected by pollen allergies	Nb. of persons—days	10 CHF
Injured and ill	Number	100,000 CHF
Deaths	Number	5 million CHF
Environment		
Biodiversity	Quality classes	No monetisation
Area of valuable biotopes	Quality classes	No monetisation

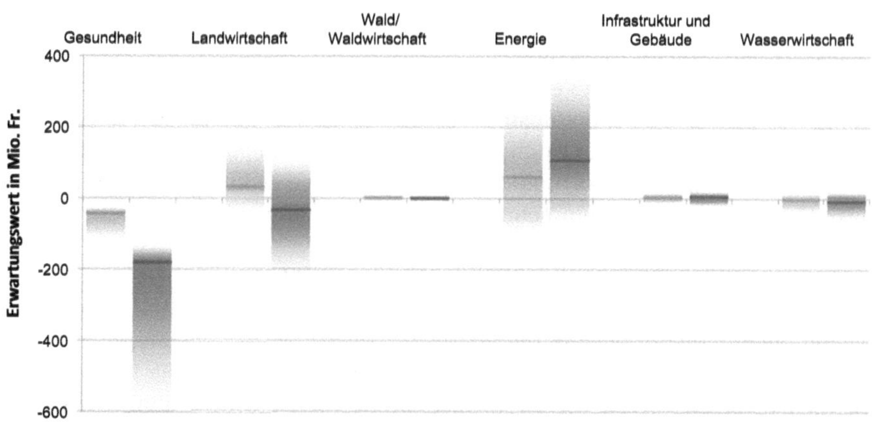

Key: Gesundheit = Health, Landwirtschaft = Agriculture, Wald/Waldwirtschaft = Forestry, Energie = Energy, Infrastruktur und Gebäude = Housing and Infrastructure, Wasserwirtschaft = Water management.

Fig. 3.3 Quantifiable climate risks and opportunities in canton Aargau, Ernst Basler + Partner (2013b), p. 55. *Key*: Erwartungswert in Mio. Fr. = Anticipated value in millions of Swiss francs

scenarios (yellow/left = 2060-moderate, red/right = 2060-severe). Expected values are marked with a dark line on the width bands to indicate uncertainties. For example, the largest costs of around 190 million CHF are expected for health impacts, albeit with a large amount of uncertainty. Uncertainties are represented

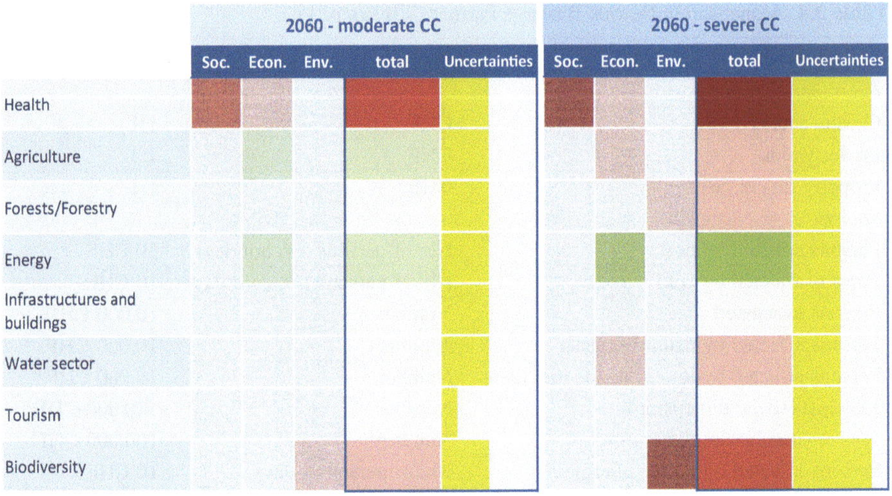

Fig. 3.4 Summary of climate risks and opportunities in the canton of Aargau, Swiss Confederation (2013), Chap. 5; Ernst Basler + Partner (2013c), p. 17

by three classes of width bands (*small*: $0.8 \leq f \leq 1.3$, *medium*: $0.5 \leq f \leq 2$ and *large*: exceeding the $0.5 \leq f \leq 2$ range at both ends), indicating that *small* uncertainties range within ∓20–30 % of expected values of impacts, *medium* uncertainties range within a range of ∓50–100 %, while large uncertainties are driven by extreme events that are difficult to predict.[8]

Non-quantifiable risks are expressed in qualitative terms only, but 'significance ratings' are provided by expert teams for each impact area in order to compare them with the quantified risks. For example, non-quantifiable impacts in agriculture such as soil contamination/fertilisation due to flood-induced sediment movements were ranked between '3' (very significant) and '1' (equally important) but were disputed among experts.

The summary risk scoring compares (1) quantifiable benefits (+values, green) and risks (−values, red-yellow) to (2) climate-risk related uncertainty classes and (3) non-quantifiable risks as in Fig. 3.4.

The main challenges that were identified in this pilot are (1) the lack of robust quantitative results given the large uncertainties; (2) the up-scaling of regional results to the national-level; and (3) ethical issues related to monetisation.[9]

[8] Examples of *small* uncertainties in terms of risk are found in relation to demography and changing agricultural and forestry land uses. *Medium* uncertainties are seen in changes of housing density and average house values, while *large* uncertainties apply to changing agricultural and energy prices (Ernst Basler + Partner 2013a, p. 25).

[9] Specifically in the monetarisation of health effects. For further information see: www.bafu.admin.ch/klimaanpassung/11529/11578/index.html?lang=de and Swiss Federal Confederation (2013), Chap. 5. Details of the pilot project can be found at: www.bafu.admin.ch/klimaanpassung/11529/11578/index.html?lang=de

3.6 Climate Risk Assessment Methods in Europe and the Added Value of COIN

Bottom-up national assessments of climate change risks in Europe can be classified according to how impacts are considered and what valuation metrics is used (see Table 3.5).

The overview of methods in Europe reveals that, despite the: different decision making contexts (local/regional/national, climate risk adaptation and disaster risk reduction) all the methodologies acknowledge the inevitability of "unquantifiable impacts". Methods for translating qualitative impacts into multi-criteria risk assessments are still under development (see CRA-CH). Risk assessments at the local or sectoral level (such as ONERC or the costing spreadsheets of UKCIP) are often selective due to limited resources or time constraints. Uncertainty is pervasive in all generic climate risk assessments, but confidence rankings are not universally applied. Often uncertainty is considered in qualitative terms such as worst case-narratives (COIN) or 'wild cards' (CCRA). Risk score cards and CBA coexists in almost all countries but are differently integrated into adaptation planning processes. Climate risk assessments (CRAs) are often formally embedded into planning procedures and are conducted more frequently while CBA is more informal and less regularly conducted in adaptation planning. The examples from Switzerland demonstrate how CRAs and CBA can be pragmatically combined in an 'order of magnitude' approach by standardised monetisation factors. In contrast, V-assessments (such as in DAS-DE) are still very much in their infancy and are rarely applied in practice.

The COIN project in Austria complements these approaches in several respects. It further refines the UK cost-benefit analysis by explicitly considering uncertainties by means of worst case narratives, not just ranges of results as in Metroeconomica (2004). It carefully defines concepts and impact chains, and applies consistent socio-economic scenarios and shared policy assumptions across sectors. It also combines observations and projections, which can be more easily communicated in national dialogues than top-down models. It specifically advances the state of the art of the assessment of cross-sectoral, indirect and macroeconomic effects. This is a key gap in many bottom-up methodologies (e.g. ONERC). Despite the short time period of the project (1 year), it delivers many insights on the cost drivers across and within sectors and, importantly, it highlights the public budget consequences of climate change in a separate assessment. It also identifies the lack of data (or access to data) in Austria and existing research needs.

There are limitations to COIN, however. Precisely because of the significant gaps in data and methods it tends to be "conservative" in its overall results. The economic costs of climate change in Austria are small in mid-range climate change scenarios. These results need to be carefully contrasted by the range of results from high climate scenarios, which imply much higher damages in the an order of 50–100 %. Quantitative results should be accompanied by the equally important qualitative results. In addition, national aggregate results should cloud our view of significant regional differences of climate change impacts in Austria.

Table 3.5 Overview of methods in Europe, modified from Steinemann and Füssler (2012)

	Impacts		Risk/Opportunities (Scores)	Cost/Benefit (€)	Vulnerability
	Quantitative and non-quantitative	Uncertainty consideration			
Metroeconomica—UK	X			X	
CCRA—UK	X	X	X		
Costing spread-sheets—UK	Selected			X	
ONERC-F	Selected			X	
DAS-DE	X				X
CRA-CH	X	X	X	X	
COIN-AU	X	X		X	

The next steps in a national effort aimed at building upon the COIN approach as presented in this book could be a parallel risk scoring effort, designed explicitly to consider the relevance of the risks to stakeholders and decision makers—as in other European countries. This being an initiative aimed at "costing of inaction", it needs to be accompanied by projects in Austria aimed at "costing of action"—obviously in the field of adaptation but also in that of mitigation. Small domestic costs of climate change do not justify inaction on mitigation as this would fail to take account of how climate change in the rest of the world will affect the national economy. The extent to which, inaction on mitigation in Austria would affect the EU's leadership role and the future outcomes of international negotiations similarly requires additional consideration.

Acknowledgements I acknowledge the comments and support of my fellow members of the COIN-Strategic Advisory Board Roger Street and Paul Watkiss in arriving at the concluding judgements on the COIN project.

References

CCRA (2012) UK climate change risk assessment. Government report. Presented to Parliament pursuant to Section 56 of the Climate Change Act 2008, London, January 2012. https://www.gov.uk/government/publications/uk-climate-change-risk-assessment-government-report

Döll S (2010) Climate change impacts in integrated assessment models: an overview. Zeitschrift für Umweltpolitik und Umweltrecht 1/2010, pp 33–58

Ernst Basler + Partner (2013a) Risiken und Chancen des Klimawandels. Methodenbericht. Bern, Mai 2013

Ernst Basler + Partner (2013b) Risiken und Chancen des Klimawandels im Kanton Aargau. Ergebnisbericht. Bern, Juni 2013

Ernst Basler + Partner (2013c) Climate change risk analysis as a basis for adaption: case study of Aargau/Switzerland. CLARR conference, Bremen, 24–25 Feb 2014

EU Commission (2009) Adapting to climate change: towards a European framework for action / *COM/2009/0147*/. Eur-LEX 52009 DC0147, Brussels

European Environment Agency (2007) Climate change: the cost of inaction and the cost of adaptation, Technical report, Copenhagen

Fisher-Vanden K, Sue Wing I, Lanzi E, Popp D (2013) Modeling climate change feedbacks and adaptation responses: recent approaches and shortcomings. Clim Change 117:481–495

Hunt A (2006) Background paper: reflections on cross-cutting issues in impact assessment. University of Bath and Metroeconomica, UK. Produced for the global forum on sustainable development on the economic benefits of climate change policies, Paris, 6–7 July 2006

Intergovernmental Panel on Climate Change (IPCC) (2013) Managing the risks of events and disasters to advance climate change adaptation. Special report. http://www.ipcc-wg2.gov/SREX/

Metroeconomica (2004) Costing of impacts of climate change in UK: overview of guidelines. UKCIP technical report, Oxford

National Observatory for the Impacts of Global Warming Climate change (ONERC) (2009a) Costs of impacts and lines of adaptation. Report to the Prime Minister and Parliament, Paris

National Observatory for the Impacts of Global Warming Climate change (ONERC) (2009b) Rapport du groupe de travail interministériel sur l'évaluation des impacts du changement

climatique, du coût des dommages et des mesures d'adaptation en France. Paris, 8 Oct 2009. http://www.developpement-durable.gouv.fr/IMG/pdf/rapport_onerc_3_ENG_vf_2.pdf
Reysset B (2012) National assessment of the cost of the impact of climate change in France. Presentation at the workshop of experiences with climate related risk and vulnerability assessments in Europe, Bern, 14 Dec 2012
Schenker O (2012) Ansätze und Herausforderungen in der integrierten makroökonomischen Bewertung von Anpassung. Vortrag beim Workshop "Ökonomische Aspekte der Anpassung an den Klimawandel. Integrierte Modellierung und angewandte Entscheidungsunterstützung, Hamburg, 11 Juni 2012
Steinemann M, Füssler J (2012) Climate related risk and vulnerability assessments. Overview of methodological approaches in countries. Workshop of experiences with climate related risk and vulnerability assessments in Europe, Bern, 14 Dec 2012
Street RB (2013) UKCIP methodology for costing of impacts of climate change. Presentation at the COIN Kick-off meeting, ZAMG Wien, 28 Jan 2013
Swiss Confederation (2013) Switzerland's sixth national communication and first biennial report under the UNFCCC, edited by Federal Office for the Environment, Bern. http://unfccc.int/files/national_reports/annex_i_natcom/submitted_natcom/application/pdf/che_nc6_opt.pdf
Umweltbundesamt (2011) Entwicklung eines Indikatorensystems für die Deutsche Anpassungsstrategie an den Klimawandel, UBA-FB 001555. www.uba.de/uba-info-medien/4230.html
Watkiss P, Hunt A (2010) The costs and benefits of adaptation in Europe: review summary and synthesis. Policy paper no. 2 of the EC 7th FWP climate cost study

Listed Projects

FAVAIA – Formal Approaches to Vulnerability Assessment that Informs Adaptation. http://www.pik-potsdam.de/research/transdisciplinary-concepts-and-methods/archiv/projects/project-archive/favaia/diva/favaia

Network Vulnerability (Germany). http://netzwerk-vulnerabilitaet.de/tiki-index.php (in German)

Part II
Evaluation at the National Level: Methodological Issues

Chapter 4
Climate Impact Evaluation at the National Level: The Interdisciplinary Consistent Framework

Karl W. Steininger, Martin König, Birgit Bednar-Friedl, and Herbert Formayer

Abstract Impact assessment at the national level requires sectoral detail, economy-wide integration, and a consistent framework and toolbox to do so. This chapter discusses the issues and derives the requirements for climate scenarios and local indicators, shared socioeconomic pathways, and economic evaluation to allow for and ensure consistent integration. Finally, a methodological check-list for national level quantitative climate impact assessment is provided.

4.1 Introduction

Research on climate impact benefits and costs is rich and has, when quantitative, so far focused mainly on an aggregated global or continental scale (i.e. aggregated countries and integrated sectors) or on the other end of the possible spectrum, single sector detailed analysis within a particular country (or region within a country). Global economic integrated assessment models estimate the economic costs of climate change, using highly aggregated economic damage functions (usually based on global temperature increase as sole aggregated climate parameter). They

K.W. Steininger (✉) • B. Bednar-Friedl
Department of Economics, University of Graz, Graz, Austria

Wegener Center for Climate and Global Change, University of Graz, Graz, Austria
e-mail: karl.steininger@uni-graz.at; birgit.friedl@uni-graz.at

M. König
Environmental Impact Assessment and Climate Change, Environment Agency Austria, Vienna, Austria
e-mail: martin.koenig@umweltbundesamt.at

H. Formayer
Institute of Meteorology, BOKU—University of Natural Resources and Life Sciences, Vienna, Austria
e-mail: herbert.formayer@boku.ac.at

K.W. Steininger et al. (eds.), *Economic Evaluation of Climate Change Impacts*,
Springer Climate, DOI 10.1007/978-3-319-12457-5_4

are applied to provide total net present values for future damage over time and to appraise the marginal social costs of carbon (the damage cost of an extra tonne of greenhouse gas (GHG) emissions). Their use to this end has been questioned for a number of reasons. Pindyck (2013) emphasises the arbitrary choice of damage function and neglect of catastrophic outcomes; Weitzman (2009, 2012) emphasises the deep-seeded uncertainty around climate sensitivity that is not reflected in the models. Furthermore, the regional sensitivity of various damaging climate parameters (precipitation patterns and intensity, storm frequency/amplitude, heat wave distribution etc.) to increasing global temperatures cannot be reflected adequately in integrated global models. Wagner and Weitzman (2015) synthesise the issues, adding, among others, inadequately high discount rates to the list.[1]

There are two main approaches that differentiate impacts specifically, and thus forgo the related criticism just mentioned. These approaches are briefly presented in Chap. 2 of the present volume, on which also this paragraph is based, and discussed in more detail in Watkiss and Hunt (2010). First, Scenario-Based Impact-Assessments combine climate model outputs with sector impact models (or functional relationships) in order to estimate physical impacts, which are then valued so as to estimate economic costs (expressed as GDP or welfare losses). However, these assessments are not able to capture cross-sectoral, economy-wide effects as they only capture impacts within the respective impact field or sector. There are a number of variations, including risk assessments, which focus on extreme (probabilistic) events such as flooding (using historical analogues or damage-loss relationships), and econometric assessments, which use historical relationships between economic production and climate and then apply these to future climate scenarios. Second, Computable General Equilibrium (CGE) models provide multi-sectoral and macro-economic analysis of the economic costs of climate change. They have the advantage of capturing cross-sectoral linkages and economy-wide effects (and metrics), and they can also look at price and international trade effects. However, they use aggregated representations of impacts and omit non-market impacts.

Major progress within the regionally detailed sectoral analyses has been achieved by methods summarised comprehensively in Metroeconomica (2004). A recent comprehensive application for the example of biodiversity services is carried out by Fezzi et al. (2011).

---

[1] The three most often applied Integrated Assessment Models (IAMs) to date are DICE (Dynamic Integrated Climate and Economy), PAGE (Policy Analysis of the Greenhouse Effect), and FUND (Climate Framework for Uncertainty, Negotiation, and Distribution), with model descriptions given by Nordhaus (1991, 2011), Hope (2006)—on which the Stern review is based (Stern 2007)—and Tol 2002a, b, respectively. The modelling aspects questioned most—for derivation of social costs of carbon by such means—include arbitrary parameter choice in social welfare functions, climate sensitivity (the temperature increase a GHG doubling implies), arbitrary and non-empirical based climate damage functions (usually a functional relationship between temperature increase and (regional) GDP loss, for FUND also distinguishing individual sectors), and neglect of consideration of possible catastrophic outcomes. For a detailed discussion see Ackerman et al. (2009), Watkiss 2011, Pindyck (2013) and Wagner and Weitzman (2015).

One major issue in combining the sectoral and economy-wide aspects is to link the bottom-up and the top-down approaches. A study that is based on the CGE side to that end, focusing on the energy sector, is Böhringer and Rutherford (2009), one that starts at the sectoral impact levels is Bosetti et al. (2006).

Relative to these developments at the global scale on the one hand and at the regional sectoral scale on the other hand, there seems to be quite a gap in the literature for the field in between, i.e. for comprehensive analysis at the country level across *all* changing climate parameters and the resulting types of climate change impacts for as broad a range of relevant impact fields as possible. Chapter 3 of this volume indicates examples of national risk evaluations on climate change, that require such inputs in a more comprehensive framework. National governments are in need to know about challenges that might hit different (economic) sectors and society simultaneously in order to prepare and to prioritise their climate change adaptation efforts. Such research can build upon and has to acknowledge both research strands mentioned earlier, yet new research challenges arise. The present chapter develops in detail one example of a consistent framework for carrying out such research. This framework has been developed by an interdisciplinary team, with its members based in 18 different research institutions across Europe. The framework is tested by applying it to one particular country, Austria, deriving specific results for this country, concluding on uncertainties and, finally, generalising results where possible beyond this particular country's results.

This chapter starts with an overview of the climate change challenges at the national level (Sect. 4.2), develops the core elements of the consistent framework for analysis in Sect. 4.3, and summarises demands in a check-list for national country level studies in the final section.

4.2 Climate Change Challenges at the National Level

Climate change impacts arise basically from two types of changes: gradual changes of climate parameter(s) and changes in the magnitude and variability of extreme events. For the economic valuation of impacts it is useful to distinguish these two classes, as they require different methods of analysis.

Different sector-specific studies exist for both types of changes, mostly for those sectors or regions within a country which face the most adverse economic effects of climate change. The first challenge for a nationwide consistent evaluation thus is to build on available sectoral findings (and the methods and tools available) and develop them further such that sectoral results become comparable in terms of costing impacts and assessing macroeconomic cost, result robustness, and concept of uncertainty.

The merit of such an approach is obvious: Climate change impacts quite often involve compensatory effects at the aggregate level, both across sectors (e.g. while agriculture may suffer from a drought, hiking tourism may benefit from the weather situation) and across regions (e.g. while the regions intensive in winter tourism may suffer, those specialised in summer tourism may benefit). For any government to be able to adequately react beyond autonomous adaptation by planned adaptation, a solid evaluation—consistent across sectors and regions—is a necessary precondition. In this evaluation, the sectoral and regional details in impact evaluation are highly relevant.

The framework we follow here starts with the identification of impact chains of climate change. Mainly based on past experience, the sensitivity of economic sectors to weather and climate conditions can be identified, related to particular climate parameters, or indices thereof. Sometimes one may be restricted to expert judgement alone. We then seek to get a handle on the quantitative relevance of each of these impact chains, both in physical terms and in economic terms (costing of impacts).

The expected costs of climate change will depend on various parameters such as economic structure, land use patterns, demographic developments or state of technology. Let us take one example to illustrate the interdependent relevance of a range of factors:

i. The cost impacts in the health system due to extended and intensified heat waves are dependent on the share of elder people, them being most sensitive for cardiovascular diseases that make them vulnerable to heat waves (calls for demographic analysis).
ii. Whether or not they are living in poorly isolated homes or whether many of them might afford air-conditioned and well-isolated/shaded homes calls for the assessment of the living conditions and/or the financial resources of this group of elder people now and in the future (calls for socio-economic analysis).
iii. If urban sprawl will further accelerate up to 2050, urban heat island effects might hit even more elder/heat vulnerable parts of the population (calls for land-use analysis).

Quantitative evaluation thus needs to be based on a scenario analysis, covering both climatic and socioeconomic dimensions. Thereby, uncertainty in future developments is not only a question of uncertainty on climate change as such, it is—at least of similar importance—a question of which socioeconomic, demographic and land-use path lies ahead.

This first step in our analysis—within sector impacts—can be analysed for each economic sector (or field of analysis) individually. For a comprehensive analysis we have to go at least two steps further, however. First, there might be knock-on effects on other sectors or activities. For example, when longer drought periods require more investment in water supply infrastructure, the financing of this infrastructure would bind public or private budgets (depending on who finances this expansion ultimately) no longer available for other purposes and change the respective demand. Second, these knock-on effects might have a feedback-effect on the sector we started our analysis with. For example, higher water sector investment might raise investment costs and thus bind investment (and increase the knock-on effect just discussed). All these impacts actually occur simultaneously.

This is the challenge at hand. To get a consistent answer that does allow for an aggregate and still meaningful evaluation of impact chains occurring simultaneously, we have to be careful about a number of dimensions. The next section is devoted to discuss them.

4.3 Core Elements of a Consistent Framework

4.3.1 Cost of Inaction: What We Are Concerned with

So far there is no international common understanding of what exactly is meant by 'inaction'. Therefore, we choose a pragmatic definition of inaction for this volume, which is drawn upon the policy needs. For the definition it makes sense to distinguish between planned measures which refer to policies at various scales (basically the National Adaptation Strategies, NAS, setting a frame for provincial, regional or municipal adaptation strategies) and autonomous adaptation which is already happening in the private sector and by individuals and will further happen in all sectors up to a certain degree (in most sectors autonomous adaptation is highly relevant). A second important distinction should be made between anticipatory measures which try to avoid or reduce damage costs before they strike and responsive or reactive measures which aim at repairing and parallel retrofitting after damages occurred (these latter measures might also be referred to as 'late measures').

Since the NAS is usually clearly focusing on planned and anticipatory measures, we refer to 'inaction' as a state of affairs omitting such adaptation.

There also is a range of concepts for "costs of climate change", yet used more consistently in the literature. The focus of this volume is an assessment for the costs of climate change without (anticipatory action in) adaptation, structured by sector (or field of activity, as they are often termed in NAS) and for different time horizons. For conceptual clarification, Stern (2007)—see Fig. 4.1—differentiates among cost concepts, with the present volume here focusing on an assessment of the costs of climate change without adaptation (top line). Note, that this figure only serves a conceptual clarification, and neither will this curve be linear nor is it solely determined by global mean temperature, but by various (regional) climate parameters' change which again activate different impact chains.

4.3.2 Sectoral Detail

The quantification of costs of inaction is thus elaborated along a sector-wise determination of the most important climate-sensitive cost drivers. For each sector, low vs. high damage case assumptions can be derived from a 'scenario family'.

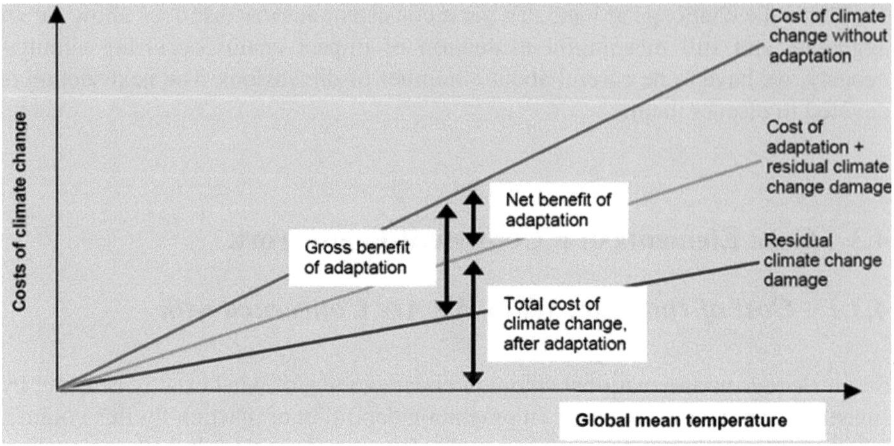

Fig. 4.1 Costs of climate change. *Source*: The Stern review © Cambridge University Press (2007)

Example for health: low fertility/high share of elderly people—low financial resources of elderly or particularly vulnerable people e.g. due to low retirement income—urban sprawl (and thus extension of urban heat islands).

These non-climate scenario settings are then coupled with the sector-wise determined cost-driving climate parameters from the climate scenarios. The individual sectors do not have to work with a full range of potential climate scenario data, but can concentrate instead on the cost-relevant climate indicators that have been agreed upon by sector experts and climatologists.

This is an innovative approach since it allows a wide-range bottom-up sector approach and incorporates a wide range of climate impacts across the fields of activity. In contrast to a top-down integrated assessment, this approach is not limited to assumptions at the aggregate level but is able to deliver sector-specific cost assessments and—where the data allows for—also on a sub-national scale (e.g. for individual climatic regions).

4.3.3 Ensuring Consistency 1: Climate Scenario(s)

In a first step, the agreed cost-relevant climate indicators are calculated for the reference period on NUTS3 level for whole Austria. This allows a regional distinguishing analysis of existing weather and climate related risk and serves as basis for all derived relative climate change signals. The core of the climate change scenarios is one realisation of a high resolution regional climate model run. This 'mid-range climate change' run is forced by the ECHAM5 GCM with the A1B emission scenario and reflects an average development till the end of the century. This run is bias corrected and localised on a 1×1 km grid and allows the calculation of the agreed indicators directly for all NUTS3 regions and time frames.

The relative changes of these calculated indicators are used to define indicator values at NUTS3 level and the time frames 2016–2045 (2030), 2036–2065 (2050) and 2071–2100 (2085). NUTS3 has been chosen as aggregated level, since most socio-economic statistics and data are available at this scale allowing for a spatially consistent coupling of climate and socio-economic data.

To assess the range of potential climate change in Austria, the monthly data for temperature and precipitation of 17 GCMs and 14 RCMs, forced by four different emission scenarios are used. For every season and time frame the warmest/coldest and driest/wettest realisations of this ensemble are defined. For all indicators that have a good correlation with monthly values (e.g. heat days versus monthly mean temperature) a transfer function is derived. This function is used to estimate the indicator value for the extreme realisations of all GCMs/RCMs.

The final outcome of the climate scenario task are consistently derived cost-relevant climate indicators on NUTS3 level for present conditions, climate change conditions of a "mid-range climate change" realisation and—if possible—also the range of the indicator values statistically derived from an ensemble of climate model runs. See Chap. 5 for more details.

4.3.4 Ensuring Consistency 2: The Common SSP

The level of the climate change burden will, however, not just be dependent on the dimensions of climate change itself, but also depend on a multitude of other factors (as indicated above). Therefore, we are confronted with an urgent need to not just look at changing climate parameters and their potential cost-relevant impacts, but to include assumptions and scenarios for developments that might lead—here excluding planned adaptation—towards either resilient or vulnerable pathways thus producing a range of plausible climate costs per sector.

These sectoral assumptions need to be developed in a consistent way across sectors. To do so, we provided all sector experts with a shared socio-economic pathway, short SSP, consistent with our climate (i.e. emission) scenario [cf. Chap. 6 for details], in which the core determinants of climate exposure and sensitivity are set to a reference development. This COIN SSP is roughly coherent with the global SSP2 ('Middle-of-the-road') narrative that has been raised within the scenario family of the IPCC process (cf. O'Neill et al. 2012). Thus, the framework for sectoral assumptions is a common reference SSP with which the central sectoral assumptions do comply.

An example for the mode in which COIN SSP results are taken up is as follows: Within the next decades, urban (sealing) areas around some Austrian cities will grow reducing the amount of arable land. In some remote areas, forest areas will increase at the expense of arable land. These trends lead to a net decrease of arable land in the northern, eastern and south-eastern parts of Austria, while forest land and urban (sealed) surface shares are increasing. The shifts among the land use classes are given by the COIN SSP, while the inner-sectoral structure of the land use (for example the cultivated crops), the (afforested) tree species composition as well as the building density assumptions are taken by the sectoral experts. For full details see Chap. 6 (Shared Socioeconomic Pathways).

4.3.5 *Ensuring Consistency 3: The Common Economic Evaluation*

As the objective is to ensure consistency and avoid double counting it is crucial that

i. impact chains are clearly assigned and/or divided between fields of activity;
ii. physical economic impacts (e.g. changed productivity) are translated into economic indicators (e.g. higher/lower unit cost of production), and that fields of activity are matched to the sectoral structure of the economy;
iii. each sector works within the same framework of economic valuation; i.e. that the work for each impact field or sector is based on the same tools and methods for assessment of costs of climate change in determining direct costs of inaction (inaction in adaptation): as is shown in Chap. 7, climate impact chains can be classified such that one of five basic economic evaluation approaches is adequate;
iv. a macroeconomic assessment is first conducted for each field of activity separately, to identify the effects of climate change impacts in the sector itself and the feedback effects to the rest of the economy: here we choose the computable general equilibrium approach (CGE), as this best allows for the evaluation of significant impacts and feedbacks, which go beyond what time series analysis could explore based on past experience;
v. a combined macroeconomic assessment is then conducted for all fields of activity jointly, to assess the overall costs of inaction.

Each of these steps requires extensive discussions and joint solutions among the involved experts. In particular, an excel based toolbox was developed for step iii. In addition to setting up a common and consistent modelling framework, it is furthermore essential to develop a common language and use of terms.

For more details on the toolbox and the macroeconomic model, see Chap. 7.

4.4 Summary or "Check-List" for Country Level Study

The economic quantification of costs of inaction (costs of climate change) thus requires

- to specify the sector's (sub-)national economic relevance (e.g. share in national/regional GDP) and its exposure (e.g. location of winter ski resorts);

- to specify socio-economic developments which drive sectoral exposure and sensitivity;
- to specify the key climate stimuli per sector (e.g. change in snow cover days), and the corresponding impact chains (e.g. shortening of winter tourism season);
- to specify one or more sensitivity indicators (e.g. reduction in overnight stays due to shorter season) which can be translated into economic costs (e.g. reduced income/revenues in winter tourism);
- to collect data on current sectoral impacts to estimate a sectoral impact function or model or alternatively to apply estimates from other countries/studies;
- to set up a value function (or again use value transfer from other studies) in order to translate physical impacts into economic costs or benefits; and
- to specify how these costs or benefits influence the economy (via changed productivity or technology, changed cost structure, changed demand; as recurrent or investment cost).

As the aim of this evaluation is to assess the cost of inaction, the adaptive capacity is assumed to remain constant at today's level. By the same argument, no planned (private or policy-induced) adaptation is accounted for. Autonomous adaptation, however, is considered within each sector to an extent that it is cost-efficient for individual actors in the respective sectors.

References

Ackerman F, DeCanio SJ, Howarth RB, Sheeran K (2009) Limitations of integrated assessment models of climate change. Clim Change 95:297–315

Böhringer C, Rutherford TF (2009) Integrated assessment of energy policies: decomposing top-down and bottom-up. J Econ Dyn Control 33(9):1648–1661

Bosetti V, Carraro C, Galeotti M, Massetti E, Tavoni M (2006) WITCH: a world induced technical change hybrid model. Energy J 27:13–38. Special issue on hybrid modeling of energy-environment policies: reconciling bottom-up and top-down

Fezzi C, Crowe A, Abson D, Bateman I, Askew T, Munday P, Pascual U, Sen A, Darnell A, Haines-Young R (2011) Evaluating provisioning service values: evaluating provisioning ecosystem service values: a scenario analysis for the United Kingdom. Report to the Economics Team of the UK National Ecosystem Assessment, UK NEA Economic Analysis Assessment

Hope C (2006) The marginal impact of CO_2 from PAGE2002: an integrated assessment model incorporating IPCC's five reasons for concern. Integr Assess 6(1):19–56

Metroeconomica (2004) Costing the impacts of climate change in the UK: overview of guidelines. UKCIP technical report. UKCIP, Oxford

Nordhaus WD (1991) To slow or not to slow: the economics of the greenhouse effect. Econ J 101 (407):920–37

Nordhaus WD (2011) Estimates of the social cost of carbon: background and results from the RICE-2011 model. National bureau of economic research working paper 17540

O'Neill B, Carter TR, Ebi K, Edmonds J, Hallegatte S, Mearns L, Riahi K (2012) Workshop on the nature and use of new socioeconomic pathways for climate change research. Meeting report boulder

Pindyck RS (2013) Climate change policy: What do the models tell us? J Econ Lit 51(3):860–872

Stern N (2007) The economics of climate change: the Stern review. Cambridge University Press, Cambridge
Tol R (2002a) Estimates of the damage costs of climate change, Part I: Benchmark estimates. Environ Resour Econ 21(1):47–73
Tol R (2002b) Estimates of the damage costs of climate change, Part II: Dynamic estimates. Environ Resour Econ 21(2):135–60
Wagner G, Weitzman ML (2015) Climate shock: the economic consequences of a hotter planet. Princeton University Press, Princeton
Watkiss P, Hunt A (2010) Review of adaptation costs and benefits estimates in Europe for the European environment state and outlook report 2010. Technical report prepared for the European Environmental Agency for the European environment state and outlook report, Copenhagen, Denmark, March 2010. European Environment Agency, Copenhagen
Watkiss P (2011) Aggregate economic measures of climate change damages: explaining the differences and implications. WIREs Clim Change 2:356–372
Weitzman ML (2009) On modeling and interpreting the economics of catastrophic climate change. Rev Econ Stat 91(1):1–19
Weitzman ML (2012) GHG targets as insurance against catastrophic climate damages. J Public Econ Theory 14(2):221–244

Chapter 5
Climate Change Scenario: From Climate Model Ensemble to Local Indicators

Herbert Formayer, Imran Nadeem, and Ivonne Anders

Abstract The aim of this task within the COIN framework is the preparation of the climatological information for all involved sectors for the past and the possible range of future developments. As a basis for the historical observations, products of the Austrian weather service (ZAMG) are used. The climate change scenarios are derived from 31 regional and global climate models forced with four different emission scenarios.

Impact relevant climate depending indicators have been developed and calculated from observational data and climate change scenario on a NUTS3 level. In total, 63 impact relevant indicators have been defined. The majority of the indicators are a kind of "peak over threshold" analyses like the temperature threshold heat day (Tmax $\geq$ 30 °C).

All climate scenarios indicate a warming within the twenty-first century. The whole ensemble indicates a warming of 0.5 up to 4 °C till 2050 and at the end of the century the warming reaches from ~2 °C up to 6 °C in winter and up to 9 °C in summer. The low border stems from models forced with the RCP 4.5 emission scenario and the high border from models forced with RCP 8.5.

The climate change signal for precipitation is not that clear. The annual sum shows no clear trend. For summer precipitation, the majority of the model indicates a decrease till −20 % and in winter an increase of the same magnitude.

The derived indicators reflect the same trends. In general, it can be said that temperature depending indicators at the middle of the century derived from the hottest realisations have a similar climate change signal as the "mid-range" scenarios at the end of the century.

H. Formayer (✉) • I. Nadeem
Institute of Meteorology, BOKU—University of Natural Resources and Life Sciences, Vienna, Austria
e-mail: herbert.formayer@boku.ac.at; imran.nadeem@boku.ac.at

I. Anders
Climate Research Section, Division Data, Methods, Modelling, ZAMG—Central Institute for Meteorology and Geodynamics, Vienna, Austria
e-mail: ivonne.anders@zamg.ac.at

K.W. Steininger et al. (eds.), *Economic Evaluation of Climate Change Impacts*, Springer Climate, DOI 10.1007/978-3-319-12457-5_5

5.1 Introduction

The aim of this task within the COIN framework is the preparation of the climatological information for all involved sectors for the past and the possible range of future developments. As a basis for the historical observations, products of the Austrian weather service (ZAMG) are used. The climate change scenarios are derived from 31 regional and global climate models forced with four different emission scenarios.

In interactions with the experts of the involved sectors, impact relevant climate depending indicators have been developed and calculated from observational data on a NUTS3 level (NUTS 2013) for Austria as a whole. The direct calculation of these indicators under climate change conditions require bias corrected and localised scenario data on a daily basis, which can only be done for one scenario.

To consider the whole range of possible future developments a statistical method was developed, to estimate the indicators from high resolution local observations and monthly anomalies. With this method, it was possible to estimate all indicators of the "peak over threshold" type (e.g. heat days) from all 31 climate scenarios for the whole twenty-first century.

5.2 Basic Climate Information

5.2.1 Basic Climate Change Scenarios

Climate models are the primary tools for investigating the climate system and its response to different driving forcings, as well as for calculating climate scenarios on various time scales. Of course, climate models have limitations and we do not have the exact information of how the human society will behave within the next decades. Several different pathways of how the anthropogenic greenhouse gas emissions will develop within the twenty-first century are possible.

Multi-model ensembles produced within international research projects such as the PRUDENCE project (Christensen et al. 2007), the ENSEMBLES project (Hewitt and Griggs 2004), or the latest CMIP5 project (Taylor et al. 2012) are nowadays standard to account for this model spread (Meehl et al. 2007; Tebaldi and Knutti 2007).

To assess the range of potential climate change in Austria, the monthly data of temperature and precipitation of 17 general circulation models (GCMs) and 14 regional climate models (RCMs), forced by 4 different emission scenarios, are used. Table 5.1 gives the acronyms of the used GCMs and RCMs and the forcing emission scenarios. The ensemble includes the whole range of the RCP emission scenarios (Meinshausen et al. 2011). This approach takes into account the uncertainties stemming from the GCMs themselves (model limitations and internal

Table 5.1 List of GCMs and RCMs used to assess the spread of the future development

Model acronym (GCMs)	Emission scenario	Model run	Model acronym (RCMs)	Emission scenario	Model run
CNRM-CM5	rcp4.5	Cmip 5	C4IRCA3_HadCM3Q16	A1B	ENSEMBLES
CNRM-CM5	rcp8.5	Cmip 5	DMI-HIRHAM5_ARPEGE	A1B	ENSEMBLES
HadGEM2-CC	rcp4.5	Cmip 5	DMI-HIRHAM5_ECHAM5	A1B	ENSEMBLES
HadGEM2-CC	rcp8.5	Cmip 5	DMI-HIRHAM5_BCM	A1B	ENSEMBLES
MPI-ESM-LR	rcp4.5	Cmip 5	ETHZ-CLM_SCN_HadCM3Q0	A1B	ENSEMBLES
MPI-ESM-LR	rcp8.5	Cmip 5	METO-HC_HadCM3Q0	A1B	ENSEMBLES
MPI-ESM-MR	rcp4.5	Cmip 5	METO-HC_HadCM3Q16	A1B	ENSEMBLES
MPI-ESM-MR	rcp8.5	Cmip 5	METO-HC_HadCM3Q3	A1B	ENSEMBLES
GFDL-CM3	rcp4.5	Cmip 5	KNMI-RACMO2_ECHAM5	A1B	ENSEMBLES
GFDL-CM3	rcp6.0	Cmip 5	MPI-M-REMO_SCN_ECHAM5	A1B	ENSEMBLES
GFDL-CM3	rcp8.5	Cmip 5	SMHIRCA_ECHAM5	A1B	ENSEMBLES
HadGEM2-ES	rcp4.5	Cmip 5	SMHIRCA_HadCM3Q3	A1B	ENSEMBLES
HadGEM2-ES	rcp6.0	Cmip 5	ICTP_RegCM3_ECHAM5	A1B	ENSEMBLES
HadGEM2-ES	rcp8.5	Cmip 5	CNRM_ALADIN_ARPEGE	A1B	ENSEMBLES
HadGEM2-AO	rcp4.5	Cmip 5			
HadGEM2-AO	rcp6.0	Cmip 5			
HadGEM2-AO	rcp8.5	Cmip 5			

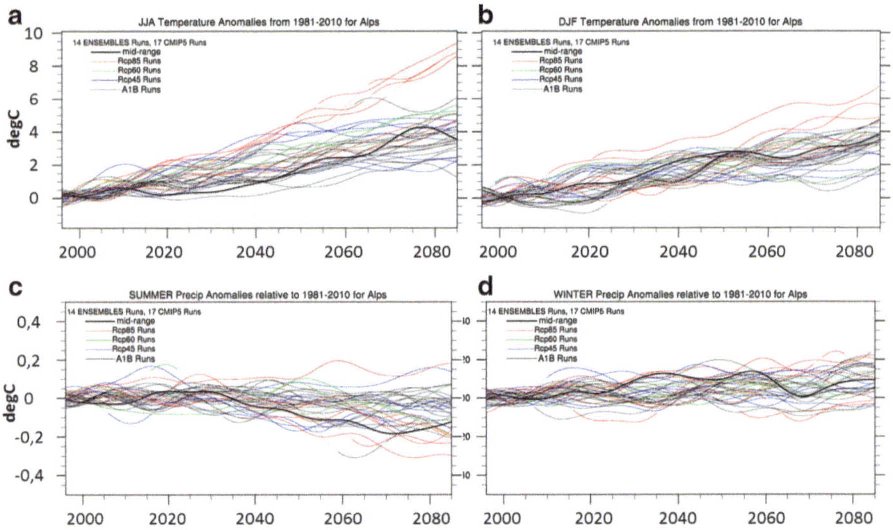

Fig. 5.1 Climate change signal of the GCM/RCM ensemble for Austria within the twenty-first century (reference period 1981–2010): (**a**) summer (JJA) temperature; (**b**) winter (DJF) temperature; (**c**) summer half year (Apr.–Sep.) precipitation, (**d**) winter half year (Oct.–Mar.) precipitation. *Bold line* indicates the high resolution "mid-range" scenario. Note that the different colours represent the simulation runs: A1B Runs (*grey*), Rcp45 Runs (*blue*), Rcp60 Runs (*green*), Rcp85 Runs (*red*) and Mid-range Run (*black bold line*)

temporal variability), the downscaling with RCMs, and different emission scenarios.

In Fig. 5.1, the smoothed climate change signals for summer and winter temperatures as well as the precipitation sum of the summer half- and winter half-year for Austria are shown. The climate change signal is the difference of the actual value and the climatological value of the reference period (1981–2010) for temperature and the quotient for precipitation. In the climate change signal, the individual model biases are removed. In summer temperature increase ranges from 1.8 up to 9 °C and in winter from 1.8 up to 7 °C. The climate change signals for precipitation are not so clear. The magnitude of the summer half year precipitation reaches from +15 to −30 % with a higher fraction of models with negative values. The magnitude of the winter half year precipitation reaches from −5 % up to +25 % with a majority of positive values.

5.2.2 *The "Mid-Range" Regional Climate Scenario*

GCM and even standard RCM results are far too coarse to resolve the complex topography of the Alps within Austria. The core of the COIN climate scenario ensemble is a very high resolution (10 km) RCM simulation of the RCM CCLM

(Meissner et al. 2009) forced with the ECHAM5 A1B, from the Austrian research project reclip:century (Loibl et al. 2010). For this model simulation all relevant meteorological variables such as temperature, precipitation, solar radiation or snow are available on a daily basis with a spatial resolution of 10 × 10 km.

Even high resolution RCMs are too coarse for a direct calculation of impact relevant indicators within the Alpine region. Additionally, the quality of the modelled variables is not sufficient for direct application. A bias correction and localisation of the RCM scenario on a daily basis is necessary.

A basis for the bias correction is the 1 × 1 km gridded observational data set on daily base for mean daily temperature and precipitation from 1977–2006 from the Austrian weather service (Schöner and Dos Santos Cardoso 2004). For daily temperature-minima and -maxima the INCA data set (Haiden et al. 2009) from 2003–2012 is used.

In a first step, the percentile of a specific threshold value (e.g. Tmax $\geq$ 30 °C) on every 1 × 1 km cell of the observational data set is calculated. For the corresponding RCM raster grid (same NUTS3 region, same elevation ±50 m) the absolute value of this percentile (e.g. 29 °C if the model has a cold bias) during the reference period is calculated. If more than one 1 × 1 km cell is within the ±50 m elevation band, the median is used. To calculate the climate change scenarios, the exceedances of these absolute values are counted. This method includes not only a bias correction based on quantiles but also localisation on the 1 × 1 km scale. It gives the same results as a classical bias correction using a quantile mapping technique (Themeßl et al. 2011), but only for the selected quantile values. This dramatically reduces the numerical calculations.

Figure 5.2 shows the resolution of the 1 × 1 km observational raster data. The steep gradients of the annual mean temperature within the mountainous regions within Austria highlight that this resolution is able to resolve the complex topography. The white lines indicate the borders of the NUTS 3 regions in Austria.

5.2.3 Climate and Climate Change Information for the NUTS3 Regions

From the climate change impact on the community, Austria is divided in sub-regions. These sub-regions highly depend on the sector or even sub-areas of a sector. Hydrologists are mostly interested in catchment areas, but agronomists which study climate change effects on grapevine focus on the wine growing regions only.

To provide climate and climate change information for all these “target regions”, we decided to prepare the climate information on a NUTS3 level. The Austrian NUTS3 regions have a size of approximately 2,500 km^2. Each NUTS3 region comprises ~25 grid cells from the high resolution RCM. Regarding climatology, these regions can be seen as homogeneous, except for elevation dependent variables

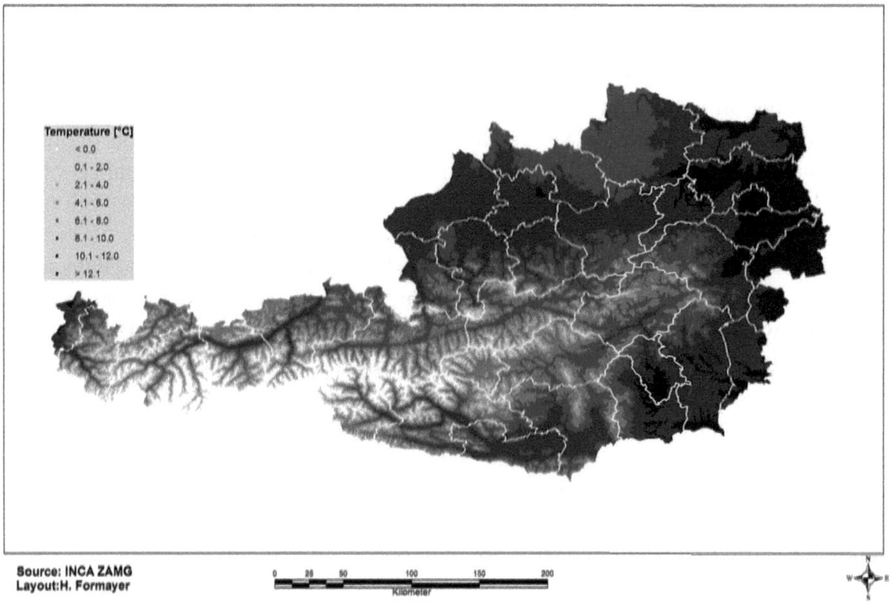

Fig. 5.2 Annual temperature distribution (2003–2012) of Austria based on the 1 × 1 km temperature data set of ZAMG-INCA and the borders of the NUTS 3 regions in Austria. The temperature distribution highlights the spatial resolution of the gridded observation data set used for bias correction

Table 5.2 List of climate related indicators calculated for COIN

Relevant meteorological parameter	Numbers
Temperature related indices	30
Precipitation related indices	19
Snow related indices	12
Global radiation	1
Relative humidity	1

like temperature or snow cover. NUTS3 regions are still small enough to capture the specific regional climate effects of the lake Neusiedel, inner Alpine valleys, or the urban effects of Vienna.

Every climate variable is calculated for the 35 NUTS3 regions, the Austrian median (based on the 1 × 1 km values), and for the river catchments of Inn, Salzach, Mur and Kamp. For temperature depending indicators, not only the median of all grid cells within a sub-region is given as for precipitation. Additionally, the median of the ten warmest grid cells and several elevation bands are calculated. The warmest grid cells are calculated from the mean annual temperature and therefore correspond mainly to the lowest areas within the sub-region. The total Austrian elevation range of 500–3,000 m is covered by 500 m steps. Elevation bands not existing in a sub-region are indicated with a missing value.

For precipitation, a different scheme is used. We assume that the daily precipitation is more or less of the same type (convective or stratiform) within a sub-region. The average condition for a NUTS3 region is defined by the spatial median value within the region. Due to the strong annual cycle of the precipitation, the indicators are calculated on a seasonal, half year, and yearly base. A table with all results of a temperature and precipitation depending indicator is included in the online supplementary material.

5.3 Impact Relevant Climate Indicators

5.3.1 General Indicators

One important benefit of the COIN approach is that the climate change indicators are defined within an interdisciplinary dialogue. In several working groups, experts of different sectors and climate modellers discussed the usability of different indicators and the possibility of calculating them in an adequate way for present and future conditions.

In total, 63 impact relevant indicators have been defined. The majority of the indicators are a kind of "peak over threshold" analyses like the temperature threshold's heat day (Tmax $\geq$ 30 °C), ice days (Tmax < 0 °C), or the length of the vegetation period. Concerning precipitation, indicators for dry conditions (e.g. consecutive dry days) and heavy precipitation (e.g. numbers of days with the amount of precipitation exceeding specific thresholds) are of interest. For snow, solar radiation, relative humidity, the inter-annual variability, and the related climate change signals are calculated. A general overview of the calculated indicators is given in Table 5.2 and a full list of indicators is given in the online supplementary material.

Some exemplary results of the spatial distribution and the climate change signal are shown in Figs. 5.3 and 5.4. They contain the comparison of average conditions of different time slices. The observation period is 1981–2010. The scenario time frames 2030, 2050 and 2085 refer to the periods 2016–2045, 2036–2065 and 2071–2100. Time slice comparisons always show a mixed signal of random decadal variability and climate change (Formayer 2010), especially when based on one model. As long as the decadal variability of a variable is not much smaller than the climate change signal, only ensemble considerations can overcome this limitation.

In Fig. 5.3, the numbers of heat days are shown for the warmest part of the NUTS3 regions. The recent warmest regions have ~15 heat days per year on average whereas in the coldest region only one heat day is observed. During the century, the number of heat days increase steadily and at the end of the century the average value of heat days exceeds 50 days in several sub-regions. Even the coldest sub-region, Lungau, where the whole area lies above 1,000 masl, 16 heat days are reached at this time.

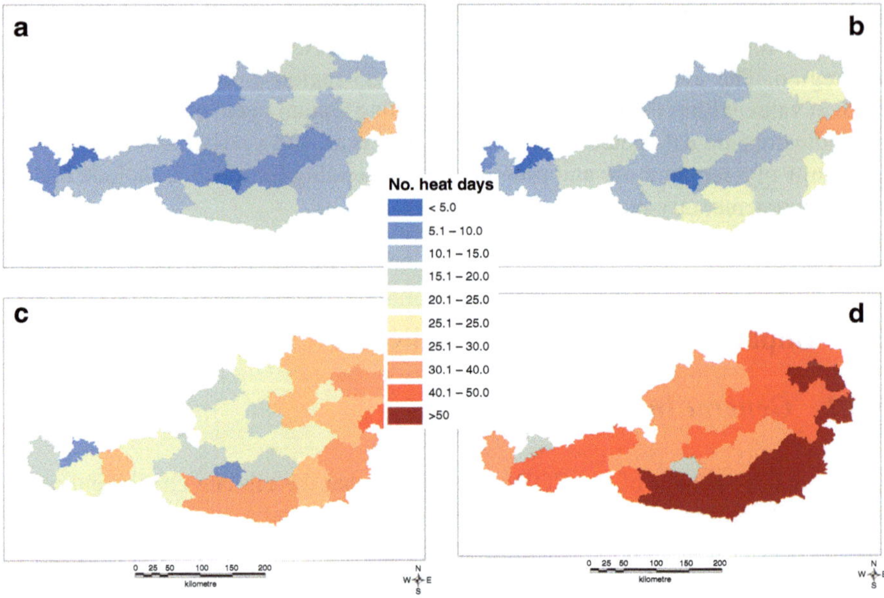

Fig. 5.3 Spatial distribution of heat days (Tmax $\geq$ 30 °C) for the observation period (**a**), 2030 (**b**), 2050 (**c**) and 2085 (**d**) for the hottest 10 km^2 per NUTS3 region and the mid-range scenario (hottest 10 km^2)

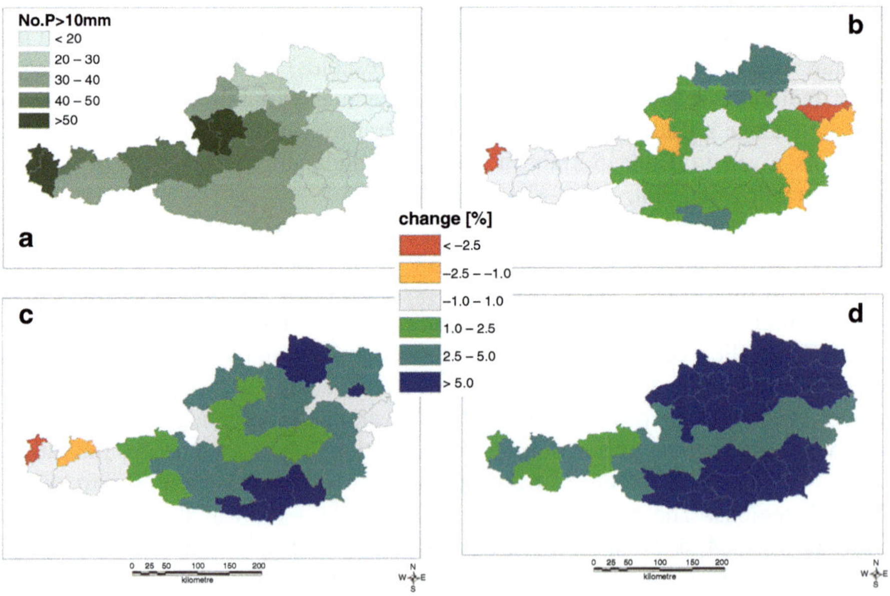

Fig. 5.4 Spatial distribution of the number of precipitation days with more than 10 mm for the observation period (**a**), and the relative changes in percent for 2030 (**b**), 2050 (**c**) and 2085 (**d**) per NUTS3 region and the mid-range scenario (median value of the NUTS 3 region)

In Fig. 5.1a, it is shown that the mid-range scenario belongs to the coldest summer realisations of the whole ensemble in the first half of the twenty-first century and at the end of the century it reflects an average warming. Thus, the heat day scenarios for 2030 and 2050 in Fig. 5.3b, c have to be interpreted as low estimates.

Figure 5.4 shows the spatial distribution of the number of days with more than 10 mm precipitation. In Fig. 5.4a, the actual values are shown. The maximum values with 40 and more than 50 days per year are the northern "Stau" regions of Austria and Vorarlberg. South of the Alpine main ridge only values below 40 days are reached and the lowest values with less than 20 days are in the dry north-eastern regions of Austria.

In the other three panels, the relative change of the number of days is shown for the three scenario periods. In the near future, some regions indicate a weak decrease in the order of a few percent. These regions are located in the eastern part and the outermost west. Till the end of the century, an increase can be observed in all regions. The relative increase is highest in the north-eastern and south-eastern part of Austria and reaches values of slightly higher than 5 %. Along the main ridge of the Alps and in the western parts of Austria the relative increase is not that high and reaches ~2.5 %.

There is no simple dependency between the average precipitation change shown in Fig. 5.1c, d and the number of precipitation days with more than 10 mm precipitation; therefore, a simple ensemble estimate is not possible. The change in the sign of the relative change during the first decades of the scenario period might indicate that inter-decadal variability dominates in this period.

5.3.2 Indicators for Special Applications

Especially for damage estimates on infrastructure specific indicators such as the amount of precipitation per day with specific return periods are calculated. Only the empirical values from the 30 year periods are used for these calculations and no extreme value statistic is applied. Therefore, only moderate extremes with return periods of 1, 2, 5 and 10 years are considered, to avoid the partly random influence of the maximum precipitation value. Figure 5.5a shows the actual spatial distribution of a 5 year precipitation event. The values range from more than 75 mm within the northern "Stau" regions and in Vorarlberg to less than 50 mm in the dry north-east.

In the near future, the climate scenario (b) shows a weak decrease of ~5 % in large areas of central and eastern Austria and a weak increase in the rest. Till the middle of the century (c), the areas with decreasing precipitation intensity for a 5 year event vanish totally in the west, north, and south of Austria. In these regions, the increase is of the magnitude of 5–10 %. At the end of the century (d), only a few regions show no to moderate increase. In large regions along the main Alpine ridge

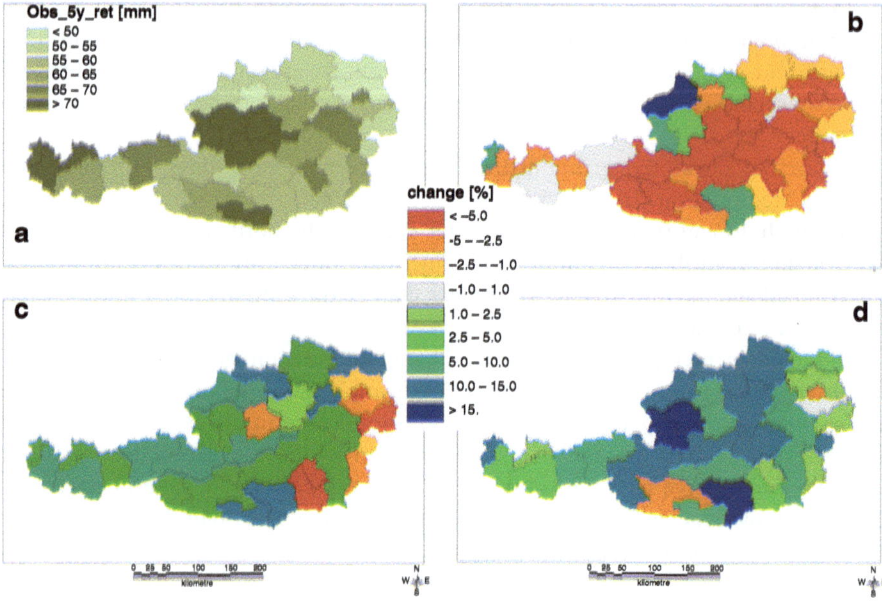

Fig. 5.5 Spatial distribution of the precipitation amount of events with 5 year return period the observation period (*up left*), 2030 (*up right*), 2050 (*down left*) and 2085 (*down right*) per NUTS3 region and the mid-range scenario (median value of the NUTS 3 region)

and north of it, the increase is in the order of 10–15 % and in two NUTS3 regions the increase exceeds 15 %.

Similar to the indicator shown in Fig. 5.4, the first half of the twenty-first century seems to be dominated by the inter-decadal variability and in the second half of the century a trend to an increasing intensity can be seen.

To assess the probability of specific amounts of precipitations in the sub-regions, a gamma distribution (Vlcek and Radan 2009) was fitted to the daily precipitation values. The parameters α, β of the gamma function, and the probability of no precipitation from the observations are calculated on a monthly basis. To get a sufficient number of precipitation events, a moving window of 3 months was used to estimate the parameter of the central month. The gamma distribution was derived for the observations and the high resolution mid-range scenario without bias correction. The gamma distribution for the median value of Austria in logarithmic version is shown in Fig. 5.6. In a standard view, the gamma distributions of the four different time slices cannot be distinguished but in the logarithmic form it can be seen that the probability of higher precipitation intensities is becoming more frequent in the scenario periods.

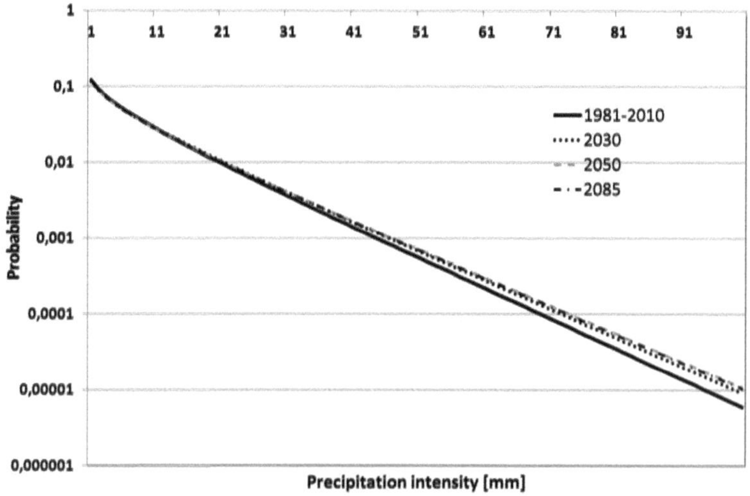

Fig. 5.6 Fit of gamma distribution function to daily precipitation for the whole of Austria for the reference period and the three scenario periods

5.4 Climate Change Signal of Impact Relevant Indicators for Austria

5.4.1 Indicator Trends for the Mid-Range Scenario

The calculated indicators include a lot of information. As Austria is separated in 35 NUTS region, the spatial differences of indicators for the present climate can be investigated. For each sub-region, the temporal development of an indicator or the elevation dependency of an indicator can be shown. For all of Austria, an analysis of the spatial dependency of the climate change signal is possible.

In the previous section, some examples of indicator distributions and climate signals were shown. In Fig. 5.7, the temporal development of heat days and frost days for the NUTS3 region Vienna are shown. According to the mid-range scenario, the heat days in Vienna will increase from 20 days at present up to ~30 in the middle of the century and ~50 at the end of the century. The frost days will decrease from actual more than 50 days, to roughly 20 in the middle of the century, and then to less than 15 days at the end of the century.

Another example is shown in Fig. 5.8. The upper left panel gives the frequency of days within a dry period of consecutive 7 days in the summer half year (April to September) for the present climate. This indicator shows a strong west/east gradient with less than 20 days in west of Austria and more than 50 days in the east. In the near future, a weak decrease can be seen in north-eastern Austria but in most parts there will be an increase up to 50 %. Till the middle of the century, this indicator is increasing throughout Austria with a maximum increase along the main ridge of the Alps up to 100 %. To the end of the century, the whole Alpine section of Austria is

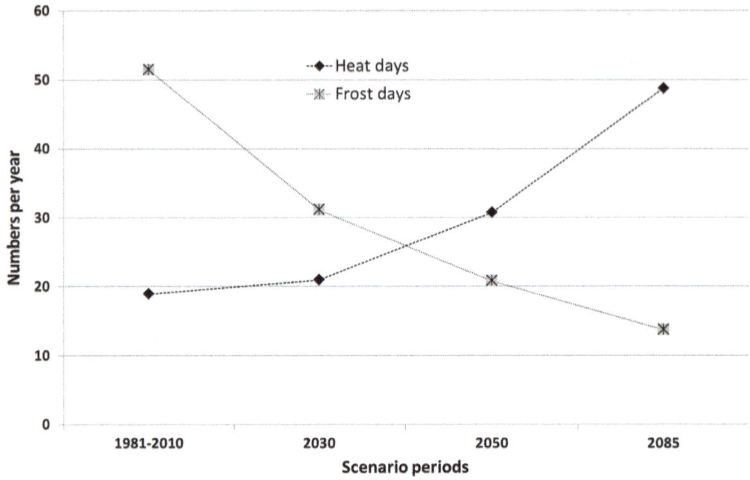

Fig. 5.7 Temporal development of heat days and frost days at the NUTS3 region Vienna within the twenty-first century (hottest 10 km^2)

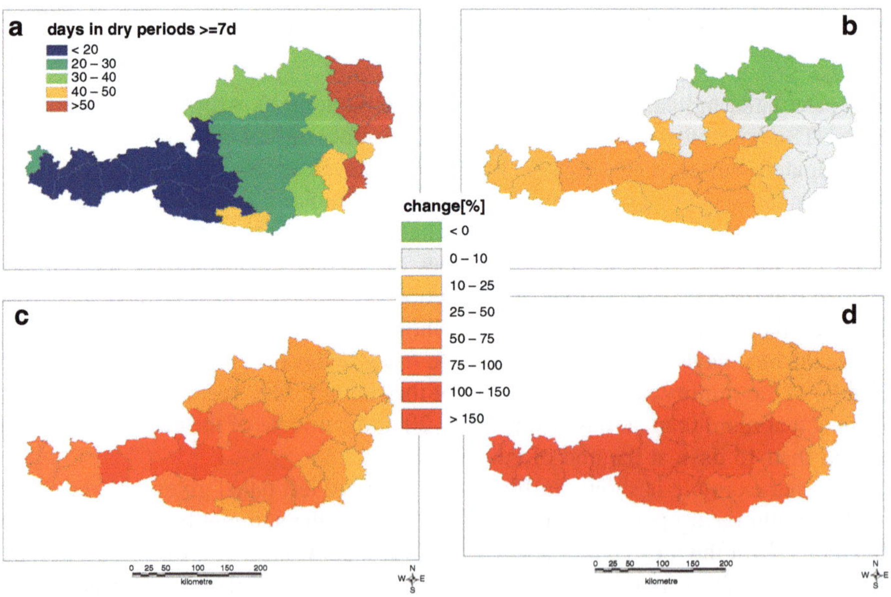

Fig. 5.8 Spatial distribution of number of days within a dry period of at least 7 days in summer in the observation period (*up left*) and the relative change (%) for the period 2030 (*up right*), 2050 (*down left*) and 2085 (*down right*) per NUTS3 region and the mid-range scenario (median value of the NUTS 3 region)

doubling the amount of days within dry periods and even in the dry parts of Austria the increase is 50 % or higher. This development fits to the climate change signal for summer precipitation in Fig. 5.1, which indicates no change in summer precipitation up to 2040 and afterwards a decrease for the mid-range scenario (thick line).

5.4.2 Deriving Indicators from Monthly Mean Values

All findings up to now are based on the mid-range scenario, but as highlighted in Fig. 5.1, a broad range of changes in temperature and precipitation are possible within the twenty-first century in Austria. This different change in the average conditions must have effects on the impact related indicators. It was not possible to bias correct and localise all 31 GCM and RCM scenarios to a 1 × 1 km raster on a daily basis within the COIN framework. Therefore, a different approach had to be found to quantify the range of the potential development of the indicator values during the twenty-first century.

To derive indicator values for the extreme realisations of the GCM/RCM ensemble, a statistical relation between monthly mean values of temperature and precipitation and indicators are necessary. To test the hypotheses of a statistical relationship between indicator and monthly mean, we used the 1 × 1 km gridded observational data set of Austria for temperature and precipitation. For all indicators of the type peak over threshold (e.g. heat days), where a single day can have the value 0 and 1, the following method is used:

In a first step, the monthly mean value for temperature or precipitation and the indicator value of every month and grid point of Austria are calculated. In a second step, the temperature or precipitation range of the monthly values is divided into bins (e.g. 1 degree steps for temperature). For every grid cell with a monthly mean falling within the bin, the number of days which exceed the threshold are counted and divided with the total number of days within this bin. This gives the probability of positive indicator values per temperature or precipitation bin.

In Fig. 5.9, an example for the temperature dependent indicator heat days is shown. The x-axis gives the bins for the mean monthly maximum temperature (Tmax) and the y-axis the probability of heat days within the bins. The crosses mark

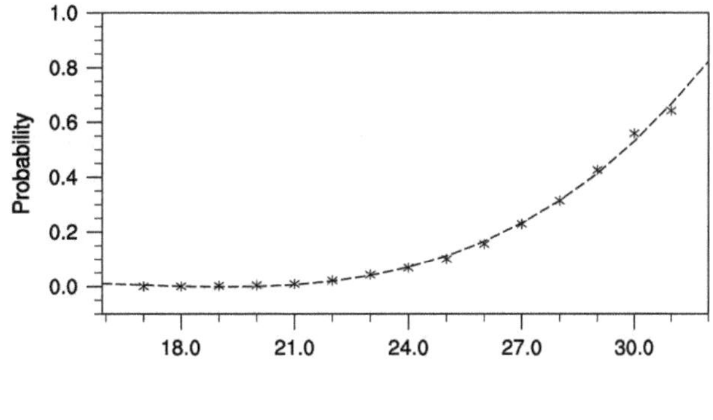

Fig. 5.9 Relation between the probability of heat days (y-axis) and monthly mean maximum temperature (x-axis; °C) in Austria based on gridded observational data and polynomial fit

Fig. 5.10 Relation between the probability of consecutive dry days (7 days) (y-axis) and monthly precipitation sums (x-axis; mm) in Austria based on gridded observational data and polynomial fit

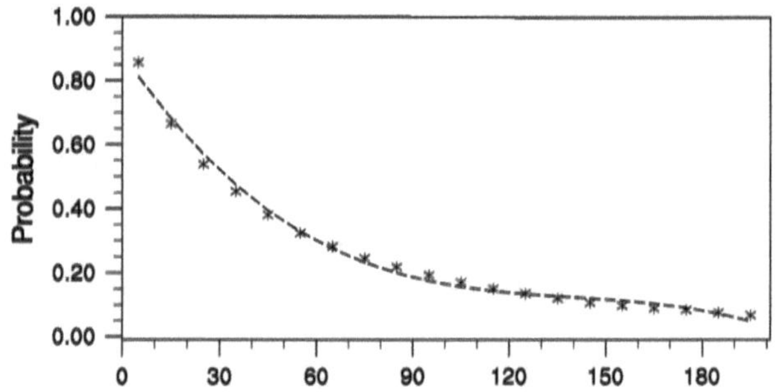

the empirical values derived from all Austrian 1 × 1 km grid points and all months. Below, a monthly mean temperature of 18 °C the probability of heat days is close to zero. For temperatures higher than 21 °C, a rapid increase of the probability of heat days occurs. For a monthly mean Tmax of 24 °C, the probability is ~10 % or 3 days per month, for 27 °C the probability is ~30 % or 9 days per month, and for months with 30 °C mean Tmax the probability is ~60 % or 18 days per month. This statistical relation between monthly mean temperature and heat days is quite clear for mean monthly maximum-temperatures above 18 °C and can be estimated by a polynomial function of the third order. For values below 18 °C, the probability is 0.

In Fig. 5.10, an example for the precipitation dependent indicator "days within a dry period of seven consecutive dry days" (DPD-7) is shown. The x-axis gives the monthly mean precipitation sum in 10 mm bins and the y-axis the probability of days within a dry period (DPD-7). Again, a clear functional dependency of the probability of days versus monthly mean sums can be seen in the range from 0 up to 200 mm. In months with a precipitation sum of less than 10 mm, the probability is close to 90 % or 27 days per month. There is a fast decline of the probability. Monthly sums between 70 and 80 mm have a probability of 25 % or 7.5 days. For higher precipitation values, the probability further decreases, but not so fast. The statistical relation can be estimated by a polynomial function of the third order.

For every peak over threshold indicator, a polynomial function for the relation between the probability of the indicator value and monthly mean values was derived. To calculate the indicator values for different extreme realisations of GCMs/RCSs and NUTS regions, the climate change signal of the monthly mean values was calculated from the climate model and added to the monthly values of the observational time series. For temperature, the absolute difference scenario period—reference period is used and for precipitation the relative change. For every time frame, extreme climate change signal and NUTS3 region a new time series for monthly mean values of temperature and precipitation is generated. To this time series, the polynomial function for the individual indicator is applied. So for every indicator of the "peak over threshold" type an estimate of the indicator value for the extreme RCM/GCM realisation is estimated.

To quantify the quality of the indicator estimate, we compared the results of the indicator values directly counted from the daily observations with the values estimated from the monthly means for all NUTS3 regions. It is assumed that the relation only depends on the mean monthly value, so no other separation (e.g. months or season) is used. Following definition of an error coefficient is used.

$$Ce = \frac{\sum |I(est, i) - I(cal, i)|}{\sum I(cal, i)}.$$

Ce = error coefficient I(est,i) = index value estimated from monthly means for the i NUTS3 region. I(cal,i) = index value calculated from daily data for the i NUTS3 region.

The absolute value of the difference between directly calculated and estimated indicator values of every NUTS3 region are summed up and divided by the sum of the calculated indicator value itself. This gives the spatial mean absolute bias normalised by the mean indicator value of all sub-regions. The error coefficient can be interpreted as the percentage of the bias relative to the indicator value itself. For regions and/or time periods where the counted index value is zero, no error coefficient is calculated. In Table 5.3, the error coefficient of selected indicators is given in the second row. In general, the Ce is higher for events that are less frequent and for precipitation dependent indicators. For heat days and Kysely days (heat wave days; Kyselý 2002), which are quite rare events, the Ce is ~10 % of the indicator value itself. More frequent temperature based events like the frost days the Ce decreases to 2 %. Estimations of precipitation depending indicators like days within dry periods of seven consecutive days have higher biases and the Ce is ~25 % and for days with precipitation higher than 10 mm, it is 8 %. Return periods of daily precipitation intensities could not be estimated with this method because this indicator is not of the "peak over threshold" type.

The quality of the estimated indicator values seems to be useful for most indicators and regions, but the quality of the estimate depends on the frequency of the investigated indicator—the higher the frequency, the better the estimate. In the interpretation of the climate change signal for estimated indicators, this factor has to be considered. As long as the error coefficient is much smaller than the relative climate change signal, the use of the estimated indicator should be without a problem. For example the relative climate change signal for Kysely days in Vienna within the twenty-first century (Fig. 5.11) is in the order of 100 % in the coldest realisation and ~1,000 % in the warmest realisation. This is much higher than the error coefficient of 10 % for this indicator.

5.4.3 Indicator Trends Including the Range of the Scenario Ensemble

All "peak over threshold" type indicators values for the extreme RCM/GCM realisations are calculated. So the whole range of all 31 available climate model

Table 5.3 Error coefficient of estimated indices from monthly data versus counted from daily observations (bold) and the climate change signal of the mid-range scenario (average, minimum and maximum of sub-region values)

Indices	Error Coef.	Min, Max, Mean (2016–2045)			Min, Max, Mean (2036–2065)			Min, Max, Mean (2071–2100)		
Heat day T > 30 (Abs)	**0.09**	0.68	4.31	2.52	4.22	17.53	11.68	15.11	42.4	30.93
Kysley (Abs)	**0.10**	−0.49	3.53	1.35	2.4	19.52	11.98	10.67	49.52	33.55
Frost day (Abs)	**0.02**	−24.9	−10.86	−21.16	−39.23	−22.96	−34.85	−57.37	−37.12	−50.57
Dry Prd 7 day sum (Abs)	**0.24**	−0.078	0.418	0.123	0.146	1.156	0.543	0.304	1.778	0.948
Precip >10 mm Ann (Rel)	**0.08**	−0.027	0.053	0.009	−0.027	0.082	0.028	0.018	0.119	0.062
Return period 5 years (Rel)	–	−0.094	0.158	−0.029	−0.09	0.149	0.038	−0.036	0.207	0.079

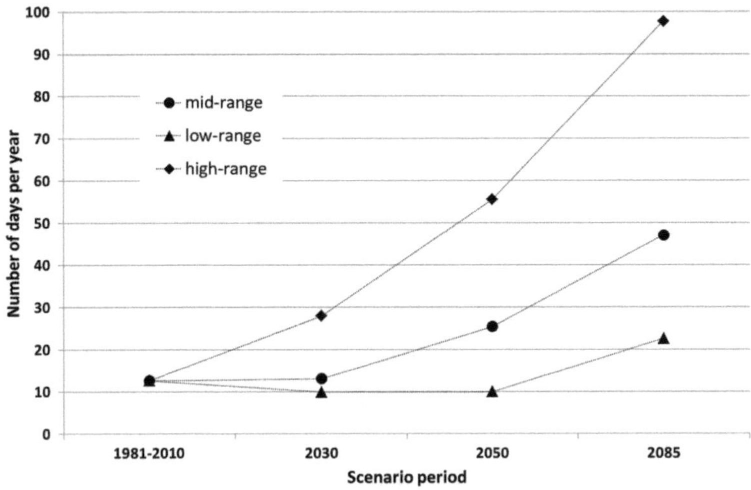

Fig. 5.11 Temporal development of Kysely days at the NUTS3 region Vienna within the twenty-first century (hottest 10 km^2) for the mid-range and the low-range and high-range ensemble member of 31 GMC/RCM realisations

realisations can be considered in the COIN analyses. Some examples are shown in the Figs. 5.11 to 5.13 for the NUTS3 region Vienna.

Figure 5.11 shows the temporal development of Kysely days (days within a heat wave) in Vienna. At present ~12 such days occur per year on average. The mid-range scenario gives ~13 Kysely days for the 2030 time frame, 25 for mid-century, and 47 for the end of the century. In the coldest realisation of each time frame, there is no increase till mid-century and a weak increase till the end of the century up to 23 days. In the warmest realisation in mid-century, the Kysely days exceed 55 days and at the end of the century are roughly 100 such days.

Figure 5.12 shows the temporal development of days within a dry period of at least seven consecutive dry days for Vienna. At present ~65 such days occur per year on average. The mid-range scenario gives no change for the 2030 time frame, and has 70 for mid-century and 74 for the end of the century. In the wettest realisation of each time frame, there is a decrease till mid- and the end of the century to less than 60 days. In the driest realisation in mid-century, the days of dry spells reach 74 days and 88 at the end of the century.

Figure 5.13 shows the temporal development of days with more than 10 mm precipitation per day for Vienna. At present, ~17 such days occur per year on average. The mid-range scenario gives a weak decrease till the end of the century. In the driest realisation of each time frame, there is a decrease to 15 events till mid-century and to ~13 days at the end of the century. In the wettest realisation, an increase of up to 20 events per year occur till the end of the century.

In general, the extreme realisations often show a slightly higher climate change signal for the mid-century than the mid-range scenario for the end of the century. At the end of the century, the extreme realisations show real dramatic developments,

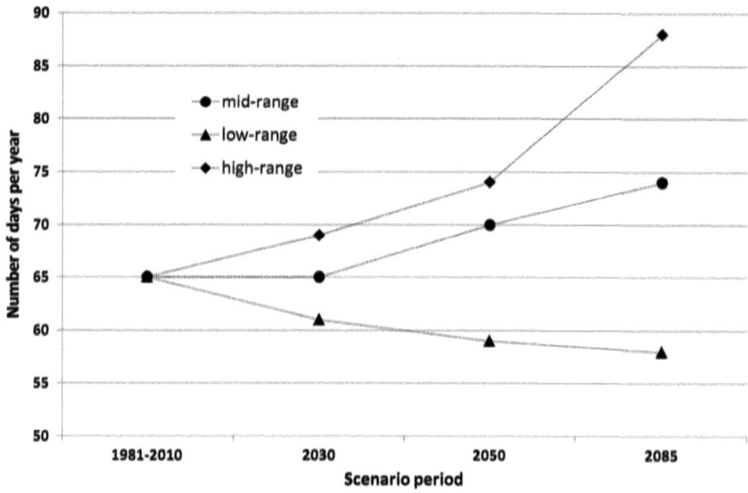

Fig. 5.12 Temporal development of consecutive dry days (7 days) at the NUTS3 region Vienna within the twenty-first century (median for the mid-range and the low-range and high-range ensemble member of 31 GMC/RCM realisations

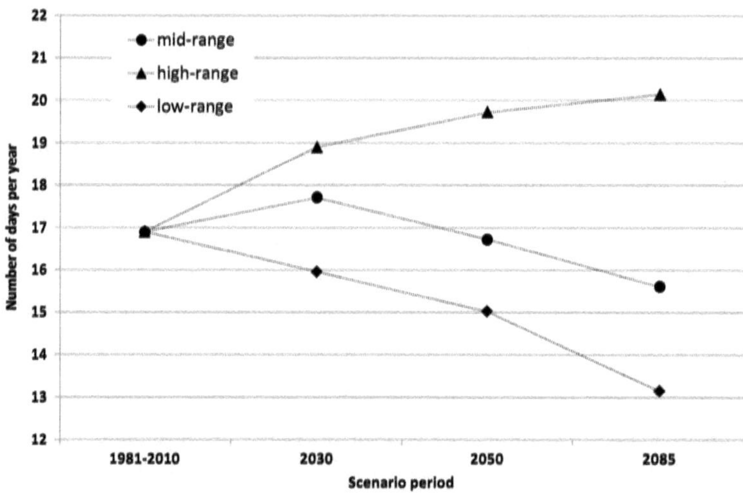

Fig. 5.13 Temporal development of the number of events with more than 10 mm precipitation amount per day at the NUTS3 region Vienna within the twenty-first century (median for the mid-range and the low-range and high-range ensemble member of 31 GMC/RCM realisations

especially for temperature indicators and drought indicators for the summer half year.

5.5 Summary and Conclusions

Within the COIN framework an ensemble of 31 climate change scenarios of different GCMs and RCMs, forced by four different emission scenarios have been prepared. The core of this ensemble is a high resolution regional model run forced with the A1B emission scenario, which were bias corrected and localised on a daily basis to a 1 × 1 km raster. This model is assumed to be the "mid-range" at the end of the twenty-first century in terms of the climate change signal for temperature and precipitation.

Local observations and the climate change scenario for all of Austria on a daily basis and on a 1 × 1 km raster have been used to calculate more than 60 impact related meteorological indicators based on temperature, precipitation, radiation, and snow. All indicators are calculated on the NUTS3 level. Temperature depending indicators were additionally calculated on elevation bands (500 m) within the NUTS3 regions.

To assess the range of the potential climate change within the twenty-first century, a statistical method was developed to estimate the "peak over threshold" type indicators (e.g. heat days) from monthly mean data. For this type of indicator, the highest and lowest indicator values could be estimated for selected time frames from the extreme scenarios of the whole ensemble.

All climate scenarios indicate a warming within the twenty-first century. The mid-range scenario belongs to the colder realisations within the first half of the century with a warming of less than 2 °C in summer and 2 °C in winter compared to the reference period of 1981–2010. At the end of the century, the warming is in the order of 4 °C in both seasons. The whole ensemble indicates a warming of 0.5 up to 4 °C till 2050 and at the end of the century the warming reaches from ~2 °C up to 6 °C in winter and up to 9 °C in summer. The low border stems from models forced with the RCP 4.5 emission scenario and the high border from models forced with RCP 8.5.

The climate change signal for precipitation is not that clear. The annual sum shows no clear trend. For summer precipitation, the majority of the model indicates a decrease till −20 % and in winter an increase of the same magnitude.

The derived indicators reflect the same trends. In general, it can be said that temperature depending indicators at the middle of the century derived from the hottest realisations have a similar climate change signal as the "mid-range" scenarios at the end of the century. The extreme warm realisations at the end of the century really show a different world. More than 100 heat days in Vienna on average are unimaginable. Precipitation depending indicators highlight a higher frequency of dry spells in summer and an increase in daily precipitation intensities.

Acknowledgement We thank the Austrian Central Institute for Meteorology and Geodynamics—ZAMG—for providing meteorological observation data. The ENSEMBLES data used in this work was funded by the EU FP6 Integrated Project ENSEMBLES (Contract number 505539) whose support is gratefully acknowledged. We acknowledge the World Climate Research Programme's Working Group on Coupled Modelling, which is responsible for CMIP, and we thank the climate modelling groups for producing and making available their model output.

References

Christensen JH, Carter TR, Rummukainenm M (2007) Evaluating the performance and utility of regional climate models: the PRUDENCE Project. Clim Change 81(1):1–6. doi:10.1007/s10584-006-9211-6

Formayer H (2010) Prognostizierte Klimaveränderungen in Österreich: Unsicherheiten und Bias. In: Österreichischer Wasser und Abfallwirtschaftsverband, Auswirkungen des Klimawandels auf Hydrologie und Wasserwirtschaft in Österreich, Eigenverlag des Österreichischer Wasser und Abfallwirtschaftsverband http://www.oewav.at/Page.aspx_param_target_is_135399.v.aspx 202 Seiten; ISBN: ISBN 978-3-902084-79

Haiden T, Kann A, Pistotnik G, Stadlbacher K, Wittmann C (2009) Integrated Nowcasting through Comprehensive Analysis (INCA) – system description. ZAMG report, p. 60. http://www.zamg.ac.at/fix/INCA_system.pdf>http://www.zamg.ac.at/fix/INCA_system.pdf

Hewitt C, Griggs D (2004) Ensembles-based predictions of climate changes and their impacts (ENSEMBLES). Eos 85:566

Kyselý J (2002) Temporal fluctuations in heat waves at Prague-Klementinum, the Czech Republic, in 1901–1997, and their relationships to atmospheric circulation. Int J Climatol 22:33–50

Loibl W, Züger J, Köstl M, Formayer H, Schicker I, Gobiet A, Suklitsch M, Schöner W, Anders I, Matulla C (2010) Reclip:century – a project conducting 21st century regional climate simulation runs focussing on the Greater Alpine Region, EGU General Assembly 2010, held 2–7 May, 2010 in Vienna, Austria, p. 11752

Meehl GA et al (2007) The WCRP CMIP3 multimodel dataset—a new era in climate change research. Bull Am Meteorol Soc 88:1383–1394

Meinshausen M, Smith S, et al. (2011) The RCP greenhouse gas concentrations and their extension from 1765 to 2500. Clim Change (Special Issue on RCPs)

Meissner C, Schädler G, Panitz H-J et al (2009) High-resolution sensitivity studies with the regional climate model COSMO-CLM. Meteorol Z 18(5):543–557. doi:10.1127/0941-2948/2009/0400

NUTS (2013) NUTS – Nomenclature of territorial units for statistics. http://epp.eurostat.ec.europa.eu/portal/page/portal/nuts_nomenclature/introduction

Schöner W, Dos Santos Cardoso EM (2004) Datenbereitstellung, Entwicklung von Regionalisierungstools und einer Schnittstelle zu den regionalen Klimamodellen. ZAMG Report for project year 1 PJ1. In: Loibl et al. (ed) Reclip:more – Research for climate protection: model run evaluation, 1. Jahresbericht. ARC systems research, Seibersdorf

Taylor KE, Stouffer RJ, Meehl GA (2012) An overview of CMIP5 and the experiment design. Bull Am Meteorol Soc 93:485–498. doi:10.1175/BAMS-D-11-00094.1

Tebaldi C, Knutti R (2007) The use of the multi-model ensemble in probabilistic climate projections. Philos Trans R Soc A 365:2053–2075

Themeßl M, Gobiet A, Heinrich G (2011) Empirical-statistical downscaling and error correction of daily precipitation of regional climate models and its impact on the climate change signal. Clim Change 112(2):449–468. doi:10.1007/s10584-011-0224-4

Vlcek O, Radan H (2009) Is daily precipitation gamma-distributed? Adverse effects of an incorrect use of the Kolmogorov–Smirnov test. Atmos Res 93:759–766. doi:10.1016/j.atmosres.2009.03.005

Chapter 6
Shared-Socio-Economic Pathways

Martin König, Wolfgang Loibl, Willi Haas, and Lukas Kranzl

Abstract Socio-economic pathways determine future climate impacts and costs thereof. Pragmatically, we have referred to a global reference socio-economic pathway (represented by SSP2 in the IPCC process) and derived figures for the core economic, demographic, land-use and (qualitatively) technological development in Austria, which again frame the sectoral development assumptions necessary to follow a scenario-based cost assessment approach.

In principal, trend projections and existing studies have been used to describe a single country, here applied for Austria, in 2030/2050 that is growing slowly in terms of population (0.27 % p.a.) and medium in terms of GDP (1.65 % p.a.) and in which forests, meadows and settlements expand in the north-east-south crescent—at the cost of arable land, within which further intensification will take place. Policy assumptions as well as technological change have been set to a medium path, at which risk zoning put forward, the EU integration 'muddles through' and no technological wonders are taken into account. A reference scenario might be regarded as least uncertain—which is not true—but we might expect more volatile developments to equilibrate over some decades.

The Austria we expose to climate change by 2050 is significantly different from nowadays: Its population is older and its public and private infrastructure density is higher—at least two factors that might influence future climate costs of inaction.

M. König (✉)
Environmental Impact Assessment and Climate Change, Environment Agency Austria, Vienna, Austria
e-mail: martin.koenig@umweltbundesamt.at

W. Loibl
Energy Department, Austrian Institute of Technology, Vienna, Austria
e-mail: wolfgang.loibl@ait.ac.at

W. Haas
Institute of Social Ecology, Alpen-Adria-Universität Klagenfurt|Wien|Graz, Vienna, Austria
e-mail: willi.haas@aau.at

L. Kranzl
Institute of Energy Systems and Electrical Drives, Vienna University of Technology, Vienna, Austria
e-mail: kranzl@eeg.tuwien.ac.at

K.W. Steininger et al. (eds.), *Economic Evaluation of Climate Change Impacts*, Springer Climate, DOI 10.1007/978-3-319-12457-5_6

6.1 Introduction

To assess the costs of future climate change in any specific country, we have to know for which type of future we carry out this assessment—climatically as well as socio-economically. The socio-economic development is structuring every countries development in various ways: It steers the building of infrastructures, drives land-use, accelerates or slows down technological development and influences the demographic features of migration and natural population growth/ageing of society et cetera.

While we have an overview of how vulnerable we are right now, we have the task to define societal vulnerability for the future to detect how much assets and which values we might expose to climate change.

We need to define the future national setting with respect to all climate sensitive assets and activities based upon these assets.

If we do so for a single country like Austria, we have to choose a top-down approach: Global trends will determine single country developments in a globalised world and set the frame, in which national policy and societies develop and how decisions are taken in future.

For example, global economic growth will allow a more prosperous Austria to invest in its infrastructure, which would alter the value of assets exposed to climate change impacts. Strong global population growth and poles apart or more convergence for global wealth will indirectly—via migration—impact on demographic structures in Austria, leaving us with different age distributions and sensitivities for heat waves. Global markets will steer the demand for certain agricultural or wood/timber products giving raise to land-use changes in Austria.

The IPCC has made significant progress in integrating socio-economic and climate scenarios and changed the mode of scenarios development from the sequential approach—first raising emission scenarios as driver for climate models and sequentially afterwards having climate impacts—to a more integrated and parallel approach, in which socio-economic development is not just steering global emission pathways but also—via different vulnerabilities—climate impacts, adaptation challenges and capacities. (cf. Moss et al. 2010)

That is why we will briefly look at the state of art for the developments of global shared socio-economic pathways, since these do matter not only for global greenhouse gas developments, but also for many factors that determine the sensitivity and exposure of people and assets in a single country. And we define our reference scenario along these assumptions taken for the global level.

To allow for range of potential future vulnerability, we have established 'diminishing' as well as 'enhancing' SSP scenario trends, wherever this was possible: E.g. for demographic as well as for economic developments plausible ranges are given in both directions leading for example to enhancing vulnerability for the demographic scenario, which results in high shares of people >65 years.

6.2 Global Socio-Economic Pathways

The IPCC has started an intensive process to develop five possible scenarios for world development in the twenty-first century as a basic set. In principal, the main variables are population and (convergent or divergent) economic growth in the major global markets, technological development, degree of globalisation/regionalisation (i.e. mainly global work share and trade interdependence), efficiency of governance and institutions with investments in future infrastructure and population (e.g. education) reducing vulnerability of societies by sufficient access to fresh water and clean energy or leverage of adaptation capacity through education (cf. O'Neill et al. 2012).

The principal difference of these new global scenarios is that they are applicable not only for the assessment of GHG emissions (as the so-called SRES scenarios were cf. Nakicenovic and Swart 2000) but also for analysing vulnerability and thus allowing hints on potential costs caused by climate change or shocks by extreme events (Fig. 6.1).

If we take a look at the five SSP narratives, we see that a global shift towards more sustainability like in SSP1 would mean that global GHG emissions would decrease. A reduction of resource consumption would also result in higher adaptation capacity, e.g. water resources would remain accessible for many people and a further spread of regenerative energy would foster self-sustaining communities with potentially much lower vulnerability. Consequently, both—adaptation and mitigation needs—would be reduced in such a world.

In contrast, SSP3—the fragmentation scenario—would leave a world of non-cooperating economic blocks with some accelerating economic growth while others fall behind. Resource depletion would be high in an SSP4 world while population grows quickly. All this would lead to high challenges for both—adaptation and mitigation.

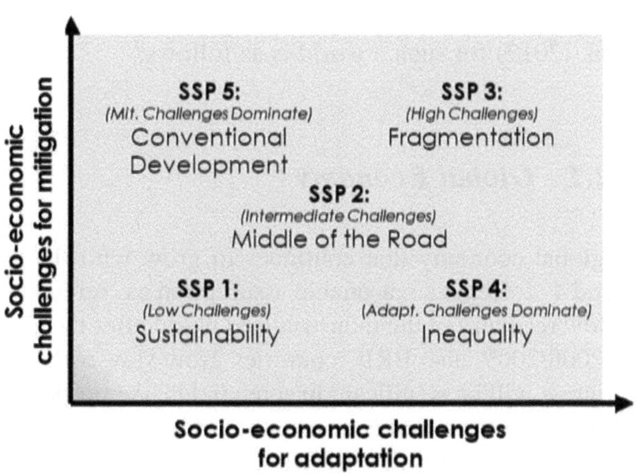

Fig. 6.1 The effects of SSPs on mitigation vs. adaptation challenges, O'Neill et al. (2012)

Yet, signals in both directions occur. Sustainable pathways come up in many parts of the world. Many of them rise bottom-up and are led by communities (Patt et al., transition towns) while others are driven by governments. In contrast to that, many global agreements (UNFCCC and Millennium Development goals) have not yet been met. Thus a likely mix of trends may result in the SSP2 reference scenario.

The core difference between the SSP4 and SSP5 worlds is disparity in welfare on a global level. While the conventional development/SSP5 argues for a convergence in economic development of transforming and developing countries in the long run, the inequality scenario/SSP4 postulates a further divergent economic growth in different world regions resulting in a rather small global elite and a vast majority of poor and vulnerable people with limited access to infrastructure, clean water and energy.

The striking difference for these two scenarios is a look at the different challenges they produce: While mitigation would be the key challenge in the conventional development/SSP5 world as a majority of world population would pursue a resource- and GHG emission intensive (and thus unsustainable!) pathway, the inequality scenario/SSP4 would create a vast majority of people remaining in poverty and—with respect to climate change—at high vulnerability, since their capacity to adapt would be very limited. In such an unequal world, the small global elite less affected by climate change might be less interested to introduce effective mitigation measures addressing the global scale quite easily while the poor majority would face climate change without being able to adapt. Consequently, the adaptation challenge in such a world would be very high.

In summary, we can expect the following trend: On the one hand, some economies in transformation countries as well as in developing countries will be catching up in terms of economic growth; on the other hand numerous countries will be left behind. Even for the ones catching up quickly, the additional national welfare is mostly gained by a rather small elite that is either controlling the natural resources (cf. Angola or Nigeria as an example) or gaining the advantages of foreign investments.

As a consequence, we refer to a reference/SSP2 world when we derive a shared socio-economic pathway for Austria. The core narrative—referring to O'Neill et al. (2012) for such a world is as follows:

6.2.1 Global Economy

A global economy that continues to grow with fluctuating growth rates between 0 and 4 % seems a reasonable assumption as 'reference': While recent trends show a slow recovery of the industrialised world after the global economic financial crisis in 2008/2009, also BRIC countries grow slower than expected. The technological progress will be significant in some fields like propulsion technology and efficiency as well as for renewable energy, but independence from fossil fuels is not expected until 2050. As a result of slowly declining demand and new exploration (especially in the Arctic) we expect a gradual global increase in energy prices until 2030 and an exponential growth only after 2030.

6.2.2 *Global Population*

The global population growth will be slower as formerly forecasted. Still, growth rates particularly in developing countries will remain high for a while. Thus, we can expect around 9 billion people around the globe in 2050 and declining global population due to ageing and a much faster demographic shift in most parts of the developing world as expected towards the end of this century. This is in line with more recent demographic studies (e.g. UN 2005) which expect that many developing countries are approaching the demographic shift (of a society with pre-industrial high birth rates but progress-triggered decreasing death rates) much faster as most industrialised countries did during the industrialisation era. In fact, decreasing birth rates show much less retention with decreasing death rates in recent industrialised societies, thus expecting less population growth.

6.2.3 *Shared Policy Assumptions*

There is no global consensus on ambitious mitigation policies. No huge transition towards a decarbonised economy occurs—neither at global nor at European or Austrian level. The achievements of the Millennium Development Goals are yet delayed leaving many people in developing countries without access to freshwater, sufficient food and energy. Vulnerability of people in developing countries will thus further increase which leads to an increase in migration within countries but also towards Europe.

6.3 Reference Scenario Assumptions for Austria

The global SSP2 as 'intermediate' scenario is selected as framing condition for the shared socio-economic pathway that we determine for our cost assessment in Austria. Shared socio-economic as well as shared policy assumptions relevant for Austria are in line with global SSP2 and shape the socio-economic conditions for the cost of inaction assessment. For demographic, economic and technological developments, we deliver country-specific plausible ranges (diminishing ← reference → enhancing), which are compliant with a global SSP2. It is important to note that 'fixing' the socio-economic conditions does not mean that there are no ranges for sectoral assumptions on their specific development. Some examples are:

- Forestry sector: The COIN reference SSP delivers as key assumptions the share of forests in Austria for the forthcoming decades and the world market price signals for timber as framing conditions, while the sector would raise the asset developments for certain (vulnerable) tree species (most adverse and most beneficial development plus best guess) within the forested land and assumptions for the development of the sawing industry in Austria and thus the import ratio for certain timber.
- Health sector: the COIN SSP delivers the share of >65 year old people (plausible range) for the forthcoming decades plus key assumptions on urban sprawl while WP6 (health) has to combine the key assumptions (e.g. share of >65 people living in non-air-conditioned flats in urban heat islands in 2030/2050) and how the development of climate sensitivity of older people in future will be to get to its exposure scenarios.
- Energy sector: the COIN reference SSP delivers the range for the price level of main energy carriers in 2030/2050, while sector energy would have to derive the energy mixes for 2030/2050 taking this overall energy price level as one boundary condition.
- Agriculture sector: the COIN reference SSP delivers the share of agricultural land use in 2030/2050 as well as world market prices for grain (ranges) plus assumptions on the level of subsidies (SPA for EU integration) while WP3 (agriculture) would derive the developments for certain (vulnerable) crop production until 2030/2050. Agriculture might be a more complex example, since crop production can be changed on a yearly basis and could thus react quickly (autonomous adaptation to climate change or changing demand/price signals). Thus the crop assets are the least fixed ones in the COIN asset exposure scenarios.

6.3.1 Shared Policy Assumptions

6.3.1.1 The EU (Dis)Integration

The future operation, intensity and budget of common EU policies, is explicitly important for the sectors agriculture (Common Agriculture Policy, CAP), energy (TEN-E), water (e.g. WFD) and transport and mobility (TEN-T policy) as well as to some extent also for ecosystem services and biodiversity (nature protection policy, e.g. protected areas like Natura2000).

As we have seen in early February 2013, the EU has reached consensus on its periodical financial framework 2014–2020 with slight decreases (960 billion euros compared to 995 billion euros for the last period 2007–2013). This agreement could be regarded as symptomatic for the EU (dis)integration. Trends in further integration especially for financial market control mechanisms and the European Stability Mechanism (ESM) are discernible. These trends might indeed empower the EU to control national budgets and to intervene in countries that are using ESM budgets.

However, since national budgets are under pressure as austerity policy prevails, we see the divergent trend of pooling less national resources on other formerly important fields of common European policy. Coherence funds as well as the budgets for CAP, TEN-E and TEN-T will most likely shrink or will at least be frozen. For all these shares of the EU budget, the national interests are diverging. E.g. France, Germany and also Austria want further protection of their farmers from major cuts in EU subsidies (via CAP), so that the CAP share will decrease to a lesser extent as many experts suggest.

To keep it simple, we expect a 'muddling through' of the EU for the coming decades. All COIN-relevant EU budgets will thus be projected as frozen to the amount of the last multiannual financial framework 2007–2013. This means in fact that

- subsidies for farmers will face a net real decrease according to the inflation rate
- major TEN-T projects will be prolonged, some might be cancelled due to budget constraints
- the share of EU protected areas (Natura2000) will remain at status quo.

6.3.1.2 National Policies

Especially for the impacts of climate-triggered natural disasters like floods, mass movements and avalanches, a rigorous risk zoning and centralisation of spatial planning policy at higher (provincial or federal) scale has to be taken into account. The degree to which this will take place until 2030/2050 is hard to project, but some proxies might be drawn from the major (national-scale) flooding events that took place in eastern Austria 2002 and western Austria in 2005.

National policies have gone through a learning process considering climate change as an issue. The ratification of the NAS in Austria could be regarded as major milestone in that respect, but we are not approaching adaptation yet as its implementation is still pending.

But the adaptation learning curve experienced will influence not only the pertinent national adaptation policies (i.e. planned adaptation, which is excluded from the COIN assessment), but will also trigger autonomous, individual adaptation, since public sector activity will inevitably foster private sector engagement for adaptation as particular tools for adaptation or at least risk management (cf. eHORA) will be available. All this will trigger public perception and policy implementation e.g. in terms of risk management for natural disasters.

E.g. after the major flood events in 2002 und 2005 the following water policies have been pushed forward:

- EU floods directive initiated by Austria

- Adjustments in the Austrian water law
- Adjustments in the Austrian law on hydraulic structures.

This indicates 'reactive' learning after natural disasters.

Pragmatically we will acknowledge a 'reactive' learning curve within public administration and policy development that is taken up by the private sector and thus lead to:

- More rigorous risk zoning
- More centralised spatial planning at provincial and federal level.

6.3.2 *Demographic and Economic Development*

6.3.2.1 Demographic and Economic Growth Assumptions

Demography is one of the key parameters that determine both biophysical and economic developments. Thus number of people and their age structure are highly relevant for consumption levels and patterns (transport, buildings, food/agriculture etc.), sensitivity of population (human health), size of labour force, dependency ratio with its financing implications for pension funds and education and so on.

According to the 2011 forecast of Statistics Austria population will still grow in future. In 2011 8.4 million people lived in Austria. For 2030 there are 9.0 million people and for 2050 9.3 million people expected (central projection; Hanika 2010). The age structure will shift towards an increased share of older people: Compared to 18 % share in population of 65 and over in 2011 this group will account for 25 % in 2030 and for 28 % in 2050. This rises the average age of population by about 4 years between now and 2050. For details on long-term assumptions for fertility, mortality and migration see Hanika (2010).

The SSP for economic development refers to Schiman and Orischnig (2012). In their economic long-term model for Austria's public finances, economic growth is driven by level of employment, its endowment by capital and technical progress. Projections are based on the central demographic projection of Statistics Austria (consistent with 'middle-of-the-road'). They state: 'Overall, the average annual trend growth is projected to be 1.65 % over the whole period (2012–2050). This increase is almost entirely accounted for by its increments of productivity, while labour input is stagnating [and capital stock only slightly increasing from 2030 onwards]'.

For both time slices, we deliver three possible scenario pathways. The 'diminishing' scenario reproduces less population and GDP growth due to less employees while the enhancing scenario reproduces a higher population, thus employee and GDP growth. In fact, the diminishing (i.e. less growth) scenario reproduces less assets, less people explicitly vulnerable to heat waves et cetera. On the other hand, one might argue that a growth scenario (here the 'enhancing' scenario as it 'produces more vulnerable people at critical age, more exposed

Table 6.1 Key economic projections for 2030, Sources cf. below

2030	Diminishing	Reference	Enhancing
Population (no.)	[a]	[b]	[c]
Total	8,926,000	9,013,000	9,385,000
<20 years	1,720,000	1,721,000	1,881,000
20 to <65 years	5,106,000	5,117,000	5,275,000
≥65 years	2,101,000	2,162,000	2,229,000
GDP growth (%)	[d]	[e]	[f]
National level	1.49 %	1.65 %	1.82 %
Employees (no.)	[g]	[h]	[i]
Total	3,443,000	3,450,000	3,557,000

[a]Statistik Austria low life expectancy projection (see also http://www.statistik.at/web_de/statistiken/bevoelkerung/index.html, cf. Hanika 2010)—midyear
[b]Statistik Austria central projection—midyear
[c]Statistik Austria growth scenario projection—midyear
[d]Based on footnote "e"; growth rate has been reduced by 10 %
[e]Schiman and Orischnig (2012) average growth rate over the whole period 2012–2050
[f]Based on footnote "e"; growth rate has been increased by 10 %
[g]Based on footnote "h" but reduced in proportion to change in population group 20 to <65 between central to low fertility scenarios
[h]Schiman and Orischnig 2012 with their assumptions on unemployment and part time work
[i]Based on footnote "h" but increased in proportion to change in population group 20 to <65 between central to growth scenario projections

assets' et cetera) calls for setting higher discount rate as the future generation has more welfare and financial power. It is thus hard to have a clear specification, which growth pathway is in fact more/less challenging in terms of climate impact costs (Tables 6.1 and 6.2).

6.3.2.2 Key Price Signals

Global market integration for energy and agricultural commodities is almost complete and world market price developments dominate supply and thus land-use shares of crops.

Simplifying the assumptions on key price signals, we opted for an evaluation of global market price projections by IEA, FAO and OECD rather than acknowledging the effect of climate change on domestic prices. This is an implicit mistake, which is due to the necessary sequential set up of (1) the climate and socio-economic scenario framework, (2) sectoral assessments and quantification of direct impact and (3) macroeconomic assessment and indirect effects (including altering price signals).

Table 6.2 Key economic projections for 2050, Sources cf. below

2050	Diminishing	Reference	Enhancing
Population (no.)	[a]	[b]	[c]
Total	9,113,000	9,334,000	10,456,000
<20	1,714,000	1,717,000	2,186,000
20 to <65	4,958,000	4,980,000	5,414,000
≥65	2,440,000	2,634,000	2,856,000
GDP growth (%)	[d]	[e]	[f]
National level	1.49 %	1.65 %	1.82 %
Employees (no.)	[g]	[h]	[i]
Total	3,435,000	3,450,000	3,751,000

[a]Statistik Austria low life expectancy projection (see also http://www.statistik.at/web_de/statistiken/bevoelkerung/index.html, cf. Hanika 2010)—midyear
[b]Statistik Austria central projection—midyear
[c]Statistik Austria growth scenario projection—midyear
[d]Based on footnote "e" growth rate has been reduced by 10 %
[e]Schiman and Orischnig (2012) average growth rate over the whole period 2012–2050
[f]Based on footnote "e" growth rate has been increased by 10 %
[g]Based on footnote "h" but reduced in proportion to change in population group 20 to <65 between central to low fertility scenarios
[h]Schiman and Orischnig (2012) with their assumptions on unemployment and part time work
[i]Based on footnote "h" but increased in proportion to change in population group 20 to <65 between central to growth scenario projections

Thus price signals for agricultural goods are important as they trigger the crop selection, shifting towards more demanded agricultural goods. Price signals for energy commodities will alter supply and demand in any given country. All this triggers sector sensitivities significantly: Different crops as well as different energy supplies lead to divergent sectoral climate sensitivities.

The assumptions for economy-wide average energy and CO_2 price development are depicted in Table 6.3 and are grounded on the assumption that even for reaching an A1B emission pathway, mitigation measures are inevitable and would require mechanisms for carbon pricing that lead to according increases in energy and CO_2 emission credit prices. This implies that current climate mitigation policies are carried forward without neither attenuation nor intensification. By this, the definition of 'inaction' in terms of mitigation must be seen as relative to current action rather than absolute.

Table 6.3 Energy and CO_2 prices in 2010 prices, after IEA WEO (2010): current policy scenario; reference scenarios from the project EISERN

	2010	2020	2030	2040	2050
Primary energy price (EUR/MWh)					
Coal	7.11	9.36	9.95	10.65	11.38
Gas	18.25	29.73	34.16	37.60	41.48
Oil	29.69	46.59	55.06	61.41	68.61
CO_2-Price (t CO_2)	15.84	21.60	26.64	33.84	41.04
Electricity wholesale price (EUR/MWh)	45.56	53.85	57.78	63.25	68.50

It is essential to stress that climate change could add on global food market prices. While climate change will most likely deepen the disparities in the global distribution of food, since most adverse effects (mainly losses in ecosystem services) will prevail in subtropical and tropical regions, cereal world market prices could rise by a maximum of 20 % under certain SRES scenarios (cf. Parry et al. 2004).

The price signals for agricultural products are hard to be projected and there is no forecast available how the price for wheat or soybean will develop over the next decades.

Uncertainty stems from various sources, where climate impacts on global yields is one among others. Nevertheless, corn markets are influenced heavily by the way that agricultural and energy markets will integrate, or in other words: How strong (and legally binding) the agrofuel mandate will be played in future. It is very likely that due to increasing demands as regards the global demographic development and the demand for agrofuels, we will face much higher volatility of the corn market (cf. Diffenbaugh et al. 2012). During the last four decades of the twentieth century, the prices decreased basically due to intensification and extension of arable land overcompensating the increasing food (and not yet energy) demand.

During the decade 2000–2010 we see the market becoming very volatile as shown by the index curves in Fig. 6.2. This depicts the increasingly complex influence on the market by energy (agrofuel) demands, increasing world population and impacts of climate change (mainly droughts in North America, Australia and Europe). On top of that, policy interventions in producer countries tend to protect their consumers by channelling more harvest shares on their domestic markets, which leads in the end to less supply on the global food market and increasing world market prices.

However, according to OECD/FAO forecasts for the most important agricultural commodities until 2020, we see almost no significant trend which would justify putting a certain index to the price development. There is only one commodity—vegetable oil—for which we see a clear increasing price level throughout this decade.

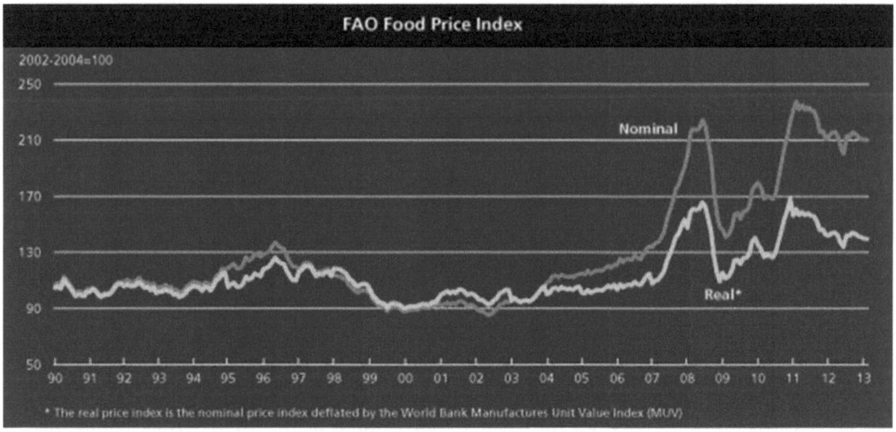

Fig. 6.2 FAO food price index since 1990, online source of FAO http://www.fao.org/worldfoodsituation/wfs-home/foodpricesindex/en

Consequently and as a best guess, we extrapolate 2020 agricultural goods prices to the decades after, acknowledging that the market will presumably be very volatile as it was in recent years. But the decreasing trend of agricultural goods prices has stopped and this has been acknowledged by the recent FAO outlooks to which we refer.

Under the assumption that the Common Agricultural Policy (CAP) would shrink in its financial intervention power (cf. above), we assume that world market prices would strike through on national food markets almost without a buffer. So, the means to decouple the EU internal market from the global market will be much less powerful in future decades (Table 6.4).

6.3.3 Land-Use Change

For the land-use change scenario, we stick to the reference scenario, because mapping ranges as giving ranges is neither possible on the existing data nor would it be easy to depict those ranges in an applicable way.

The land cover map reflects the kind and intensity of land-use. Land cover mapping is based on the European CORINE Land Cover project (http://www.umweltbundesamt.at/umwelt/raumordnung/flaechennutzung/corine/). Land use projection refers to the land use class shares and concentrates on settlement areas, carried out using population, employment and housing projections.

The map in the book is depicted in grey scale. A coloured image can be found on the web-site, accompanying the book (http://extras.springer.com) or on the Environmental Agency Austria's—CORINE website listed above. The spatial and the land use classes—especially those addressing settlements—are shown rather coarse

Table 6.4 Real prices projected for 2020

Agricultural commodity	2020 Price in EUR/t or EUR/hl (for ethanol and biodiesel)
Wheat	178
Coarse grain	150
Rice	365
Oil seeds	354
Protein meals	242
Vegetable oils	805
Raw sugar	302
Beef and veal	3,547
Pig meat	1,894
Poultry meat	1,937
Sheep meat	2,629
Butter	2,762
Cheese	3,032
Skim milk powder	2,534
Whole milk powder	2,659
Whey powder	726
Casein	6,237
Ethanol	49
Biodiesel	106

Source: OECD-FAO Agricultural Outlook 2011; exchange rate 1.35 has been applied for USD/EUR

in Fig. 6.3. As a large part of Austria is covered by mountain and thus forested, the large settlements are concentrated in the plane areas in the north-east-south 'crescent' of Austria, shaped by Pleistocen glaciers as well as the Danube and further in some wider Alpine valleys along the rivers Rhine, Inn, Salzach, Enns, Drau, Möll, Glan, Lavant, Mur and Mürz.

For COIN this CORINE land cover subset has been intersected with the Austrian NUTS-3 regions to allow a more detailed spatial relation of sector-related land uses to different exposure patterns using tabular statistics.

The following map depicts the land use intensity with respect to housing and settlement area through population density per km^2 which allows identifying the population exposure variation to climate change effects and the related hazard risk through NUTS-3 regions. The darker gray (dark red to purple) shades show a higher population and thus settlement density (Fig. 6.4).

6.3.3.1 Assumptions on Future Land Use Change

The prior land use change and population change observations allow the following assumptions on trends. Two directions in spatial development can be identified: peripheral rural regions are expected to continue losing population and thus

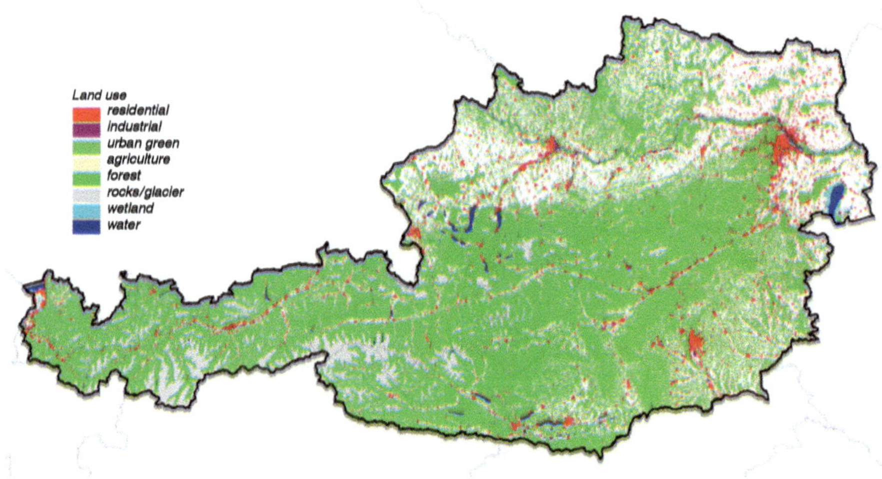

Fig. 6.3 CORINE Landcover 2006, Environment Agency Austria (2014)

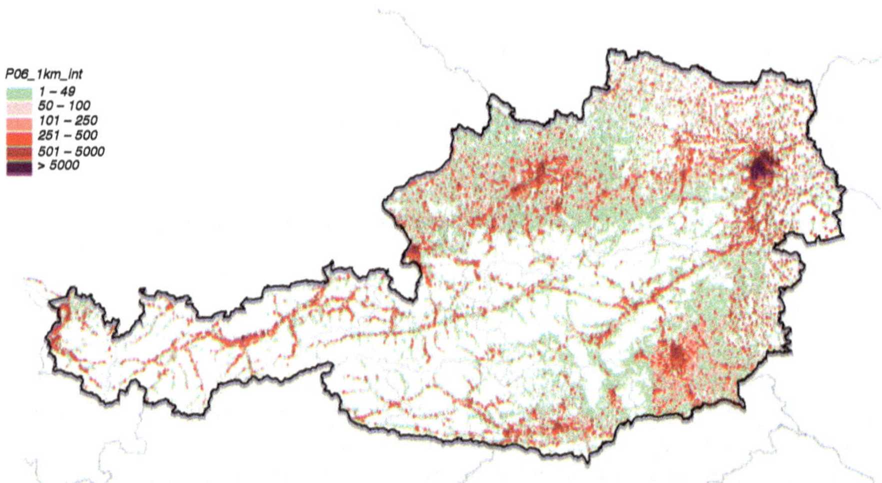

Fig. 6.4 Population distribution per km^2 2010, AIT—spatial allocation of population numbers related to CORINE land use distribution based on Statistik Austria population data

economic attractiveness, liveability and social cohesion (The reasons refer to their demographic structure and to migration patterns more or less triggered by economic expectations). Central regions with larger core cities are usually prosperous areas with migration surplus and growing businesses, all demanding additional building land.

6.3.3.2 Settlement Area

Settlement areas in rural peripheral regions with less economic activity will remain more or less as they are. As less people will be living in these areas, more houses will stay unoccupied or will be, in best case, occasionally used as holiday homes by the children of those living there before or rented out as guest houses during holiday seasons. Buildings in bad condition can be expected to be demolished. Buildings in areas which have been observed to be affected by natural hazards (floods, avalanches) will occasionally get abandoned.

Settlements in rural areas in attractive landscapes with higher touristic potential will grow—new houses will be built for guest accommodation or as holiday homes. In areas where no sufficient residential land is available, pressure to policy makers will turn up, to zone new building land. If there is no land available in areas which are free of risk to be damaged through climate induced natural hazards, areas judged to let expect little natural hazard risk may be zoned as new building land.

In urbanised regions pressure on land consumption will increase due to in-migration (either from the national periphery or from the European periphery which are both less prosperous). Settlements in urban and peri-urban areas are expected to grow because of housing demand from increasing population numbers, increasing household numbers (which still grow, even when population declines, because of decreasing household size) and because of increasing economic activities due to the establishment of new businesses or enlargement of existing ones demanding additional commercial area and related land uses like logistics, parking lots, transportation network and technical infrastructure supply. Depending on planning policy guidelines one can expect either densification with controlled growth in sub-centres or urban sprawl in the outskirts of the cities. Decline of climate comfort in densely populated and thus densely built up urban cores leads again to an urged trend of households which can afford single family homes to move to the city outskirts building houses in green environment and fresh air with sufficient nocturnal cooling.

The website accompanying the book contains a figure which shows regions in red where population numbers are expected to grow causing a growing settlement area and those regions in blue where population numbers will shrink leading to less growth pressure in those settlement areas. The grow-areas are all capital regions of Austria's provinces while the shrinking areas are those in less prosperous Alpine valleys like mountainous Carinthia, mountainous Styria and in the north-western peripheral region of Lower Austria (the "Waldviertel") and finally some parts of the southern Burgenland.

The following Table 6.5 gives an estimation of a "reference scenario" (REF) for further settlement area growth. The change rates are based on the changes between 2001 and 2010. Till 2020 the change rates are assumed to remain stable. For the decade 2021–2030 the change rates are reduced by 50 %. For the decades 2031–2050 the change rates are reduced to 25 % assuming declining building land demand due to a saturation of population numbers, household size averages and

Table 6.5 Building and transportation infrastructure land: absolute number, change rates and estimations for 2020–2050, Statistik Austria—ÖROK Scenarios, AIT

Building land and transportation infrastructure (km^2)					Growth 2001–2010			
Province	2001	2009	2010	2011	Absolute (km^2)	Relative (in % of 2001)	Reference scenario 2030	Reference scenario 2050
Burgenland	264	312	315	317	51	19	412	453
Kärnten	361	393	397	399	36	10	458	481
Niederösterreich	1,147	1,291	1,304	1,309	157	14	1,584	1,694
Oberösterreich	711	789	797	806	86	12	947	1,006
Salzburg	216	229	229	230	13	6	250	258
Steiermark	694	788	792	794	98	14	968	1,037
Tirol	270	297	299	302	29	11	349	368
Vorarlberg	117	121	122	126	5	4	130	133
Wien	191	196	194	194	3	2	200	206
Austria	3,971	4,416	4,449	4,477	478	12	5,299	5,622

workplace numbers. For Vienna the change rates are kept stable for the entire time range because of the in-migration expectations.

The projections show a slight saturation of settlement area growth after 2030 because of the saturation in population growth and thus economic activity increase. Until 2030, a 20 % increase of settlement area can be expected, till 2050 a 28 % increase can be expected (base year 2010).

The numbers above are target numbers derived from the ÖROK scenarios (see Hiess et al. 2009a, b). A further disaggregation into smaller regions is necessary to allow considering the regional variation of socio-economic trends. The following section documents the way how the projections have been carried out for the NUTS3-regions.

As the households are the major drivers for change in residential areas, projections of household numbers (Hanika 2010) are used as proxy for estimating the additional demand for settlement area for housing: From 2010 till 2030 an increase of 12.5 % can be expected, till 2050 the increase will results in 20 % since 2010. The increase will vary between stagnation in some NUTS3-regions and growth till 2030 of up to 25 % and till 2050 of up to 40 % (e.g.in the Vienna region outside the core city). The increase of household numbers in the cities will not require the same increase of building land but will be substituted to a certain extent through densification.

Growing household numbers represent just a part of the growing apartment numbers—a certain additional amount refers to second homes, holiday homes, reconstruction demand as well as a turnover demand for migration, renovation etc. (Windisch 2005). The reference household projections refer to the ÖROK household projections till 2030 (Hanika 2005) and the households' main residences projection 2011–2030 (Hanika 2010). The housing projections have been extended for the reference scenario till 2050. Therefore the regional growth rates for the decades till 2050 are taken from growth rates for decade 2021–2030 and have been reduced by the factor 0.8 assuming less apartment surplus for the future. Only for Vienna no reduction was considered, otherwise the apartment surplus share would have been reduced from 12.5 % (in 2011) down to 4 % which is assumed to be too low. Now the surplus share shrinks only to 6.7 %.

The remaining building land increase is due to increase of area demand for public infrastructure, for production, logistics, services and retail as well as for transportation infrastructure. To estimate the demand for building land for business development, the change in workplace numbers were used as proxy. As no workplace numbers are available but employee projections, we assume with some uncertainty (considering static employment ratios and neglecting changes in commuter patterns) that the workplace numbers will follow the workplace demand estimated through demographic projections and related employment quotas. Employee numbers (Kytir et al. 2010) have thus been taken as a further proxy to identify new workplace demand as new demand for business development area. From 2010 till 2030 some workplace increase of up to 4 % can be expected for the provinces Lower Austria, Vorarlberg and Vienna, till 2050 an increase ranging between 3 and 16 % can be expected for the provinces Lower Austria, Tiyrol,

Vorarlberg and Vienna. The NUTS-3-specific change rates show much more variation—e.g. up to 26 % in Vienna's outskirts.

Workplace increase projections are not available in detail for Austria. But there are projections based on NUTS-2 regions for EU27. An available projection comes from the PLUREL project (www.plurel-org.eu) based on GDP-projections of the NEMESIS economic model using energy price expectations to project GDP- and population up to 2025 (Boitier et al. 2008). These NUTS2-projections were taken to extend the 2010–2020 trend till 2050. The growth rates were reduced by a factor of 0.982 to achieve the 4.25 million workplace total as defined by the reference scenario. The NUTS2 workplace numbers were disaggregated to NUTS3 regions based on the 2010 shares of the NUTS-3 entities on the NUTS-2 regions.

Then the housing and workplace change ratios were taken as final proxy for extrapolating the urban fabric land use classes. The following Table 6.6 presents the numbers for built up land distinguished as CORINE landcover classes CLC 11 (urban residential) and CLC 12 (industrial, commercial, transportation infrastructure) for 2010, 2030 and 2050.

6.3.3.3 Agricultural and Forest Area

A study carried out by Alterra (Pérez-Soba et al. 2010) on land use change in Europe predicts a quite strong abandonment of agriculture in Western Europe's more marginal mountainous areas.

In peripheral rural regions, the share of arable land is expected to decline due to lacking cultivation, given up by prior farming families now commuting to nearby industrial or service centres for work or leaving the rural areas to move to urban regions for better education, better working and better living conditions. Some land will be first transformed into grass land used as pastures or fodder resource for cattle (or sheep in high mountain areas) by nearby farmers and will later evolve into forested area.

So, agriculture is quite under pressure loosing on the one hand arable land, transformed into residential and commercial land in prosperous, easy to be cultivated central regions and on the other hand leading to abandoned areas—both affected by climate threats endangering harvest volume and thus income. Climate and market conditions lead also to changing crop cultivation shares—e.g. increasing crops for biofuel production, sensitive to drier and hotter climates, and decreasing food crop shares as far as EU Biofuel policies may influence the biofuel cultivation shares.

Haberl et al. (2003) have estimated changes in land use due to policy response of agricultural, forestry production till 2020. A TREND scenario extrapolates effects of current common agricultural policy CAP into the future. In the GLOB (globalisation) scenario, agricultural subsidies and market regulations will be reduced leading to a concentration on agricultural products which can be produced in Austria competitively. No additional biomass production is assumed here. In the MAX (maximum liberalisation) scenario, the highest possible increase of biomass

Table 6.6 Settlement area projections for NUTS3-regions, Processing: AIT based on CORINE Land Cover classes 2006: Urban residential, industry (incl. commercial, transportation infrastructure), and Statistik Austria projections for households, apartments and workplaces (ÖROK Scenarios)

NUTS-3	Urban-residential 2006	Industrial 2006	Urban-residential 2030	Industrial 2030	Urban-residential 2050	Industrial 2050
AT111—Mittelburgenland	3,778	30	4,387	37	4,604	38
AT112—Nordburgenland	10,074	512	11,958	634	12,774	640
AT113—Südburgenland	7,744	251	8,814	311	9,126	314
AT121—Mostviertel-Eisenwurzen	10,288	620	12,086	768	12,814	774
AT122—Niederösterreich-Süd	15,097	1,451	17,622	1,796	18,582	1,812
AT123—Sankt Pölten	7,873	304	9,225	376	9,756	380
AT124—Waldviertel	14,237	227	15,877	281	16,176	283
AT125—Weinviertel	13,468	232	15,795	287	16,729	290
AT126—Wiener Umland/Nordteil	23,295	1,705	28,967	2,111	31,922	2,129
AT127—Wiener Umland/Südteil	15,620	3,201	18,512	3,963	19,727	3,997
AT130—Wien	19,387	3,620	22,527	4,569	23,935	4,653
AT211—Klagenfurt-Villach	13,627	1,136	15,833	1,424	16,624	1,445
AT212—Oberkärnten	7,815	134	8,980	168	9,358	171
AT213—Unterkärnten	6,929	208	7,852	261	8,092	265
AT221—Graz	15,072	1,536	17,688	1,903	18,708	1,921
AT222—Liezen	5,388	287	5,990	356	6,083	359
AT223—Östliche Obersteiermark	10,022	1,170	10,707	1,450	10,525	1,463
AT224—Oststeiermark	13,116	206	15,744	255	16,942	258
AT225—West- und Südsteiemark	10,425	380	12,344	471	13,150	475

(continued)

Table 6.6 (continued)

NUTS-3	Urban-residential 2006	Industrial 2006	Urban-residential 2030	Industrial 2030	Urban-residential 2050	Industrial 2050
AT226—Westilche Obersteiermark	5,555	825	6,016	1,022	5,982	1,032
AT311—Innviertel	10,368	219	12,702	271	13,859	273
AT312—Linz-Wels	17,780	3,547	21,113	4,385	22,526	4,420
AT313—Mühlviertel	9,911	1	12,563	1	14,022	1
AT314—Steyr-Kirchdorf	7,400	343	8,802	424	9,406	427
AT315—Traunviertel	11,183	114	13,291	141	14,193	142
AT321—Lungau	1,395	1	1,664	1	1,783	1
AT322—Pinzgau-Pongau	8,708	64	10,619	81	11,551	82
AT323—Salzburg und Umgebung	10,181	632	12,275	798	13,238	812
AT331—Außerfern	2,862	29	3,575	37	3,951	38
AT332—Innsbruck	9,059	413	11,037	523	11,984	534
AT333—Osttirol	2,354	120	2,787	152	2,969	155
AT334—Tiroler Oberland	6,106	64	7,713	81	8,589	83
AT335—Tiroler Unterland	11,348	123	14,265	156	15,827	159
AT341—Bludenz-Bregenzer Wald	6,162	95	7,546	118	8,226	119
AT342—Rheintal-Bodenseegebiet	10,811	398	13,402	494	14,721	499
Total km^2	354,448	24,198	420,279	30,108	448,456	30,443
Change factor, 2006 = 100 %	100.0	100.0	118.6	124.4	126.5	125.8

Table 6.7 Land cover shares: 1995, 2008, scenarios 2020, reference scenario (reference), 2030

			TREND	MAX	GLOB	Reference
	1995 (km^2)	2008 (km^2)	2020 (km^2)	2020 (km^2)	2020 (km^2)	2030 (km^2)
Urban and infrastructure area	3,967	4,360	5,191	5,191	5,191	5,300
Cropland, gardens	14,670	*26,449*	13,614	14,436	11,664	13,700
Grasslands (in use)	11,131		9,898	9,077	7,295	7,500
Alpine grasslands (in use)	8,525	8,552	8,616	8,616	5,773	6,500
Forest, woodland	38,400	36,343	39,375	39,375	42,014	42,500
Natural areas, rivers, lakes	7,164	7,622	7,164	7,164	7,164	6,500
Unused Alpine pastures	–	–	–	–	3	25
Grassland (unused)	–	–	–	–	757	757
Grassland succession	–	–	–	–	420	420
Old field succession	–	–	–	–	736	736
	83,857[a]	83,326[b]	83,858[a]	83,859[a]	81,017[a]	83,938[c]

[a]Haberl et al. (2003)
[b]Environment Agency Austria—BEV numbers
[c]AIT-estimation

production is assumed—a scenario which considers the possible impact of agricultural markets' liberalisation. Table 6.7 presents the numbers from Haberl et al. (2003), adds the observation numbers for 2008 and reference scenario numbers located between the TREND and the GLOB scenario, taking the settlement and infrastructure area from the estimations shown above. The total area numbers are varying slightly due to estimation and rounding errors.

The numbers are rough estimates based on available statistics and sources.

The areal change for non-urban land use classes is presented below. The maps give some hints on the regional differences in change of agricultural area. The numbers per NUTS-3 regions will serve as basis for the disaggregation of the changes in agricultural/forestry land cover sub-classes. Intra-regional details on patterns and shares are subject to sectoral investigations (Figs. 6.5, 6.6 and 6.7).

6.3.4 Technology and Innovation Path

In contrast to the other parts of the SSP, technological development is strongly sector specific, e.g. in terms of the breeding progress in the agricultural sector or in terms of cost-reduction of certain renewable energy technologies like photovoltaics. On the other hand, there are relevant aspects like technological progress in material science or in computing and communication technologies which have strong cross-sectoral characteristics. In this frame it is not possible to identify all

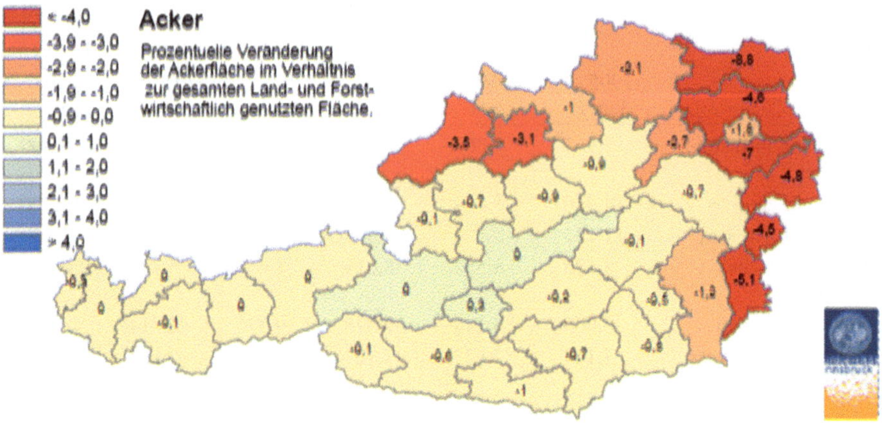

Fig. 6.5 Change of arable land 2011–2020, Rüdisser et al. (2011)

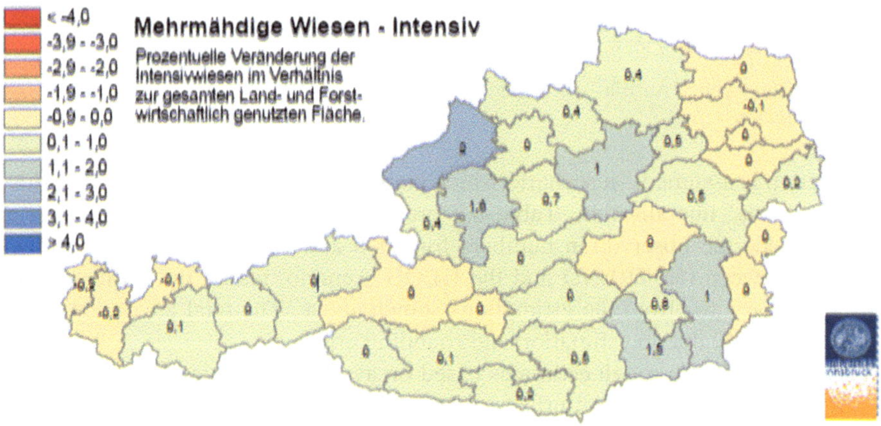

Fig. 6.6 Change of intensive grassland 2011–2020, Rüdisser et al. (2011)

relevant cross-sectoral technologies, to investigate the impacts of these technologies on the different sectors and to develop well based scenarios for the development of these cross-sectoral technologies.

Therefore, we suggest three basic qualitative storylines which are compliant with the definition of the three scenarios:

- Enhancing: Compared to the technological developments of the past years and decades, the technological development significantly decreases. No major progress is achieved and the innovations mainly refer to small improvements of single components.
- Reference: Remarkable, steady progress is achieved for single technologies, however, with no principle major change on the overall structure and efficiency of material use and sector specific technologies. No new technological

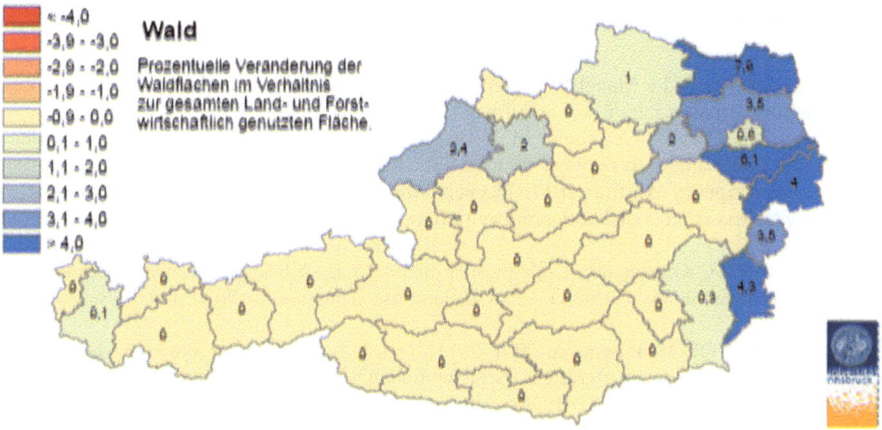

Fig. 6.7 Change of forest area 2011–2020, Rüdisser et al. (2011)

revolutions take place (compared to the technological revolutions in computing, telecommunication or internet which occurred in the last decades). This refers both to cost decrease of innovative technologies and development of new materials, components and technology systems.

– Diminishing: Substantial technological development occurs, in particular in material science (with corresponding impact e.g. on energy storage systems), computing and telecommunication. This technological development has a considerable impact on the sector-specific technologies. However, no "wonders" occur (i.e. nuclear fission still is not applied broadly and of course the physical laws of thermodynamics are still in place). The (even substantial) technological development does not lead to discontinuous jumps in the availability and cost structure of technologies. For pragmatic reasons, even in this storyline we do not consider wild cards (although we know that unexpected technology development may happen).

6.4 Uptake by Sectoral Assumptions

In part III's sectoral assessments and their socio-economic assumptions, various parts of this SSP have been uptaken. Some important ones are:

- Demographic scenario (age structure): uptaken by health chapter
- Demographic scenario (total population): uptaken/aligned by natural disasters and catastrophe management
- Land-use scenario: uptaken by agriculture, forestry, construction and buildings as well as urban green chapters
- Economic growth assumptions and key commodities price assumptions: uptaken by macro-economic, agriculture and energy chapters
- Technology/innovation pathways: uptaken by energy chapter.

References

Boitier B, Da Costa P, Le Mouel P, Zagam P (2008) Calculation of land use price and land use claims for agriculture, transport and urban land use at national level. hal-01044651

Diffenbaugh NS, Hertel TW, Scherer M, Verma M (2012) Responses of corn market to climate volatility under alternative energy futures. Nat Clim Change 2:514–518

Environmental Agency Austria (2014) Flächennutzung in Österreich. http://www.EnvironmentAgencyAustria.at/umweltsituation/umweltinfo/opendata/oed_landbedeckung. Accessed 20 May 2014

Haberl H, Erb K-H, Krausmann F, Adensam H, Schulz N-B (2003) Land-use change and socio-economic metabolism in Austria – Part II: land-use scenarios for 2020. Land Use Policy 20 (1):21–39

Hanika A (2005) ÖROK-Prognosen 2001–2031; Teil 2. Haushalte und Wohnungsbedarf nach Regionen und Bezirken Österreichs. Schriftenreihe/Österreichische Raumordnungskonferenz (ÖROK); 166/2 (including CD-ROM), Wien

Hanika A (2010) Kleinräumige Bevölkerungsprognose für Österreich 2010–2030 mit Ausblick bis 2050 ("ÖROK-Prognosen") Teil 1: Endbericht zur Bevölkerungsprognose. ÖROK, Wien. http://www.oerok.gv.at. Accessed 15 Mar 2014

Hanika A (2010) Kleinräumige Bevölkerungsprognose für Österreich 2010 bis 2030 mit Ausblick bis 2050 ("ÖROK-Regionalprognosen") Teil 3: Endbericht zu den Modellrechnungen zur regionalen Haushaltsentwicklung. ÖROK, Wien. http://www.oerok.gv.at. Accessed 15 Mar 2014

Hiess H et al (2009a) Szenarien der Raumentwicklung Österreichs 2030. Regionale Herausforderungen und Handlungsstrategien. Schriftenreihe/Österreichische Raumordnungskonferenz 176/II, Wien

Hiess H et al (2009b) Szenarien der Raumentwicklung Österreichs 2030. Materialienband, Schriftenreihe/Österreichische Raumordnungskonferenz 176/II, Wien

IEA (2010) World energy outlook 2010. http://www.worldenergyoutlook.org/media/weo2010.pdf

Kytir J, Biffl G, Wisbauer A (2010) Kleinräumige Bevölkerungsprognose für Österreich 2010–2030 mit Ausblick bis 2050 ("ÖROK-Regionalprognosen") Teil 2: Endbericht zur Erwerbsprognose. ÖROK, Wien. http://www.oerok.gv.at. Accessed 25 Mar 2014

Loibl W, Köstl M, Steinnocher K (2009) RUR-typology: list of generic rural-urban region types. Classification of the major European rural-urban regions at NUTS3-aggregate level. Deliverables D2.1.2; D2.1.3 Version V3.1. PLUREL project. Peri-urban land use relationships – strategies and sustainability assessment tools for urban-rural linkages. http://www.plurel.net

Moss RH, Edmonds JA, Hibbard KA, Manning MR, Rose SK, van Vuuren DP, Carter TR, Emori S, Kainuma M, Kram T, Meehl GA, Mitchell JFB, Nakicenovic N, Riahi K, Smith SJ, Stouffer RJ, Thomson AM, Weyant JP, Wilbanks TJ (2010) The next generation of scenarios for climate change research and assessment. Nature 463:747–756. doi:10.1038/nature08823

Nakicenovic N, Swart R (eds) (2000) The IPCC's special report on emissions scenarios (SRES). Cambridge University Press, Cambridge

O'Neill B, Carter TR, Ebi K, Edmonds J, Hallegatte S, Mearns L, Riahi K (2012) Workshop on the nature and use of new socioeconomic pathways for climate change research. Meeting report Boulder. https://www2.cgd.ucar.edu/sites/default/files/iconics/Boulder-Workshop-Report.pdf

Parry ML, Rosenzweig C, Iglesias A, Livermore M, Fischer G (2004) Effects of climate change on global food production under SRES emissions and socio-economic scenarios. Glob Environ Chang 14:53–67

Pérez-Soba M, Verburg PH, Koomen E, Hilferink MHA, Benito P, Lesschen JP, Banse M, Woltjer G, Eickhout B, Prins A-G, Staritsky I (2010) Land use modelling – implementation. Preserving and enhancing the environmental benefits of "land-use services". Final report to the European Commission, DG Environment. Alterra Wageningen UR, Geodan, Object Vision, BIOS, LEI and PBL

Rüdisser J, Schönhart M, Schmid E, Sinabell F, Tappeiner U, Tasser E (2011) Szenarien für Österreichs Landwirtschaft 2020: Ergebnisse einer integrierten Modellierung. University of Innsbruck, BOKU, Joanneum Research, Wien, Graz

Schiman S, Orischnig T (2012) Coping with potential impacts of ageing on public finances in Austria; the Demography-based Economic Long-Term Model for Austria's Public Finances (DELTA-BUDGET) assumption report. Federal Ministry of Finance, working paper 1/2012, Wien

UN (2005) World population prospects. The 2004 revision. http://www.un.org/esa/population/publications/WPP2004/2004Highlights_finalrevised.pdf

Windisch P (2005) Prognosen für Österreich Teil II: Haushalte und Wohnbautätigkeit – regionale trends bis 2031. Österreichischer Sparkassenverband (Hg.), working paper, 1030 Wien. https://www.sparkasse.at/erstebank/.../0901481b80045b87.pdf

Chapter 7
Economic Evaluation Framework and Macroeconomic Modelling

Gabriel Bachner, Birgit Bednar-Friedl, Stefan Nabernegg, and Karl W. Steininger

Abstract The first step in an economic assessment of climate change impacts at the country level is the identification of so-called "impact fields". These fields can be either single economic sectors, parts of sectors or aggregates of sectors. For the case of Austria that is explored in this book, 12 impact fields are identified and investigated regarding climate change impacts and the resulting economic costs and benefits. As impact fields are often of very different character, the mechanisms of climate change impacts are different and, therefore, also the costing methods to obtain costs and benefits of climate change are diverse. Hence, depending on the impact field, one or several of the following costing methods are applied: Changes in production technology and subsequent production cost structure, changes in productivity, changes in final demand, changes in investment, changes in public expenditures, and, finally, level of replacement cost. By applying these methods we obtain the direct costs by impact field.

As a modern economy is characterised by a strong specialisation across activities and sectors, there are strong interdependencies between different economic sectors (e.g. the food sector relies heavily on agriculture). For that reason, indirect effects on other sectors may contribute to total costs (or benefits) for the economy as well. A framework is needed which is able to capture these interactions between economic sectors. For that reason we here employ a computable general equilibrium (CGE) model as it depicts linkages between economic sectors as well as agents and is therefore able to cover interaction between different climate impacts occurring in different sectors. Relevant model outputs are changes in welfare, changes in sectoral activity (output), changes in value added and GDP, as well as in public budgets.

G. Bachner (✉) • S. Nabernegg
Wegener Center for Climate and Global Change, University of Graz, Graz, Austria
e-mail: gabriel.bachner@uni-graz.at; stefan.nabernegg@uni-graz.at

B. Bednar-Friedl • K.W. Steininger
Wegener Center for Climate and Global Change, University of Graz, Graz, Austria

Department of Economics, University of Graz, Graz, Austria
e-mail: birgit.friedl@uni-graz.at; karl.steininger@uni-graz.at

K.W. Steininger et al. (eds.), *Economic Evaluation of Climate Change Impacts*,
Springer Climate, DOI 10.1007/978-3-319-12457-5_7

7.1 Introduction

In the process of evaluating the economic impacts of climate change, often a monetary quantification is sought (i.e. costs and benefits), which is not always an easy task to accomplish. The possibility of monetisation depends to a high degree on the type of impacts which are analysed, but also on the availability of (market) prices. For a better understanding of this problem, climate induced costs are often categorised along the following sets of dimensions, tangible/intangible and direct/ indirect (e.g. Merz et al. 2010; Jonkman et al. 2008).

Tangible costs are relatively easy to measure, as the respective assets or goods are traded on markets and thus price tags exist, whereas intangible damages are not traded on markets and thus hard to transform into monetary values. The question whether damage is tangible or intangible can lead to deep controversies among authors, as it comes down to ethical principles (e.g. the pricing of human life or psychological trauma).

In contrast, the distinction between direct and indirect costs is more straightforward. Costs which are caused by climatic impacts themselves are direct costs (e.g. the destruction of a road due to flooding and the related costs of replacement and repair), whereas indirect costs are induced by the consequences of the damage event. Thus indirect costs are not induced by the physical impacts themselves but by their further consequences, i.e. negative effects which affect the whole economy via intersectoral linkages or the loss of services (EEA 2007; Merz et al. 2010).

Whereas the direct costs of climate change impacts are in principle relatively straightforward to approximate, e.g. by investigating expenditures for reconstruction and repair activities (if data is available), the indirect costs are more difficult to determine. In general the indirect costs depend very much on the macroeconomic importance and embedding of the regarded sectors or agents. A necessary first step to later analyse this mutual embedding is therefore to identify the key economic "impact fields". For the case of Austria we take them as represented by the impact fields in the Austrian National Adaptation Strategy (NAS).

The first aim of this chapter is to provide an overview of the economic costing methods applied to elicit the direct costs and benefits within each of these impact fields (i.e. of the respective impact chains there). Second, the macroeconomic model which is used to assess the indirect effects, i.e. cross-sectoral effects to other sectors as well as the effects for welfare, GDP, employment, and public budgets is described.

7.2 Economic Assessment Framework: The Direct Costs

7.2.1 Sectoral Aggregation and Representation of Impact Fields

The costs of inaction are assessed by distinguishing the following different impact fields:

- Agriculture *(agr)*
- Forestry *(for)*
- Ecosystem Services: Pest Control and Pollination *(ess)*
- Human Health *(hea)*
- Water Supply and Sanitation *(wat)*
- Electricity *(ele)*
- Buildings: Heating and Cooling *(h&c)*
- Transport *(trn)*
- Manufacturing and Trade: Labour Productivity Losses *(m&t)*
- Cities and Urban Green *(cug)*
- Catastrophe Management: Riverine Flooding *(cam)*
- Tourism *(tsm)*

In the macroeconomic assessment, climate change impacts in 10 out of these 12 impact fields will be assessed regarding their macroeconomic consequences. In the impact field Ecosystem Services there is no robust impact function available that could be used to assess changed pollination and pest control due to changes in climatic parameters and hence the resulting consequences for agricultural productivity and output cannot be estimated as a functional quantitative relationship. In the impact field Human Health, a robust impact function is available to estimate the consequences of changed temperature and humidity for morbidity and mortality, yet there is a multitude of options (but no agreement across them) how to monetise these climate change impacts (and even more so the indirect consequences thereof). In all other impact fields, climate change impacts could be assessed in terms of their direct costs and hence could also be included in the macroeconomic assessment. But again, the coverage of impacts differs across impact fields. The limitations of the current assessment are described in more detail in the respective chapters for each impact field (see Chap. 8—Agriculture to Chap. 19—Tourism) as well as in Chap. 21 (Macroeconomic Evaluation) where all impact fields are compared and assessed jointly.

To characterise these impact fields within an economic framework national input-output tables—which are often classified by NACE codes (French: "Nomenclature statistique des activités économiques dans la Communauté européenne")—provide a good starting point. In general, there are four different options for the characterisation of impact fields:

- Option 1: a single (aggregated) NACE sector represents an impact field (e.g. Forestry)
- Option 2: several, distinct NACE sectors represent an impact field (e.g. Manufacturing and Trade consists of several manufacturing sectors)
- Option 3: a new sector is extracted from several, distinct NACE sectors (e.g. Tourism is represented by respective fractions of sectors accommodation, entertainment, travel agencies, sport)
- Option 4: a single NACE sector is disaggregated into two or more subsectors (e.g. Agriculture could be disaggregated into crop and livestock)

The economic data provided by the national input-output table for Austria for the year of 2008 (Statistics Austria 2013) is our starting point. The original sectoral structure of 75 NACE sectors was aggregated to a total of 40 sectors (see Table 7.1 for the sectoral aggregation and the respective model code. Table 7.1 in the Supplementary Material for this chapter provides information about the respective NACE codes).

7.2.2 Costing Methods

To elicit the direct costs and benefits for each of the impact fields, the following types of costs can be distinguished:

1. Changes in production cost structure (i.e. changed input structure or technology)
2. Change in productivity (i.e. changed output for the same amount of input)
3. Change in final demand (e.g. demand shift as response to less snow availability)
4. Replacement costs (i.e. investment to reinstall damaged assets and infrastructure)
5. Preventive expenditures (i.e. for replacing a natural service, such as the protective function of a forest, by a technical measure)
6. Change in public expenditure (transfers and subsidies e.g. for damage compensation)

Figure 7.1 provides a compact overview of the applied costing methods in all of the impact fields under investigation (except for Ecosystem Services and Human Health which are not included in the macroeconomic assessment).

All costs have to be assessed both for the baseline scenarios up to 2030 and up to 2050 (relative to the average of a defined base period, e.g. 1981–2010) and again for the climate change scenarios for 2030 (representing climatic period 2016–2045) and 2050 (climatic period 2036–2065). The difference between the two constructed scenarios (climate change versus baseline scenario) yields the climate change impact in future periods. For details on the specific methods employed for each impact field, we refer to the Chaps. 8–19. Note that within each impact field one or several costing methods are applied, e.g. replacement costs jointly with change in

Table 7.1 Sectoral structure and model codes

Code	Activity/industry	Code	Activity/industry
AGRI	Agriculture	LTRA	Land transport and pipelines
FORE	Forestry	WTRA	Water transport
REXT	Rest of extraction	ATRA	Air transport
FOOD	Food and tobacco	STRA	Warehousing and transport support, post
WOOD	Wood and wood products	ACCO	Accommodation and restaurants
BEVE	Beverages	FINA	Financial service activities
PAPE	Paper and paper products	INSU	Insurance, reinsurance and pension funding
COKE	Coke and refined petroleum	AFIN	Auxiliary financial services and insurance activities
CHEM	Chemicals and chemical products	REAL	Real estate
PHAR	Pharmaceutical products and preparations	ARCH	Architectural and engineering; technical testing and analysis
PLAS	Rubber, plastic and other non-metallic mineral products	TRAV	Travel agency, tour operator and related activities
META	Metals and metal products	PUBL	Public administration and defense; compulsory social security
MACH	Machinery and equipment	HEAL	Human health activities
RMAN	Rest of manufacturing	ENTE	Creative, arts and entertainment activities
ELEC	Electricity, gas, steam and air conditioning supply	CULT	Libraries, archives, museums and other cultural activities
WATE	Water collection, treatment and supply	SPOR	Sports activities and amusement and recreation activities
WAST	Sewerage and waste	RECR	Recreational services
CONT	Construction	TELE	Telecommunication and computer services
MOTO	Wholesale and retail trade, repair of motor vehicles and motorcycles	SCIE	Scientific and professional services
TRAD	Wholesale and retail trade except for motor vehicles and motorcycles	RSER	Rest of services and education

productivity (e.g. in Forestry) or change in productivity and change in production cost structure (e.g. in Agriculture).

7.3 Macroeconomic Model Description

7.3.1 Methodology

To assess the macroeconomic effects of climate change impacts there are several options. The three main approaches are econometric analysis, input-output analysis

Impact field	Costing method		Exposure unit		Costing unit
Agriculture	**Change in agricultural productivity**	=	Change in average annual crop and grassland forage yields	x	Average commodity prices
Agriculture	**Change in production costs**	=	Change in management techniques	x	Average variable production costs (e.g. fertiliser costs)
Agriculture	**Change in public expenditures (change in agricultural policy premiums)**	=	Change in management techniques	x	Average agricultural policy premiums
Forestry	**Change in forest productivity**	=	Change in annual timber volume production	x	Average timber prices
Forestry	**Change in productivity (increase of bark beetle damage in production forests)**	=	Change in salvaged area	x	Average afforestation costs
Forestry	**Change in productivity (increase of bark beetle damage in production forests)**	=	Change in salvaged timber volume	x	Average reduction in contribution margin
Forestry	**Replacement cost (increase of damaged area in protection forests)**	=	Change in damaged area	x	Average cost required to restore protection functionality
Water Supply and Sanitation	**Change in final demand**	=	Change of demand for water and associated number of assets that need to be adapted	x	Costs of water production and adaptation of assets
Water Supply and Sanitation	**Change in production cost**	=	Number of assets that need to be adapted due to changes in water availability, quality and volume of wastewater	x	Cost of adaptation of assets and changed operating expenditure
Water Supply and Sanitation	**Replacement cost**	=	Change in number of infrastructure damage events	x	Average repair cost per infrastructure damage event
Buildings: Heating and Cooling	**Change in final demand**	=	Change in final energy demand for heating and cooling by energy carrier	x	Price of energy carrier used for heating and cooling
Buildings: Heating and Cooling	**Change in final demand**	=	Change in number of buildings equipped with AC by building category	x	Average specific investment costs of AC per building by respective building categories
Electricity	**Change in final demand**	=	Change in final demand for electricity for heating and cooling	x	Average electricity price
Electricity	**Change in production cost**	=	Change in electricity generation mix	x	Cost of respective electricity generation technologies
Transport and Mobility	**Replacement cost**	=	Change in number of damage events	x	Average repair cost per damage event
Manufacturing and Trade Services	**Change in labour productivity**	=	Change in labour productivity of (part of) total labour force	x	- Average wage per employee - Hourly GDP per employee
Cities and Urban Green	**Preventative expenditure**	=	Change in required park area [km^2] to compensate for increased heat	x	Investment and maintenance cost of park area per km^2
Catastrophe Management	**Replacement costs (losses due to catastrophes)**	=	Affected number of buildings or infrastructure	x	Average damage according to return period of floods (damage function)
Tourism	**Change in final demand**	=	Change in overnight stays by season and region	x	Average expenditure of visitor per overnight stay by season and region

Fig. 7.1 Applied costing method by all impact fields

and computable general equilibrium (CGE) modelling. Econometric models use statistical methods to analyse past time series on the economic performance of a region (typically on a yearly base) in order to project economic activity into the future. To do so a dependent variable is explained by a function of one or more independent (explanatory) variables. This approach is used extensively for sectoral assessments (see for example de Cian et al. (2013) for the case of energy demand changes under climate change) but lacks the ability to cover cross-sectoral and economy-wide feedback effects. Furthermore, the historical data—on which econometric models are based—are unlikely to represent the economic behaviour during climate change.

When cross-sectoral feedback effects are of interest there are basically two options left: input–output (I–O) analysis and CGE modelling. I–O analysis is widely used to analyse the economic costs of climate change. Such models are based on input-output tables of economies which depict interdependencies between economic sectors. In such a table there is a row and a column for every sector, representing the sector's supplied and demanded goods/services to and from other sectors. Hence the technical relations of production are depicted implicitly in the structure of an I–O table and therefore shocks to a single (or more) sector(s) ripple through the whole economy. The advantages of this method are: relatively good availability of data (I–O tables of countries are often easy to get at national statistical offices), transparency and its straightforwardness (if data is available). Nevertheless the main disadvantage of I–O models is their rigidity, as there is no possibility for price induced substitution effects which are essential in the long run (but the long run is the context relevant for answering questions of climate change). Nevertheless this approach is popular and widely used in cost assessments and has been refined in many ways. Examples for the application of I–O models for hazard loss estimation are the Indirect Economic Loss Module of the HAZUS loss estimation methodology (FEMA 2001) or the adaptive I–O model in Hallegatte (2008). See Rose (2004) for further examples of refinement.

Another powerful modelling framework is CGE modelling (Shoven and Whalley 1992), which has become more attractive for impact and cost assessment more recently. For instance, CGE models have been used to assess the macroeconomic costs of climate change impacts, such as changed productivity in agriculture, sea level rise, tourism, energy demand or health (e.g. Aaheim et al. 2010; Mechler et al. 2010; Bosello et al. 2011; Ciscar et al. 2011, 2012). Yet, most of these studies are applied at the EU level and cover only selected impact fields like sea-level rise, agriculture and electricity while many effects on smaller spatial scales are neglected or cancelled out due to aggregation. In the present study, we therefore focus exclusively on a single country, Austria. We combine earlier CGE modelling exercises on tourism (Schinko et al. 2013), agriculture (Schönhart et al. 2011) as well as electricity (Bachner et al. 2013), extend the sectoral coverage towards the remaining impact fields, and conduct a cross-sectoral assessment of climate change costs and benefits. This national modelling has the advantage on the one hand of paying sufficient attention to national specifics (e.g. importance of certain sectors, degree of federalism); on the other hand we face the disadvantage of depicting international trade relations in less detail and therefore climate change impacts via international price mechanisms are covered poorly (e.g. changes in global food prices and their implications at the national level).

The backbone of CGE models is a table similar to an I-O table (it is an I–O table enriched by Systems of National Accounting data) which gives a static picture of the economy and describes intersectoral dependencies and demands: a so-called "social accounting matrix" (SAM). In CGE models multiple markets, producers and consumers with simultaneous optimising behaviour are simulated, while certain budget and cost constraints must hold. Thus, allowing for responses to possible climate change induced price effects, this method is well suited to analyse the stated research questions of costs of climate change and is therefore used in this analysis. Note that as the model allows for endogenous optimal responses to shocks and resulting price changes, autonomous adaptation to climate change (which is still in the scope of "inaction") is possible in such a model environment. Allowing for such endogenous adjustments of perfectly informed and rational agents, the modelled impacts and the resulting costs may be underestimated compared to assessments with less flexibility.

As in every model, some elements of the real world have to be neglected for the sake of simplification and therefore we face limitations in modelling climate change impacts. CGE models are an excellent tool to assess economy wide or "global" feedback effects of "local" shocks, such as sectoral impacts by climate change. Nevertheless, the intangible dimension (e.g. health, welfare costs of ecosystem service changes) is depicted poorly in model environments like this. Therefore the modelling of climate change impact chains is restricted to the model structure and its elements. Another drawback of the application of CGE models is their inability to capture changes in stocks. By definition such models are based on annual flows within the economic system and as such, stocks are not captured. Results like impacts on GDP (flow) therefore show solely the impact of climate change on the economy's ability to produce goods and services, but do not tell anything about the impacts on stocks, which might be much higher. The absence of economic stocks in CGE models together with the fact that those models are usually applied on a yearly basis implies that catastrophic events can be captured only poorly by this class of models, if not adjusted adequately.

7.3.2 General Model Description

Austria's economy is modelled as a static, small open economy with 2008 as the base year (date of latest available input-output table for Austria). The model comprises 40 economic sectors (see Table 7.1) according to their relevance for the identified impact fields (forward and backward linkages) and major Austrian sectoral activities. Austria is one regional entity (NUTS-0 level) and the rest of the world is represented by respective trade flows to and from Austria. Regarding international trade, Austria is modelled as a small open economy and trade is assumed to comply with the Armington assumption. Final demand is distinguished for households and government consumption.

Figure 7.2 gives a diagrammatic overview of the applied CGE model and shows flows of goods and services as well as production factors (monetary flows run in the respective opposite directions). The representative private household (*privHH*) is endowed with the production factors labour (*L*) and capital (*K*) and obtains transfers from the government (*GOV*). The production factors are used in domestic production (*X*) together with intermediate inputs to produce goods and services which are either used domestically or exported (*EX*). According to Armington (1969) goods and services produced in different world regions are not perfectly substitutable, thus every region treats its imports (*IM*) and goods from domestic production (*D*) differently. Therefore the so-called "Armington aggregate" bundles goods and services coming from domestic production and imports, which can be substituted for each other with sector specific elasticities. Goods and services from the Armington aggregate are then used by private households and the government as final demand (i.e. consumption) or as intermediate input for production. The government collects taxes which are levied on *L* and *K* (input taxes) as well as taxes on production and consumption (output taxes) (not depicted in Fig. 7.2). The algebraic model formulation can be found in the Online Supplementary Material for Chap. 7.

The production structure of *X* is shown in Fig. 7.3. A nested constant elasticity of substitution (CES) production function is applied: On the top level of production of a commodity *i*, a capital-labour-energy composite (*(KL)E*) can be substituted for an intermediate material composite (*INT*) with the sector specific elasticity of substitution *top*. On the second level of the nesting structure there are two branches: First, *(KL)E* is produced by a capital-labour composite (*KL*) and an energy composite *E* (consisting of sectors COKE, ELEC and REXT) which can be substituted with a sector specific elasticity *kle*. Second, *INT* is produced by intermediate inputs coming from all economic sectors (and imports) except COKE, ELEC and REXT (G_i to G_k). The intermediate inputs can be substituted against each other with the

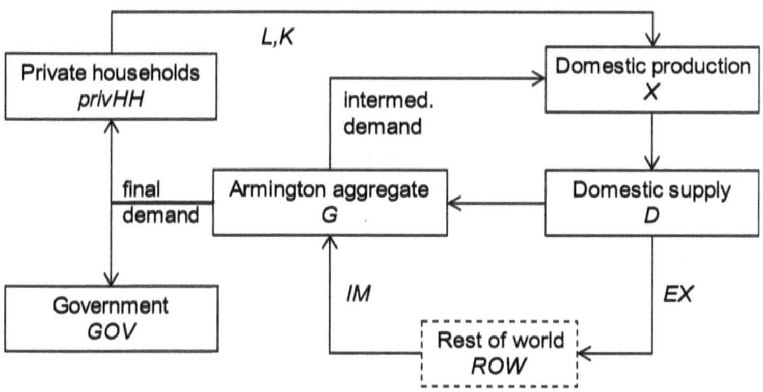

Fig. 7.2 Diagrammatic overview of the CGE model

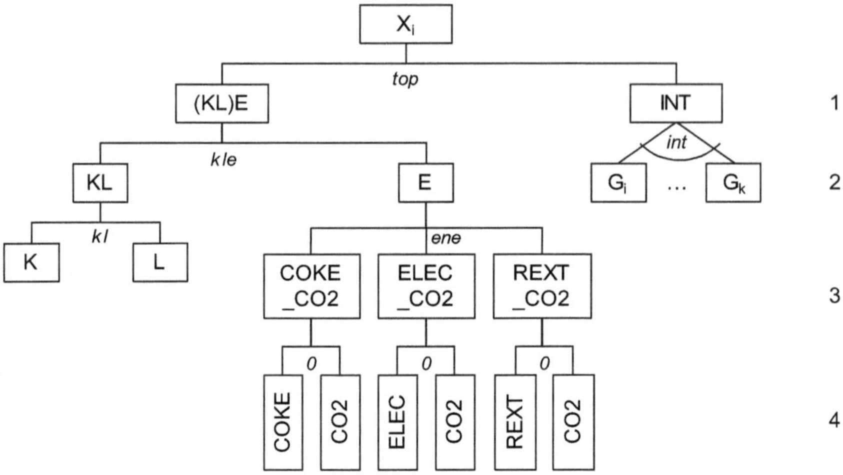

Fig. 7.3 Production structure of domestic production with four nesting levels

sector specific elasticity *int*. On the third nesting level the composite *KL* is composed of *K* and *L*, whereas the composite *E* is composed of inputs from the energy sectors COKE, ELEC and REXT, with an elasticity of substitution of *kl* and *ene*, respectively. On the lowest nesting level the energy sectors are coupled with a sector specific coefficient for CO_2 emissions (with an elasticity of 0). Only those sectors which are covered by the current EU emissions trading scheme (ELEC, COKE, META, PLAS and PAPE) are confronted with the resulting additional costs for CO_2 emissions (i.e. permits).

Concerning final demand, the aggregate demand function (composite *W*) is depicted in Fig. 7.4. On the top level a non-energy composite (*NE*) can be traded off for the energy composite *E* with an elasticity of substitution of *s*. Similar to the production structure of domestic production the *NE* composite is produced using commodities G_i to G_k but with a different elasticity of substitution (*nene*). The energy composite *E* consists of inputs from COKE, ELEC and REXT (but without CO_2, as final demand is not covered by the EU Emission Trading Scheme [ETS]).

Regarding public budgets, the following taxes and transfers are distinguished:

- Production tax: Sector specific output tax (or subsidy)
- Labour tax: Production input tax for labour (including social insurance)
- Capital tax: Sector specific capital returns tax
- Value added tax: Consumption tax
- Export tax
- Unemployment benefits
- Transfers to households

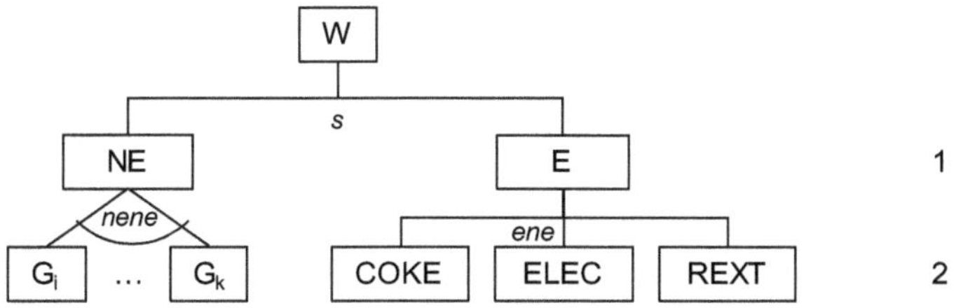

Fig. 7.4 Final demand structure of private households and government with two nesting levels

7.3.3 Implementation of Baseline Scenario

In order to implement the socioeconomic and political scenario assumptions several exogenous assumptions are introduced in the macroeconomic model which are described briefly in the next paragraph (see Chap. 6, SSP for more details).

Economic growth is given exogenously by a given GDP growth rate of 1.65 % p.a. until 2050 (Schiman and Orischnig 2012). All sectors are assumed to grow at the same rate. Land use and demographic changes are indirectly covered by the assumed GDP growth rates. Regarding international markets, changes in prices of fossil energies (oil, gas, coal) as well as agricultural products are implicitly taken account of by the sectoral models for impact fields Electricity (Chap. 14) and Agriculture (Chap. 8). Climate policy is fixed at those levels which were agreed in 2013. The CO_2 permit price is exogenously given according to the shared socioeconomic pathways (see Chap. 6), in particular according to IEA (2010). Regarding agricultural policy, changes in subsidies and taxes on land/capital/labour are implicitly taken into account by the sectoral models applied in the impact field Agriculture (see Chap. 8). Concerning Water Supply and Sanitation, reduction in subsidy rates are implemented (see Chap. 12 for more details). Changes in the energy structure of Austria are implicitly taken into account by the impact fields Buildings: Heating and Cooling (see Chap. 13) as well as Electricity (see Chap. 14).

In addition, for each impact field, specific baseline assumptions are implemented (covering implicitly also technological change). These assumptions are described in more detail in the respective chapters (see Chaps. 8–19). Note, that in the baseline scenario the economic structure remains mostly the same as in the base year, disregarding the mentioned sector specific baseline assumptions. Hence, the economy structurally looks mostly the same in 2030 and 2050 as in the base year, but is larger in size as it grows at assumed GDP growth rates. Implicitly this assumption also covers population growth which is translated to labour force as well as capital accumulation in the CGE model.

7.3.4 Inclusion of Climate Change Impacts

As already mentioned, the impacts of climate change on a specific impact field can be divided into direct and indirect effects. The former are triggered by the climatic stimuli themselves, whereas the latter emerge from the direct effects in the long run and by economy-wide interaction through price signals. The direct effects which were calculated according to the methods described in Sect. 7.3.3 are included in the CGE model in the following ways:

- *Change in production cost structure*: Climate change may force certain sectors to use different technologies for production (e.g. a change in the electricity generation mix) or require that costs are redirected towards maintenance (for some types of recurring replacement costs or preventative costs). This in turn leads to a change in the production cost structure, as the sectoral composition of intermediate inputs, capital and labour changes.
- *Change in productivity*: Due to climate change there may be positive or negative changes in productivity of certain sectors. In the case of productivity losses, more input is needed to achieve the same amount of output than without climate change (or equivalently, less output is possible with the same amount of input). In the case of productivity gains, the reverse holds. Productivity changes are either modelled for a single factor (e.g. labour) which becomes more or less productive or via multi-factor productivities (MFPs).
- *Change in final demand*: Final demand of private households as well as the government may shift to other goods and services under climate change (e.g. more demand for water because of higher temperatures or less tourism demand due to lower snow availability). As the benchmark endowment of private households and the government is limited, final demand is also limited to this amount. Therefore overall final demand can not change but can be shifted to other commodities.
- *Change in investment*: Both replacement cost and preventative expenditure in most cases require additional investment or changed investment patterns. As a consequence, the affected sectors may have to (autonomously) adjust their capital stock, leading to the emergence of higher or lower investments (e.g. to meet higher peak loads due to cooling demand, the electricity sector has to build new power plants) or depreciation (e.g. shorter depreciation periods due to a faster deterioration of water and sewerage infrastructure). We assume fixed savings which determine the overall level of investments. Therefore overall investment expenditures can not be changed but can be shifted to other sectors.
- *Change in public expenditures*: Regarding government expenditures, climate change may trigger changes in transfers to private households but also in tax revenues and subsidy expenses (e.g. higher expenditures for catastrophe management).

Table 7.2 provides an overview of how the mentioned effects are implemented in the CGE model for each impact field and gives a short explanation of the respective

Table 7.2 Overview of the implementation of climate change impacts by impact field

Impact field	Change in production cost structure	Change in productivity	Change in final demand	Change in investments	Change in public expenditures	Short explanation of impact(s) and respective modelling
Agriculture	X	X				In agriculture climate change triggers productivity gains due to longer growing seasons. These are modelled via multi-factor productivities (MFPs). Furthermore the production cost structure changes as, due to higher yields, more livestock farming is possible and more fertilisers are used (modelled as a shift within the production cost structure)
Forestry	X	X		X	X	Climate change triggered effects in the forestry sector induce higher investment in protective forests (modelled as shifts within investment expenditures), lower productivity in commercial forests (modelled via capital productivity), as well as higher costs for silvicultural measures (modelled as lower labour productivity) and reduced contribution margins in commercial forests (modelled via capital productivity) due to bark beetle infestation
Ecosystem Services: Pest Control and Pollination	Not quantified within the CGE model					
Human Health	Not quantified within the CGE model					
Water Supply and Sanitation			X	X		In the water supply and sanitation sector there are climate change triggered effects on drinking water demand (modelled as a shift within consumption expenditures) as well as changes in investments (modelled as shifts within the investment expenditures), as more extreme

(continued)

Table 7.2 (continued)

Impact field	Change in production cost structure	Change in productivity	Change in final demand	Change in investments	Change in public expenditures	Short explanation of impact(s) and respective modelling
						weather events lead to more frequent damages and faster deterioration of water infrastructure
Buildings: Heating and Cooling	X		X			Due to climate change final and intermediate energy demand for heating is reduced, whereas energy demand for cooling increases (modelled as a shift within consumption expenditures). When combining heating and cooling effects triggered by climate change, the reduction in heating is stronger than the increase for cooling (measured in expenditures)
Electricity	X		X	X		First, there is a change in the electricity generation mix (shift within the production cost structure) which is triggered by climate change (more gas fuelled power plants to compensate shortages from water power). Second, the net effect of changes of electricity demand for heating and cooling increases final demand for electricity (modelled as a shift within consumption expenditures). Third, more investment expenditures for new power plants are needed because of higher air conditioning demand and thus higher peak loads (modelled as shifts within the investment expenditures)

Transport				X		Climate change triggered damages to road infrastructure increase and therefore additional investments are necessary (modelled as a shift within the investment expenditures). In addition capital becomes less effective, as capital which is used to rebuild damaged infrastructure does not increase the capital stock, but solely holds it constant
Manufacturing and Trade: Labour Productivity Losses		X				Due to hotter and more humid climatic conditions in summer, labour productivity is subject to decrease
Cities and urban green					X	The preventative costs for climate change triggered effects are assumed to arise for the public sector and are modelled as reduction in government budget
Catastrophe Management: Riverine Flooding			X	X	X	Damages due to additional flooding events are modelled as an increase in final demand of private households, which are covered parthly by additional transfers from the government. Furthermore, reconstruction of damaged buildings is modelled as additional investment into the construction sector
Tourism			X			Tourism demand is reduced due to climate change triggered effects (modelled as a shift within consumption expenditures). The reduction of foreign tourism demand is modelled as a reduction in available capital endowment

impacts modelled. As described in Table 7.2 many of the impacts lead to shifts within the model only (e.g. between investment categories), thus leading to moderate total economic impacts only, which might be positive or negative. The parameter values applied are given in the Supplementary Material to Chap. 21 (Macroeconomic Evaluation).

7.3.5 Scenario Simulations: Representation of Long Run Effects of Investment and Demand Changes on Capital Accumulation and Public Budgets

Simulations are undertaken for a baseline scenario reflecting reference socioeconomic development assumptions up to 2030 and 2050 (see Chap. 6, Shared Socioeconomic Pathways), and for the mid-range climate change scenario 2016–2045 and 2035–2065 (see Chap. 5, Climate).[1] As we use a static CGE model which represents annual monetised flows, it is necessary to convert investments into annual expenditures and thereby to take investment cycles, the size of the capital stock (which is not depicted explicitly in the model) as well as depreciation periods into account. Furthermore, we distinguish whether costs arise temporarily (one time) or permanently (each year).

In addition to changed investment costs, it is necessary to take knock-on effects which emerge over time into account, because investments are diverted from other fields of investment to investment for repair of climate change damages (mostly investment for construction activities). As we use a static model, we account for these effects by higher investment requirements, depreciation and hence capital services increase in the respective impact fields.

Regarding shifts in demand, we assume that the savings rate and subsequently also the consumption rate of private and public households is fixed. Therefore changes in the structure of demand (e.g. higher expenditures on construction and buildings) have to be compensated for by demand changes for all other expenditure categories. A key assumption here is the CES structure of demand. As a consequence of fixed savings, the total level of investment has to adjust to this level and hence the basic assumption here is that investments are savings driven.

In addition to accounting for which costs arise in which sector, it is necessary to specify who is bearing these costs, e.g. whether costs are covered in the respective sector and shifted (partly) via higher prices to consumers, or whether government subsidies/expenditures are increased (as a form of autonomous adaptation, e.g. when subsidies are based on the level of infrastructure investment in the sector).

---

[1] For all impact fields, simulations are undertaken for the reference socioeconomic development and the mid-range climate change scenario. For most impact fields, also an impact diminishing and enhancing socioeconomic development as well as a low-range and high-range climate scenario are analysed.

Regarding public budgets, the default option is to assume a constant net budget balance.

7.3.6 Model Outcomes

For each impact field, results are displayed by comparing the effects of the baseline to the climate change scenarios, both for 2030 (representing the average of 2016–2045) and 2050 (2036–2065). Table 7.3 summarises key macroeconomic output variables of the CGE model which are described in more detail in the respective chapters. Next to these variables we also investigate changes on the labour market as well as effects on tax revenues and government budgets.

7.4 Summary and Outlook

This chapter gave a short overview of the methods employed for the costing of climate change impacts in each of the 12 impact fields at the national level, here applied for the case of Austria. While these methods enable the assessment of the direct costs of climate change, not only the direct costs but also the indirect costs which arise through cross-sectoral linkages matter from the perspective of the economy. We therefore briefly reviewed different methods for eliciting these indirect costs (econometric, input-output, and CGE modelling) and described the CGE approach which is used in the remainder of this book in more detail. After the model overview, the different options for implementation of the direct costs of climate change into the macroeconomic model were described. Finally, we also described how the difference between the climate and socioeconomic scenarios is used to elicit the economic impacts of climate change.

In the following chapters, the impacts of climate change are evaluated first for each impact field separately, implying that we assume that climate change effects are only present there. This exercise is useful to get an understanding of the direct and the indirect effects triggered by climate change in the respective field. Thereafter, impact fields are assessed jointly to provide an estimate of the total economic costs of climate change for Austria.

For both the assessment by impact field as well as for the total assessment, it needs to be kept in mind that the coverage of impact chains is partial due to data limitations but also due to lack of knowledge. As a consequence, all numbers presented have to be interpreted with care and the total cost estimate is likely to be a lower bound estimate due to these knowledge gaps.

Table 7.3 Characterisation of macroeconomic results

Variable	Description	Measurement
Output quantity	Output quantities are interpreted as physical units of traded goods and services, which are—for the sake of comparability—multiplied by prices of the benchmark year (in this case 2008)	Quantities (Q)
Output value	Sectoral and spill-over effects are measured in changes of "output values" of the respective sector. When output quantities are multiplied by prices which emerge in a counterfactual solution of the model (for instance a "future" price of a scenario run), so-called "output values" are obtained	Price times quantities (P*Q)
Intermediate Demand	Intermediate demand reflects the costs (or value) of all sectoral intermediate inputs which are necessary for production	P*Q
Value added	Intermediate inputs are used together with capital and labour to create goods and services which are valued higher than the sum of all intermediate inputs. Hence, the used capital and labour represents the additional value which is created by an activity. Sectoral value added is therefore obtained by subtracting intermediate demand from output value. When sectoral value added is increasing in a model simulation run, we speak of "gains", whereas "losses" emerge when value added is decreasing	P*Q
Gross domestic product	Gross domestic product (GDP) of a country is the value added of all produced goods and services within a year. GDP can be determined by the sum of all sectoral value added. Note that GDP only measures flows and therefore gives no information about the development of (natural) stocks	P*Q

(continued)

Table 7.3 (continued)

Variable	Description	Measurement
Welfare change	The change of welfare is measured as the difference between consumption in the climate change case priced with baseline prices and the consumption of the baseline case at baseline prices (so-called Hicksian equivalent variation). Alternatively the welfare change can be interpreted as the amount of money that is needed to be added to (or subtracted from) the household's benchmark income in order to keep its utility at the same level as in the benchmark. The standard welfare measure is corrected by forced consumption which is not welfare enhancing (e.g. additional water consumption which is necessary to provide the same service level as without climate change). Note that this welfare measure ignores non-market values such as environmental quality or health	Baseline price times climate change quantity minus baseline price times baseline quantity $(P_{BL}*Q_{CC}) - (P_{BL}*Q_{BL})$

References

Aaheim A, Dokken T, Hochrainer S, Hof A, Jochem E, Mechler R, van Vuuren DP (2010) National responsibilities for adaptation strategies: lessons from four modelling frameworks. In: Hulme M, Neufeld H (eds) Making climate change work for us: European perspectives on adaptation and mitigation strategies. Cambridge University Press, Cambridge

Armington P (1969) A theory of demand for products distinguished by place of production. Staff Pap Int Monet Fund 16:159. doi:10.2307/3866403

Bachner G, Bednar-Friedl B, Birk S, Feichtinger G, Gobiet A, Gutschi C, Heinrich G, Kulmer V, Leuprecht A, Prettenthaler F, Rogler N, Schinko T, Schüppel A, Stigler H, Themessl M, Töglhofer C, Wagner T (2013) Impacts of climate change and adaptation in the electricity sector – the case of Austria in a continental European context (EL.ADAPT). Wegener Center Scientific Report 51-2013. Wegener Center Verlag, Graz. ISBN: 978-3-9503112-8-0

Bosello F, Eboli F, Parrado R, Nunes PALD, Ding H, Rosa R (2011) The economic assessment of changes in ecosystem services: an application of the CGE methodology. Economía Agraria y Recursos Naturales 11(1):161–190

Ciscar J-C, Iglesias A, Feyen L, Szabó L, Van Regemorter D, Amelung B, Nicholls R, Watkiss P, Christensen OB, Dankers R, Garrote L, Goodess CM, Hunt A, Moreno A, Richards J, Soria A (2011) Physical and economic consequences of climate change in Europe. Proc Natl Acad Sci USA 108(7):2678–2683

Ciscar J-C, Szabó L, van Regemorter D, Soria A (2012) The integration of PESETA sectoral economic impacts into the GEM-E3 Europe model: methodology and results. Clim Change 112(1):127–142

de Cian E, Lanzi E, Roson R (2013) Seasonal temperature variations and energy demand: a panel cointegration analysis for climate change impact assessment. Clim Change 116(3–4):805–825

European Environment Agency (EEA) (2007) Climate change. The cost of inaction and the cost of adaptation. EEA Technical report No. 13/2007. Copenhagen

Federal Emergency Management Agency (FEMA) (2001) Earthquake loss estimation methodology (HAZUS). National Institute of Building Sciences, Washington, DC

Hallegatte S (2008) An adaptive regional input-output model and its application to the assessment of the economic cost of Katrina. Risk Anal 28(3):779–799

International Energy Agency (IEA) (2010) World energy outlook 2010. OECD/IEA, Paris

Jonkman S, Bočkarjova M, Kok M, Bernardinini P (2008) Integrated hydrodynamic and economic modelling of flood damage in the Netherlands. Ecol Econ 66(1):77–90

Mechler R, Hochrainer S, Aaheim A, Salen H, Wreford A (2010) Modelling economic impacts and adaptation to extreme events: insights from European case studies. Mitig Adapt Strateg Glob Chang 15(7):737–762

Merz B, Kreibich H, Schwarze R, Thieken A (2010) Assessment of economic flood damage. Nat Hazards Earth Syst Sci 10:1697–1724

Rose A (2004) Economic principles, issues, and research priorities in hazard loss estimation. In: Okuyama Y, Chand SE (eds) Modeling spatial and economic impacts of disasters. Springer, Berlin

Schiman S, Orischnig T (2012) Coping with potential impacts of ageing on public finances in Austria; the demography-based economic long-term model for Austria's public finances (DELTA-BUDGET) assumption report. Working Paper 1/2012. Federal Ministry of Finance, Vienna

Schinko T, Köberl J, Bednar-Friedl B, Prettenthaler F, Töglhofer C, Heinrich G, Gobiet A (2013) Climate change and the Austrian tourism sector: impacts, adaptation and macroeconomic spillover effects. Under review

Schönhart M, Schmid E, Schneider UA (2011) CropRota—a crop rotation model to support integrated land use assessments. Eur J Agron 34:263–277

Shoven JB, Whalley J (1992) Applying general equilibrium. Cambridge University Press, Cambridge

Statistics Austria (2013) Input-output tabelle 2008. Statistics Austria, Vienna

Part III
Fields of Impact

Chapter 8
Agriculture

Hermine Mitter, Martin Schönhart, Ina Meyer, Klemens Mechtler, Erwin Schmid, Franz Sinabell, Gabriel Bachner, and Birgit Bednar-Friedl

Abstract Agriculture is highly exposed to climate change. The severity of impacts on agricultural systems usually varies by geographic, natural, and socioeconomic factors. We match results from a bio-physical process model with the climate change scenario of the COIN (Cost of Inaction) project to derive climate induced yield impacts on major crops and permanent grassland in Austria. An economic calculation is applied to estimate average annual changes of production values and costs for the periods 2016–2045 and 2036–2065. Results feed into a computable general equilibrium (CGE) model to assess economy-wide effects. Uncertainties are addressed in the study and are mainly due to high spatial and sectoral aggregation as well as the unknown autonomous adaptation behaviour of farmers. Our

H. Mitter (✉)
Institute for Sustainable Economic Development, BOKU—University of Natural Resources and Life Sciences, Vienna, Austria

Doctoral School of Sustainable Development, BOKU—University of Natural Resources and Life Sciences, Vienna, Austria
e-mail: hermine.mitter@boku.ac.at

M. Schönhart • E. Schmid
Institute for Sustainable Economic Development, BOKU—University of Natural Resources and Life Sciences, Vienna, Austria
e-mail: martin.schoenhart@boku.ac.at; erwin.schmid@boku.ac.at

I. Meyer • F. Sinabell
WIFO—Austrian Institute of Economic Research, Vienna, Austria
e-mail: ina.meyer@wifo.ac.at; franz.sinabell@wifo.ac.at

K. Mechtler
Institute for Sustainable Plant Production, Austrian Agency for Health and Food Safety, Vienna, Austria
e-mail: klemens.mechtler@ages.at

G. Bachner
Wegener Center for Climate and Global Change, University of Graz, Graz, Austria
e-mail: gabriel.bachner@uni-graz.at

B. Bednar-Friedl
Wegener Center for Climate and Global Change, University of Graz, Graz, Austria

Department of Economics, University of Graz, Graz, Austria
e-mail: birgit.friedl@uni-graz.at

K.W. Steininger et al. (eds.), *Economic Evaluation of Climate Change Impacts*, Springer Climate, DOI 10.1007/978-3-319-12457-5_8

analysis indicates moderately higher outputs and value added at the sector level. This results in a positive impact on the rest of the Austrian economy. The aggregated results conceal adverse regional and farm type specific impacts.

8.1 Introduction

Agriculture is highly exposed to climate change, in particular due to regional temperature increases and shifts in precipitation patterns (Rosenzweig et al. 2007; Peltonen-Sainio et al. 2010; Olesen et al. 2011). The scope of impacts varies by geographic, natural, and socioeconomic factors. Productivity gains are likely in temperate zones whereas sub-tropical and arid regions are expected to face losses even in the short run (Iglesias et al. 2011; Lobell et al. 2011; World Bank 2012). Non-climatic factors such as farm management, demographic and land use change can alter the exposure to climate change (Ziervogel and Ericksen 2010). Our analysis complements the findings on global scale by exploring the implications of the COIN (Cost of Inaction) climate change scenario for agriculture in Austria. The analysis also accounts for climate heterogeneity within the country as well as for economy-wide impacts including feedback effects with other sectors. The following questions are guiding our analysis: What are potential impacts of climate change on the agricultural sector in Austria? What are the costs induced by climate change at sector level and economy-wide if there is no planned adaption (i.e. long-term adaptations implying major structural changes)? Which uncertainties are to be considered because of spatial and structural heterogeneity?

The ADAPT.AT project (Adaptation to Climate Change in Austria) and the CAFEE project (Climate change in agriculture and forestry: an integrated assessment of mitigation and adaptation measures in Austria)[1] complement our study. Among the results of these projects are detailed data sets which are used in the present economic analysis.

We estimate crop and grassland forage yield responses to the COIN climate change scenario based on statistical meta-models. Yield responses are merged with a dataset on national land use, valued in monetary terms and fed into a computable general equilibrium (CGE) model to assess the macroeconomic impacts.

The chapter is organised as follows: In Sect. 8.2, we describe the sensitivity and exposure of agricultural production to climatic and non-climatic factors. In Sect. 8.3, we develop a scenario of potential future exposure to climate change, assess the impacts in a quantitative and qualitative manner and discuss uncertainties. In Sect. 8.4, we wrap up the findings, present our conclusions, and identify topics of future research.

[1] Both projects have been funded by the Austrian Climate Research Programme (ACRP).

8.2 Sensitivity and Exposure of the Agricultural Sector to Climatic and Non-climatic Factors

8.2.1 Sensitivity of the Agricultural Sector to Climatic and Non-climatic Factors

Temperature and precipitation patterns (among other climatic parameters) strongly influence plant development. Changes in temperature affect growing season lengths, late and early frosts, extreme heat waves and—in combination with precipitation—dry periods. Precipitation determines occurrence of wet periods, hail and extreme rainfall events. Consequently, not only changes in average climate conditions but also inter- and intra-seasonal patterns determine agricultural production. Crops, for instance, exhibit threshold responses to the climatic environment which affect their development, growth, and yield (Porter and Semenov 2005; Sánchez et al. 2014). Rising CO_2 levels in the atmosphere stimulate crop yields by the CO_2 fertilisation effect (Chavas et al. 2009; McGrath and Lobell 2013). Global warming is also related to detrimental effects such as deteriorating soil resources (Klik and Eitzinger 2010), affecting permanent crops (Weber 2009), altering balances between pest and predators (Garrett et al. 2013), and changing pollination by insects (see Chap. 10).[2]

While most natural production factors are highly affected by climate conditions, man-made factors seem to be less climate-driven. The following (non-climatic) developments are likely to prevail in the coming decades:

- Economic growth although at lower levels than in the past: loss of agricultural land and soil fertility; increasing demand for agricultural products such as food, renewable energy, bio-pharmaceuticals, green bio-refinery products, safe and high quality food; increased administrative burden for producers.
- Further investments in research and development in agriculture although at lower levels than in the past: increasing use of new cultivars and crops (e.g. bio-technology in plant breeding); input saving agri-technological advances (e.g. precision farming, digital agriculture); more efficient and animal welfare friendly livestock production systems.
- Uncertainties about future conditions: preferences and behaviour of consumers (e.g. diets, food waste); policy responses (e.g. production restrictions); public support of certain production systems (e.g. mountain farming, organic farming, small-scale farming and agri-environmental programs); technological progress (e.g. methane free milk production, nutrient accumulating crops).

[2] More details on the 'Sensitivity of Austrian agricultural regions to climatic factors' are provided in the Supplementary Material to this chapter.

8.2.2 Past and Current Exposure to Climate Change

Yields of major crops in Europe and Austria have increased steadily for decades. This is mainly due to advances in agricultural technologies such as agro-chemical measures, and plant breeding (Hafner 2003; Ewert et al. 2005). The slowing average yield growth of many crops, which can be observed since the early 1990s, may result from decreasing success in crop breeding and technology development as well as from farmers' participation in extensification programs (Darnhofer and Schneeberger 2007; Finger 2010), but is not necessarily the result of global warming. However, the European heat wave and drought in 2003 illustrates potential negative effects of extreme weather conditions. In Austria, grain maize and winter wheat yields were about 30 % below the average of 1997–2002 in north-eastern regions. There, mean temperatures were about 4 °C above the long-term means and annual precipitation sums amounted to around 350 mm only (Soja et al. 2005a, b). The year 2003 was not the only one with exceptional conditions during the recent past. According to our statistical analysis there were 5 years in which crop and grassland forage yields deviated significantly (positive and negative) from the expected values during the last two decades, i.e. 2003, 2008, 2011, 2012 and 2013.[3]

8.2.3 Potential Large-Damage and Beneficial Combinations

Some combinations in climatic and other natural and man-made production factors may lead to considerable devastating or beneficial situations for the agricultural sector. Here, we give an overview on potential impacts and their economic effects.

Heat periods and drought: Heat periods, i.e. daily mean temperature >30 °C for several days in combination with insufficient water supply can severely damage or diminish the harvest. If occurring at large spatial scales, short supply leads to price hikes as food demand is usually inelastic. Thus farmers are often partially compensated for poor harvests but consumers usually bear the full additional costs.

Heavy precipitation events at sensitive points in crop production: The time of seedbed preparation and early crop development stages are very sensitive. Changes in sub-daily rainfall intensities and land use change lead to an increase in surface run-off and soil water erosion (Mullan 2013). For instance, long-term field experiments in Austria revealed that 2–5 % of the rainfall events produce soil losses of more than 50 t/ha (Klik 2003) which is far above the average annual soil formation rate and thus diminishing long term productivity. Such extreme events typically occur at local or regional scale. They hardly affect market prices but diminish farm incomes in the affected regions.

[3] A detailed account of recent years with exceptional climate conditions is presented in the Supplementary Material 'Bio-physical impacts up to now'.

Higher temperatures and atmospheric CO_2 levels: The combined effect of increasing temperatures and CO_2 levels stimulates plant growth in water-rich regions. The effects are crop specific and contingent upon other production factors. For instance, winter cereals are likely to gain more from longer vegetation periods (daily mean temperature >5 °C) than summer cereals. Positive effects of CO_2 fertilisation are expected to be higher for C_3-crops like wheat and barley than for C_4-crops like maize (e.g. Ainsworth et al. 2002; Högy et al. 2009). However, farmers can only benefit from increased agricultural productivity if market conditions are favourable as well.

Adaptive capacity and social capital: Climate change impacts on agriculture can be alleviated by adaptation. However, the adaptive capacity depends on knowledge and education, management skills, the availability of technologies, timely information (such as climate services) and liquidity as well as the flexibility of the farming system (Rodriguez et al. 2011). Insufficient farm income and a lack of funds for investments are major obstacles to climate change adaptation. Not only farmers but also agri-businesses and governments have to introduce or promote adaptation measures at field, farm, regional, national, and international levels in a co-ordinated manner (Smit and Skinner 2002; Gupta et al. 2010).

8.2.4 *Impact Chains in Agricultural Production*

Table 8.1 provides an overview on the most important climate change impact chains on Austrian agriculture. Its last column indicates those parameters and impact chains which are analysed in the quantitative assessment presented in the next section.

8.3 An Exploration of Future Exposure to and Impacts of Climate Change

In this section, we present the steps of our climate change impact assessment. An overview is given in Fig. 8.1. The economic impacts are driven by changes in crop and grassland forage yields as well as yield-driven changes in production costs, which are induced by the scenario on future climate conditions. The yield changes of grain maize, winter wheat, winter rape, soybean as well as temporary and permanent grasslands[4] are assessed with the bio-physical process model EPIC. Statistical meta-models of crop (grassland forage) yield responses are developed for three (one) climate region(s) in Austria (see Fig. 8.2). Autonomous adaptation at farm level is taken into account by adjusting time schedules for seeding, plant

[4] Alpine meadows and pastures are not considered due to data and model restrictions.

Table 8.1 Impact chains in agricultural crop and grassland production

Climate change parameter	Impact chain	Quantified in this analysis
Temperature		
Increase in mean annual temperature	Prolonged growing season → changes in sowing and harvesting dates → increase in biomass production → increase in farm income	Yes
	Higher evapotranspiration rate → increase in water demand → yield decrease in dry regions	Yes
	Optional: higher evaportranspiration rate → increase in irrigation costs → change in farm income	No
	Desertification of porous soils → decrease in biomass production → decrease in farm income	Yes
	Higher infestation pressure of pests, diseases, and weeds → increase in variable production costs, e.g. pesticide inputs → decrease in farm income	No
Increase in maximum daily temperature in summer; heat waves	Higher evapotranspiration rate → increase in water demand → water as limiting factor for plant growth → change in farm income	Yes
	Optional: higher evapotranspiration rate → increase in water demand → increase in irrigation costs → change in farm income	No
	Heat stress → water as limiting factor for plant growth → change in farm income	Yes
Cold waves, severe frosts in winter/autumn/spring	Frost damage → drop in yield quality and in the value of the harvested crops → decrease in farm income	No
	Frost damage → decrease in biomass production → decrease in farm income	Yes
Precipitation		
Increase (decrease) in annual precipitation sums	In currently drier regions: increase (decrease) in plant available water → increase (decrease) in biomass production → increase (decrease) in farm income	Yes
	In currently wetter regions: increase (decrease) in surface runoff → increase (decrease) in soil water erosion → decline in (improving) soil structure and soil fertility in the long run → reduction (increase) in biomass production → decrease (increase) in farm income	Yes

Shift in the seasonal precipitation distribution and dry period in summer	Limited water availability during the growing season → decrease in biomass production → decrease in farm income	Yes
	Optional: limited water availability during the growing season → increase in irrigation costs → change in farm income	No
Wet period in summer	Damages to crops and soil structure	Yes
(Sub-daily) heavy precipitation events	Increase in surface runoff → increase in soil water erosion → decline in soil structure and soil fertility in the long run → reduction in biomass production → decrease in farm income	No
	Flooding and stagnant surface water in agricultural fields → reduction in biomass production → decrease in farm income	No
Hail events	Crop damages, drop in crop quality, and reduction in biomass production → reduction in the value of the harvested crops → decrease in farm income	No

Δ Climate signal	
Model & data	Statistical climate change model (ACLiReM) COIN-climate change data (COIN-CCD)
Indicators	Ø Temperature (°C) Ø Precipitation (mm)

↓

Bio-physical impacts	
Models	Bio-physical process model EPIC Meta-models of yield responses
Results	Ø Yields at NUTS-3 level (changes in %)

↓

Economic impacts on the agricultural sector	
Calculation	Economic gross margin calculation
Results	Ø Production values at NUTS-3 level (changes in %) Ø Production costs at NUTS-3 level (changes in %) Ø Productivity changes

↓

Macro-economic impacts	
Model	Computable general equilibrium (CGE) model
Results	Ø Welfare (changes in M€) Ø GDP (changes in M€) Ø Unemployment rate (changes in % points) Ø Public budget (changes in % points)

Fig. 8.1 Methodology to assess climate change impacts in agriculture

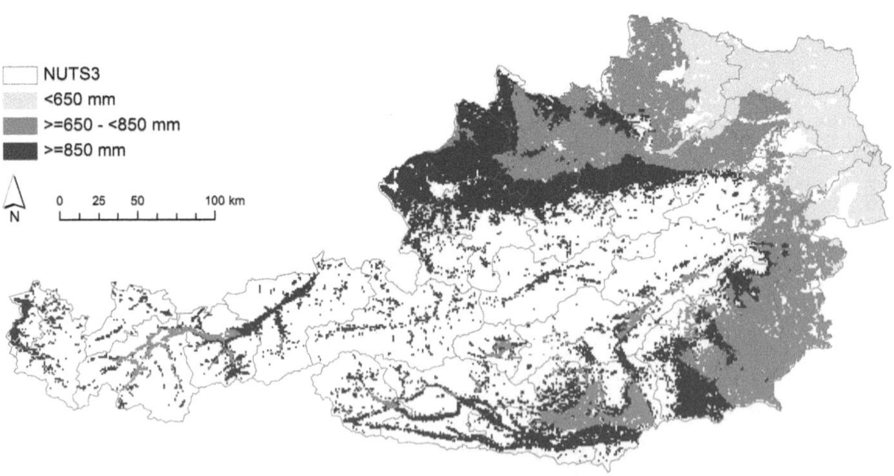

Fig. 8.2 Long-term historical mean annual precipitation sums on cropland

protection, fertilisation and harvest. Such adjustments require considerable farm agronomic skills and flexibility in farm management (see Rodriguez et al. 2011) but they do not increase variable costs assuming sufficient mechanisation capacity. Adaptation in land use (i.e. cultivar choices, land use types), mechanisation and farm management are not considered.

Expected yield changes directly affect farm revenues and production costs. Assumptions on real agricultural prices are based on OECD-FAO (2013). In our analysis, most production costs are assumed to be independent from yield levels, e.g. soil mechanisation and insurance. An exception is fertilisation, which depends on changing crop yield potentials in the economic calculation. Revenue and cost changes are calculated at the NUTS-3 level.

The results on changed production values and costs are aggregated to the agricultural sector level and used as input in a CGE model to assess indirect effects on outputs of other sectors' as well as effects on welfare, GDP, employment and public budget.

8.3.1 Climate Change Data for Agriculture

In our analysis, we use two different sets of climate change data, (1) the mid-range COIN climate change data (COIN-CCD), and (2) five climate change scenarios (SCEN) derived from a statistical climate change model for Austria (ACLiReM) (see Table 8.2 for an overview). The second data set is necessary because the bio-physical simulations require daily weather data such as those provided by ACLiReM, i.e. SCEN. We develop a procedure to match the bio-physical simulation outputs of SCEN to the COIN-CCD in order to be consistent with the other sector analyses.[5]

8.3.2 Impact Calculation

8.3.2.1 Bio-physical Process Model EPIC

The EPIC (Environmental Policy Integrated Climate) model simulates bio-physical processes, which respond to daily weather, topographic information (i.e. elevation and slope), crop and soil characteristics, crop management measure, and atmospheric CO_2 concentration (Williams 1995; Izaurralde et al. 2006). EPIC has already been applied for site, regional, and national investigations of future crop yield changes in Austria (e.g. Strauss et al. 2012; Koland et al. 2012; Eitzinger et al. 2013b;

[5] A detailed description of the climate change data and the matching procedure is provided in the Supplementary Material 'Climate change data for agriculture'.

Table 8.2 Overview on climate change data sets

	COIN climate change data (COIN-CCD)		Climate change scenarios (SCEN)
Models	RCM simulation of the RCM CCLM (Meissner et al. 2009) forced with the ECHAM5 A1B, from the Austrian research project reclip: century (Loibl et al. 2010)		Statistical climate change model for Austria (ACLiReM) using historically observed data from the period 1975–2007 (Strauss et al. 2013a)
Base period	1981–2010		1975–2005
Scenario period(s)	2016–2045	2036–2065	2010–2040
Changes in mean annual temperature compared to the base period	+1.05 °C	+2.02 °C	+1.5 °C
Changes in mean annual precipitation sums compared to the base period	+1.5 %	−2.3 %	SCEN01: ±0 %
			SCEN05: +20 %
			SCEN09: −20 %
			SCEN13: ±0 % (shift to winter)
			SCEN17: ±0 % (shift to summer)

Kirchner and Schmid 2013; Stürmer et al. 2013; Mitter et al. 2014). In this analysis, it provides outputs for a base period (1975–2005) and five climate change scenarios for the period 2010–2040 (see Sect. 8.3.1 and Supplementary Material) at 1 km pixel resolution, i.e. 40,244 cropland and 46,525 grassland pixels. For each climate change scenario and pixel, average annual dry matter crop yields and grassland forage yields have been simulated for various crop management measures including alternative fertilisation rates (high, moderate, and low nitrogen and phosphorus inputs), and irrigation (in combination with high fertilisation intensity). In a simplified validation procedure, EPIC outputs are compared to data from field experiments on research stations. The results indicate that EPIC overestimates average crop yields, especially for soybean.[6] In this study economic estimations are based on simulated crop yield changes, which likely are more robust than absolute yield level results.

8.3.2.2 Statistical Meta-models of Yield Responses

Statistical meta-models of yield responses have been developed in order to aggregate detailed EPIC results. Such an approach is deemed useful to merge large data

[6] The validation procedure and results are discussed in the Supplementary Material 'Validation of the bio-physical process model EPIC'.

sets (Ciscar et al. 2011; Iglesias et al. 2012), accelerates data processing and thus provides short-cut solutions for subsequent economic analysis (Wei et al. 2009; Finger et al. 2011). We apply multiple linear regression models with dummy explanatory variables to estimate the impact of five climate change scenarios (SCEN) and four crop management measures on average 30-years crop and grassland forage yields. The meta-models have been developed for five arable crops (i.e. grain maize, winter wheat, winter rape, soybean, and temporary grassland) and for permanent grassland. In order to account for regional differences, the meta-models for cropland have been developed for three climate regions which are classified according to long-term historical mean annual precipitation sums (Fig. 8.2): drier regions in eastern Austria which are partly irrigated (<650 mm), areas in western Austria where water is usually not limiting for crop growth (>850 mm), and the remaining areas which do not suffer from water stress at present but may experience water limitations if dry weather conditions increase in the future (650–850 mm; see Strauss et al. 2013b). For permanent grassland, Buchgraber and Gindl (2004) identified mean annual precipitation sums below 800 mm as critical for optimal plant growth. Because of the rather low share of permanent grassland in regions with precipitation sums below 800 mm, we do not distinguish climate regions for grassland and consider one average value for Austria.

The meta-models of yield responses are specified as follows:

$$\hat{Y}_{i,c} = \beta_{0i,c} + \beta_{1i,c} SCEN_{i,c} + \beta_{2i,c} MANA_{i,c} + u_{i,c} \tag{8.1}$$

where index i refers to climate regions and c to crops. $\hat{Y}$ is representing estimated annual dry matter crop and grassland forage yields (t/ha), *SCEN* refers to the base period and climate change scenarios, and *MANA* refers to alternative crop management measures (high, moderate, and low fertilisation intensity as well as irrigation). The base period and climate change scenarios as well as crop management measures enter the regression as dummy variables whereby the base period and high fertilisation intensity serve as reference. The β_0 is the intercept value; the β_1 and β_2 are parameter vectors for the climate change scenarios and the crop management measures, respectively. The $u_{i,c}$ is the residual term. The estimations are performed using the OLS estimator. In total, 16 regression models (for five arable crops—including temporary grassland—and three climate regions as well as for permanent grassland) have been run.

8.3.2.3 Economic Calculation of Bio-physical Climate Change Impacts

The meta-model estimates on regional crop and grassland forage yield changes, i.e. a weighted linear combination of the climate change scenarios (SCEN) that best match COIN-CCD (see Sect. 8.3.1 and Supplementary Material), serve as proxy for climate induced yield changes of major Austrian crops and grassland at NUTS-3

Table 8.3 Modelled and assigned crops, relative share (%) of total cropland in 2011

Modelled crop	Crop code	Assigned crops	% of cropland
Grain maize	CORN	All corn production systems, sugar beets, potatoes, spring cereals	37
Winter wheat	WWHT	All winter cereals	34
Winter rape	WRAP	Winter rape	4
Soybean	SOYB	Faba beans, field peas, sunflowers, soybeans, lupines, summer rape	6
Perennial grass	PGRS	All temporary grasslands	12
		Total	93

Note: Shares estimated from cropland data based on BMLFUW (2012)

level. The five crops modelled in EPIC (including temporary grassland) represent 46 % of total Austrian cropland in 2011. They are considered to be representative for other crops that are actually planted in Austria in order to extend the coverage of our analysis to 93 % of total Austrian cropland (Table 8.3).

Relative changes in crop yields are converted to absolute changes based on crop yield statistics. In a next step, absolute average annual yield changes for the periods 2016–2045 and 2036–2065 are attributed to a land use share of each crop at NUTS-3 level. Therefore, data are required about current and future land use. We build on land use and livestock scenarios for 2008 (base year) and 2020 provided by the ADAPT.AT project (Schönhart et al. 2014). It results from the bottom-up optimisation agricultural sector model PASMA and takes future conversion of agricultural land into non-agricultural land use, OECD-FAO price forecasts, different management variants, as well as the recent Common Agricultural Policy (CAP) reform into account. The land use scenario for 2020 is assumed to be valid for the periods 2016–2045 and 2036–2065. Future productivity changes and land use losses are assumed to cancel out each other beyond 2020. In a next step, absolute aggregated crop yield changes at NUTS-3 level are valued by market prices for each crop to derive the production value. Average output prices and production costs are based on projections for 2022 by OECD-FAO (2013). They are converted to real prices (base year 2008) with an assumed inflation rate of 2 % and are assumed constant in real terms for the periods 2016–2045 and 2036–2065. The base year was chosen to be consistent with the CGE analysis. The climate induced changes in crop specific production value are aggregated to crop groups similar to those in the economic accounts of agriculture (EAA) classification system. We weight the changes of each crop group by its production share to obtain aggregated national production values. Additional permanent grassland forage yields are monetised via its conversion to dairy yields due to the high competitiveness of dairy farming in Austria. Outputs of milk and beef increase proportionally.

Average annual climate change impacts are compared to the base year 2008 and two baseline scenarios without climate change for the periods 2016–2045 and

2036–2065. For the base year 2008, statistical crop and grassland forage yields are attributed to the 2008 land use scenario and valued by statistical crop prices of the period 2007–2009. Livestock production outputs are valued by the same price data set. At the aggregated national level, these results refer to and are validated by the EAA production values. The procedure to estimate the baseline scenarios for the periods 2016–2045 and 2036–2065 is identical to that described in the previous paragraph but takes statistical crop yields without climate change impacts into account.

With respect to production costs, we only take variable costs of plant and livestock production into account and separate among yield-independent and yield-dependent costs. Valuation of costs is based on the PASMA data base for gross margins, nutrient demand and nitrogen fixation. Fertiliser costs are based on 2007–2009 observations and projections for 2022 (OECD-FAO 2013; real prices from the base period 2008 with an assumed inflation rate of 2 %). Additional fertiliser demand from climate induced yield increases on permanent grassland is assumed to be covered by additional organic fertiliser supply of larger cattle herds. Additional variable costs for increasing livestock numbers are accounted for.

8.3.2.4 Macroeconomic Assessment

For the macroeconomic assessment, we use a CGE model for Austria which distinguishes 41 sectors (see Chap. 7 for details on this model). With this model, we simulate both a baseline scenario (reference socioeconomic development without climate change) and a climate change scenario (reference socioeconomic development and mid-range climate change) for the two future periods, i.e. 2016–2045 and 2036–2045.[7] The model's base year is 2008.

The quantified impact chains from Table 8.1 are implemented in the CGE model in two different ways.[8] First, productivity gains are implemented such that the required increase of production quantities is replicated in the CGE model. Second, changes in the production cost structure in the agricultural sector (i.e. due to production technology) are implemented in the CGE model. Compared to the base year 2008, the cost structure in the baseline scenario is altered to meet projected increases in demand for meat and dairy products. Adding climate change impacts, this cost structure is changing again. Due to higher grassland forage yields more livestock farming is possible and fertiliser inputs increase. Finally, regarding the sector-specific socioeconomic developments, we take account of the Common Agricultural Policy by implementing a reduction in subsidy rates for the agriculture sector.

---

[7] In the CGE model, the period 2016–2045 is represented by the year 2030 and the period 2036–2065 is represented by the year 2050.

[8] In the CGE model, all impacts are considered in relative terms (in % relative to the base year) because the economic gross margin calculation and the CGE model use different databases.

Table 8.4 Changes in mean annual crop and grassland forage yields with high fertilisation intensity in % for the periods 2016–2045 and 2036–2065, compared to the base period

Scenario period	2016–2045			2036–2065		
PRCP (mm)	**<650**	**≥650 to <850**	**≥850**	**<650**	**≥650 to <850**	**≥850**
CORN	−6.0	−3.7	5.7	−9.7	−4.1	6.0
WWHT	0.3	3.6	6.7	−2.3	3.6	7.1
WRAP	9.3	4.1	3.6	5.6	3.9	3.9
SOYB	3.4	4.2	9.3	0.0	3.6	9.2
PGRS	6.8	9.7	7.4	5.5	9.7	7.6
GRAS	n.a.	23.1	n.a.	n.a.	23.0	n.a.

Notes: Climate regions are based on long-term historical mean annual precipitation sums (PRCP in mm). Crop codes are defined in Table 8.3. For permanent grassland (GRAS), we do not differentiate between climate regions and consider an average value for Austria

8.3.3 *Results*

8.3.3.1 Bio-physical Impacts

Based on the results of the statistical meta-models of yield responses, Table 8.4 gives an overview on potential changes in mean annual crop and grassland forage yields with high fertilisation intensity for the climate regions and two scenario periods.[9] Crop yield changes range from −10 to +9 %, whereby increases and decreases are often higher in the second scenario period, i.e. variability increases. Decreasing crop yields are projected for grain maize and winter wheat in climate regions with mean annual precipitation sums below 650 mm indicating that these regions are particularly vulnerable to changing climatic conditions. Highest crop yield increases are received for soybean in climate regions with mean annual precipitation sums above 850 mm. Grassland forage yields (temporary and permanent grassland) are projected to rise between 7 and 23 % in the first and between 6 and 23 % in the second period.[10]

8.3.3.2 Direct Sector Impacts

Table 8.5 presents the aggregated benefits for Austrian agriculture from the estimated crop and grassland forage yield impacts. For the whole sector, climate change is estimated to increase the average annual production value by 193 million euros (M€) (real values 2008) in the period 2016–2045 compared to the baseline scenario. In the period 2036–2065, the average annual production value increases

[9] Note that change rates of mean annual crop and grassland forage yields refer to periods of 30 years and do not refer to annual changes.

[10] More detailed results are presented in the Supplementary Material 'Results of statistical meta-models of yield responses'.

Table 8.5 Average annual climate change-triggered economic impacts in crop and grassland production in the future compared to the baseline in million euros (M€)

Scenario period	2016–2045	2036–2065
Change in production costs	+77	+74
Change in production value	+193	+180
Net effect from climate change	+116	+106

by 180 million euros (real values 2008). This is equivalent to 4 % of the production value in 2008 of the agricultural products considered. These results are driven by mixed effects for both, crops and regions.

Productivity gains from climate change on temporary and permanent grasslands are an important driver of increasing production value. Forage gains on permanent grassland are assumed to be utilised by additional dairy cows. Increasing herds also increase livestock production costs. Furthermore, higher yields increase fertilisation costs. Total average annual production costs under the climate change scenarios are 77 and 74 million euros above the baseline scenarios for the periods 2016–2045 and 2036–2065.

8.3.3.3 Macroeconomic Effects

A summary on macroeconomic effects is shown in Table 8.6. All effects are given as average changes of annual values in million euros (M€) between the baseline and the climate change scenario. We see positive effects on gross output value and gross value added in the agricultural sector in both periods, i.e. 2016–2045 and 2036–2065.[11] Productivity gains in agriculture also lead to positive consequences for downstream sectors like food products (lower prices). The resulting increase in households' purchasing power is leading to an increase in typical final demand for goods and services coming from the sectors real estate and construction (living) or trade (including e.g. supermarkets). The emerging rise in final and intermediate demand for many other goods and services (such as real estate, construction and trade) leads to higher outputs in these sectors and also to slight demand-driven price increases. Compared to the baseline scenario, GDP thus rises by 248 million euros p.a. in the period 2016–2045 and by 441 million euros p.a. on average in the period 2036–2065. This corresponds to an average GDP effect of +0.07 % in the first period and +0.09 % in the second. Note that in both periods price effects are responsible for the major share of GDP increases.[12]

[11] Gross output value is the sum of sectoral intermediate demand and gross value added. Summing up gross value added across sectors and correcting for indirect taxes gives Gross Domestic Product (GDP).

[12] More detailed results are presented in the Supplementary Material 'Additional macroeconomic effects: GDP, welfare, and public budgets'.

Table 8.6 Sectoral and total effects of quantified climate change impacts in the agricultural sector, average annual effects relative to baseline (for periods 2016–2045 and 2036–2065)

Scenario period	Ø 2016–2045			Ø 2036–2065		
Changes in M€ p.a. (relative to baseline)	Gross output value	Inter-mediate demand	Gross value added	Gross output value	Inter-mediate demand	Gross value added
Gaining sectors	**+268**	**+17**	**+251**	**+579**	**+135**	**+444**
Agriculture	+3	−12	+15	+12	−8	v20
Real estate	+80	+23	+56	+125	+37	+89
Trade	+57	+22	+35	+102	+39	+62
Construction	+44	+28	+16	+75	+47	+28
Food products	−49	−59	+10	−52	−64	+12
All other gaining sectors	+134	+15	+119	+317	+85	+232
Losing sectors	**−11**	**−7**	**−3**	**−5**	**−3**	**−2**
Total effect (all sectors)	**+257**	**+9**	**+248**	**+574**	**+132**	**+441**
GDP at producer price			+0.07 %			+0.09 %
...thereof price effect			+0.06 %			+0.10 %
...thereof quantity effect			+0.01 %			−0.01 %

Note: baseline scenario = reference socioeconomic development without climate change; climate change scenario = reference socioeconomic development and mid-range climate change

8.3.4 Discussion of Qualitative (Non-monetised) Impacts

According to our results, average crop and grassland forage yields at the aggregated national level will increase over the period considered in our simulations. Despite the accordance of trends in yield changes to the scientific literature, one has to acknowledge important assumptions and uncertainties that determine the results of crop and grassland forage yields.[13] Economic results concerning land use and livestock as well as the development of output prices and production costs are based on specific assumptions as well. We apply land use and livestock scenarios from the ADAPT.AT project for the base period 2008 and a baseline scenario in 2020 including major elements of the 2013 CAP reform. The baseline scenario in 2020 is assumed to represent land use and livestock production in the periods 2016–2045 and 2036–2065 without adaptation of land use (i.e. land use types, cultivar

[13] A more detailed discussion is presented in the Supplementary Material 'Sector-specific uncertainties'.

choices) to changing climate conditions. There is one exception: Higher yields of grassland forage will be used to feed larger cow herds. We do not consider planned adaptation in order to follow the guiding principle of limited adaptation in COIN. According to our reasoning farmers adjust smoothly to higher crop and grassland forage yields (see definition of 'autonomous adaptation' at the beginning of Sect. 8.3). It is consistent with our assumptions that farmers will harvest all of their grassland yields.

Due to data limitations, we do not take into account climate change impacts on product quality and livestock productivity as well as on certain land uses such as fruit and wine production, horticulture and vegetables. These are important economic subsectors of agriculture, which may be crucially affected by climate change in positive and negative ways. However, in many cases, fruit, vegetable and wine producers are able to control weather impacts better than other sub-sectors, e.g. by irrigation, hail nets, or greenhouses.

With respect to production costs, we only take fertilisation costs and additional variable livestock costs into account. Climate change may also impact plant protection costs due to new or more intense weeds, pests, and diseases. Furthermore, we assume constant real costs to account for the observed divergence between input and output prices over time. We are aware of the literature that prices of farm commodities and thus land uses may change significantly in the future.[14] Our assumptions to keep many variables constant is made deliberately in order to isolate the effect of changes of crop and grassland forage yields in the climate change scenario.

A frequently raised issue in climate change impact research is the occurrence of extreme weather events (see e.g. Mitter et al. 2014), which may increase in magnitude and frequency during climate change. In this study extreme weather events are taken account of in EPIC as long as they are represented by historical daily weather input data. Their magnitude and frequency is assumed to be similar in future periods compared to the base period. Nevertheless, it includes heat waves, droughts, and cold periods but to a lesser extent extreme precipitation events within single days. In the Box 'Potential Impacts of Droughts on Crop Production' we provide an overview on possible impacts of specific drought scenarios on crop yield changes and agricultural production value. However, in order to deal with extreme events in a coherent manner, the economic calculation would need to be adapted as well. It would require an economic model in order to better account for dynamic and stochastic effects.

[14] See e.g. Special Issue of Agricultural Economics, Volume 45, Issue 1, January 2014.

Potential Impacts of Droughts on Crop Production

Hermine Mitter, Erwin Schmid, Martin Schönhart, Herbert Formayer, Imran Nadeem

Meteorological and agricultural droughts can reduce average crop yields as already experienced during the Central European drought and heat waves in 2003 and 2013.[15] Due to climate change, frequency and duration of drought events are expected to increase in the next decades and — in some cases and in the absence of adaptation — may lead to social, environmental, and economic impacts (Olesen et al. 2011; Dai 2013; IPCC 2014). We present potential impacts of three selected drought scenarios (Strauss et al. 2013) on crop yield changes of grain maize, winter wheat, winter rape, and soybean as well as the implications on the agricultural production value until 2040. The four crops are chosen to represent a number of crops that are currently cultivated in Austria (see Table 8.3 in the main text). Accordingly, this analysis covers about 81 % of total Austrian cropland.

Dry matter crop yields have been simulated with the bio-physical process model EPIC for three drought scenarios and three fertilisation intensities at 1 km pixel resolution. The drought scenarios are provided by a statistical climate change model for Austria and cover the period 2010–2040. In the reference scenario (S1), dry days are distributed similarly as in the past (1975–2007). In the other two drought scenarios (S2, S3), the probability that more than 60 % of the Austrian territory does not experience precipitation events within a random day rises from 38 % in S1 to 50 % and 59 % in S2 and S3, respectively (Strauss et al. 2013). In comparison to S1, S2 (S3) shows decreasing mean precipitation sums between 11 and 15 % (23 and 25 %) during the most dry sensitive crop development stages, i.e. July and August for grain maize and soybean, May and June for winter wheat, and April and May for winter rape. Such dry anomalies represented in S2 (S3) are exceeded approximately every fourth (tenth) year in the reference period S1. Analysing the ensemble of 31 climate models[16] indicates that the probability of dry anomalies within the most dry sensitive crop development stages of grain maize and soybean corresponding to the S2 scenario will increase by 15 % until 2030. In 2050, roughly every second year is likely to show a dry anomaly similar to the projections of S2 for grain maize and soybean. For winter wheat, the 31 climate models indicate no changes in precipitation means during the dry sensitive crop development stages until 2030 and an increase of 4 % until 2050. For winter rape, the climate models indicate a dry anomaly reduction of 6–7 % until 2030 and 2050. For S3 type anomalies, the

[15] More details on the 'Bio-physical impacts up to now' are provided in the Supplementary Material.

[16] More details on the ensembles of 31 climate models are presented in Chap. 5.

31 climate models indicate similar changing rates as for S2. The differences between the four crops stem from the differences in drought sensitive months.[17]

The three fertiliser intensities are assumed to be applied equally likely across Austria and are thus equally weighted in the analysis. Changes in land use are not taken into account, i.e. we assume that the crop shares remain at the same level as in the past. Real prices and land use are used from the database presented in Sect. 8.3.2.3 in the main text in order to estimate potential changes in the agricultural production value. Variable production costs are not taken into account, because they are either yield independent (e.g. planting and harvesting costs) or arise in advance according to the yield expectation (e.g. fertiliser costs).

Compared to the reference scenario S1, model results show that, on national average, yields of winter crops are hit hardest in scenarios S2 (−13.2 % for winter rape and −8.8 % for winter wheat) and S3 (−27.2 % for winter rape and −21.8 % for winter wheat) whereas soybean shows the smallest decline in crop yields (−3.0 % in S2 and −9.5 % in S3; Table 8.7). However, the impacts differ by cropland regions due to heterogeneous agronomic, topographic, and climate conditions. The average annual agricultural production value is computed to decrease by 56 million euros in S2 and by 137 million euros in S3 compared to S1 (Table 8.7).[18]

Table 8.7 Changes in average annual crop yields (in %) and agricultural production value (in M€) of the drought scenarios (S2 and S3) compared to the reference scenario (S1)

Modelled crop	S2	S3
Grain maize	−5.9 %	−14.4 %
Winter wheat	−8.8 %	−21.8 %
Winter rape	−13.2 %	−27.2 %
Soybean	−3.0 %	−9.5 %
Agricultural production value	−56 M€	−137 M€

Our integrated assessment aims to contribute to the discussion on how to model extreme weather events, analyse drought impacts, assess effective adaptation measures, and derive drought policy recommendations.

[17] More details on the 'likelihood' of such drought scenarios are provided in the Supplementary Material of Chap. 5.

[18] Note that changes in crop yields and agricultural production value cannot be directly compared to the results in Sect. 8.3.2.3 in the main text due to differences in data inputs.

References

IPCC (2014) Fifth assessment report, Working group II, Chap 7. Food Security and Food Production Systems

Dai A (2013) Increasing drought under global warming in observations and models. Nat Clim Change 3:52–58. doi:10.1038/nclimate1633

Olesen JE, Trnka M, Kersebaum KC et al (2011) Impacts and adaptation of European crop production systems to climate change. Eur J Agron 34:96–112. doi:10.1016/j.eja.2010.11.003

Strauss F, Moltchanova E, Schmid E (2013) Spatially explicit modeling of long-term drought impacts on crop production in Austria. Am J Clim Change 02:1–11. doi:10.4236/ajcc.2013.23A001

8.4 Summary of Climate Change Impacts on Agriculture and Conclusions

Climate change as projected by the mid-range scenario used in the COIN project (COIN-CCD), leads to moderate increases in agricultural production in the period 2036–2065 at the aggregated national level due to moderate temperature increases and rather stable precipitation sums and inter-annual distribution. Compared to the base period and considering high fertilisation intensity, crop yields are simulated to change between −10 and +9 % and grassland forage yields (temporary and permanent grassland)[19] between +6 and +23 % in the period 2036–2065. This is in line with previous studies. For example, Easterling et al. (2007) show average yield increases for maize, wheat, and rice with temperature increases between 1 and 2 °C in their IPCC meta-study. Trnka et al. (2010) and Eitzinger et al. (2013a) analyse crop growing conditions for Central Europe and expect increasing production potentials on average despite challenges for the Pannonian basin and the Mediterranean region. Similarly, Henseler et al. (2009) model productivity increases from climate change in 2020 for cereals, fodder crops, and grassland for the western parts of Central Europe along the Danube watershed. Furthermore, Ciscar et al. (2011) simulate moderate increases in crop yields at European average with temperature increases of 2.5 °C. For Austria, Alexandrov et al. (2002) model climate change impacts on winter wheat and soybean in the Weinviertel region. Their results show mixed effects which are driven by water constraints and CO_2 fertilisation. Thaler et al. (2012) analyse winter wheat production in the Marchfeld region with similar results, i.e. changes between −18 and −3 % which are moderated or even reversed if CO_2 fertilisation effects are considered.

[19] Note that crop and temporary grassland forage yield changes are calculated for three climate regions whereas permanent grassland forage yield changes are calculated at the national level.

Our economic analysis indicates moderate gains in the average annual agricultural production value (+180 million euros) and variable costs (+74 million euros) resulting in benefits of +106 million euros on average in the agricultural sector up to the period 2036–2065. Increasing costs are stimulated by productivity increases from climate change. Farm and public level adaptation such as shifts in crop mixes or investments in irrigation systems may even increase benefits. Macroeconomic effects are also positive. Welfare and GDP are higher in the climate change scenario, compared to the baseline. However, these results must be taken with care as effects on international markets (e.g. higher prices for food imports) are not taken into account. Furthermore, aggregated effects at the national level conceal a mixed picture of climate change effects at regional and farm level. Our current state of knowledge indicates productivity increases in regions with sufficient precipitation while regions with relevant crop production but limited precipitation at present may face increasing water stress in the future. The important driver for the increasing production values are productivity gains on grasslands. Such stylised results must not undermine international and national climate change mitigation efforts due to the fact that major global agricultural production regions and regions with high dependency on domestic food production will face considerable challenges from climate change (see Iglesias et al. 2011). While Austrian agriculture may gain on average in the coming decades under moderate climate change, impacts may reverse signs once certain climate thresholds are passed. For example, Easterling et al. (2007) show decreasing yields for maize and wheat for mid to high latitudes and temperature changes above 2–3 °C in their meta-analysis. Similarly, Schlenker and Roberts (2009) find that yield growth of corn and soybean rises gradually up to maximum temperatures of 29 and 30 °C but decreases considerably above these thresholds. Such temperature increases and levels are likely to be realised in Austria in the second half of the twenty-first century and call for further research with a focus on crop-specific temperature sensitivities.

It is often assumed that agriculture can easily and quickly adapt to changes in climate conditions. However, increasing inter-annual yield variability (as demonstrated in Balkovič et al. (2013) and Iglesias et al. (2012)) may enhance the need for strategic adaptations, which refer to major structural changes at farm level. There are several examples of new approaches, e.g. development of adequate insurance instruments in order to smooth income variations, adaptation of crop mixes in order to diversify farm income, or innovations in soil and water management.

Acknowledgements This research has been supported by the research project COIN—Cost of Inaction funded by the Austrian Climate and Energy Fund within the Austrian Climate Research Programme as well as by the Doctoral School of Sustainable Development (dokNE) at the University of Natural Resources and Life Sciences, Vienna (Austria). The work on this chapter benefited from scientific input of the JPI-FACCE project MACSUR—Modelling European Agriculture with Climate Change for Food Security (http://www.macsur.eu) and the Federal Ministry of Agriculture, Forestry, Environment and Water Management of Austria (Contract No. 100875). We are especially thankful to the project leaders Karl Steininger and Martin König, to the scientific advisory board established within the project COIN, i.e. Reimund Schwarze, Roger Street, and Paul Watkiss, and to two anonymous reviewers for helpful comments.

References

Ainsworth EA, Davey PA, Bernacchi CJ et al (2002) A meta-analysis of elevated [CO2] effects on soybean (Glycine max) physiology, growth and yield. Global Change Biol 8:695–709

Alexandrov V, Eitzinger J, Cajic V, Oberforster M (2002) Potential impact of climate change on selected agricultural crops in north-eastern Austria. Global Change Biol 8:372–389

Balkovič J, van der Velde M, Schmid E et al (2013) Pan-European crop modelling with EPIC: implementation, up-scaling and regional crop yield validation. Agric Syst 120:61–75

Buchgraber K, Gindl G (2004) Zeitgemäße Grünlandbewirtschaftung, 2nd edn. Leopold Stocker Verlag, Graz

Chavas DR, Izaurralde RC, Thomson AM, Gao X (2009) Long-term climate change impacts on agricultural productivity in eastern China. Agric Forest Meteorol 149:1118–1128

Ciscar J-C, Iglesias A, Feyen L et al (2011) Physical and economic consequences of climate change in Europe. Proc Natl Acad Sci 108:2678–2683

Darnhofer I, Schneeberger W (2007) Impacts of voluntary agri-environmental measures on Austria's agriculture. Int J Agric Resour Govern Ecol 6:360–377

Easterling WE, Aggarwal PK, Batima P et al (2007) Food, fibre and forest products. In: Parry ML, Canziani OF, Palutikof JP et al (eds) Climate change 2007: impacts, adaptation and vulnerability. Contribution of working group II to the fourth assessment report of the intergovernmental panel on climate change. Cambridge University Press, Cambridge, UK, pp 213–272

Eitzinger J, Trnka M, Semerádová D et al (2013a) Regional climate change impacts on agricultural crop production in Central and Eastern Europe – hotspots, regional differences and common trends. J Agric Sci 151:787–812. First View

Eitzinger J, Thaler S, Schmid E et al (2013b) Sensitivities of crop models to extreme weather conditions during flowering period demonstrated for maize and winter wheat in Austria. J Agric Sci 151:813–835

Ewert F, Rounsevell MDA, Reginster I et al (2005) Future scenarios of European agricultural land use: I. Estimating changes in crop productivity. Agric Ecosyst Environ 107:101–116

Finger R (2010) Evidence of slowing yield growth – the example of Swiss cereal yields. Food Policy 35:175–182

Finger R, Hediger W, Schmid S (2011) Irrigation as adaptation strategy to climate change—a biophysical and economic appraisal for Swiss maize production. Clim Change 105:509–528

Garrett KA, Dobson ADM, Kroschel J et al (2013) The effects of climate variability and the color of weather time series on agricultural diseases and pests, and on decisions for their management. Agric Forest Meteorol 170:216–227

Gupta J, Termeer C, Klostermann J et al (2010) The adaptive capacity wheel: a method to assess the inherent characteristics of institutions to enable the adaptive capacity of society. Environ Sci Policy 13:459–471

Hafner S (2003) Trends in maize, rice, and wheat yields for 188 nations over the past 40 years: a prevalence of linear growth. Agric Ecosyst Environ 97:275–283

Henseler M, Wirsig A, Herrmann S et al (2009) Modeling the impact of global change on regional agricultural land use through an activity-based non-linear programming approach. Agric Syst 100:31–42

Högy P, Wieser H, Köhler P et al (2009) Effects of elevated CO2 on grain yield and quality of wheat: results from a 3-year free-air CO2 enrichment experiment. Plant Biol 11:60–69

Iglesias A, Quiroga S, Diz A (2011) Looking into the future of agriculture in a changing climate. Eur Rev Agric Econ 38:427–447

Iglesias A, Garrote L, Quiroga S, Moneo M (2012) A regional comparison of the effects of climate change on agricultural crops in Europe. Clim Change 112:29–46

Izaurralde RC, Williams JR, McGill WB et al (2006) Simulating soil C dynamics with EPIC: model description and testing against long-term data. Ecol Model 192:362–384

Kirchner M, Schmid E (2013) Integrated regional impact assessment of agricultural trade and domestic environmental policies. Land Use Policy 35:359–378

Klik A (2003) Einfluss unterschiedlicher Bodenbearbeitung auf Oberflächenabfluss, Bodenabtrag sowie auf Nährstoff- und Pestizidausträge. Österreichische Wasser- und Abfallwirtschaft 55:89–96
Klik A, Eitzinger J (2010) Impact of climate change on soil erosion and the efficiency of soil conservation practices in Austria. J Agric Sci 148:529–541
Koland O, Meyer I, Schönhart M et al (2012) Regionalwirtschaftliche Auswirkungen von Maßnahmen zur Anpassung und Minderung des Klimawandels im Agrarsektor. WIFO-Monatsberichte 2:131–149
Lobell DB, Schlenker W, Costa-Roberts J (2011) Climate trends and global crop production since 1980. Science 333:616–620
Loibl W, Züger J, Köstl M et al (2010) Reclip: century – a project conducting 21st century regional climate simulation runs focussing on the Greater Alpine Region, EGU General Assembly 2010, held 2–7 May, 2010 in Vienna, Austria, p 11752
McGrath JM, Lobell DB (2013) Regional disparities in the CO2 fertilization effect and implications for crop yields. Environ Res Lett 8:014054
Meissner C, Schädler G, Panitz H-J et al (2009) High-resolution sensitivity studies with the regional climate model COSMO-CLM. Meteorologische Zeitschrift 18:543–557
Mitter H, Kirchner M, Schmid E, Schönhart M (2014) The participation of agricultural stakeholders in assessing regional vulnerability of cropland to soil water erosion in Austria. Reg Environ Change 14:385–400
Mullan D (2013) Soil erosion under the impacts of future climate change: assessing the statistical significance of future changes and the potential on-site and off-site problems. CATENA 109:234–246
OECD-FAO Organisation for Economic Co-operation and Development-Food and Agriculture Organization of the United Nations (2013) OECD-FAO agricultural outlook 2013–2022. OECD Publishing, Paris
Olesen JE, Trnka M, Kersebaum KC et al (2011) Impacts and adaptation of European crop production systems to climate change. Eur J Agron 34:96–112
Peltonen-Sainio P, Jauhiainen L, Trnka M et al (2010) Coincidence of variation in yield and climate in Europe. Agric Ecosyst Environ 139:483–489
Porter JR, Semenov MA (2005) Crop responses to climatic variation. Philos Trans R Soc Lond B Biol Sci 360:2021–2035
Rodriguez D, deVoil P, Power B et al (2011) The intrinsic plasticity of farm businesses and their resilience to change. An Australian example. Field Crops Res 124:157–170
Rosenzweig C, Casassa G, Karoly DJ et al (2007) Assessment of observed changes and responses in natural and managed systems. In: Parry ML, Canziani OF, Palutikof JP et al (eds) Climate change 2007: impacts, adaptation and vulnerability. contribution of working group II to the fourth assessment report of the intergovernmental panel on climate change. Cambridge University Press, Cambridge, pp 79–131
Sánchez B, Rasmussen A, Porter JR (2014) Temperatures and the growth and development of maize and rice: a review. Global Change Biol 20:408–417
Schlenker W, Roberts MJ (2009) Nonlinear temperature effects indicate severe damages to U.S. crop yields under climate change. Proc Natl Acad Sci 106:15594–15598
Schönhart M, Mitter H, Schmid E et al (2014) Integrated analysis of climate change impacts and adaptation measures in Austrian agriculture. German J Agri Econ 63:156–176
Smit B, Skinner MW (2002) Adaptation options in agriculture to climate change: a typology. Mitig Adapt Strateg Global Change 7:85–114
Soja G, Soja AM, Eitzinger J et al (2005a) Analyse der Auswirkungen der Trockenheit 2003 in der Landwirtschaft Österreichs – Vergleich verschiedener Methoden. Teilprojekt von StartClim2004.C, Analysen von Hitze und Trockenheit und deren Auswirkungen in Österreich
Soja G, Eitzinger J, Schneider W, Soja AM (2005b) Auswirkungen meteorologischer Extreme auf die Pflanzenproduktion in Österreich. In: Kämpf A, Claupein W, Graeff S, Diepenbrock W (eds) Gesellschaft für Pflanzenbauwissenschaften: 48. Jahrestagung d. Ges. f. Pflbw., 27–

29 September, Wien, Mitteilungen der Gesellschaft für Pflanzenbauwissenschaften, Wasser und Pflanzenbau – Herausforderungen für zukünftige Produktionssysteme 17. Heimbach, Stuttgart, pp 229–230

Strauss F, Schmid E, Moltchanova E et al (2012) Modeling climate change and biophysical impacts of crop production in the Austrian Marchfeld region. Clim Change 111:641–664

Strauss F, Formayer H, Schmid E (2013a) High resolution climate data for Austria in the period 2008–2040 from a statistical climate change model. Int J Climatol 33:430–443

Strauss F, Moltchanova E, Schmid E (2013b) Spatially explicit modeling of long-term drought impacts on Austrian crop production. Am J Clim Change 2:1–11

Stürmer B, Schmidt J, Schmid E, Sinabell F (2013) Implications of agricultural bioenergy crop production in a land constrained economy – the example of Austria. Land Use Policy 30:570–581

Thaler S, Eitzinger J, Trnka M, Dubrovsky M (2012) Impacts of climate change and alternative adaptation options on winter wheat yield and water productivity in a dry climate in Central Europe. J Agric Sci 150:537–555

Trnka M, Eitzinger J, Dubrovský M et al (2010) Is rainfed crop production in central Europe at risk? Using a regional climate model to produce high resolution agroclimatic information for decision makers. J Agric Sci 148:639–656

Weber RWS (2009) Betrachtung möglicher Auswirkungen des Klimawandels auf Schadpilze im Obstbau am Beispiel von Fruchtfäuleerregern an Äpfeln. Erwerbs-Obstbau 51:115–120

Wei Y, Davidson B, Chen D, White R (2009) Balancing the economic, social and environmental dimensions of agro-ecosystems: an integrated modeling approach. Agric Ecosyst Environ 131:263–273

Williams JR (1995) The EPIC model. In: Singh VP (ed) Computer models of watershed hydrology. Water Resources Publications, Highlands Ranch, CO, pp 909–1000

World Bank (2012) Turn down the heat, Why a 4 °C warmer world must be avoided. World Bank, Washington, DC

Ziervogel G, Ericksen PJ (2010) Adapting to climate change to sustain food security. Clim Change 1:525–540, Wiley Interdisciplinary Reviews

Chapter 9
Forestry

Manfred J. Lexer, Robert Jandl, Stefan Nabernegg, and Birgit Bednar-Friedl

Abstract A warmer climate with reduced summer precipitation will affect the biomass productivity in Austrian forests. In mountain forests with sufficient precipitation, an extended growing season will lead to productivity increases. In eastern and north-eastern lowlands and in inner-Alpine basins, extended and more frequent drought periods will result in reduced production. The two most relevant disturbance factors in Austrian forests are bark beetles and storms. Disturbance regimes from temperature driven agents such as bark beetles will intensify under all available climate change scenarios. Abiotic disturbance factors such as storms, snow and late frost events have the potential to cause damage, however information about future development of these drivers is highly uncertain. Forest fires will likely occur more often in Austrian forests, however, the scale of fires is comparably small and of local importance only. Disturbances will impact on contribution margins of timber production and increase regeneration costs. Other ecosystem services like recreation and CO_2 sequestration might be negatively affected as well by an intensified disturbance regime. A particularly important service in Austria is protection against gravitational hazards (snow avalanches, rockfall, mudflow). We have used data of the National Forest Inventory and available results from earlier climate change related studies to quantify the effect of climate change on timber production and on the bark beetle disturbance regime in Norway spruce forests. At the end of the century (2070–2100) conservative

M.J. Lexer (✉)
Institute of Silviculture, BOKU—University of Natural Resources and Life Sciences, Vienna, Austria
e-mail: mj.lexer@boku.ac.at

R. Jandl
Department of Forest Ecology and Soils, Austrian Forest Research Centre, Vienna, Austria
e-mail: robert.jandl@bfw.gv.at

S. Nabernegg
Wegener Center for Climate and Global Change, University of Graz, Graz, Austria
e-mail: stefan.nabernegg@uni-graz.at

B. Bednar-Friedl
Wegener Center for Climate and Global Change, University of Graz, Graz, Austria

Institute of Economics, University of Graz, Graz, Austria
e-mail: birgit.friedl@uni-graz.at

K.W. Steininger et al. (eds.), *Economic Evaluation of Climate Change Impacts*, Springer Climate, DOI 10.1007/978-3-319-12457-5_9

estimates indicate mean annual losses of 2.43 million euros p.a. due to impacts on productivity. Bark beetle damages in production forests may cause damages as high as 141 million euros p.a. (period 2074–2100). To substitute for losses in protective capacity, mean annual investment costs between 85 million euros (period 2014–2035) and 189 million euros (period 2074–2100) have been estimated. The macroeconomic effects triggered by these three impact chains are found to be negative, both for welfare and GDP. Welfare is found to decline by up to −0.10 % by mid-century, compared to a baseline scenario without climate change. The lion's share of macroeconomic impacts is due to the investment requirements to maintain or restore protective forests after disturbances, followed by the damages due to bark beetle disturbances in timber production. Future storm damages as one potential implication of climate change were beyond the scope of this study and were not considered.

9.1 Introduction

Forest ecosystems play an important role in the global biogeochemical cycles, acting as both sources and sinks of greenhouse gases, and in doing so they have significant influence on the earth's climate. At the same time, forests are particularly sensitive to climate change, because the long life-span of trees does not allow for rapid adaptation to environmental changes. This may hamper the sustainable provision of ecosystem services from forests. Consequently, climate change is at the centre of attention of science and policy, with a major focus on mitigating the anthropogenic interference with the climate system. Yet it becomes increasingly clear that alongside intensified mitigation efforts, adaptation will be necessary to cope with the already inevitable adverse effects of climate change.

Stern (2009) showed that for vulnerable systems timely and well-designed adaptation measures are economically favorable over delayed or no adaptation. Due to long time lags before actions become effective, proactive adaptation strategies in forest management indeed need to be timely implemented in order to sustain the provision of multiple goods and services under changing future conditions. While a number of conceptual studies on adapting forest management to altered climate have been presented, recent surveys among policy makers and practitioners revealed large prevailing gaps between planned and implemented adaptation measures. Demonstrated how the wide variety in personal beliefs about climate change influences decisions on adaptation of individual forest owners. Furthermore, a recent survey in Austria documented large differences in adaptation to climate change between ownership categories. A common denominator in these surveys among practitioners is that strategic management decisions such as adaptation, often associated with an investment of considerable intellectual, human and economic capital, need to be robust with regard to future uncertainties, including climate change. Making information about the cost of inaction available may be valuable support for decision makers.

9.2 Dimensions of Sensitivity to Climate Change

The impact of climate change on forests depends on the interplay of state of forests, climate related disturbance factors, changes in climate as such, and how forest management affects these interrelationships.

9.2.1 Climatic Factors

Tree species and forests respond directly and indirectly to climatic factors (Table 9.1).

9.2.2 Non-climatic Factors

Despite the diverse demands by society, only few products and services (timber, berries, mushrooms, hunting) are marketable. Payments for other ecosystem services are still debated. Timber production and game management are major drivers of forest management and affect species composition, biomass accumulation levels and forest structure. Browsing of naturally regenerating trees due to high population levels of red and roe deer often impairs the establishment of mixed forests. Recently, the demand for bioenergy from tree biomass has stimulated harvesting activities (Neumann 2012). Demographic changes and an increasing share of urban forest owners are reducing the relevance of timber production as source of income. Nevertheless, the Austrian forest based wood sector (saw mills, pulp and paper, plywood) has an annual value creation of 12 billion euros and has created an annual trade surplus of up to 3.8 billion euros recently. Austria is a major global exporter of conifer saw wood.

Temperate forests are naturally N-limited systems. Nitrogen enrichment from atmospheric deposition is increasing forest productivity and is altering the competitive relationship between tree species. However, the capacity of absorbing N is limited and adverse effects on forest growth are foreseen (Butterbach-Bahl and Gundersen 2011; Leip 2011). The direct effects of increasing levels of atmospheric CO_2 on forest growth are uncertain in the long run. However, short to midterm impacts on net assimilation may be positive and increase growth rates by about

Table 9.1 Sensitivity of tree species and forests with regard to climatic factors

Climate factor	Impact mechanism
Increase in temperature	Increased growth through higher net photosynthesis, at most sites in Austria trees currently grow at sub-optimal temperature conditions
	Increasing respiratory losses, may outbalance increased efficiency in assimilation
	Higher evapotranspiration losses and subsequently shortage in water supply, may lead to reduced growth
	Poikilothermic insects may increase population densities quickly and increase tree mortality
	Predisposition for storm damage increases due to shorter periods with frozen soils
	Increased fire risk due to reduced moisture content in forest litter
Longer growing season	Increased productivity; may also lead to de-synchronisation of insect-host relationships and hence reduce damage potential
Late frost	Dieback of young shoots and subsequently reduced growth
Limited water supply, drought periods	Reduced growth, strong dependence on species specific sensitivity
	Limitation to tree regeneration due to dieback of seedlings (particularly sensitive to limited water supply)
Storms	Increase of storm damages as a consequence of climate change is controversial, huge uncertainties related to storm frequency and intensity particularly regarding to local storm events
Increased CO_2-content of atmosphere	Fertilisation effect increases efficiency of assimilation; affects also quality of diet for insects

10 %. A potentially relevant indirect effect of CO_2 concentration in the atmosphere is an increase in water use efficiency which may partly counterbalance decreasing summer precipitation (Curtis and Wang 1998).

9.2.3 Identification of Potential Large-Damage Combinations

In Austrian forestry there are two major areas particularly prone to large-scale damages related to a warming climate.

1. The promotion of Norway spruce (Picea abies) outside its natural range at sites naturally supporting mixed broadleaved forests has resulted in large areas of secondary spruce forests at low elevations which are vulnerable to an array of insects and pathogens (Gschwantner and Prskawetz 2005; von Teuffel et al. 2005).
2. Norway spruce dominated conifer forests in mountain regions where so far insect damages have been negligible are increasingly susceptible to bark beetle damages.

The interplay of drought periods affecting the vitality of host trees during the summer season and high temperatures favouring the fast build-up of high insect population densities lead to increased disturbance intensities. Increasing temperatures extend the area of potential damage into mountain forests. Large scale storm events damage forests and create huge supplies of suitable breeding material for bark beetles, thus having the potential to further fuel disturbance regimes and thus affect damage levels. Beyond direct damages to timber values large-scale disturbances further affect timber markets and result in decreasing timber prices also outside the damaged regions.

9.3 Exposure to Climatic Stimuli and Impacts Up to Now

9.3.1 Past and Current Climatic Exposure and Physical Impacts

Forest damages have increased at the continental scale (Nabuurs et al. 2013; Seidl et al. 2011; Schelhaas et al. 2003). Austrian statistics show a series of large-scale storm damages, followed by increased levels of damage from bark beetles damages in the last 20 years (Tomiczek and Schweiger 2012). However, the climate change interacts with other drivers of disturbances such as management and aging of forests (Seidl et al. 2011). The most relevant climatic factors are summarised in Table 9.2.

9.3.2 Impact Chains up to Socio-economic System

Beside climate change the disturbance regimes (frequency and severity of droughts, storms and insect infestations) affects forest productivity and tree mortality (Table 9.3). Natural disturbances are a main driver of ecosystem development and succession. However, in managed forests they interfere with management objectives. Enforced harvests of immature forests and premature tree mortality

Table 9.2 Recent climate change exposure of Austrian forests

Climatic factor	Change over time	Consequence
Temperature increase	Slow consistent trend, huge interannual variation	Longer growing season for trees, favourable conditions for insect development
Summer droughts	Increased frequency	Reduced tree vitality and increased susceptibility to insects and pathogens
Storms	No clear long-term trend apparent in historic analysis	Direct damage through overtoppled and broken trees, supply of breeding habitat for bark beetles

Table 9.3 Impact chains in the forest sector

Climate change parameter	Impact chain	Quantified in the COIN cost estimates
Precipitation: Lower precipitation, change of seasonal distribution, less summer precipitation in eastern/southern foothills; little change in total annual precipitation	*Drought periods*	
	→ In areas with shallow and coarse-textured soils the water storage is insufficient. The consequence is limited water supply and a reduction in growth rate	Yes
	→ Increases the susceptibility of Norway spruce stands to bark beetle attacks and an array of pathogens	Yes (bark beetles)
Storm events: Increase of frequency and/or intensity is highly uncertain	*Increase of storm events*	
	→ Physical strain on forests; higher volume of damaged trees, insufficient control on timber price due to high supply; little option for reducing the supply; increased harvesting costs; in COIN storm damage is not considered based on the assumption that no increase of storm events will happen	No
Temperature: Increase of mean values and heat waves increasing length of growing season	*Drought periods*	
	→ Impact on forest productivity (see above)	Yes
	Tree species change	
	→ Change of competitive balance between tree species; need for new management strategies	No
	Pests and pathogens will lead to intensified disturbance regimes	
	→ Higher population density of bark beetle; invasion of new pests and pathogens → Will impair protective functions of mountain forests against gravitational hazards	Yes (bark beetles)
	Annual tree growth	
	→ Longer growing season → increasing productivity especially in not-drought-prone mountain forests) → potentially lower timber quality due to lower wood density as a trade-off of faster tree growth/higher productivity	Yes No

lead to economic loss, reduced protective function of forests, and declining carbon stocks. Warmer temperatures and summer droughts will increase the risk from biotic disturbance agents. The extent of future storm damages is uncertain. Based

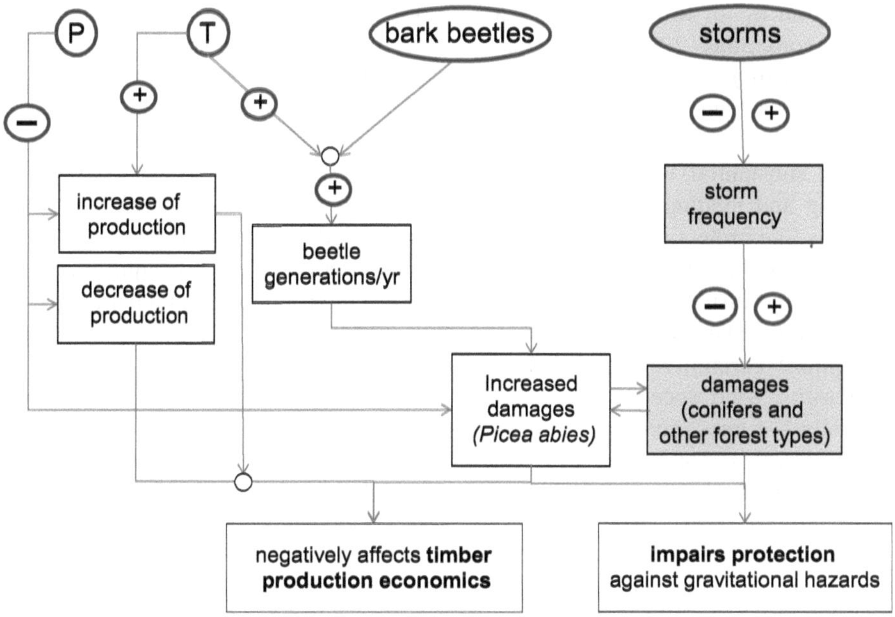

Fig. 9.1 Climate change related impact chains for Austrian forestry. *Chains* represented by *white boxes* are considered in the COIN assessment of the forest sector

on an analysis of historic storm events we assumed no significant increase of storm damages. Figure 9.1 displays the major impact chains and the interaction of drivers.

Temperature (1) affects potential number of bark beetle generations, and (2) reduces the winter mortality of beetles. **Precipitation** increases the drought stress on trees. Thus, on one hand there is a negative impact of drought on forest growth, on the other hand the combination of increased insect pressure and reduced tree vitality may result in increased disturbance intensity and tree mortality. Unplanned harvests of bark beetle infested trees increase harvesting costs and decrease revenues. Also, replanting on damaged sites creates additional costs.

In protective forests **larger gaps and decreased canopy cover and stem density resulting from disturbances** will call for restorative measures such as artificial regeneration or protective constructions.

Increasing temperatures will in create more **favourable temperature regimes for net primary production**, where the water supply for trees is not limited. Forest productivity may increase in mountain regions and decrease at sites with **limited precipitation**.

9.3.3 Physical and Economic Impacts Up to Now

Storms are irregular with respect to extent and periodicity (see supplementary material, Sect. 9.3). Salvage of storm damages fluctuated between annually 1 and

11 million m^3 of timber in 1990–2012, corresponding to 4–50 % of the annual cut (Anonymous 2005–2013). Peak years due to winter storms were 1990 (7 million m^3), 2002 (6 million m^3), 2007 (9 million m^3) and 2008 (11 million m^3). Other salvage was due to small-scale, endemic disturbances. However, these low-intensity small-scale damages make management plans obsolete and accumulate to substantial losses in timber value and additional costs for harvesting, planting and tending. Windthrow also negatively affects other forest ecosystem services like protection of infrastructure from rockfall and avalanches, soil erosion and mudflow (see Chap. 15), drinking water production or carbon sequestration.

The wind disturbance regime is driven by the interplay of forest characteristics and weather (Dale et al. 2000). Beyond wind speeds of 45 m/s large scale damages are almost certain, regardless of stand conditions (Gardiner et al. 2010). With decreasing wind speed the effect of tree and stand characteristics on damage intensity becomes more apparent.

The frequency of storm events may increase in a warmer climate, partly due to a poleward shift in extratropical storm tracks (Haarsma et al. 2013; IPCC 2012; Gardiner et al. 2010; Usbeck et al. 2010). The 20th century showed no significant storm increase in Central Europe (Matulla et al. 2007) and future dynamics are uncertain. An intensifying wind disturbance regime may positively feed back on bark beetle disturbances through the provision of abundant breeding habitat and in consequence to higher greenhouse gas emissions from forest soils (Schirimeier 2013) (see supplementary material, Sect. 9.1).

However, the effect of management activities is slow (Thom et al. 2013). Timber prices in the wake of large scale disturbances reflect the disturbance regime and wood needs to be brought to the market quickly after disturbances because the storage capacity of unprocessed wood is limited and the wood protection methods are expensive.

In 2003 some Austrian mountain forests could benefit from higher temperatures and increased their increment by approx. 10 %. In large parts of the country a reduction of the productivity was observed (Kindermann and Neumann 2011; Eastaugh et al. 2011).

9.4 Future Exposure to and Impacts of Climate Change

9.4.1 Mid-Range Climatic Scenario for Forestry

A mid-range climate change scenario was defined to provide a harmonised base for impact assessments in the different sectors (see Chap. 5). The original scenario data were derived from an ensemble of general circulation models, regional climate models and emission scenarios. The spatially coarse resolution of the scenarios was aggregated to a spatial resolution were few political districts are combined (NUTS

3 regions). The climate scenario is described in Sect. 9.2 of the supplementary material.

For the quantitative analysis of impacts in the forest sector earlier studies of the authors have been used. The impacts on productivity relate to the A1B scenario in Jacob et al. (2008). The scenario reflects a temperature increase of 4 °C until 2100 with increasing summer precipitation increases until 2050 and subsequent decreases. Bark beetle damages were taken from Seidl et al. (2009a) who used the B1 scenario from Mitchell et al. (2004) with increasing mean annual temperatures of +2.4 °C in 2090–2099. Precipitation in the Mitchell et al. scenario is similar to Jacobs et al. (2008).

9.4.2 High and Low Range Climatic Scenarios for Forestry

A decreasing water availability in spring and summer will definitely have negative impacts on forest growth and will increase the susceptibility to insects and pathogens. Vanishing winter frosts may affect the frost hardening processes of trees which in turn could increase their vulnerability to frost events. Seed production may also be affected by impairing the establishment of flower buds (Sykes et al. 1996).

Late frost events are difficult to predict because the accumulation of cold air depends on small-scale physiogeographic conditions and wind speed. Such events can be decisive for the presence and absence of tree species. Even at increasing average temperatures tree development species may be impaired by single frost events. The elongation of the growing season can make trees even more vulnerable to frost damage that increasingly fall in the period when vulnerable juvenile shoots already have developed. On the other hand, late frost events in a warming climate may in turn negatively affect the swarming activities and early development stages of bark beetles (Wermelinger 2004).

9.4.3 Specific Method(s) of Valuation and Their Implementation Steps

9.4.3.1 Productivity Changes

Estimates of climate change related productivity changes were taken from simulations of the KLIMADAPT project (Schörghuber et al. 2010) were used. PICUS 3G was run under the A1B IPCC scenario as provided by Jacob et al. (2008). Climate parameters for the NPP simulations were monthly values of temperature, precipitation, global radiation and vapor pressure deficit. The data were aggregated at the province level. Figure 9.2 shows the contrasting simulated climate change effects

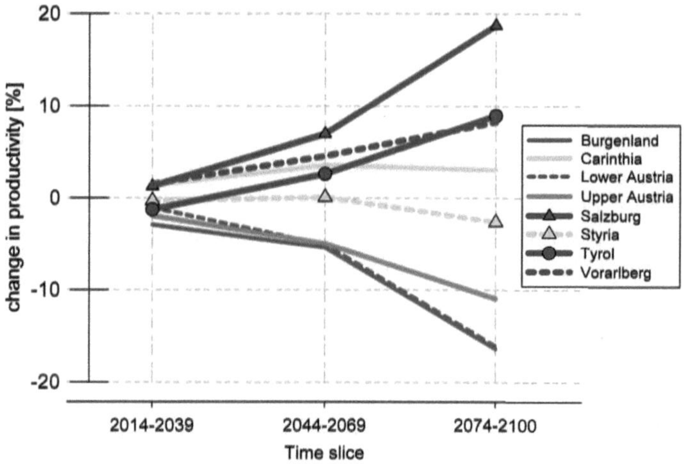

Fig. 9.2 Simulated climate change impacts on forest growth under an A1B climate change scenario relative to historic climate (1961–1990), Schörghuber et al. (2010) and Jacobs et al. (2008)

on forest growth for eight Austrian provinces. The mountain forests in the provinces of Salzburg, Vorarlberg and Tyrol are mostly benefitting from climate change, mainly due to longer growing seasons. The forests in provinces with a higher proportion of lowland forests show reduced growth due to more frequent drought periods during the summer season. Due to large areas of forests in Lower and Upper Austria which contribute a large share of Austria's timber production the overall trend of the country is a growth reduction.

9.4.3.2 Bark Beetle Damages in Production Forests

Based on simulations with an enhanced FISCEN scenario model, Seidl et al. (2009a, b) has estimated damage from spruce bark beetles. The damages were available in 5-year periods for the Austrian provinces. These data were aggregated to periods which are very similar to the COIN analysis periods (see Sect. 9.4.3.1). Regarding temperature and precipitation changes the underlying climate change signal is weaker compared to the COIN mid-range scenario (for details see Supplementary Material Chap. 9, Section 2).

Based on data of the Austrian Forest Inventory 2007/2009 (http://www.waldinventur.at) we estimated the standing stock of Norway spruce production forests at province level which are vulnerable to bark beetle attacks. We focused on stands older than 40 years (age classes IV to VIII) and assumed that 20 % of the expected damage occurs as stand-replacing disturbance (i.e. the entire forest stand is destroyed and 100 % of standing Norway spruce stock is salvaged). Eighty percent of the expected damage was assumed to happen as partial damages with a mean damage intensity of 30 % of standing stock. Damage quantities and mean

Table 9.4 Estimates of damaged area (ha) in Norway spruce stands under a B1 climate change scenario [change in mean annual temperature at the end of the twenty-first century: +2.4 °C, change in precipitation: slight increase or decrease (<5 %) depending on province, Seidl et al. (2009a)]

		Affected area (ha) in time periods		
Province	Damage intensity	2014–2039	2044–2069	2074–2100
Burgenland	Full damage	782	1,235	1,809
	Partial damage	9,391	14,841	21,731
Carinthia	Full damage	2,365	4,196	6,326
	Partial damage	28,483	50,539	76,200
Lower Austria	Full damage	6,451	8,608	12,903
	Partial damage	80,742	104,495	156,642
Upper Austria	Full damage	3,576	5,756	8,171
	Partial damage	44,292	71,298	101,213
Salzburg	Full damage	1,409	2,333	3,087
	Partial damage	18,038	29,875	39,517
Styria	Full damage	4,680	7,282	11,963
	Partial damage	57,973	90,205	148,190
Tyrol	Full damage	1,437	2,342	3,258
	Partial damage	19,089	31,103	43,264
Vorarlberg	Full damage	144	182	305
	Partial damage	1,821	2,305	3,867

standing stock were used to estimate the area of affected Norway spruce stands (Table 9.4).

9.4.3.3 Impacts on Protection Functionality

In Austria 820,000 ha are labeled as protection forest (National Forest Inventory; ANFI). ANFI differentiates between productive and unproductive protection forests. For protective forests we distinguished two situations. Protective forests that are managed like regular forests ('productive protection forests') we assumed the damage scenario as in Section 9.4.3.2 for commercial production forests. Unproductive protective forests are mostly located at high elevations outside the present habitat of bark beetles. Even under warmer climates the infestation pressure may remain relatively low. We assumed that the vulnerability of these forests is lower than that of productive protective forests at lower elevations and that the share of damaged area is only 50 % of the rate in productive protection forests.

Table 9.5 shows the estimated potential damage from bark beetles by province in Austrian protection forests. From the total damaged area 20 % were considered a complete damage (i.e. protection function requirements no longer met). In the case of damage the protective function needs to be immediately restored by technical measures. In the remaining damaged area we have assumed partial damage where

Table 9.5 Area damaged by bark beetle disturbances in Austria's protection forests

Province	Total damaged area in productive protection forests (SiE) (ha/period)			Total damaged area in non-productive protection forests (SaE) (ha/period)		
	2014–2039	2044–2069	2074–2100	2014–2039	2044–2069	2074–2100
Burgenland	85	135	197	0	0	0
Carinthia	642	1,140	1,718	383	679	1,024
Lower Austria	1,059	1,370	2,054	381	493	740
Upper Austria	857	1,380	1,959	536	862	1,224
Salzburg	695	1,152	1,524	720	1,192	1,577
Styria	1,095	1,704	2,799	817	1,272	2,089
Tyrol	1,138	1,854	2,579	1,029	1,677	2,333
Vorarlberg	117	148	249	138	175	293

SiE = productive protection forest, SaE = non-productive protection forest

silvicultural measures (i.e. planting of seedlings) and soft technical measures are considered sufficient (e.g. deadwood logs felled in the contour line against snow creeping).

9.4.4 Range of Sectoral Socio-economic Pathway Parameters that Co-determine Climate Impact

The peculiar role of Norway spruce forests is described in Sect. 3 of the Supplementary Material

9.4.5 Monetary Evaluation of Impacts

9.4.5.1 Direct Sector Impacts (Costs and Benefits) Without Feedback Effects from Other Sectors

Productivity

The direct economic impact of this decrease in production was estimated by assuming an average price per cubic metre timber of 61. This mean timber price was calculated from 5-year average prices for roundwood, industrial wood and fuelwood combined with 10-year mean shares of these assortment groups in the total annual cut in Austria. Table 9.6 shows the estimated mean annual costs from reduced volume growth in Austrian forests.

Table 9.6 Natural and monetary losses due to simulated decreasing timber productivity in Austrian forests under a A1B climate change scenario (Jacob et al. 2008) representing COIN midrange climate change conditions

	Time period		
Cost category	2014–2039	2044–2069	2074–2100
Loss in timber production (m^3/year)	−136,300	−234,069	−1,052,461
Mean annual economic loss (million euros/year)	−0.33	−0.56	−2.43

A mean mixed price of 61 €/m^3 was used to calculate the monetary loss

Table 9.7 Estimated annual cost of bark beetle disturbances in Austrian production forests under an A1B climate change scenario (Jacob et al. 2008) representing COIN midrange climate change conditions

	Time period		
Cost category	2014–2039	2044–2069	2074–2100
Afforestation	6	9	13
Reduced CM1	58	89	127
Total cost	64	98	141
Increase in total costs relative to baseline without climate change	+28 %	+54 %	+70 %

CM1 = contribution margin 1 (timber revenues minus harvesting cost), in million euros/year

Bark Beetle Damages

A further major source of potential costs from climate change impacts on forestry are damages and related costs from natural disturbances. In COIN bark beetle damages in Norway spruce were used as a major factor. The monetary valuation of bark beetle damages in commercial forests was based on the assumption that fully damaged stands had to be regenerated by planting which results in average costs of 10,000 €/ha. The salvaged timber caused higher harvesting costs and lower revenues due to reduced quality after beetle infestation and eventual delayed removal resulting in a net loss in contribution margin of 25 €/m^3 (Table 9.7).

To isolate climate change impacts from other developments such as changes in age structure of the forest stand, total costs in Table 9.7 need to be compared to a baseline development without climate change (historic climate 1971–2000). In the climate change scenario, costs from bark beetles are estimated about 30 % higher in period 2014–2039 and about 70 % higher in 2074–2100.

Protection

The implications of damaged area from bark beetle infestations (see Sect. 4.3.2) followed the assumptions: (a) the totally damaged forests require technical construction measures to immediately restore protective functionality while the

Table 9.8 Cost estimates (M€/year) to maintain and restore the protective functionality in Austria's protective forests under mid-range climate change conditions, for assumptions see text

	Time period		
Protective forest type	2014–2039	2044–2069	2074–2100
Unproductive (SaE)	35	56	78
Productive (SiE)	50	78	111
Total	85	134	189

partially damaged forests can be restored by afforestation measures. The restoration costs in protective forests are higher than in production forests due to the limited accessibility of the site. Costs for a complete replacement of a damaged forest stand by advanced technical measures may be as high as 700,000 €/ha, whereas afforestation measures combined with simpler technical measures to support the establishment and early growth of planted seedlings may require investment costs of about 100,000 €/ha. For the current COIN analysis a weighted average restoration cost of 220,000 €/ha was used. Table 9.8 shows the estimated costs of maintaining or restoring the protective function against gravitational hazards in Austria's protective forests. The general hypothesised underlying relationships of the employed approach appear plausible. However, it has to be noted that these cost figures bear huge uncertainties and should be seen as a frame for potential economical implications of climate change impacts on protective forests.

9.4.5.2 Macroeconomic Effects

In the macroeconomic model (see Chap. 7), three impact chains are quantified representing mid-range climate change conditions: a changed productivity in commercially used forests due to changed climatic conditions, a negative impact of bark beetle disturbances, and the investment necessary in protective forests to restore protective functions after negative disturbance impacts (as a proxy for the damages of climate change to the capacity of protective forests). For details on the implementation of the impact chains as well as the baseline scenario regarding the age structure of forest stands see Sect. 9.4 of the supplementary material.

Assessing the three climate change impacts in the macroeconomic model gives changes in output and value added as shown in Table 9.9. All numbers display average annual changes in million euro (M€) compared to the baseline scenario without climate change impacts in the respective period. The macroeconomic effects are calculated as effects in real prices (of 2008) for 2030 (representative for climatic period 2016–2045) and 2050 (period 2036–2065). Regarding sectoral effects, some sectors are affected positively while others are affected negatively in terms of changes in output and value added. In particular, the construction sector can increase its gross output value on average by 80 million euros/year in 2016–2045, mainly caused by the higher investment in this sector. In the first period, gross value added increases (by 7 million euros) due to higher demand for capital and

Table 9.9 Sectoral and total effects of quantified climate change impacts in sector Forestry, average annual effects relative to baseline (for periods 2016–2045 and 2036–2065)

	Ø 2016–2045			Ø 2036-2065		
Changes in M€ p.a. relative to baseline	Gross output value	Inter-mediate demand	Gross value added	Gross output value	Inter-mediate demand	Gross value added
Gaining sectors	**+80**	**+40**	**+40**	**+124**	**+60**	**+64**
Construction	+80	+47	+33	+122	+71	+50
Forestry	+0	−7	+7	+2	−11	+13
All other gaining sectors	+0	+0	+0	+0	+0	+0
Losing sectors	**−564**	**−288**	**−276**	**−940**	**−474**	**−466**
Trade	−69	−27	−41	−116	−46	−69
Real estate	−52	−16	−36	−89	−27	−62
Rest of manufacturing	−28	−7	−21	−47	−12	−36
All other losing sectors	−416	−238	−178	−689	−390	−299
Total effect (all sectors)	**−484**	**−248**	**−237**	**−816**	**−414**	**−402**
GDP at producer price			−0.07 %			−0.08 %
...thereof price effect			−0.01 %			−0.01 %
...thereof quantity effect			−0.06 %			−0.07 %

Note: baseline scenario = reference socioeconomic development without climate change; climate change scenario = reference socioeconomic development and mid-range climate change; quantified climate impact chains: bark beetle infestation, productivity losses due to unfavourable climatic conditions, investment in protective forest

labour (because of reduced timber growth and more silvicultural measures) which is compensated by lower demand for other inputs (intermediate demand). As a consequence, the gross output value in forestry stays about the same. Because of sectoral linkages and changes in relative prices, sectors like trade, real estate and rest of manufacturing have a lower gross value added in the climate change scenario relative to the baseline without climate change. The net effect of gaining and losing sectors is given in GDP (as the sum of gross value added across sectors), which is reduced by 0.07 % in total. The reduction of GDP is dominated by reduced output quantities (86 % of this effect) while lower prices have a much smaller effect (14 %). The same effects occur in both time horizons, but they are up to three times larger in the second period.

Compared to the impact on GDP of −0.07 % in the first period and −0.08 % in the second, the effect on welfare is slightly larger (−0.08 and −0.10 %). This stronger negative effect on welfare is driven by two channels: on the one hand, the

Table 9.10 Effects of quantified climate change impacts in sector Forestry on government budget, average annual effects relative to baseline (for periods 2016–2045 and 2036–2065)

Changes in M€ p.a. relative to baseline	Ø 2016–2045	Ø 2036–2065
Revenues	**−181**	**−297**
Production tax	−92	−145
Labour tax	−49	−82
Capital tax	−5	−10
Value added tax	−35	−59
Other taxes	−1	−1
Expenditures	**−181**	**−297**
Unemployment benefits	+57	+93
Transfers to households net of other taxes	−237	−390
Government budget in baseline (p.a.)	**148,949**	**206,113**
Climate change impact on government budget	−0.12 %	−0.14 %

Note: baseline scenario = reference socioeconomic development without climate change; climate change scenario = reference socioeconomic development and mid-range climate change; quantified climate impact chains: bark beetle infestation, productivity losses due to unfavourable climatic conditions, investment in protective forest

government pays compensation for restoration of protective forests and hence households receive less government transfers. On the other hand, unemployment rises slightly (by 0.0 %-points and 0.08 %-points) because of in total negative macroeconomic effects and hence household incomes decline, also lowering welfare.

In Table 9.10 the effects on government expenditures and revenues are listed in average million euros/year. The lower revenues in (net) production taxes of –92 million euros in period 2016–2045 arise due to the higher subsidies for construction of additional protective forests, as well as for a general decrease in the domestic production and hence lower tax revenues. The latter effect is also the reason for the lower prices of capital and labour, and therefore reduced revenues from capital and labour taxes. As shown before, GDP goes down, and consequently also value added taxes decline as well as some minor other taxes. The quantified effect on these taxes is even higher in period 2036–2065.

On the expenditure side the government payments for unemployment increase with corresponding higher unemployment rates than in the baseline without climate change. The net transfers to households have to compensate these changes in government budget, as the assumption in the CGE model is a balanced government budget, and therefore transfers to households need to decline. The balanced changes in revenues and expenditures affect the total government budget by a share of −0.12 % in the first and −0.14 % in the second time horizon.

9.4.6 Sector-Specific Uncertainties

The disturbance regimes in forest ecosystems in a future warmer climate are subject to substantial uncertainties with regard to disturbance agents as well as with regard to disturbance intensity. In the current study we considered the most detrimental biotic disturbance agent in Austrian forests, the European spruce bark beetle (*Ips typographus*). In contrast, we did not include storm damage in the assessment. Major reasons were the lack of economically evaluated historic storm events and the controversy whether or not the frequency and the intensity of storms will increase in Central Europe in course of the twenty-first century. Based on Gardiner et al. (2010) storm damage related uncertainties can be summarised as follows: (1) huge variation among climate models regarding storm intensity, (2) shifts of storm tracks resulting in changes of affected areas, (3) warmer winter temperatures and thus reduced time of frozen soils leading to less anchoring resistance of trees, and (4) uncertainty regarding future management which may affect mean standing stock in forests. Large-scale storm damages also affect the timber price and thus add complexity to economic analysis.

The development of bark beetle generations is primarily driven by temperature. Thus, quicker build-up of population densities is expected in the future which may lead to larger damages from bark beetles. The development of disturbance regimes in future climatic conditions in forests at higher altitudes and marginal sites is even more difficult to estimate. Reasons are that forest composition and structure are not so well known as in commercial productive forests and thus stratification and more specific assumptions are difficult to implement. Also the extrapolation of today's bark beetle driven disturbance regimes in commercial forests to extensively managed subalpine forests bears substantial uncertainties. The eventually required investments for the restoration of protective forests affected by large scale disturbances are potentially considerable. Although protective forests cover only about 20 % of the total forest area the expected costs are 50 % of the total costs arising in the forestry sector. The higher costs are a consequence of the difficult terrain, the limited accessibility and the need to immediately restore the required protective function through technical measures.

Highly relevant factors such as singular extreme events featuring late frost episodes, canopy damages from heavy snow in spring, and erosion due to heavy rainfall events are ignored. An additional factor is the potentially increasing pressure from exotic pests and pathogens. Each of these impact chains can have enormous implications for the forestry sector and may dwarf the quantified impact chains. However, due to high uncertainties, both on the extent of climate change and the general agreement on its consequences, no robust economic assessment is currently possible. While methodology for robust estimates of productivity under climate change conditions is available and has been used within COIN, disturbances and how they may be affected by a change in climate, forest conditions and forest management remain a major challenge in forest science. Moreover, it is important to note that society demands an array of other ecosystem services from forests as well which are not marketed at the moment. The protective function against gravitative hazards can be monetarised via the cost of re-establishing the desired vegetation structure or an equivalent technical structure such as rockfall nets. The value of carbon storage can be calculated from physical C sinks and a market-based price per ton of C. However, C markets so far did not prove as particularly effective instrument in stimulating increased C storage. The value of services such as landscape beauty to support tourism could be quantified via willingness-to-pay or contingent valuation approaches. However, transferability and generality of such studies is still a matter of debate (Tacconi 2012). To indicate the relevance of non timber services Vacik et al. (2008) estimated the value of non-timber products and ecosystem services in Austria as 220 million euros in 2005. The sensitivity to climatic changes, however, has not been assessed so far.

The forestry sector is facing a challenging future. The globalisation of trade is an effective vector for pests and pathogens and the requirement for monitoring activities will increase. Climate change effects are overlaid by a rapidly changing society. Societal changes manifest themselves in different expectations towards forest ecosystems and services they should provide and by a declining workforce available for forestry. Only timber and a few minor products and services are marketable and the dependence on subsidies and payments for ecosystem services may increase.

9.5 Summary of Climate Costs for Forestry and Conclusions

The forestry sector in Austria will partially benefit from climate change because in the immediate future an elongation of the growing season will lead to higher growth rates. However, this positive effect will be balanced by an increasing pressure from pests and pathogens and possibly by drought stress so that negative effects may dominate in the long run. The consequences are moderate declines in the revenue in the timber sector (assuming timber prices of today) and increased harvesting costs (at technology and prices of today). The considered impact chains that are the basis for our results are elaborated. The main drivers for the economic performance of the forest sector are timber production estimates under climate change conditions and the pressure from bark beetle on commercially used Norway spruce forests.

The three modeled impact chains lead to macroeconomic effects equal to a reduction of 0.10 % in welfare, a reduction of 0.07 % in GDP and an increase of 0.05 %-points in unemployment in period 2016–2045 (all values relative to a baseline scenario without climate change). This negative effect on welfare and

GDP is the net effect of on the one hand higher output and value added in construction, which is triggered by the investment to restore protective forests. On the other hand, investment is redirected from other sectors to construction, leading to output losses in sectors such as trade, real estate and manufacturing. The forestry sector is affected by higher value added but lower gross output value. Comparing across impact chains, the main cost trigger is investment in restoration of protective forests, followed by bark beetle damages, while changes in timber productivity play only a minor role. Regarding changes of impacts over time, effects are generally found larger for period 2036–2065 compared to period 2016–2045, up to a factor of three.

Based on these findings for forestry it appears to be useful to design and implement robust forest management concepts. In this context one strategy is to foster mixed species stands. As in most cases this will come at the cost of Norway spruce shares, in relation to current timber prices this may result in a decrease of profitability of commercial timber production. However, long-term future development of timber markets (several decades) under climate change are difficult to project and highly uncertain. Particularly in climate—sensitive regions inaction is not the adequate option in forestry. Leaving managed forests with high shares of conifer tree species at sites naturally supporting broadleaved species to natural ecosystem dynamics will very likely result in severe biotic damages.

While there is high confidence in this conclusion it must also be noted that our assessment has several shortcomings. We have used the climate scenario A1B to assess climate change related impacts and cost of inaction-related to production and a B1 for bark beetle disturbances. However, currently the global emissions are not yet sufficiently reduced and even stronger warming trends are possible as those represented in the A1B and particularly in B1 scenario storylines. Under such conditions the pressure from bark beetles on Norway spruce will be even stronger and will manifest itself also earlier. A second gap is the ignorance of eventually elevated damages due to storms. The decision not to consider storms was justified with the lack of unanimous scientific evidence and the poor data availability. However, in the case of more frequent and more intense storms in the future the economic situation of the forestry sector will be less favorable and costs due to negative impacts on other ecosystem services such as protection against gravitational hazards will increase.

Acknowledgements Thanks to Hubert Siegel, Austrian Federal Ministry of Agriculture, Forestry, Environment and Water Management, for insightful comments on the costs of restoration of high-elevation protection forests.

References

Butterbach-Bahl K, Gundersen P (2011) Nitrogen processes in terrestrial ecosystems. The European nitrogen assessment, vol 6. Cambridge University Press, Cambridge, pp 99–125
Curtis PS, Wang X (1998) A meta-analysis of elevated CO2 effects on woody plant mass, form, and physiology. Oecologia 113(3):299–313
Dale VH, Joyce LA, McNulty S, Neilson RP (2000) Climate change and forest. Sci Total Environ 262:201–204
Eastaugh CS, Potzelsberger E, Hasenauer H (2011) Assessing the impacts of climate change and nitrogen deposition on Norway spruce (Picea abies L. Karst) growth in Austria with BIOME-BGC. Tree Physiol 31(3):262–274
Gardiner B, Blennow K, Carnus JM, Fleischner P, Ingemarson F, Landmann G, Lindner M, Marzano M, Nicoll B, Orazio C, Peyron JL, Reviron MP, Schelhaas M, Schuck A, Spielmann M, Usbeck T (2010) Destructive storms in European forests: past and forthcoming impacts. Final Report to European Commission – DG Environment
Gschwantner T, Prskawetz M (2005) Sekundäre Nadelwälder in Österreich. BFW-Praxisinformation 6:11–13
Haarsma RJ, Hazeleger W, Severijns C, de Vries H, Sterl A, Bintanja R, van Oldenborgh GJ, van den Brink HW (2013) More hurricanes to hit western Europe due to global warming. Geophys Res Lett 40:1783–1788
IPCC (2012) In: Field C, Barros V, Stocker T, Qin D, Dokken D, Ebi K, Mastrandrea M, Mach K, Plattner G-K, Allen S, Tignor M, Midgley P (eds) Managing the risks of extreme events and disasters to advance climate change adaptation. A special report of working groups I and II of the intergovernmental panel on climate change. Cambridge University Press, Cambridge, p 582
Jacob D, Göttel H, Kotlarski S, Lorenz P, Sieck K (2008) Climate change/Klimaauswirkungen und Anpassung in Deutschland – Phas 1: Erstellung regionaler Klimaszenarien für Deutzland. Max-Planck-Institut für Meteorologie (MPI-M), Hamburg. ISSN:1862–4359
Kindermann G, Neumann M (2011) Zuwachsreaktionen auf Witterungsextreme. BFW Praxisinfo 24:25–28
Leip A (2011) Integrating nitrogen fluxes at the European scale. In: Sutton MA et al (eds) The European nitrogen assessment, vol 16. University Press, Cambridge, pp 345–376
Matulla C, Schoener W, Alexandersson H, von Storch H, Wang XL (2007) European Storminess: late 19th century to present. Clim Dyn 31(2–3):125–130
Mitchell TD, Carter TR, Jones PD, Hulme M, New M (2004) A comprehensive set of high-resolution grids of monthly climate for Europe and the globe: the observed record (1901–2000) and 16 scenarios (2001–2100). Tyndall Centre working paper 55, p 25
Nabuurs GJ, Lindner M, Verkerk PJ, Gunia K, Deda P, Michalak R, Grassi G (2013) First signs of carbon sink saturation in European forest biomass. Nat Clim Change 3:792–796
Neumann M (2012) Biomasse aus traditioneller forstlicher Nutzung. BFW Praxisinformation 28:9–10
Qin J (2013) China battles army of invaders. Nature 503:450–451
Schelhaas MJ, Nabuurs GJ, Schuck A (2003) Natural disturbances in the European forests in the 19th and 20th centuries. Glob Chang Biol 9:1620–1633
Schirimeier Q (2013) Wild weather can send greenhouse gas emissions spiralling. Nature 496:147
Schörghuber S, Seidl R, Rammer W, Kindermann G, Lexer MJ (2010) KlimAdapt - Ableitung von prioritären Maßnahmen zur Adaption des Energiesystems an den Klimawandel - Arbeitspaket 3: Biomasse Bereitstellung, Technical report, BOKU, Vienna
Seidl R, Schelhaas MJ, Lexer MJ (2011) Unraveling the drivers of intensifying forest disturbance regimes in Europe. Glob Chang Biol 17(9):2842–2852

Seidl R, Schelhaas MJ, Lindner M, Lexer M (2009) Modelling bark beetle disturbances in a large scale forest scenario model to assess climate change impacts and evaluate adaptive management strategies. Reg Environ Change 9(2):101–119
Stern N (2009) A blueprint for a safer planet. How to manage climate change and to create an era of progress and prosperity. The Bodley Head, London
Sykes MT, Prentice IC, Cramer W (1996) A bioclimatic model for the potential distribution of north European tree species under present and future climates. J Biogeogr 23:203–233
Tacconi L (2012) Redefining payments for environmental services. Ecol Econ 73(1):29–36
von Teuffel K, Baumgarten M, Hanewinkel M, Konold W, Sauter UH, Spiecker H, von Wilpert K (2005) Waldumbau für eine zukunftsorientierte Waldwirtschaft. Springer, New York, p 422
Thom D, Seidl R, Steyrer G, Krehan H, Formayer H (2013) Slow and fast drivers of the natural disturbance regime in Central European forest ecosystems. Forest Ecol Manag 307:293–302
Tomiczek C, Schweiger C (2012) Assessment of the regional forest protection risk in Austria on forest district level. Forstschutz aktuell 54:2–4
Usbeck T, Wohlgemuth T, Dobbertin M, Pfister C, Bürgi A, Rebetez M (2010) Increasing storm damage to forests in Switzerland from 1858 to 2007. Agric Forest Meteorol 150:47–55
Vacik H, Wolfslehner B, Ruprecht H (2008) Identifizierung von Dienstleistungen, Holz- und Nichtholzprodukten aus nachhaltiger Waldbewirtschaftung. Steckbriefe für Nichtholzprodukte und Dienstleistungen.Berichte aus Energie- und Umweltforschung 17b/2008, 80, bmvit, Wien
Wermelinger B (2004) Ecology and management of the spruce bark beetle it Ips typographus – a review of recent research. Forest Ecol Manag 202:67–82

Chapter 10
Ecosystem Services: Pest Control and Pollination

Klaus Peter Zulka and Martin Götzl

Abstract Among the several ecosystem services delivered by biodiversity, natural pest control and pollination are comparatively well understood and highly relevant for ensuring food provision. We describe the potential impacts of climate change, in particular the effects of increasing temperatures, on pest antagonists and pollinators, and evaluate the relevance for Austria's agricultural ecosystems. Temperature changes lead to species range shifts, causing a reshuffling of assemblages and a decoupling of community interactions, followed by an impairment of pest control and pollination services. The effects are strongly modulated by socio-economic factors, particularly the development of semi-natural elements in agricultural landscapes. An enlargement of semi-natural area might mitigate the effects of climate change; a reduction in semi-natural area might exacerbate the climatic effects by impeding migration to track temperature changes even further. We calculated the value of pest control in Austria to be approximately 255 million euros or 8.5 % of the total agricultural plant product value in 2008. Pollination in Austria is worth 298 million euros, corresponding to 9.9 % of the total agricultural plant product value. We distinguish and discuss four possible climate impact scenarios; a scenario describing a moderate reduction of these values emerged as the most likely one.

10.1 Introduction

Studies have compared the impact of climate change on biodiversity with the five mass extinctions in earth history (Barnosky et al. 2011). While overly pessimistic outlooks have recently been scrutinised (He and Hubbell 2011), most scholars agree that massive climate effects can be expected in ecosystems and biota in the future (reviewed in Bellard et al. 2012); some of these effects have already been documented (Thomas et al. 2006).

Aspects of ecosystem functioning that benefit humans have been referred to as ecosystem services (Millennium Ecosystem Assessment 2005; Maes et al. 2011).

K.P. Zulka (✉) • M. Götzl
Biodiversity and Nature Conservation, Environment Agency Austria, Vienna, Austria
e-mail: peter.zulka@umweltbundesamt.at; martin.goetzl@umweltbundesamt.at

K.W. Steininger et al. (eds.), *Economic Evaluation of Climate Change Impacts*, Springer Climate, DOI 10.1007/978-3-319-12457-5_10

Table 10.1 Typology of ecosystem services, Maes et al. (2011)

Provisioning services: the goods or products obtained from ecosystems
Food
Water
Raw materials
Genetic resources
Medicinal resources
Ornamental resources
Regulating services: the benefits obtained from an ecosystem's control of natural processes
Air quality regulation
Climate regulation
Moderation of extreme events
Regulation of water flows
Waste treatment
Erosion prevention
Maintenance of soil fertility
Pollination
Biological control
Habitat services: services supporting the provision of others by providing habitat
Nursery habitat
Gene pool protection
Cultural services: the nonmaterial benefits obtained from ecosystems
Aesthetic information
Opportunities for recreation and tourism
Inspiration for culture, art and design
Spiritual experience
Information for cognitive development

Traditionally, a distinction is made between provisioning services, regulating services, habitat services and cultural services (Table 10.1). Ecosystem services are usually dependent on a variety of species. In some circumstances, higher species richness provides for more ecosystem services (Tylianakis et al. 2008) and, through redundancy, protects ecosystem functions against failure under altered conditions. Climate change might act directly on ecosystem services or indirectly by changing the species composition in a particular area (Schröter et al. 2005).

The relationship between climate change and specific provisioning services is covered in Chap. 8 (Agriculture) and Chap. 9 (Forestry). Cultural ecosystem services are difficult to quantify. For many of the ecosystem services listed in Table 10.1, the only data available, if at all, are data related to a specific context. By contrast, the importance of pollination and natural pest control for the functioning of agricultural ecosystems is unquestioned, and methods for the quantification of these services in monetary terms have been provided (Losey and Vaughan 2006; Gallai et al. 2009). In this chapter, we thus focus on the economic value and its climate-induced alteration of these two ecosystem services. Clearly, the total economic impacts of climate change impinging on biodiversity may be much larger, in particular if modelling projections about climate change leading to the extinction of nearly one third of all species come true (Thomas et al. 2004). For the time being, however, it is not possible to envision all the ramifications of such a disastrous species decline, let alone estimate its monetary implications. Pest control and pollination directly determine agricultural yields, consequently, their causal relationships, their economic importance and its climate-induced change are comparatively easier to assess.

It is expected that climate effects will be modulated by future socio-economic developments, which include the modification of the structure and composition of agricultural landscapes. In this paper, we will thus discuss the implications of climate change in the context of socio-economic developments.

Firstly, we describe the most important effects of climate change on biodiversity in general. Comprehensive reviews of this topic are already available in the literature (Bellard et al. 2012); here, we restricted our description to the most important aspects. Secondly, we evaluated the implications of these effects on species assemblages that provide pollination and pest control services in Austria. Thirdly, we attempted to estimate the change in ecosystem services based on these impacts. Given the limitations of knowledge and data available, a numerical calculation of this change factor was not possible. Instead, we provided four combined scenarios to describe climate and land use change effects on pollinators and pest control agents and associate them with ecosystem service change factors. In a fourth step, we related these factors to the calculated monetary value of pest control and pollination in Austria.

10.2 Dimensions of Sensitivity to Climate Change

10.2.1 Climatic Factors

Like all organisms, pest control antagonists and pollinators require a range of ecological conditions in which they can live and reproduce, referred to as their niche (Schoener 1989). Temperature, humidity and their extremes are important niche dimensions, which limit the geographical distribution of a species. In open agricultural landscapes, prolonged drought in hot summers may lead to stress in a number of Central European species. Warm winters lead to elevated metabolic rates and to the depletion of body fat reservoirs in overwintering pest antagonists. As Dillon et al. (2010) show, metabolic rates are the ultimate driver of species shifts, and in warmer regions (for which a smaller temperature increase has been predicted) even small temperature increases might cause a physiological response as strong as in the arctic regions.

The number of sunny days and the maximum day temperature, combined with soil moisture, are important parameters responsible for the larval development of some bee species which develop in sand or soil. Total annual rainfall and the distribution of precipitation during the growing season have an influence on soil moisture, which is of relevance for 70 % of the European wild bee species which are soil nesters (Müller et al. 1997).

Long dry periods in combination with open soil (e.g. after harvest) can lead to a population breakdown for many generalist predator species inhabiting agricultural landscapes (e.g. spiders, ground beetles or rove beetles) that require moist conditions.

10.2.2 Socio-economic Factors

In arable fields, the resource supply for pollinators and pest control organisms is highly variable and changes from fields in full bloom to barren fields after harvest. Semi-natural landscapes, e.g. road verges, set-asides, hedges, field margins, flower strips or dry grassland patches, provide a habitat buffer for parasitoids, generalist predators and pollinators from which arable fields can be repopulated. Compared to uniform arable fields, semi-natural landscape elements show higher plant species richness values and higher spatial heterogeneity. They thus provide a variety of food resources and are able to support sufficiently stable populations of pest antagonists and pollinators (Hänke et al. 2009). Semi-natural landscape elements are important hibernation sites for generalist predators (Pfiffner and Luka 2000), such as rove beetles (D'Hulster and Desender 1982) and spiders (Bayram and Luff 1993).

The availability of semi-natural landscape elements is strongly influenced by socio-economic factors, e.g. by incentive systems designed to enhance farmland biodiversity, so-called agri-environment schemes (Kleijn et al. 2006). At times of high food prices, incentive payments may no longer compete with market prices of agricultural products. Additionally, farming on unused landscape elements with poor soils may become viable. By contrast, reductions in the global competitiveness of Austrian agriculture will lead to the abandonment of more agricultural areas. While, in the short run, such a reduction in agricultural activity will result in higher biodiversity levels (van Buskirk and Willi 2004), shrub and bush encroachment will lead, over time, to the spread of forest species, with a limited potential for pest control and pollination (Kleijn and Báldi 2005).

Another socio-economic factor is the development of settlement and transportation infrastructures, of which area requirements restrict the proportion of non-used arable land in agricultural landscapes.

10.2.3 *Interaction Between Climatic and Socio-economic Factors*

The largest reductions in the provision of agricultural ecosystem services are to be expected if severe climate change with long lasting extreme periods coincides with the abandonment and fragmentation of semi-natural landscape elements. The latter reduces not only habitat area and available resources for pest control and pollination agents, but also the density of stepping stone habitats that allows migrating species to track suitable climatic conditions. Consequently, climatic and socio-economic effects not only add on each other, but interact to some degree.

10.3 Exposure to Climatic Stimuli and Impacts up to Now

We first describe overall climate change effects on biodiversity. Second, we relate them to pest control and pollination services. Finally, we infer the impact on Austria's agricultural production systems.

10.3.1 *Impacts of Climate Change on Biodiversity in General*

10.3.1.1 Range Shifts

Climate change and global warming will lead to range shifts of a large proportion of the world's species (Bellard et al. 2012). Examples of poleward and, in the case of mountains, upslope migration are already numerous and well-documented (Parmesan and Yohe 2003; Walther et al. 2005). Current warming rates require range shifts of several kilometres per year (Diffenbaugh and Field 2013); average migration speeds of 6.1 km per decade have already been measured (Parmesan and Yohe 2003). Range shifts have been recorded in a wide range of animal taxa (Hickling et al. 2006). Species' responses to global warming often coincide with responses to land use change (Mair et al. 2012), as shown by the range contractions experienced by the Large Heath butterfly *Coenonympha tullia* in Britain, which were attributed to habitat degeneration rather than climate change (Franco et al. 2006). Climate change and habitat loss compound their effects: A meta-analysis of 1,319 papers on habitat loss and fragmentation showed that the effect of habitat loss on species and their survival was most destructive in dry climates with high warming rates (Mantyka-Pringle et al. 2012). Range shifts will lead to an elevated degree of local species turnover. This may lead to a disruption in predator–prey systems (see below), and to more frequent pest outbreaks (Stireman III et al. 2005).

10.3.1.2 Phenology

Phenology describes the seasonal plant and animal activity driven by environmental factors. Phenological shifts have already been documented in many species and can help species keep synchrony with cyclical abiotic factors. However, they can also be disruptive as they may increase asynchrony in predator–prey and insect–plant systems (Parmesan 2006). In flowering plants and insect pollinators, this leads to mismatches between plant flowering and pollinator activity. Likewise, modifications of interspecific relationships between prey and predator or host and parasite have to be expected as a consequence of climate change (Bellard et al. 2012).

At the end of the last century, the phenology of both wild and cultivated plants in Europe changed much more than in the decades before (Scheifinger et al. 2002). Phenological changes in Europe have also been observed in vertebrates (Dunn and Winkler 2010) and several insect taxa (Stefanescu et al. 2003; Roy and Sparks 2000; Dell et al. 2005).

10.3.1.3 Voltinism

The number of insect generations per year is called voltinism. The length of the favourable season influences the number of generations insects can have in a year (Stearns 1992). An extended length of the favourable season enables some insect species in central Europe to produce a second generation, which allows them to enlarge their populations. Multi-voltinism in herbivorous pests leads to an increased yield loss in agriculture and forestry. Additionally, voltinism may have negative effects because existing interactions between parasites and pathogens become decoupled. Modelling studies predict a temperature-related production of a second generation for some species (Tobin et al. 2008; Jönsson et al. 2009). However, empirical data on current changes in voltinism resulting from climate change and affecting a whole community are still scarce (Altermatt 2010).

10.3.1.4 Extinctions

With geographic range shifts of species, new areas along the colder northern border of species' ranges are colonised, but these expansions are accompanied by extinctions near the southern border. Species with limited dispersal capacity, a fragmented habitat and a high degree of local genetic adaptation (Harte et al. 2004) might not be able to keep track of climate shifts at all, and could become extinct across their entire geographical distribution (Opdam and Wascher 2004; Hodgson et al. 2012).

Using the species-area relationship, Thomas et al. (2004) predicted climate change-related extinctions of more than one third of the global fauna in a worst-case scenario. Their modelling approach has recently been questioned (He and

Hubbell 2011) and the overall global extinction risk is difficult to gauge at the moment. To date, few cases of climate-related extinctions have been recorded, but this might be due to poor documentation and data deficiencies rather than a small impact of climate change. As shown by Sinervo et al. (2010), a high percentage of species with a narrow distribution range and limited dispersal capacity, such as Mexican lizards, may indeed become extinct by 2080 because of their inability to keep track of climatic changes. Neither the local habitat heterogeneity (e.g. southern and northern slopes; Clusella-Trullas and Chown 2011) nor genetic population adaptation will allow species to compensate for the predicted global warming effects (Sinervo et al. 2011).

10.3.2 Impact on Pest Control Organisms

Natural pest control services on agricultural land are provided by a number of generalist predator taxa, such as birds, ground beetles, rove beetles, spiders, harvestmen or hoverflies. Additionally, parasitoids (often small Hymenoptera) play a major role in keeping pest organisms in check and avoiding pest outbreaks. The availability of these organisms in agro-ecosystems depends on the area of non-farmed landscape elements and semi-natural patches, which ensure a baseline population density of antagonists (Hänke et al. 2009). A quantitative prediction of the effects of climate change is difficult; a large-scale modelling approach including the climatic envelopes of many species would be required. Civantos et al. (2012) modelled the effects of global warming on pest control agents in Europe; however, because of data deficiencies, the study was restricted to vertebrates. It predicts major negative effects on species survival and pest control services in Mediterranean countries, whereas in Central Europe, species shifting their ranges northward will largely be replaced by thermophilous species from the South and the East. In the case of parasitoids, the effects of climate warming depend on the migration speed of the host, the migration speed of the parasitoid and the host specificity (Jeffs and Lewis 2013). Asynchrous migration will lead to a decrease in parasitoid–host interactions and thus to "an increase in the frequency and intensity of herbivore outbreaks" (Stireman III et al. 2005). Similar effects can be expected from a disruption of the relationship between insect predators and herbivores, even if the average host specificity is lower within these communities (Table 10.2).

10.3.3 Impacts on Pollinators

10.3.3.1 Climate Impacts at the Global Scale

Honey bees, wild bees, hoverflies, butterflies, flies, moths, midges, thrips, and beetles are the main insect species providing animal-mediated pollination to wild

Table 10.2 Impact chains related to pest control and pollination in Austria

Climate change parameter	Impact chain
Precipitation: Lower precipitation or change in seasonal distribution	*Impediment of soil nesters (wild bees) in the dry season* → Decrease of soil moisture → deterioration of conditions for pollinators which are soil nesters or hibernating in the soil → reduced population strength → reduced pollination service → decreased agricultural yield
	Changes in the amount of snow fall → Reduced thickness of snow cover → earlier timing of snowmelt → earlier flowering time → decoupling of species interactions → reduced pest control and pollination services → decreased agricultural yield
	Long drought periods Local extinction of species depending on Central European and oceanic climatic conditions → expansion of species of Eastern or Mediterranean distribution ranges into Austria → decoupling of predator–prey systems → reduced pest control services → decreased agricultural yield
Temperature: Increase in mean values and in the frequency heat waves	*Changes in range and phenology; local extinction of species* → Increased decoupling of species interactions → reduced pest control and pollination service → decreased agricultural yield
	Emergence of multi-voltinism → Longer growing season → enlargement of populations → increase in the abundance of herbivorous pests → decreased agricultural yield
	Impediments to insect larval development (wild bees) → Increase in soil and sand temperature → increased (or total) larval mortality → decreased agricultural yield

and cultivated plants. Generally, pollination services depend on the species richness of pollinators and on pollinator abundance. There is a clear link between pollinator diversity and sustainable crop pollination (Carvalheiro et al. 2011; Garibaldi et al. 2011).

Impacts of climate change may affect all organisational levels, from the individual level to the community level. At the individual level, raised temperatures have a direct influence on the behaviour and the physiology of pollinators and determine the insects' foraging activity. Avoidance of extreme temperatures could significantly influence pollen removal and deposition efficiency of bees. With increasing temperatures, pollinators are at risk of overheating (Reddy et al. 2012).

Climate change can alter the quality and composition of the floral environment, define new bee distribution ranges of pollinators and give rise to new competitive relationships among species and among their parasites and pathogens (Le Conte and Navajas 2008).

Temporal mismatches of flowering plants and pollinating insects may lead to a disruption of plant–pollinator interactions. Direct temperature responses and the occurrence of mismatches in pollination interactions may vary among species and regions (Memmott et al. 2007). As many pollinators visit plants quite opportunistically, such phenological decoupling may result in the emergence of novel plant–pollinator interactions (Hegland et al. 2009).

Alien species can both compensate for the negative effects of climate change and amplify them. On the one hand, alien species might affect functional pollinator community compositions by intruding into native plant–pollinator communities where they partly take over and sustain pollination services. On the other hand, alien pollinators may not compensate for morphological and behavioural mismatches because of a lack of specialist morphology. This would lead to lower pollination rates. However, the net effects of direct and indirect interactions among native and alien species are difficult to predict (Schweiger et al. 2010).

On the whole, generalist species are expected to be more resilient to interactions between pollinators and plants under climate change. Consequently, non-random novel communities with an over-representation of generalist species are expected as a result of climate change-induced shifts in species' distributions and phenology (Gonzáles-Varo et al. 2013). Pollinator communities might become progressively poorer in species and dominated by mobile habitat generalists (Vanbergen et al. 2013).

10.3.3.2 Model Calculations of Climate Impacts on Pollinators at the European Scale

Future global warming may also have severe impacts on the ranges of butterflies, as described in the 'Climatic Risk Atlas of European butterflies', which is based on model calculations (Settele et al. 2008). Model-based predictions of climate change effects along a pan-European transect revealed declines in wild bee species richness in warmer climates, whereas the species richness of hoverflies showed a slight increase (Dormann et al. 2008). Further results indicate that soil properties (soil bulk density, pH, soil moisture) most likely affect the development of the larval stages of terrestrial arthropod species which are bound to the soil. This is important because about 70 % of the European bee species are soil nesters (Müller et al. 1997) and a large proportion of mobile insects such as hoverflies hibernate in the soil.

In simulations of phenological shifts, floral resource availability (429 plant species) was reduced for 17–50 % of 1,419 pollinator species (Memmott et al. 2007). Consequently, half of the original activity period of the animals would coincide with times where no food plants were available. The pollinators most likely to be left with no food at all were specialists using the flowers of a small number of plant species. But generalist pollinators, which feed on the flowers of many plants, were affected by the reduced flower availability as well. As the data on changes in flowering phenology in response to climate warming were based on reports about phenologial shifts over the past century from many northern temperate sites in the USA and the UK, the results might also be valid for European countries.

10.3.3.3 Obstacles to Estimating Impacts on Pollination Services in Austria

Bees (honey bees and wild bees, including bumble bees) are the most important pollinators for many wild and cultivated plant species in Europe (Hegland et al. 2009). Most of the Austrian crop plants are pollinated by foraging habitat generalist bees (polylectic bees). As generalist insects have been shown to be less sensitive to altered climatic parameters, it is assumed that there will be no dramatic decrease in the pollination services provided to agricultural production in Austria. However, this assumption cannot be applied to the pollination of wild plant species, which often depend on specialist insects for their reproduction. The relevant impact chains are listed in Table 10.2.

Up to now, there are no detailed analyses of climate change impacts on insect-mediated pollination services sustaining agricultural production in Austria. Therefore, no sound estimate of the degree to which climate change will influence the pollination services provided to Austrian crops is possible (see also discussion in Sect. 10.4.7).

10.3.4 Impacts Observed up to Now

10.3.4.1 Pest Control

High summer temperatures, prolonged dry periods, large-scale inundations and warm winters are already impinging on species providing pest control services. Warm summers have led to pest outbreaks that could no longer be controlled by antagonists. For example, the exceptionally warm and dry summer in 2003 triggered pest outbreaks and reductions in agricultural yield in Austria. Among the pest organisms favoured by hot and dry conditions are the European Corn Borer *Ostrinia nubilialis*, the Four-Spotted Sap Beetle *Glischrochilus quadrisignatus*, the European Wheat Stem Sawfly *Cephus pygmaeus*, the Wheat Bugs *Eurygaster maura* and *Aelia acuminata*, all of which attacked crops in Austria and caused substantial yield losses in 2003 (Grünbacher et al. 2006).

10.3.4.2 Pollination

Recent climate change has caused shifts in butterfly species distributions in Great Britain (Hickling et al. 2006). Range retractions and elevated extinction risks have been observed for 16 mountain butterfly species in central Spain (Wilson et al. 2005). Parallel declines in pollinators and insect-pollinated plants have been reported for Britain and the Netherlands (Biesmeijer et al. 2006). Corresponding studies for Austria that describe climate change induced range shifts or declines in pollinators have not been published yet, but similar trends can be also expected for Austria.

10.4 Future Exposure to and Impacts of Climate Change

10.4.1 Climatic Scenarios

Climate mid-range scenarios indicate an increase in average air temperatures across Austria by about 1 °C between 1981 and 2010 and from 2016 to 2045, with slightly higher increases in the agricultural regions of eastern Austria. Temperature increases will be higher in summer than in winter. The number of heat days will increase steadily over the twenty-first century. A small reduction in dry periods in Eastern Austria up to 2030 will be followed by an increase in dry periods as in all other parts of Austria throughout the century. In the low-range scenario, heat days will increase only in the second half of the century, in the high-range scenario, heat days will increase much faster over the century than in the mid-range scenario [see Chap. 5 (Climate)].

10.4.2 Socio-Economic Scenarios

According to the reference socio-economic scenario, farmland area in Austria will have experienced reductions of about 100,000 ha by 2030, while the area used for settlement and transportation infrastructures will have increased by about 90,000 ha at the same time. The area available for unused semi-natural landscape elements will thus remain almost unchanged.

In the enhancing scenario, European legislation requires that a certain percentage of agricultural land will have to be set aside to comply with "greening" practices (van Zeijts et al. 2011). A target of 7 % of non-used arable land has been under discussion; this percentage could be increased even further.

The diminishing scenario would assume a large-scale loss of semi-natural landscape elements. Prices for agricultural products showed a negative trend for decades, but experienced an inflection point in the beginning of the twenty-first century (von Witzke 2008). High prices could render farming on semi-natural landscape elements viable and lead to the large-scale cultivation of semi-natural areas.

10.4.3 Combined Scenarios Including Climatic and Socio-economic Factors

From the evidence described above, it is not possible to calculate a numerical impact function that would describe the change in pest control and pollination service value against climate change and socio-economic trends. Instead, we set up four combined scenarios for possible developments of ecosystem services, and compare them according to their likelihood and economic impacts (Table 10.4).

The boundaries for ecosystem service change given in Table 10.4 are not derived from scientific evidence; they only provide a practical delineation of the four scenarios.

Combined Scenario 1: Increase of ecosystem services by 0–20 %. Climate change leads to the immigration of new species. Species adapted to a hotter and drier climate complement existing species assemblages of pollinators and pest antagonists and thus enhance ecosystem services, in particular if additional arable land is set-aside at the same time.

Combined Scenario 2: Despite a change in climatic parameters and socio-economic factors, pest control and ecosystem services stay at their current levels. This result is not very likely, since at least a substantial species turnover (Thuiller et al. 2004) has to be expected in Central Europe. It is, however, possible if the original ecosystem services are adequately substituted by the services provided by the immigrant species.

Combined Scenario 3: Climate change and socio-economic factors lead to a moderate (0–20 %) decrease in the supply of ecosystem services in agricultural landscapes. The supply of pest control and pollination services is reduced (Jeffs and Lewis 2013; Voigt et al. 2003; Zarnetske et al. 2012; van Grunsven et al. 2010). Local extinctions of species are not entirely counterbalanced by species immigration because of migration impediments (Hodgson et al. 2012). Novel predator–prey and pollinator–plant relationships work less efficiently than the finely tuned community interactions which had become established over long periods of time.

Combined Scenario 4: Climate change and socio-economic factors lead to a severe reduction (more than 20 %) in the supply of ecosystem services. Species loss due to climate change is aggravated by the loss of semi-natural habitat. Agricultural landscapes provide neither habitat nor migration structures for pollinators or pest control agents. Natural ecosystem services need to be replaced to a large degree by artificial pollination and pest control measures.

10.4.4 *Methods of Valuation and Their Implementation Steps*

10.4.4.1 Pest Control Services

For the USA, Losey and Vaughan (2006) assess the value of pest control ecosystem services supplied by predators and parasitoids using the following formula:

$$V = (NC - CC) \times P_i$$

with V being the value of the pest control service, NC being the cost of damage incurred without any natural pest control services, CC being the cost of damage incurred at current levels of pest control (natural antagonists in place) and P_i the proportion of pest control services offered by antagonist insects and parasites as compared to other control mechanisms.

Using previously published data on eight major crops in North America (Yudelman et al. 1998), Losey and Vaughan (2006) estimate CC to be 17.7 % of the total yield, if the costs for pesticides are included. On the basis of estimates of damage caused by invasive species, the authors conclude that 65 % of the pest organisms are being suppressed, in other words, 65 % of the potential damage costs are currently avoided. NC thus amounts to 50.5 % and the present total value of pest control is NC–CC = 32.9 % of the current total yield. Assuming that only 39 % of the costs are caused by native pests, which are under control by native antagonists, Losey and Vaughan (2006) conclude that 32.9 % × 39 % = 12.8 % of the yield is saved through natural control services, of which one-third (4.3 %) is delivered by native insects. If we assume that the importance of the services of predators and parasitoids from taxonomic groups other than insects are is similar to those of insects, we end up with a proportion of 8.5 % of the total agricultural yield currently safeguarded by natural pest control. Clearly, these numbers are based on multiple assumptions and may just roughly define an order of magnitude. However, with these limitations in mind, they seem transferable to European and Austrian conditions.

10.4.4.2 Pollination Services

To estimate the total economic value of the pollination service provided by insects for agricultural production in Austria, only those pollinator-dependent agricultural products (see Table 10.3) were selected which are used for human food directly. A further selection criterion was the availability of published figures relating to production outputs and producer prices (Statistik Austria 2009).

The economic value of each of the products was calculated according to a formula used by Gallai et al. (2009): Economic value = production output × production price. In order to assess the pollinator-dependent production output and the corresponding producer prices, we attributed an average pollination dependence coefficient to each of the products, as suggested by Klein et al. (2007) (Table 10.3).

10.4.5 Monetary Evaluation of Impacts

Using the proportions derived in Sect. 10.4.4, we calculate the monetary value of pest control and pollination. We then apply the combined scenarios (Sect. 10.4.3) to estimate the effect of climate change on these values.

10.4.5.1 Value of Pest Control Services

In Austria, the calculated value of agricultural plant products in 2008 was 3,001 million euros (Statistik Austria 2009). Applying the rationale for estimates as described in Sect. 10.4.2 and using a proportion of 8.5 % of the agricultural product value as guaranteed by pest control services currently in place, we obtain a current value of pest control services of 255 million euros.

10.4.5.2 Value of Pollination Services

For the crop plants selected, the total output produced was about 1.3 million metric tons (Table 10.3) in 2008, with producer prices amounting to 564 million euros. The total producer price of pollinator-dependent output is about 298 million euros (economic value of insect pollinators). This corresponds to 9.9 % of the economic value of all the Austrian agricultural plant crops produced in 2008 (3,001 million euros).

Table 10.3 Production output, pollination dependency and value of pollinator-dependent products presented for selected fruits, vegetables, edible oil crops and pulses

	Production output (t) in 2008[e]	Producer prices (M€) in 2008[f]	Average pollination dependency coefficient[g]	Value of pollinator dependent production output (M€)
Fruits[a]	831,203	404.5	0.25/0.65	239.8
Vegetables[b]	101,569	36.6	0.5/0.25/0.65/0.95	14.7
Edible oil crops[c]	262,169	95.5	0.25/0.95	38.6
Pulse[d]	107,745	27.5	0.5/0.25	4.9
Total	1,302,686	564.1		297.9

t tons, M€ million euros

[a,f]Apple (0.65), Pear (0.65), Plum (0.65), Peach (0.65), Apricot (0.65), Sweet and Sour Cherry (0.65), Red and Black Currant (0.25), Strawberry (0.25), Raspberry (*0.25*), Elderberry (0.25)

[b,g]Fennel (0.65), Cucumber (0.65), Aubergine (0.25), Melon (0.95), Pumpkin (0.95), Tomato (0.05), Zucchini (*0.25*), String Bean (0.05)

[c,g]Pumpkin seeds (0.95), Sunflower (0.25), Oilseed Rape (0.25)

[d,g]Runner Bean (*0.05*), Field Bean (0.05), Pea (*0.05*), Soybean (0.25)

[e]Sources of figures for output produced: Statistik Austria (2009)

[f]Sources of figures for producer prices: Statistik Austria (2009)

[g]This coefficient provides an estimate of the extent to which the output produced depends on insect-mediated pollination. The figures represent the average dependency on pollinators as published by Klein et al. (2007). Figures in italic represent assumed dependency coefficients

10.4.5.3 Value of Pest Control and Pollination Services in the Future

Our assessment revealed that pest control and pollination services together accounted for more than 18 % of the total economic value of agricultural plant products in 2008. Forecasts made by the OECD-FAO (2013) predict a decline in many (though not all) of the agricultural output prices for the years up to 2022. The economic value of pest control in 2022 will remain constant at 255 million euros, since gains in producer prices for some agricultural plants will be compensated by declines in others. By contrast, the value of pollination will decrease to 245 million euros, since strong declines in producer prices for fruits and pulses will not be offset by gains in producer prices for vegetables and edible oil crops. These values are within an order of magnitude which is still comparable to economic values in 2008 (input prices were corrected by an assumed inflation rate of 2 %). The prices of 2022 (based on 2008, OECD-FAO 2013) are assumed to remain constant up to 2030 and 2050 (for further explanation of these assumptions see Chap. 8 (Agriculture). The assumption that both pest control and pollination service will remain as important as they are today (also from an economic point of view) is therefore plausible.

10.4.6 Evaluation of Impacts on Pest Control and Pollination Driven by Climate Change and Habitat Loss

Having determined the total value of pest control and pollination services, we may estimate how the four combined scenarios from Sect. 10.4.3 translate into future losses and gains (Table 10.4).

Even if all four scenarios are possible, scenario 3 appears to be best in line with existing evidence, as laid out in Sect. 10.3. Decoupling of interspecific interactions (Sects. 10.3.1–10.3.3) has been predicted repeatedly and, in some cases, already been observed (Sect. 10.3.4). A reduction in ecosystem service efficiency has thus to be expected. This scenario would imply a reduction in the value of pest control and pollination to about one-fifth of the present value and reduce the value of agricultural plant products marketed in Austria by about 3.7 %. The other scenarios are less likely but still possible, and they are in line with some of the predictions provided in the literature.

10.4.7 Sector-Specific Uncertainties

The main sources of uncertainty are the relationship between climate change and species assemblage responses and the relationship between species assemblages and ecosystem services. Both of these relationships have not been studied explicitly

Table 10.4 The four combined scenarios, their delineation and economic effects

Scenario	1	2	3	4
Characteristics	Increase in species richness and semi-natural habitat elements	Both species numbers and semi-natural habitat elements remain constant	Small changes in species numbers and semi-natural habitat area, but with decoupling of community relationships and increased habitat fragmentation	Large species losses and loss of semi-natural habitat
Ecosystem services	Increase	Stable	Decrease	Strong decrease
Likelihood	Unlikely	Rather unlikely	Likely	Possible, but unlikely
Ecosystem service change	0 to +20 %[a]	No change	0 to −20 %[a]	<−20 %[a]
Losses in agricultural output value	None	None	0–100 M€	>100 M€
Gains in agricultural output value	0–100 M€	None	None	None

M€ million euros
[a]Arbitrary values; selected to provide a delineation of the four scenarios

for the Austrian landscapes, and possible responses can only be inferred from circumstantial evidence. Considering the relationship between species assemblages and pollination service, scientific evidence indicating a reaction to altered climatic conditions is available only for a few pollinator species. These studies often focus on single species or on a low number of pollinators which are analysed independently of each other. Studies on the simultaneous impact of climate change, habitat loss and agricultural intensification on pollinating insects are lacking.

Civantos et al. (2012) have modelled the distribution of vertebrate pest antagonists in Central Europe, but analogous simulations for invertebrate pest antagonists (such as ground beetles, spiders, hoverflies or rove beetles) are not available due to insufficient data. Even if many of the species of these taxa are capable of flying, their dispersal capacity might not match the dispersal capacity of large mammals, let alone birds. Little is known of the dispersal capacity of parasitoids and their migration potential in heavily transformed human-dominated landscapes. Methodological attempts to combine climatic envelope modelling with landscape modelling are just beginning (Hodgson et al. 2012).

It is unclear how climate-induced extinction processes will unfold. Substantial extinction debts might be incurred (Kuussaari et al. 2009), meaning that some populations might survive for decades even if the overall environmental and climatic factors are no longer suitable for their long-term existence. The spatial resolution of climatic envelope models is limited; small-scale microclimatic heterogeneity might safeguard local populations despite the overall change in climatic conditions. However, the relevance of this effect is in dispute (see e.g. Clusella-Trullas and Chown 2011; versus Sinervo et al. 2011).

A second source of uncertainty is the relationship between species richness and ecosystem services. A positive yet asymptotical relationship has been assumed for some time (rivet popper or redundancy hypothesis, Ehrlich and Ehrlich 1981, see also Naeem 1998). However, recent analyses have shown that in a diverse landscape, ecosystem services might increase with species richness over large gradients (Tylianakis et al. 2008). Srivastava and Vellend (2005) argue that the direct effects of environmental stressors will have a much greater influence on ecosystem services than the indirect effects of altered assemblage structures or reduced species richness. There are no sound methods to estimate the degree to which a reduced provision of pollination services (due to the loss of some insect species) will be compensated by other pollinator species. There is also a lack of comprehensive analyses which might reveal the relative contribution of different insect species to the pollination of all relevant crops in Austria.

10.5 Summary of Climate Costs for Ecosystem Services and Conclusions

Climate change will likely lead to newly assembled pollinator and pest control communities. Range shifts and asynchrony between parasitoid and pest cycles can be expected with a certain level of reliability. This will lead to a reduction in ecosystem services. It is less clear whether species relevant for pest control and pollination services will become extinct and to what degree a possible reduction in species richness will impair ecosystem services.

Land use changes and combinations of stressors could produce a similar or even stronger negative influence on both of these services. On a regional level, a diverse landscape including semi-natural habitats is the most important background supporting sustainable pollination of wild and cultivated plants, and also an assemblage of pest antagonists capable of keeping pest outbreaks in check. Losses of appropriate habitats due to land use intensification, fragmentation of landscapes and excessive use of fertilisers and pesticides lead to a deterioration of foraging and nesting conditions for most of the insects responsible for ecosystem services. These pressures seem to impair insect community networks at least to the same extent as do changing climatic parameters.

Decline and local extinctions have been observed in the case of some pollinator species (butterflies, hoverflies, honey bees and wild bees including bumble bees) in some European countries. However, it is still unclear to what extent climate change (rather than other pressures) is responsible for these species declines. Neither has it been possible to quantify the losses in agricultural production, which might be attributed to a change in climatic parameters.

The value for pest control services in Austria as estimated in this chapter amounts to 8.5 % of the agricultural plant product value or 255 million euros and the value for pollination is 9.9 % or 298 million euros. In the most likely scenario, climate change will lead to a moderate reduction of these values. Unfortunately, an exact reduction function cannot be provided. Obstacles to such quantifications are: a lack of knowledge about (1) species reactions to altered climatic conditions, (2) compensation capacity within a species community (including alien species), (3) the importance of various insect species (compared to other species) in providing pest control and pollination to agriculture, and (4) the level of combined impacts on ecosystem services. At present, it is not possible to predict a temporal trajectory for these impacts, although most of the trends already known may continue throughout the century.

To reduce the remaining uncertainties, more niche modelling analyses, in particular those addressing species that are responsible for pest control and pollination services, are needed. Invertebrate distribution data are often incomplete and poorly suited for such analyses. The relationship between species richness and ecosystem services should be further analysed. Exclusion experiments might help to elucidate the role of natural pest control in maintaining yields.

Acknowledgements We appreciate feedback by Paul Watkiss, Reimund Schwarze and Martin König on a first manuscript draft, the comments of three anonymous reviewers and Birgit Bednar-Friedl on a second draft; linguistic corrections by Brigitte Read, inspiring discussions with Wolfgang Rabitsch and the encouragement of Maria Stejskal-Tiefenbach.

References

Altermatt F (2010) Climatic warming increases voltinism in European butterflies and moths. Proc R Soc B 277:1281–1287

Barnosky AD, Matzke N, Tomiya S et al (2011) Has the earth's sixth mass extinction already arrived? Nature 471:51–57

Bayram A, Luff ML (1993) Winter abundance and diversity of lycosids (Lycosidae, Araneae) and other spiders in grass tussocks in a field margin. Pedobiologia 37:357–364

Bellard C, Bertelsmeier C, Leadley P et al (2012) Impacts of climate change on the future of biodiversity. Ecol Lett 15:365–377

Biesmeijer J, Roberts S, Reemer M et al (2006) Parallel declines in pollinators and insect-pollinated plants in Britain and the Netherlands. Science 13:351–354

Carvalheiro L, Veldtman R, Shenkute A et al (2011) Natural and within-farmland biodiversity enhances crop productivity. Ecol Lett 14:251–259

Civantos E, Thuiller W, Maiorano L et al (2012) Potential impacts of climate change on ecosystem services in Europe: the case of pest control by vertebrates. BioScience 62:658–666

Clusella-Trullas S, Chown SL (2011) Comment on "Erosion of lizard diversity by climate change and altered thermal niches". Science 332:537

Dell D, Sparks T, Dennis R (2005) Climate change and the effect of increasing spring temperatures on emerge dates on the butterfly Apatura iris (Lepidoptera: Nymphalidae). Eur J Entomol 102:161–167

D'Hulster M, Desender K (1982) Ecological and faunal studies of Coleoptera in agricultural land. III. Seasonal abundance and hibernation of Staphylinidae in the grassy edge of a pasture. Pedobiologia 23:403–414

Diffenbaugh NS, Field CB (2013) Changes in ecologically critical terrestrial climate conditions. Science 341:486–492

Dillon ME, Wang G, Huey RB (2010) Global metabolic impacts of recent climate warming. Nature 467:704–707

Dormann C, Schweiger O, Arens P (2008) Prediction uncertainty of environmental change effects on temperate European biodiversity. Ecol Lett 11:235–244

Dunn P, Winkler D (2010) Effects of climate change on timing of breeding and reproducing success in birds. In: Moller A (ed) Effects of climate change on birds. Oxford University Press, Oxford, pp 113–128

Ehrlich P, Ehrlich A (1981) Extinction: the causes and consequences of the disappearance of species. Random House, New York

Franco AMA, Hill JK, Kitschke C et al (2006) Impacts of climate warming and habitat loss on extinctions at species' low-latitude range boundaries. Glob Chang Biol 12:1545–1553

Gallai N, Salles J, Settele J et al (2009) Economic valuation of the vulnerability of world agriculture confronted with pollinator decline. Ecol Econ 68:810–821

Garibaldi L, Steffan-Dewenter I, Kremen C et al (2011) Stability of pollination services decreases with isolation from natural areas despite honey bee visits. Ecol Lett 14:1062–1072

Gonzáles-Varo J, Biesmeijer J, Bommarco R (2013) Combined effects of global change pressures on animal-mediated pollination. Trends Ecol Evol 28:524–530

Grünbacher EM, Kromp B, Formayer H, Hann P (2006) Einflüsse des Klimawandels auf landwirtschaftliche Schädlinge und Nützlinge im Biologischen Landbau Österreichs, Endbericht zum Projekt StartClim2005.C3a. Bio Forschung Austria, Wien. http://www.austroclim.at/fileadmin/user_upload/reports/StCl05C3a.pdf. Accessed 27 Aug 2013

Hänke S, Scheid B, Schaefer M et al (2009) Increasing syrphid fly diversity and density in sown flower strips within simple vs. complex landscapes. J Appl Ecol 46:1106–1114

Harte J, Ostling A, Green JL et al (2004) Climate change and extinction risk. Nature 430:6995

He F, Hubbell SP (2011) Species-area relationships always overestimate extinction rates from habitat loss. Nature 473:368–371

Hegland S, Nielsen A, Lázaro A et al (2009) How does climate warming affect plant-pollinator interactions? Ecol Lett 12:184–195

Hickling R, Roy D, Hill J et al (2006) The distributions of a wide range of taxonomic groups are expanding polewards. Glob Chang Biol 12:450–455

Hodgson JA, Thomas CD, Dytham C et al (2012) The speed of range shifts in fragmented landscapes. PLoS One 7:e47141

Jeffs CT, Lewis OT (2013) Effects of climate warming on host-parasitoid interactions. Ecol Entomol 38:209–218

Jönsson A, Appelberg M, Harding G et al (2009) Spatio-temporal impact of climate change on the activity and voltinism of the spruce bark beetle, *Ips typographus*. Glob Chang Biol 15:486–499

Kleijn D, Báldi A (2005) Effects of set-aside land on farmland biodiversity: comments on Van Buskirk and Willi. Conserv Biol 10:963–966

Kleijn D, Baquero RA, Clough Y et al (2006) Mixed biodiversity benefits of agri-environment schemes in five European countries. Ecol Lett 9:243–254

Klein A, Vaissière B, Cane J et al (2007) Importance of pollinators in changing landscapes for world crops. Proc R Soc B 274:303–313

Kuussaari M, Bommarco R, Heikkinen RK et al (2009) Extinction debt: a challenge for biodiversity conservation. Trends Ecol Evol 24:564–571

Le Conte Y, Navajas M (2008) Climate change: impact on honey bee populations and diseases. Rev Sci Tech Off Int Epizoot 27:499–510

Losey JE, Vaughan M (2006) The economic value of ecological services provided by insects. BioScience 56:311–323

Maes J, Paracchini M, Zulian G (2011) A European assessment of the provision of ecosystem services. European Commission Joint Research Centre, Luxembourg
Mair L, Thomas CD, Anderson BJ et al (2012) Temporal variation in responses of species to four decades of climate warming. Glob Chang Biol 18:2439–2447
Mantyka-Pringle C, Martin TG, Rhodes JR (2012) Interactions between climate and habitat loss effects on biodiversity: a systematic review and meta-analysis. Glob Chang Biol 18:1239–1252
Memmott J, Craze P, Waser N et al (2007) Global warming and the disruption of plant-species interactions. Ecol Lett 10:710–717
Millennium Ecosystem Assessment (2005) Ecosystems and human well-being. Biodiversity synthesis. World Resources Institute, Washington, DC
Müller A, Krebs A, Amiet F (1997) Bienen—Mitteleuropäische Gattungen, Lebensweisen, Beobachtungen. Weltbild Verlag, Augsburg
Naeem S (1998) Species redundancy and ecosystem reliability. Conserv Biol 12:39–45
OECD-FAO Organisation for Economic Co-operation and Development-Food and Agriculture Organization of the United Nations (2013) OECD-FAO Agricultural Outlook 2013–2022. OECD Publishing, Paris
Opdam P, Wascher D (2004) Climate change meets habitat fragmentation: linking landscape and biogeographical scale levels in research and conservation. Biol Conserv 117:285–297
Parmesan C (2006) Ecological and evolutionary responses to recent climate change. Annu Rev Ecol Evol Syst 37:637–669
Parmesan C, Yohe G (2003) A globally coherent fingerprint of climate change impacts across natural systems. Nature 421:37–42
Pfiffner L, Luka H (2000) Overwintering of arthropods in soils of arable fields and adjacent semi-natural habitats. Agric Ecosyst Environ 78:215–222
Reddy P, Verghese A, Rajan V et al (2012) Potential impact of climate change on honeybees (*Apis* spp.) an their pollination services. Pest Manag Hort Ecosyst 18:121–127
Roy D, Sparks T (2000) Phenology of British butterflies and climate change. Glob Chang Biol 6:407–416
Scheifinger H, Menzel A, Koch E et al (2002) Atmospheric mechanisms governing the spatial and temporal variability of phenological phases in central Europe. Int J Climatol 22:1739–1755
Schoener TW (1989) The ecological niche. In: Cherrett JM, Bradshaw AD, Goldsmith FB et al (eds) Ecological concepts. The contribution of ecology to an understanding of the natural world. Blackwell, Oxford, pp 79–113
Schröter D, Cramer W, Leemans R et al (2005) Ecosystem service supply and vulnerability to global change in Europe. Science 310:1333–1337
Schweiger O, Biesmeijer J, Bommarco R et al (2010) Multiple stressors on biotic interactions: how climate change and alien species interact to affect pollination. Biol Rev 85:777–795
Settele J, Kudrna O, Harpke A et al (2008) Climatic risk atlas of European butterflies. Pensoft, Sofia
Sinervo B, Méndez-De-La-Cruz F, Miles DB et al (2010) Erosion of lizard diversity by climate change and altered thermal niches. Science 328:894–899
Sinervo B, Miles DB, Martínez-Méndez N et al (2011) Response to comment on "Erosion of lizard diversity by climate change and altered thermal niches". Science 332:537–538
Srivastava DS, Vellend M (2005) Biodiversity-ecosystem function research: is it relevant to conservation? Annu Rev Ecol Evol Syst 36:267–294
Statistik Austria (2009) Statistik der Landwirtschaft. Statistik Austria, Wien
Stearns S (1992) The evolution of life histories. Oxford University Press, Oxford
Stefanescu C, Penuelas J, Filella I (2003) Effects of climatic change on the phenology of butterflies in the northwest Mediterranean Basin. Glob Chang Biol 9:1494–1506
Stireman JO III, Dyer LA, Janzen DH et al (2005) Climatic unpredictability and parasitism of caterpillars: implications of global warming. Proc Natl Acad Sci 102:17384–17387
Thomas CD, Cameron A, Green RE et al (2004) Extinction risk from climate change. Nature 427:145–148

Thomas C, Franco A, Hill J (2006) Range retractions and extinction in the face of climate warming. Trends Ecol Evol 21(8):415–416

Thuiller W, Araújo MB, Pearson RG et al (2004) Uncertainty in predictions of extinction risk. Nature 430. doi: 10.1038/nature02716

Tobin P, Nagarkatti S, Loeb G et al (2008) Historical and projected interactions between climate change and insect voltinism in a multivoltine species. Glob Chang Biol 14:951–957

Tylianakis JM, Rand TA, Kahmen A et al (2008) Resource heterogeneity moderates the biodiversity-function relationship in real world ecosystems. PLoS Biol 6:947–956

Van Buskirk J, Willi Y (2004) Enhancement of farmland biodiversity within set-aside land. Conserv Biol 18:987–994

Van Grunsven RHA, van der Putten WH, Bezemer TM et al (2010) Plant-soil interactions in the expansion and native range of a poleward shifting plant species. Glob Chang Biol 16:380–385

Van Zeijts H, Overmars K, van der Bilt W et al (2011) Greening the common agricultural policy: impacts on farmland biodiversity on an EU scale. PBL Netherlands Environmental Assessment Agency, The Hague

Vanbergen A, Baude M, Biesmeijer J et al (2013) Threats to an ecosystem service: pressures on pollinators. Front Ecol Environ 11:251–259

Voigt W, Perner J, Davis AJ et al (2003) Trophic levels are differentially sensitive to climate. Ecology 84:2444–2453

von Witzke H (2008) Agriculture, world food security, bio-energy and climate change: some inconvenient facts. Quart J Int Agric 47:1–4

Walther G, Berger S, Sykes MT (2005) An ecological 'footprint' of climate change. Proc R Soc Lond B 272:1427–1432

Wilson R, Gutiérrez D, Gutiérrez J et al (2005) Changes to elevation limits and extent of species ranges associated with climate change. Ecol Lett 8:1138–1146

Yudelman M, Ratta A, Nygaard D (1998) Pest management and food production: looking into the future. International Food Policy Research Institute, Washington, DC

Zarnetske PL, Skelly DK, Urban MC (2012) Biotic multipliers of climate change. Science 336:1516

Chapter 11
Human Health

Willi Haas, Ulli Weisz, Philipp Maier, and Fabian Scholz

Abstract There are manifold pathways by which climate change affects human health. Most directly, temperature increases will bring fewer deaths from cold on a global scale. However, despite temperature increases, single cold events might occur at the same time, mainly threatening humans in regions like Southern Europe, which are not well adapted to cold conditions. For Austria, cold-related deaths play a minor role. In contrast, the risk of dying due to increasing temperatures and heat waves in summer is growing significantly in the future. Estimates for Austria for three climate and three socioeconomic scenarios excluding adaptation forecast roughly 600–3,000 deaths or 7,000–32,000 "years of life lost due to premature mortality" for the 2050s. Impacts are three times more sensitive to varying climatic than to varying socioeconomic scenarios. Amongst indirect health effects, Salmonellosis cases have been declining steadily due to EU-wide programs. It is highly uncertain if and to what extent climate change will slow down the effectiveness of these programs in future. Allergy as another indirect health effect will be enhanced by climate change due to the increasing spread of Ragweed, which is a potent source of allergens. A study shows that the increase in treatment costs is about ten times higher than the implementation costs of appropriate management plans in the case of Austria up to 2050.

11.1 Introduction

The fact that climate conditions and their variability have wide-ranging impacts on human health (e.g. Haines et al. 2006; McMichael 2011; WHO 2009) is not disputed. Furthermore, substantial evidence indicates that human-induced climate change already causes adverse health effects (Allison et al. 2009; Confalonieri et al. 2007). As climate change becomes more severe, impacts on human health will steadily intensify in the foreseeable future. This has led to a growing interest within

W. Haas (✉) • U. Weisz • P. Maier • F. Scholz
Institute of Social Ecology, Alpen-Adria-Universität Klagenfurt|Wien|Graz, Vienna, Austria
e-mail: willi.haas@aau.at; ulli.weisz@aau.at; phmaier@edu.uni-klu.ac.at; fabian.scholz@gmx.at

K.W. Steininger et al. (eds.), *Economic Evaluation of Climate Change Impacts*, Springer Climate, DOI 10.1007/978-3-319-12457-5_11

and beyond the public health community in better understanding which mechanisms linking climate change and health are potentially at work (Haines et al. 2006). Such insights will help to improve assessments of the future disease burden induced by climate change and subsequently of related monetary effects (Stern 2006).

Interlinkages between climate change and health are considered to be wide-ranging and complex (Ciscar et al. 2010; Costello et al. 2009). Direct negative effects comprise temperature-related morbidity and mortality and the impacts of extreme weather events (i.e. heat waves, droughts, heavy precipitation, floods, storms and tropical cyclones even outside the directly exposed areas; IPCC 2012, cf. Schubert et al. 2008). Indirect health effects include climate-sensitive communicable diseases (such as water-, food- and vector-borne diseases), non-communicable diseases (e.g. allergies or diseases caused by air pollution associated with ground level ozone (Confalonieri et al. 2007) or ozone layer depletion) or food and water shortages and further health consequences in the aftermath of catastrophic events (e.g. post-traumatic stress disorder) (see Fig. 11.1).

Before identifying the potentially most severe health effects to be discussed in this chapter, the possibly beneficial health effect of reduced cold-related deaths deserves attention too. The mechanisms linking cold stress and mortality are more complex than those linking heat stress and mortality and furthermore are not fully understood (e.g. Laschewski and Jendritzky 2002). Within international literature, different aspects of cold-related health effects are discussed in the context of climate change. According to AR 4 IPCC, there is a high confidence that climate change will result in fewer deaths from cold on a global scale (Confalonieri et al. 2007, p. 393), due to decreasing number of cold days and nights (IPCC 2013, p. 3). Regarding single extreme episodes [i.e. cold events, which will

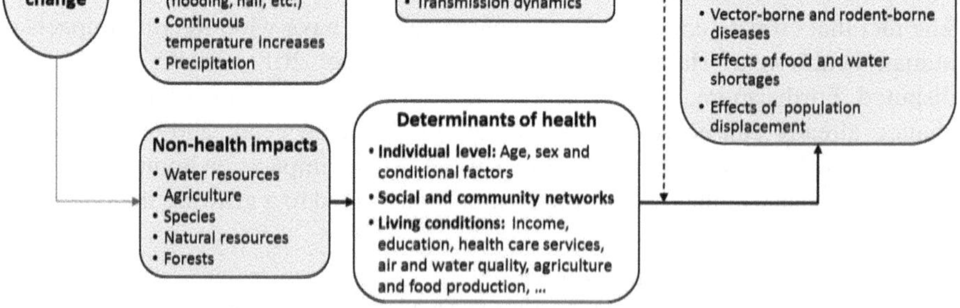

Fig. 11.1 Pathways through which climate change may affect health, modified from Confalonieri et al. (2007), McMichael et al. (2004) and Patz et al. (2000) (for a comprehensive list of known effects of weather and climate variability see Sect. 11.1 of the supplementary material)

Table 11.1 Number of people killed per 10,000 due to extreme weather events and wild fire for Northern, Western Europe and all European regions (1980–2011), EEA (2012) based on EM-DAT and World Bank

	Flood and wet mass movement (including landslides)	Cold event	Heat wave	Storm	Wildfire
Northern Europe	0.10	0.12	0.34	0.41	0.00
Western Europe	0.27	0.06	18.76	0.37	0.02
All European regions	2.41	2.68	41.24	1.16	0.22

Western Europe: Austria, Belgium, France, Germany, Luxembourg, the Netherlands and Switzerland

Northern Europe: Denmark, Estonia, Finland, Iceland, Ireland, Latvia, Lithuania, Norway, Sweden, United Kingdom

Total Europe: eu27 countries plus Albania, Bosnia and Herzegovina, Croatia, former Yugoslav Republic, Iceland, Macedonia, Montenegro, Norway, Serbia and Turkey

continue to occur; IPCC (2013, p. 18), cf. IPCC (2012)], it is argued that mortality will be highest within countries that are not well adapted to cold conditions (cf. Eurowinter 1997, cf. Healy (2003) for a comparison of winter death rates among European countries). Furthermore, there is evidence that cold waves can lead to increased mortality even in countries well adapted to cold conditions, if sufficient heating is not provided (e.g. if heating systems fail (Confalonieri et al. 2007), or where heating is unaffordable for poor population groups). Increasing rates in cold deaths in some European countries such as the UK can be understood in this context (Llyod 2013). Poor socioeconomic and infrastructural conditions (i.e. obsolete and inefficient heating systems, including poor insulation of buildings) put mainly the elderly already suffering from chronic diseases like respiratory problems or cardiovascular diseases at risk (Llyod 2013). Possibly due to better building and heating standards, Austria is not affected by this kind of negative cold-related health effect. Table 11.1 shows the minor role of cold-deaths in Western Europe in the past (including Austria).

There is broad consensus that the overall balance of effects on health will be negative (Confalonieri et al. 2007; Haines et al. 2006; WHO 2009). Primarily poorer regions of the world and poorer population groups within wealthier regions will be hit hardest due to their limited adaptive capacities (Haines et al. 2006; Schubert et al. 2008). However, wealthy states will also be challenged. The diverse health impacts "... will have economic consequences through incurring medical treatment costs and health protection costs, the potential loss of work productivity,[1] as well as welfare changes" (Ciscar et al. 2010, p. 6; see also Confalonieri et al. 2007; Watkiss and Hunt 2012).

"The experience of the 2003 heat wave in Europe shows that high-income countries may also be adversely affected" (Haines et al. 2006, p. 585). The 2003

[1] "Loss of work productivity" is discussed in Chap. 16 (Manufacturing), not in this chapter.

heat wave has been described as one of the worst heat wave events in recent history (UNEP 2004), resulting in around 70,000 heat-related deaths, mainly in western and central Europe (Barriopedro et al. 2011). In Eastern Europe and large parts of Russia, the exceptionally warm summer of 2010 caused adverse impacts that exceeded even the amplitude and spatial extent of the previous hottest summer of 2003. "Mega-heat waves" such as the 2003 and 2010 events likely broke the 500-year-long seasonal temperature records over approximately 50 % of Europe. Further analysis showed that the probability of a summer experiencing "mega-heat waves" will increase by a factor of 5–10 within the next 40 years (Barriopedro et al. 2011).

Such past impacts and the present understanding of future developments shape the focus of this chapter. Amongst the numerous possible pathways by which climate change might affect human health, heat waves seem to be by far the most severe threat to human health in Western and Southern Europe, especially when employing the assumption that adaptation to climate change will not take place. In the case of extreme weather events, Table 11.1 gives an overview of fatal incidents in Europe for the period from 1980 to 2011 (EEA 2012).

Therefore, one focus of this chapter is on temperature-related mortality caused by heat waves. For this climate change-induced health effect, we estimate the death toll and burden of diseases according to varying assumptions regarding climate change and socioeconomic development.

Further choices of the most severe health effects are far more difficult to make since they are highly determined by research gaps, lack of comparability between studies, and data availability. Against this backdrop, in Sect. 11.4.8, we discuss examples of indirect health effects with reference to literature. These comprise climate-sensitive communicable diseases (especially the case of food-borne diseases involving salmonella infection) and allergies caused by aeroallergens (pollen) as an example for climate-sensitive non-communicable diseases; here we focus on the well-studied accelerating spread of Ragweed. There are two reasons for this selection: The examples are representative of a wide range of possible impacts related to indirect pathways, while rough estimates that can be used to discuss the situation in Austria are available in literature. In the complex interaction between climate change and health, it remains unclear whether these health effects will be significant or if other pathways might become more relevant in future (see Fig. 11.1).

11.2 Dimensions of Sensitivity to Climate Change

11.2.1 Climatic Factors

Heat waves in Western and Southern Europe (especially France) in 2003 and in Eastern Europe (especially Russia) in 2010 have demonstrated what impacts high temperatures can have on daily mortality. Thus, heat waves as consecutive days with high maximum and minimum temperatures are crucial climatic factors. Besides heat waves, continuous temperature increases in summer months also cause adverse health effects.

Increased temperature interferes with human health due to heat stress, heat stroke, dehydration, skin eruptions, heat fatigue, heat cramps or heat syncope. If the body heat load from increased temperature is not reduced, it may cause heat stroke, which can lead to damage to cellular structures. Heat stroke, for example, can cause adult respiratory distress syndrome, kidney failure or liver failure (Koppe et al. 2004).

11.2.2 Non-climatic Factors

Climate sensitivity is influenced by non-climatic factors like age, sex and conditional factors at the individual level, but also by the embeddedness of a person in social and community networks as well as by living conditions like basic environmental conditions, income and access to good health services [based on the concept of health determinants: Dahlgren and Whitehead (1991)]. Applying this concept shows that there are several factors that make humans especially sensitive to climate change. One factor is old age; based on past events, there is ample evidence that people aged 65 and above are especially affected (e.g. Baccini et al. 2008; Kovats et al. 2011; Moshammer et al. 2006). Consequently, the demographic composition of a population matters when it comes to climate sensitivity (for Austria see Table 11.4).

Another factor concerns weak health conditions. Thus, chronic illness, the use of medication (adversely affecting thermoregulation) or previous hospital admission (low fitness level) increase susceptibility to heat-related mortality or morbidity. Particularly pre-existing health problems as mental illness, cardiovascular or respiratory diseases and being overweight increase the risk of heat-related health hazards (Koppe et al. 2004). Actual evidence from different countries is available, but due to the complexity and diversity of the evidence base on case level, no impact function at country level has been developed so far.[2]

---

[2] For a discussion of further social factors and potential large-scale damage combinations including urban heat islands please refer to Sect. 11.2 of the supplementary material.

11.3 Exposure to Climatic Stimuli and Impacts to Date

11.3.1 Past and Current Climatic Exposure and Physical Impacts

Europe during the last three decades experienced several rather severe heat waves, with approximately 19 deaths per 10,000 people in Western Europe (see Table 11.1).

Robine et al. (2008) investigated daily mortality during summer 2003 for 16 European countries at the NUTS 2 level. The study compared the data for the number of daily deaths observed during the summer and the average number of deaths, noted on the same day, during the 5 years in the 1998–2002 reference period.

During the period from 3 to 16 August, 39,000 additional deaths were recorded in 12 European countries. In nine of these 12 countries, the ratio between the additional deaths compared to the mortality figures for the same period in the years 1998–2002 exceeded 10 % (see Table 11.2). According to Robine et al. (2008), Salzburg and Burgenland were hit the strongest out of nine provinces in Austria.

Triggered by the 2003 heat wave, three studies were undertaken specifically to investigate the situation in Austria. Matzarakis et al. (2011) carried out a biometeorological evaluation of heat-related mortality in Vienna by analysing the period from 1970 to 2007. This study clearly shows the increases of relative heat-related mortality over the duration of heat waves. Two further studies performed by different teams but with Moshammer as the same lead author (2006 and 2007) provided statistical analyses of mortality data; one for Vienna (cf. Hutter et al. 2007) and one for Upper Austria. For the period from 1990 to 2004 they identified an excess mortality of 1,510 deaths for the entire period and 7.33 deaths for a day during heat waves. For the purpose of comparison, the researchers calculated the daily mortality figures during waves of influenza, with an excess mortality of 9.00 per day (Moshammer et al. 2006).

Table 11.2 Excess mortality ratio (expressed as a percentage) for 3–16 August 2003 compared to same period in 1998–2001, Robine et al. (2008, p. 174)

Selected countries	Excess mortality ratio (%)
France	96.5
Portugal	48.9
Italy	45.4
Spain	41.2
Luxemburg	40.8
Germany	28.9
Switzerland	26.7
Belgium	21.6
Austria	12.6

Although the short-term effects of high ambient temperatures on mortality have been well documented, there is a lack of consistent data and related analysis about the effects on morbidity. While episode analyses of heat waves have documented a comparatively higher impact on mortality than on morbidity, morbidity (Michelozzi et al. 2009) is still a factor that deserves special attention. An evaluation study of hospital admissions by Michelozzi et al. (2009) between April and September in 12 European cities carefully discusses these effects. The study performed a time series analysis by looking at the relationship between maximum apparent temperature with a time lag of 0–3 days and daily hospital admissions for cardiovascular, cerebrovascular and respiratory causes by age and for an overall population of 25 million people. There was only a positive association for respiratory admissions, with some heterogeneity between cities. For a 1 °C increase in maximum apparent temperature above a threshold, respiratory admissions increased by 4.5 % in Mediterranean and 3.1 % in northern continental cities for the age group of 75 years and older. In contrast to respiratory causes, the association between cardiovascular and cerebrovascular admissions and high temperatures did not reach statistical significance.

Despite a poor understanding of the mechanisms in the relation of temperature and morbidity, scholars are confident in assuming that the impact of extreme heat events on respiratory admissions is expected to increase in European cities as a result of both progressive population aging and higher surface temperatures in polluted regions, which is associated with peak levels of ground level ozone and particular matter of less than 2.5 μm (Confalonieri et al. 2007; IPCC 2013).

11.3.2 Impact Chains up to Socioeconomic System

In the first section we described the diverse pathways by which climate change might affect human health. Here we want to identify what kind of damage climate change might cause to both quality of life and the economy.

The health effects we quantitatively investigate in this chapter are temperature-related diseases caused by heat waves. Salmonellosis as a food–borne disease and allergy-related health effects are not calculated, although potential impacts are discussed on the basis of literature (see Sect. 11.5).

In general, health effects might lead to morbidity or mortality (or a combination of both). Mortality can be expressed in recorded deaths and in "Years of Life Lost due to premature mortality" (YLL). The latter is a component of the so-called "Disability Adjusted Life Years" (DALYs). Due to a lack of data regarding morbidity effects caused by heat events, the results presented in this chapter refer only to excess mortality and related YLL.

Within these constraints, it is also possible to derive monetary values considering three components (Kovats et al. 2011; see also Fig. 11.2):

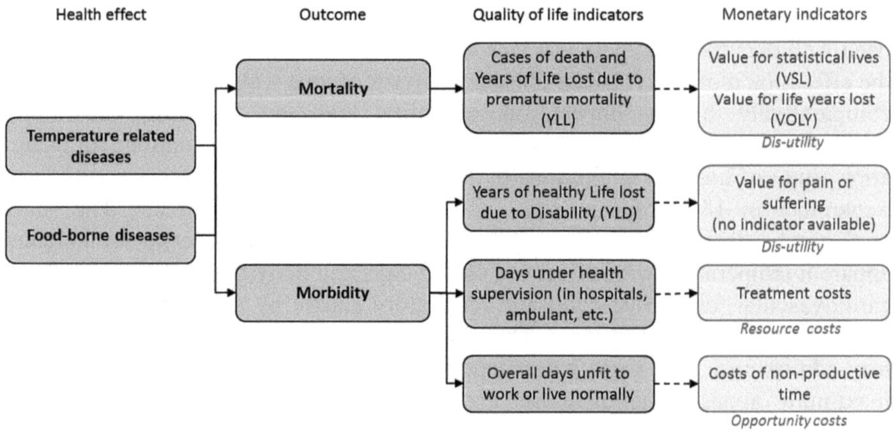

Fig. 11.2 Impact chain from health effects to impacts on the economy

- Resource costs i.e. medical treatment costs;
- Opportunity costs, in terms of lost productivity; and
- Dis-utility i.e. pain or suffering, concern and inconvenience to family and others.

Since this chapter discusses heat waves related to the mortality of old-aged people only, an economic valuation could neither calculate resource nor opportunity costs, but could only focus on the non-market related dis-utility. Here the valuation methods are rather uncertain and, depending on the method used, it provides results differing by a factor 10 (authors' estimates). We therefore decided to focus our evaluation on cases of death and "Years of Life Lost due to premature mortality" (YLL).

11.4 Future Exposure to and Impacts of Climate Change

This section mainly presents methods for and results of estimating health impacts of heat waves by considering different climate and socioeconomic scenarios for Austria for an average year during the two periods 2016–2045 and 2036–2065 (here we refer to these periods as 2030s and 2050s).

11.4.1 High and Low Range Climatic Scenarios for Human Health

For a better understanding of the range of climate change impacts on human health in Austria, three different climate scenarios were developed (low-, mid- and high-range). Regarding past periods the estimates are based on the regionally most

Table 11.3 Prognosticated Kyselý days for Austria: three different climate scenarios for 2030s and 2050s: low-, mid- and high-range; further Kyselý days for the base period 2003–2012, model calculations see Chap. 5

	Climate scenario		
	Low-range	Mid-range	High-range
2030s Kyselý days	8.2	7.1	14.1
2050s Kyselý days	8.1	16.0	26.7
Base period 2003–2012	5.6		

appropriate and most up-to-date study by Moshammer et al. (2006) which relates past heat waves to mortality in a traceable manner. This study used so-called 'Kyselý days' as a definition for heat waves[3]; for methodological compatibility we used the same definition. Jan Kyselý (2004) developed the definition for a long-term study of heat wave occurrence in the Czech Republic. According to his work, heat waves are defined as consecutive periods of at least three days during which the daily maximum temperature is $\geq$30 °C. The heat wave persists for as long as the average maximum temperature of the whole period remains above 30 °C and the daily maximum temperature never drops below 25 °C.

Table 11.3 illustrates the different climate scenarios and their future developments. The number of Kyselý days increases strongly over the next two decades in the low-range climate scenario but then tend to stagnate. In contrast, the mid-range climate scenario assumes a less marked increase in Kyselý days over the next two decades but a significant acceleration after 2030. The low-range scenario envisages an increase compared to the base period. The numbers are presented as the Austrian average. Due to Austria's very diverse topography, they vary significantly on NUTS 3 level.

11.4.2 Mid Range Climatic Scenario for Human Health

Figure 11.3 highlights the prognosticated Kyselý days (2030s and 2050s) for the mid-range climate scenario. The increase in Kyselý days is highly dependent on the geographical position of the observed area. Compared to the period 2003–2012, in an average year during the period 2036–2065, Eastern and Southern Austria experience an additional 10 or more Kyselý days. At the same time, areas that have previously not faced a significant number of heat waves might be exposed more severely in future: *Tiroler Oberland* and *Bludenz-Bregenzer Wald* might face an increase of more than 5–7 Kyselý days in the period 2036–2065, compared to the period 2003–2012.

[3] For a discussion of further heat wave definitions see supplementary material Sect. 11.3.

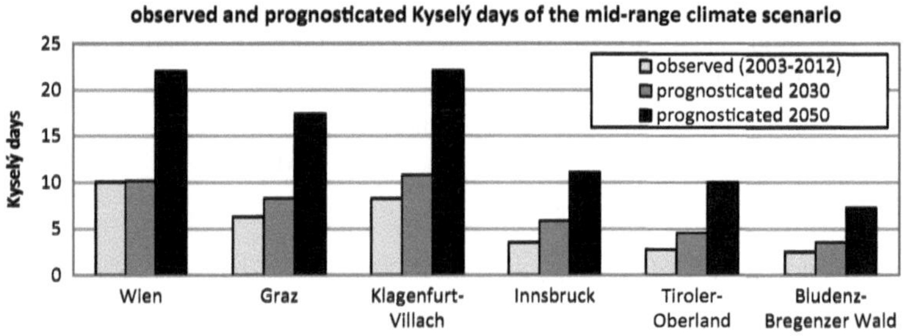

Fig. 11.3 Observed and prognosticated Kyselý days for the mid-range climate scenario for selected NUTS-3 areas, see Chap. 5

11.4.3 Range of Sectoral Socioeconomic Pathway Parameters that Co-determine Climate Impact

For the development of socioeconomic pathways, we focus on demographic structure and the related number of people aged 65 and above. Furthermore, we consider the diffusion of air-conditioning among this age group. Finally, we look into heat alert systems as adaptation measures that have already been identified (additional adaptation measures are excluded due to the overall assumption of the book).

In Austria, elderly people as the most sensitive group regarding climate change-related health risks are increasing in both number and share of the overall population. While in 2011 there were about 1.5 million people aged 65 and older, by 2050 this number will have increased to 2.6 million people in the main variant forecast of Statistics Austria (2012) (cf. Hanika 2011) (Table 11.4).

Since the diffusion of air conditioners can significantly mitigate death tolls, we formulate assumptions for the different socioeconomic scenarios. In the enhancing scenario for elderly people, no additional air conditioners are assumed. This can be seen as a consequence of unfavorable economic developments that reduce the economic capacity of this vulnerable group. In the diminishing scenario, it is assumed that 20 % of the population over 65 have access to air conditioning and hence reduce their risk by 50 %. Experts for infrastructure and buildings who have contributed to this book confirm the specific assumption related to the diffusion of air conditioners. The assessment of effectivity of air conditioning in preventing deaths caused by temperature increases is based on discussions in literature (e.g. Kovats et al. 2011; Ostro et al. 2010). For the reference scenario, we assumed that air conditioning was available to 10 % of the population over 65, with the same effectivity as in the diminishing scenario.

Spontaneous adaptation, which includes household air conditioning, reduced outdoor activities or changes in social and cultural habits as well as planned adaptation measures such as early warning systems can significantly alter the impacts of heat waves on human health. In Austria no superior national heat alert

Table 11.4 Comparison of Statistics Austria's demographic forecast with main variant, low and high variant of life expectancy for 2011, 2030 and 2050 with special attention to the age-group 65 years and over, Statistic Austria (2012)

	2011	2030	2050
High variant of life expectancy (enhancing scenario)			
Total	8,420,900	9,067,150	9,513,990
≥65	1,486,441	2,218,806	2,798,942
Main variant (reference scenario)			
Total	8,420,900	9,000,007	9,330,904
≥65	1,486,441	2,162,221	2,633,751
Low variant of life expectancy (diminishing scenario)			
Total	8,420,900	8,911,970	9,105,925
≥65	1,486,441	2,094,023	2,433,155

system has been implemented yet, but two provinces provide heat alerts since 2013.[4] However, these very recent developments are not considered in our estimates, since there is no assessment of the effectivity of these systems and analyses of past periods do not cover the years when they were already installed.

11.4.4 Specific Method(s) of Valuation and Their Implementation Steps

The following steps were applied to arrive at the results:

1. The estimates presented here are based on Moshammer et al. (2006). Hence we used the Austria-specific postulated (investigated) interrelationship between Kyselý days and excess mortality. The statistical analysis of Moshammer et al. revealed an excess mortality of 7.33 deaths per Kyselý day. By use of model calculation they concluded that about 50 % of these deaths are due to continuous temperature increases (Moshammer et al. 2006, p. 41,[5] see Table 11.5).

   A verification of the factors with the study for Upper Austria (Moshammer et al. 2007), which covers rural areas, revealed high accordance; on this basis, factors were applied for the whole of Austria.
2. **Projected Kyselý days** were taken from climate model results (see Chap. 5—Climate) for the observed period (2003–2012) and prognosticated data for 2030s and 2050s (see Table 11.6). Climate data were used on NUTS 3 level in accordance with the population assumptions. Thus special regional differences in temperature-related health effects can be considered, which is important because of the high altitude diversity in Austria. Three climate scenarios

[4] For further information on heat alert systems in Austria see Sect. 11.4 of the supplementary material.

[5] An Austrian study conducted after the 2003 heat wave investigated Vienna during the period from 1990 to 2004.

Table 11.5 Average excess mortality (cases per day) during so-called Kyselý days and the part of it that can be ascribed to continuous temperature increases resulting in Vienna during the period 1990–2004, Moshammer et al. (2006)[a]

	Heat wave (Kyselý day)	Continuous temperature increase only
Increase excess mortality (absolute)	7.33[b]	3.63[c]
Increase excess mortality (percentage)	15.75 %	7.80 %

[a]For a table of relative risk of a person dying on a Kyselý day please see supplementary material Sect. 11.5

[b]The statistical analysis was performed for 206 Kyselý days in the period 1990–2004. On these days in average 53.91 persons died, whereas on the remaining summer days (June to August) only 46.58 persons died. Thus, the difference of 7.33 per Kyselý day was taken as excess mortality (see Moshammer et al. 2006)

[c]Excess mortality per Kyselý day

(low-, mid- and high-range) as presented in Chap. 5 (Climate) were taken into account. Temperature-related health impacts were only calculated from June to August.

3. *Population*: As discussed above, we only considered the population fraction aged above 65. The applied data on NUTS 3 level (population 65+ between 1990 and 2004 as well as prognosticated population 65+ for 2030 and 2050) were taken from the Statistics Austria database (see Table 11.6). Three different population assumptions were take into account. The growth scenario implies a high fertility rate (1.95) and a high absolute number of the population aged 65 years and over. The low population scenario calculates on the contrary using a medium fertility rate (1.53) and a comparable lower share of population aged 65 years and above. The reference scenario combines the most common assumptions of Austria's population development and lies in between the two scenarios.
4. *Mortality by Kyselý day and increased temperature*: Two different calculations were performed: *Firstly,* excess mortality because of Kyselý days including continuous temperature increase and *secondly,* excess mortality solely because of continuous temperature increase. The first is more relevant to outlining climate change-related health impacts since it shows the overall effect, for which we present the excess mortality and the "Years of Life Lost due to premature death" (YLL).
5. *Life expectancy, excess mortality and YLL*: We used mortality tables for different ages to calculate the average number of life years lost for the group aged 65 years and over. Assumptions for future life expectancy were calculated once (main variant—referring to the Statistics Austria database) and applied to all different population scenarios. By means of the relation between impacts due to

Kyselý days on excess mortality, we calculated the specific YLL for 2030s and 2050s—respectively for each point in time adjusted with the corresponding life expectancy.[6]

6. *Calculations*: Each calculation was performed for an average year in both the 2030s and 2050s. The results table includes three different socioeconomic and three different climatic scenarios. Thereby we cover a wide range of possible future developments and provide a differentiated decision-making basis for policy makers. Physiological adaptation was not considered but will be (qualitatively specified) discussed within uncertainties. Socioeconomic scenarios include the different population scenarios as well as technical adaption.

Morbidity caused by heat waves was not calculated. Impacts can be expected to be much smaller than for mortality and uncertainties to be much higher. So far, no study for Austria has performed such estimates [Moshammer et al. (2006) collected data but did not analyse these]. Michelozzi et al. (2009) investigated 12 European cities including Ljubljana, Budapest and Paris. Results were not appropriate to enable estimates for Austria.

11.4.5 Evaluation of Health Impacts

Table 11.6 and Fig. 11.4 provide the range of results for all scenarios. They show that the number of deaths and the "Years of Life Lost due to premature death" (YLL) for 2030s, low-range climate assumption, are higher than for climate mid-range scenario. This is caused by the different climate scenarios. The low-range climate scenario implies a higher increase in temperature during the next two decades (thus more heat-related deaths) than the climate mid-range scenario. After 2030s it tends to stagnate, while the climate mid-range scenario tends to increase strongly, which is also shown in the results. Such a counter-intuitive effect is mainly created by climate variability, and the climate mid-range scenario turning out "mid-range" only at mid to end century, but cool before.

Altogether, results show that the climate scenario is far more significant than the socio-economic scenario: climate scenarios vary roughly by factor 3.7, socio economic scenarios by factor 1.3. The scenario combination with the lowest impact (low old-aged population + mid-range climate scenario + reduced mortality through technical adaptation) still reaches a significant value of 370 deaths or 3,600 YLL in 2030s. The highest impact is with the high-range climate and the enhancing

[6] e.g.: excess mortality 65+ for 2030: 638; we assumed a distribution of age groups amongst the death cases and calculated a remaining life expectancy at time of death of 9.66 years in average. Therefore excess mortality equals 6,159 "Years of Life Lost due to premature death" for 2030.

Table 11.6 2030 and 2050 estimates for annual deaths and "Years of Life Lost due to premature death" (YLL) regarding health effects due to heat waves for three climate and three socio-economic scenarios

			Annual effects	No climate change	Climate scenarios Low-range	Mid-range	High-range
2030	Socio-economic scenarios	Enhancing[a]	Deaths	380	640	430	1,200[2)]
			YLL	3,600	6,200	4,200	11,500
		Reference[b]	Deaths	360	580	400	1,100
			YLL	3,500	5,600	3,800	10,500
		Diminishing[c]	Deaths	350	540	370[1)]	1,010
			YLL	3,400	5,200	3,600	9,800
2050	Socio-economic scenarios	Enhancing[a]	Deaths	490	830	1,200	2,960[4)]
			YLL	5,200	8,800	12,800	31,500
		Reference[b]	Deaths	450	730	1,060[3)]	2,610
			YLL	4,800	7,700	11,300	27,800
		Diminishing[c]	Deaths	420	640	920	2,280
			YLL	4,500	6,800	9,800	24,200
Base period 2003–2012[d]			Deaths	240			
			YLL	2,300			

For comparison, a calculation that does not alter the climate signal but socioeconomic assumptions over future periods is added. Further health impacts for the baseline period are given as reference, authors' estimates

[a]Neither physiological acclimatisation nor technical adaptation are considered

[b]No physiological acclimatisation is considered. Technical adaptation: we assume that 10 % of the population aged 65+ reduce their risk by 50 % due to air conditioning

[c]No physiological acclimatisation is considered. Technical adaptation: we assume that 20 % of the population aged 65+ reduce their risk by 50 % due to air conditioning

[d]Since there is no country-wide account on excess mortality for Austria, the same impact function as for future scenarios is used for calculating the baseline period

socioeconomic scenario, which provides death tolls and YLL that are about three times higher than the lowest impact [see superscripts 1) and 2) in Table 11.6].

The "worst case scenario" implies a temperature-related mortality rate in 2030 which is nearly the same as calculated in the "average scenario" for 2050 [see superscripts 2) and 3) in Table 11.6]. It claims that the heat wave related mortality effect will cause 1,200 deaths and 11,500 YLL in 2030s. In 2050s, the value increases to 2,960 deaths [see superscript 4) in Table 11.6] and 31,500 YLL, which might pose a huge burden for Austria's socioeconomic system.[7]

[7] For an estimate of deaths on Kyselý days that can be ascribed to continuous temperature increase only please refer to supplementary material Sect. 11.6.

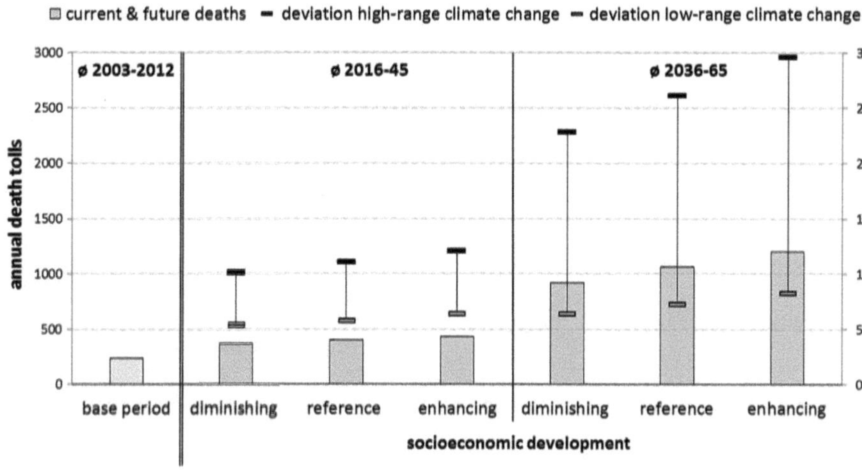

Fig. 11.4 2030s and 2050s estimates for annual deaths due to heat waves combining three climate and three socio-economic scenarios are shown. The base period shows the already existing excess mortality in the period 2003–2012, authors' estimates

11.4.6 Sector-Specific Uncertainties

Results need to be seen in the light of uncertainties involved. Therefore, Table 11.7 summarises the reliability of assumptions and approaches, which have been applied to carry out these estimates (see Risbey and Kandlikar 2007).

Although the different population scenarios are widely accepted in the scientific community, migration as a specific issue deserves some attention. Austrian in-migration depends highly on EU migration policies and migration pressure, which in turn is contingent on conditions in countries from which migrants come (economic and political developments as well as catastrophic and slow-onset disasters). While a huge scale of in-migration is not very likely, it is still a possible scenario that is not reflected in our population assumptions. This reduces the level of confidence.

Given the definition used, prognosticated Kyselý days are only dependent on temperature values, which amongst other climate signals are considered to be relatively reliable (for uncertainties of the climate scenarios used, please refer to Chap. 5).[8]

Estimates for technical adaptation are connected with higher uncertainties. Planned or spontaneous technical adaptation to heat stress through better housing insulation, air conditioning, new urban planning policies or the promotion of retirement homes (to reduce the number of elderly living alone) can mitigate health impacts of temperature increases. Since exploiting technical potential depends

[8] For discussion of methodological underestimation of effects by using Kyselý days and overestimation due to neglecting physiological adaptation see supplementary material Sect. 11.7.

Table 11.7 Level of confidence and agreement for different assumptions and concepts used in the estimation for health effects due to heat waves

Terminology	Level of confidence	Level of agreement	Sources
Population assumptions	Middle	High	Statistics Austria
Prognosticated Kyselý days	High/ Middle	Middle	See Chap. 5
Kyselý days as lead indicator	Middle	Middle	Kyselý (2004) and Moshammer et al. (2006)
Physiological adaptation	Low	Low	Dessai (2003), Watkiss et al. (2009) (peseta-study), empirical hints but no reliable evidence
Extrapolation of case studies for Austria	Middle	Middle	Socioeconomic and cultural aspects: not based on specific studies; climate aspects: climate aspects: model results at regional level (NUTS-3)
Technical adaptation	Low	Middle	Chapter infrastructure and buildings (problematic because it depends highly on uncertain economic developments in future)

highly on factors at individual as well as societal level, which are uncertain in themselves, assumptions here are rated with a low confidence level.

Finally we want to compare our estimations with results from other European studies. Watkiss and Hunt (2012) provide summaries for two temperature-related impact functions. This comparison excludes heat extremes. While our results related to temperature increases only (excluding Kyselý days) range roughly between 260 and 560 deaths for 2030s, specific figures provided by Paul Watkiss are between 180 and 560 deaths depending on climate, socioeconomy and acclimatisation assumptions for Austria for 2011–2040. Another study from Kovats et al. (2011) provides estimates for 2020s of 385 deaths and for 2050s of 1,414 deaths, compared to our results of 400 deaths for 2030s and 1,060 deaths for 2050s (reference and mid-range scenario combination, see Table 11.7). Despite different methods, these results can be seen to be similar in range and differences might stem from differences in climatic and socioeconomic scenarios as well as how physiological adaptation is treated.

11.4.7 Estimating Extremes

To estimate the effect of extremely hot years, years with a return period of 20 years (95 %) of a mid-range climate change scenario have been selected for both periods. In such hot years the number of Kyselý days increases to 59 for the period 2015–2045 (before 8–14) and to 77 for the period 2036–2065 (before 8–27). Further it was estimated, how causalities change, if additionally to the group of people 65 years and over chronically ill are included (i.e. people suffering from chronic respiratory and cerebrovascular diseases cf. Moshammer et al. 2006; Baccini et al. 2008). Assuming that 10 %[9] of the population aged 20–64 are chronically ill and have the same sensitivity regarding heat stress than old persons, the number of deaths will more than double for extremely hot years of the period 2036–2065 compared to about 3,000 deaths of an average year of this period. It should be stressed that these estimates do neither account for prolonged nor for more intense heat waves. Each of these changes are regarded to lead to even more severe effects (D'Ippoliti et al. 2010, WHO and WMO 2012). Since no reliable impact functions are available for Austria no estimates are included here.

11.4.8 Indirect Health Effects of Climate Change

In this section we give insight into indirect effects on human health: climate-sensitive communicable diseases and here especially food-borne ones with Salmonella infection as a case-study and allergies caused by aeroallergens (pollen) as an example for climate sensitive non-communicable diseases (NCD). Here we refer to the well-studied accelerating spread of Ragweed. The section is based on literature, highlighting qualitative and quantitative aspects.

11.4.8.1 Climate-Sensitive Communicable Diseases: Food-Borne Diseases

Today it is widely anticipated that a changing climate will impact the spread of communicable diseases and in many cases these will pose new threats to public health (ECDC 2010). Estimations of future impacts (incidence rates) are based on epidemiologically derived functions for mortality and morbidity (impact relationships) of the respective disease. However, there is a lack of quantification studies for many health impacts, including food-borne diseases. Thus, they are limited to selected endpoints (see Watkiss and Hunt 2012; Kovats and Lloyd 2010).

Disease vectors (e.g., mosquitoes, sand-flies and ticks) as well as pathogens causing food- and water- borne disease are sensitive to climatic factors such as temperature and humidity (ECDC 2010).[10]

Many food-borne pathogens increase their growth rate at higher temperatures (ECDC 2010). Campylobacter and Salmonella are among the most important

---

[9] Due to lack of more specific data the assumption can be seen as rough estimation only. It is based on BMG (2009) and Statistics Austria (2008).

[10] For an overview of disease groups caused by pathogens and their possible links to climate variables please refer to supplementary material Sect. 11.8.

causes of food-borne diseases (gastrointestinal infections) in developed countries (see ECDC 2012; Kornschober et al. 2009; Kovats et al. 2004, 2005). Incidence rates of Salmonellosis have been correlated to increasing temperatures showing a seasonal pattern with peaks in the summer months (Kovats et al. 2004) whereas the main driver of seasonality of Campylobacter remains elusive (Kovats et al. 2005). Due to this fact, in the following section we focus on Salmonellosis.

Salmonellosis

According to the European Centre for Disease Control and Prevention, European Salmonellosis remained the second most commonly identified gastrointestinal disease across the European Union (after Campylobacter, which has increasing incidence rates). Its occurrence has been declining steadily since 2004, mainly due to EU programs (e.g. control programs in poultry farms).[11] However, Salmonella continues to be the source of many outbreaks, both within and between countries (ECDC 2012) and there are estimates for European countries which show an increase of cases of up to 50 %.[12]

In Austria, Salmonellosis cases show remarkable decline due to the introduction of vaccination for laying hens and broilers, together with intensified outbreak investigation efforts (Kornschober et al. 2009; NRCS 2013). Climate change could slow down these improvements, however, to a degree that is difficult to predict since both future effectiveness of programs and climate change-induced effects on Salmonella cases for Austria are highly uncertain.

11.4.8.2 Climate Sensitive Non-communicable Diseases: Allergies

Pollen quantity and seasonality depend on climatic variables. The observed earlier onset of the spring pollen season in the northern hemisphere as well as the introduction and further spread of new invasive plants species (neophytes) with highly allergenic pollen (aeroallergens), in particular Ragweed (Ambrosia artemisiifolia), present important health risks which are attributable to climate change (Confalonieri et al. 2007). Further, as pollen is an important trigger for some types of asthma, the global increase of asthma incidents[13] is discussed in the context of a changing climate (Beggs and Bambrick 2005).

---

[11] For a graph on trends and numbers of reported confirmed cases of Salmonellosis in EU/EEA countries see supplementary material Sect. 11.9.

[12] For a summary of causes of Salmonellosis and an estimate of future climate change-induced increases of Salmonellosis cases in Europe see supplementary material Sect. 11.10.

[13] In 2008 respiratory diseases, including asthma and chronic obstructive pulmonary disease (COPD), were the third leading cause of NCD deaths worldwide and a major cause of disability (WHO 2011, cf. Lozano et al. 2012).

Ragweed

A rapid spread of Ragweed is observed in several parts of the world, including Europe and Austria (Confalonieri et al. 2007; Essl et al. 2009). Its pollen is highly allergenic; today 4–5 % of Europeans are sensitised, a trend which is expected to worsen in future. A recently published study by Richter et al. (2013) simulated the future spread of this plant in Austria and Bavaria (Southern Germany) under different climate assumptions. The results for the more extreme climate scenario show that without adaptation mean treatment costs for allergy range from about 290–365 million euros annually up to 2050. This would offset the annual adaption costs of about 30 million euros by a factor 10 (cf. Beggs and Bambrick (2005) for Switzerland).

11.5 Summary of Climate Impacts for Health and Conclusions

Pathways by which climate change affects health are considered to be both wide-ranging and complex. Furthermore, the specific emergence of phenomena depends highly on geographic area, population density and degree of industrialisation.

In this chapter we focussed on Europe, using Austria as a case study for estimations and discussions of heat waves which seem to be the most severe threat to human health. However, estimating future impacts requires some simplifications to be made, depending on the availability of cases analysed. In our estimates we focussed on people aged 65 and older as well as on mortality. Estimates for morbidity were not possible. Both the empirical data base and estimation methods still lack the requisite level of maturity. Against this background, results presented in Table 11.6 are quite significant, with 640 to almost 3,000 deaths and 6,800–31,500 life years lost (YLL), depending on the scenario combination for 1 year in the 2050s. Results show that the considered variation in the climate signals matters far more than the considered socioeconomic assumptions (given our assumptions the result is about three times more sensitive to climate than to socioeconomic conditions).

An estimate for extremely hot years with an extension of the vulnerable groups to chronically ill the number of deaths will more than double for the period 2036–2065 compared to about 3,000 deaths in an average year of this period.

In the case of temperature-related mortality of old-aged people, monetisation is highly uncertain, since the calculation depends only on dis-utility (value of a life year lost or of a life lost). Besides ethical concerns, results depend highly on the specific valuation; compared methods show a difference by factor 2, different assumptions reveal a difference by factor 14.

However, given the potential health effects, monitoring of temperature-related mortality and morbidity is required. Further to general heat information systems, more tailored approaches need to be developed to care for potentially vulnerable people ranging from old-aged persons to people with poor health conditions, especially when accompanied by weak ties to social networks. Furthermore, outdoor and indoor heat islands need to be both identified and mitigated.

There are many other pathways by which health might be affected. We limited our discussion to climate-sensitive communicable diseases with an emphasis on food-borne diseases (Salmonellosis) and on allergic health effects (caused by Ragweed).

Against the background that effective control and prevention programs have reduced numbers of cases significantly in Austria within one decade, it is likely that these will lead to further decreases in future. Climate change could slow down these improvements, however, this is to a degree difficult to predict since both the future effectiveness of programs and climate change-induced effects on Salmonella cases for Austria are highly uncertain. Studies are required to better understand the potential effects of climate change in Austria.

Allergies as non-communicable disease are strongly related to quantity and seasonality of pollen, which themselves are dependent on climatic variables. Due to climate change, a rapid spread of highly allergenic Ragweed is observed in several parts of the world, including Europe and Austria. A recently published study estimates mean treatment costs for allergy within a range of about 290–365 million euros annually up to 2050. This would offset the annual adaption costs of about 30 million euros annually by factor 10.

References

Allison I (2009) The Copenhagen diagnosis. Updating the world on the latest climate science. The University of New South Wales, CCRC, Sydney

Baccini M et al (2008) Heat effects on mortality in 15 European cities. Epidemiology 19(5):711–719

Barriopedro D, Fischer EM, Luterbacher J, Trigo RM, García-Herrera R (2011) The hot summer of 2010: redrawing the temperature record map of Europe. Science 332:220

Beggs H, Bambrick J (2005) Is the global rise of asthma an early impact of anthropogenic climate change? Environ Health Perspect 113:915–919

BMG (ed) (2009) Gesundheitsbericht Österreich 2009. Verfasst von Gesundheit Österreich GmBH, Wien

Ciscar JC et al (2010) Physical and economic consequences of climate change in Europe. In: Schellnhuber H-J (ed) PNAS (Early edition). Potsdam Institute, Potsdam

Confalonieri U, Menne B, Akhtar R, Ebi KL, Hauengue M, Kovats RS, Revich B, Woodward A (2007) Human health. Climate change 2007: impacts, adaptation and vulnerability. In: Parry ML, Canziani OF, Palutikof JP, van der Linden PJ, Hanson CE (eds) Contribution of working group II to the fourth assessment report of the intergovernmental panel on climate change. Cambridge University Press, Cambridge, pp 391–431

Costello A et al (2009) Managing the health effects of climate change. Lancet 373:1693–1733

Dahlgren G, Whitehead M (1991) Policies and strategies to promote social equity in health. Institute for Future Studies, Stockholm

Dessai S (2003) Heat stress and mortality in Lisbon Part II. An assessment of the potential impacts of climate change. Int J Biometeorol 48:37–44

D'Ippoliti D et al (2010) The impact of heat waves on mortality in 9 European cities: results from the EuroHEAT project. Environ Health 9:37

ECDC European Center for Disease Control and Prevention (2010) Climate change and communicable diseases in the EU Member States Handbook for national vulnerability, impact and adaptation assessments. Technical document. ECDC, Stockholm

ECDC European Center for Disease Control and Prevention (2012) Annual epidemiological report 2012. Reporting on 2010 surveillance data and 2011 epidemic intelligence data. Surveillance report. ECDC, Stockholm

EEA European Environment Agency (2012) Climate change, impacts and vulnerability in Europe 2012 – an indicator-based report. EEA, Copenhagen

Essl F, Dullinger S, Kleinbauer I (2009) Changes in the spatio-temporal patterns and habitat preference of Ambrosia artemisiifolia during its invasion of Austria. Preslia 80:119–133

Eurowinter (1997) Cold exposure and winter mortality from ischaemic heart disease, cerebrovascular disease, respiratory disease, and all causes in warm and cold regions of Europe. Lancet 349:1341–1346

Haines A, Kovats RS, Campbell-Lendrum D, Corvalan C (2006) Climate change and human health: impacts, vulnerability and public health. Public Health 120:585–596

Hanika A (2011) Kleinräumige Bevölkerungsprognose für Österreich 2010 bis 2030 mit Ausblick bis 2050 („ÖROK-Regionalprognosen") Teil 3: Endbericht zu den Modellrechnungen zur regionalen Haushaltsentwicklung. ÖROK, Wien. http://www.oerok.gv.at

Healy JD (2003) Excess winter mortality in Europe: a cross country analysis identifying key risk factors. J Epidemiol Community Health 57:784–789

Hutter HP, Moshammer H, Wallner P, Leitner B, Kundi M (2007) Heatwaves in Vienna: effects on mortality. Wiener klinische Wochenschrift 119(7–8):223–227

IPCC (2012) Managing the risks of extreme events and disasters to advance climate change adaptation. In: Field CB, Barros V, Stocker TS, Qin D, Dokken DJ, Ebi KL, Mastrandrea MD, Mach KJ, Plattner GK, Allen SK, Tignor M, Midgley PM (eds) A special report of working groups I and II of the intergovernmental panel on climate change. Cambridge University Press, Cambridge

IPCC (2013) Summary for policymakers. In: Stocker TF, Qin D, Plattner G-K, Tignor M, Allen SK, Boschung J, Nauels A, Xia Y, Bex V, Midgley PM (eds) Climate change 2013: the physical science basis. Contribution of working group I to the fifth assessment report of the intergovernmental panel on climate change. Cambridge University Press, Cambridge

Koppe C, Kovats S, Jendritzky G, Menne B (2004) Heat-waves: risks and responses. Health and global environmental change, vol 2. World Health Organization, Copenhagen

Kornschober C, Mikula C, Springer B (2009) Salmonellosis in Austria: situation and trends. Wiener Klinische Wochenschrift 121(3–4):96–102

Kovats RS, Edwards SJ, Charron D, Cowden J, D'Souza RM, Ebi KL, Gauci C, Gerner-Smidt P, Hajat S, Hales S, Hernández Pezzi G, Kriz B, Kutsar K, McKeown P, Mellou K, Menne B, O'Brien S, van Pelt W, Schmid H (2005) Climate variability and campylobacter infection: an international study. Int J Biometeorol 49(4):207–14

Kovats RS, Edwards SJ, Hajat S, Armstrong BG, Ebi KL, Menne B (2004) The effect of temperature on food poisoning: a timeseries analysis of salmonellosis in ten European countries. Epidemiol Infect 132(3):443–53

Kovats RS, Lloyd S (2010) Health effects of climate change: a review of impact studies. ClimateCost project. London School of Hygiene and Tropical Medicine, London

Kovats S, Lloyd S, Hunt A, Watkiss P (2011) Technical policy briefing note 5: the impacts and economic costs on health in Europe and the costs and benefits of adaptation, results of the EC RTD ClimateCost project. In: Watkiss P (ed) The ClimateCost project, vol 1. Stockholm Environment Institute, Sweden

Kyselý J (2004) Mortality and displaced mortality during heat waves in the Czech Republic. Int J Biometeorol 49:91–97
Laschewski G, Jendritzky G (2002) Effects of the thermal environment on human health: an investigation of 30 years of daily mortality data from SW Germany. Clim Res 21:91–103
Llyod J (2013) Cold enough. Excess winter deaths, winter fuel payments and the UK's problem with the cold. The Strategic Society Center SSC, London
Lozano et al (2012) Global and regional mortality from 235 causes of death for 20 age groups in 1990 and 2010: a systematic analysis for the Global Burden of Disease Study 2010. Lancet 380 (9859):2095–128
Matzarakis A, Muthers S, Koch E (2011) Human biometeorological evaluation of heat-related mortality in Vienna. Theor Appl Climatol 105:1–10
McMichael A (2011) Insights from past millennia into climatic impacts on human health and survival. PNAS 109(113):4730–4737
McMichael AJ, Campbell-Lendrum D, Kovats S, Edwards S, Wilkinson P, Wilson T, Nicholls R, Hales S, Tanser F, Le Sueur D, Schlesinger M, Andronova N (2004) Global climate change. In: Ezzati M, Lopez A, Roders A et al (eds) Comparative quantification of health risks. Global and regional burden of disease attributable to selected major risk factors. World Health Organization, Geneva, pp 1543–1650
Michelozzi P, Accetta G, De Sario M, D'Ippoliti D, Marino C, Baccini M, Biggeri A, Anderson HR, Katsouyanni K, Ballester F (2009) High temperature and hospitalizations for cardiovascular and respiratory causes in 12 European cities. Am J Respir Crit Care Med 179(5):383–389
Moshammer H, Geresdorfer A, Geresdorfer T, Hutter H-P, Formayer H, Kromp-Kolb H, Leitner B (2007) Abschätzung der Auswirkung von Hitze auf die Sterblichkeit in Oberösterreich
Moshammer H, Hutter HP, Frank A, Geresdorfer T, Hlava A, Sprinzl G, Leitner B (2006) Einflüsse der Temperatur auf Mortalität und Morbidität in Wien. StartClim2005.A1a
NRCS Austrian National Reference Centre for Salmonella (2013) Jahresbericht. Nationale Referenzzentrale für Salmonellen. AGES IMED, Graz
Ostro O, Rauch S, Green R, Malig B, Basu R (2010) The effects of temperature and use of air conditioning on hospitalizations. Am J Epidemiol 172(9):1053–1061
Patz JA, McGeehin MA, Bernard SM et al (2000) The potential health impacts of climate variability and change for the United States: executive summary of the report of the health sector of the US national assessment. Environ Health Perspect 108:367–376
Richter R, Berger UE, St D, Essel F, Leitner M, Smith M, Vogl G (2013) Spread of invasive ragweed: climate change, management and how to reduce allergy costs. J Appl Ecol 50:1422–1430
Risbey JS, Kandlikar M (2007) Expressions of likelihood and confidence in the IPCC uncertainty assessment process. Clim change 85:19–31
Robine JM, Cheung SLK, Le Roy S, Van Oyen H, Griffiths C, Michel JP, Herrmann FR (2008) Death toll exceeded 70,000 in Europe during the summer of 2003. Epidemiol C R Biol 331:171–178
Schubert R, Schellnhuber HJ, Buchmann N, Epiney A, Grießhammer R, Kulessa M, Messner D, Rahmstorf S, Schmid J (2008) Climate change as a security risk. German Advisory Council on Global Change (WGBU)/Earthscan, London/Sterling, VA (German edition published 2007)
Statistics Austria (2008) Gesundheitsbefragung 2006/2007. Chronische Erkrankungen und Gesundheitsprobleme. www.statistik.at/web_de/statistiken/gesundheit/gesundheitszustand/chronische_krankheiten/index.html. Accessed 20 May 2014
Statistics Austria (2012) Vorausberechnete Bevölkerungsstruktur für Österreich 2011–2075. http://www.statistik.at/web_de/statistiken/bevoelkerung/demographische_prognosen/bevoelkerungsprognosen/index.html#index1. Accessed 14 Sept 2012
Stern N (2006) Stern review on the economics of climate change. Cambridge University Press, Cambridge
UNEP United Nations Environment Programme (2004) Early warning on emerging environmental threats: impacts of summer 2003 heat wave in Europe. UNEP, Nairobi

Watkiss P, Horrocks L, Pye S, Searl A, Hunt A (2009) Impacts of climate change in human health in Europe. PESETA-Human health study. JRC scientific and technical reports
Watkiss P, Hunt A (2012) Projection of economic impacts of climate change in sectors of Europe based on bottom up analysis: human health. Clim Change 112:101–126
WHO World Health Organization (2009) Protecting health from climate change. Connecting science, policy and people. WHO, Geneva
WHO World Health Organization (2011) In: Alwan A (ed) Global status report on non-communicable diseases 2010. WHO, Geneva
WHO and WMO (2012) Atlas of health and climate change. World Health Organization and World Meteorological Organization: WMO-No 1–98. WHO Press, Geneva

Chapter 12
Water Supply and Sanitation

Roman Neunteufel, Reinhard Perfler, Dominik Schwarz, Gabriel Bachner, and Birgit Bednar-Friedl

Abstract The Water Supply and Sanitation (WSS) sector is a complex system that involves all, specific and geographically bound natural water resources; vast and diverse technical infrastructure; and a strong nexus to lifestyle and consumer behaviour. Therefore the sensitivity to changes, including climate changes, originates from many levels.

We consider a baseline scenario that reflects changes due to socioeconomic and demographic changes as well as a climate change scenario that reflects additional changes due to climate change. Based on changes of units like changes in final demand, new built assets, enlargements, or replacement of assets we attempt to give cost estimates for the WSS sector until 2050 (under the differentiation of the causal nexuses and exemplarily based on empirical data). Based on the estimated costs for the WSS sector macroeconomic effects are calculated, including spill-over effects to other sectors, as well as effects on welfare, GDP and public budgets. Note that both scenarios are subject to various assumptions and considerably high uncertainties and therefore the underlying results must be interpreted with care.

We show that an increase of infrastructure damages in the WSS sector will be mainly caused by floods or landslides due to intense precipitation events. Even higher impacts will originate from changed production costs (e.g. treatment effort, operation and maintenance etc.) due to climate change as more assets and labour will be needed to provide the same service as today or to meet an additional climate change induced consumer demand. In total, the adaptation to socioeconomic and demographic changes will be the bigger challenge than the adaptation to climate

R. Neunteufel (✉) • R. Perfler • D. Schwarz
Institute of Sanitary Engineering and Water Pollution Control, BOKU—University of Natural Resources and Life Sciences, Vienna, Austria
e-mail: roman.neunteufel@boku.ac.at; reinhard.perfler@boku.ac.at; dominik.schwarz@boku.ac.at

G. Bachner
Wegener Center for Climate and Global Change, University of Graz, Graz, Austria
e-mail: gabriel.bachner@uni-graz.at

B. Bednar-Friedl
Wegener Center for Climate and Global Change, University of Graz, Graz, Austria

Institute of Economics, University of Graz, Graz, Austria
e-mail: birgit.friedl@uni-graz.at

K.W. Steininger et al. (eds.), *Economic Evaluation of Climate Change Impacts*, Springer Climate, DOI 10.1007/978-3-319-12457-5_12

changes. However, the costs of climate change will only add up to the total costs for each customer. Despite of all uncertainties involved, investigations on the effects of climate change suggest that there will be hardly any benefits but a lot of different costs for the WSS sector.

In order to adapt the long-living assets of WSS sector in an efficient way, more and early information on the impacts and their magnitudes on the sector will be needed.

12.1 Introduction

Besides the unquestioned influence of socioeconomic and demographic parameters the influence of weather, and thus of climate change, on the WSS sector is manifold.

For example the natural water resources—especially some types of springs—are very sensitive to changes in the distribution of precipitation. This is with regard to quantity and quality. A lack of precipitation during the summer season can lead to lower yields of spring water. Furthermore a long duration of dry periods can cause drying cracks in the soil enabling pollution to enter deeper soil layers and finally to get into the water body when precipitation sets in. The water supply demand side has a strong nexus to weather and climate. Seasonal characteristics—the actual weather and its variability—have a clear influence on the consumer's behaviour. On the sanitation side an increased variability of the weather conditions—with regard to increased frequency or intensity of heavy precipitation events—will lead to an increased number of Combined Sewer Overflow (CSO) events and thus to a higher total pollution of the environment.

Furthermore water supply and sanitation infrastructures are sensitive to damages due to extreme events and natural disaster. Damage, including cleaning (removal of sediments from blocked sewer systems), disinfection of water supply assets, damage of machinery and electronic equipment etc. is following mainly two mechanisms: (i) flooding after intense, long-lasting precipitation events and (ii) landslides after extreme heavy precipitation events.

The combination of changing weather/climate parameters together with physical assets of the WSS sector as well as influences of human behaviour results in an extensive matrix of quantifiable, semi-quantifiable and non-quantifiable impact chains.

The Austrian water supply sector consists of a very high number of municipal (about 2,000) and very small co-operative (in total another 3,500) utilities. Approximately 900,000 inhabitants (about 10 %), mainly in remote areas, are supplied by onsite single water supply systems (household wells or springs). The wastewater sector is organized in a similar way whereas organizational units are often a bit bigger due to some cooperation of neighboring municipalities, resulting in about 600 bigger (>2,000 population equivalent) wastewater treatment plants and still about 1,000 smaller treatment plants (>50 population equivalent). The number of very small and private onsite wastewater treatment plants is estimated to be at 200,000 serving approximately. This is with regard to approximately 1.2 million inhabitants not being connected to a public sewer system. The whole WSS sector is predominately publicly owned and operated and pricing is not under central regulation. As municipal policy sometimes tries to keep prices as low as possible rather minimum prices than price-caps are applied in some regions in order to achieve proper cost recovery.

However, water pricing in most of Europe is subject to significant regulation that may introduce a distortionary effect from the true opportunity cost of the resource. Furthermore some additional risks on water quality have a variety of impacts on water ecosystem services and human welfare that are generally not captured in market prices.

12.2 Dimensions of Sensitivity to Climate Change

The water cycle of the WSS sector starts at the water resources like groundwater, spring water or surface water, includes abstraction, treatment to drinking water quality, transmission, storage and supply, consumption, collection and transport of wastewater, wastewater treatment including treatment sludge disposal and finally discharge of the water into the environment and thus back to the water resources, usually into rivers.

The WSS sector is a complex system that involves all, very specific and geographically bound natural water resources; vast and diverse technical infrastructure; and a strong nexus to lifestyle and consumer behaviour. Therefore the sensitivity to changes, including climate changes, originates from many levels. To give at least some examples: the natural water resources are sensitive to changes in the total amount and in geographical and temporal distribution of precipitation; technical infrastructure is sensitive to damages due to extreme events and natural disaster; treatment plant performance is sensitive to quality or quantity changes; and consumer behaviour is sensitive to the actual weather as well as to the weather forecast.

In any case, the main issue is the complexity of impact chains but still, also the monetary estimations of the effects are subject to great uncertainty.

12.2.1 Climatic Factors

Thinking about climatic influences one has to distinguish between short-term and long-term parameters. Short-term factors are within the timeframe of the variation of the actual weather up to the timeframe of the seasons whereas the long-term factor is the gradual shift of the mean values of temperature, precipitation, evapotranspiration and variation of the weather.

For example, according to Schöner et al. (2010), climate change could lead to a reduced groundwater recharge due to less precipitation during the winter in the south and in some eastern regions of Austria. Furthermore the water quality of surface water could be affected by decreased dilution; groundwater chemistry could change due to temperature increase and its influence on interaction of surface- and groundwater.

Besides the resource situation, the demand side has a strong nexus to weather and climate. Seasonal characteristics as well as the days of the week or time of the day show a clear influence on the consumer's behaviour. This includes the actual weather and its variability of temperature and precipitation. A change of frequency or duration of specific weather situations due to climate change, especially an increase of dry hot periods, will result in more frequent and intense peak demands.

On the side of the wastewater treatment an increase of variation of the weather will lead to an increased number of Combined Sewer Overflow (CSO) events and thus to a higher total pollution of the environment. On the other hand, higher temperatures will increase the performance of the waste water treatment process by ways of an enhanced activity of microorganism responsible for degradation of pollutants within the treatment plant.

12.2.2 Non-climatic Factors

The non-climatic factors mainly involve socio-economic and demographic changes and technological development at consumer's level as well as on the side of service providers.

Furthermore structural parameters like housing types (with or without private gardens) and population density have a clear influence on the efficiency and thus performance of service provision in the WSS sector.

On the other hand one must keep in mind the deterioration of natural water resource due to increasing use with respect to quality and quantity. Quality aspects mainly come up with point (any kind of discharge) or non-point pollution (percolation of e.g. fertilizers or pesticides into the groundwater). Quantity aspects—especially with regard to groundwater—can either be related to depletion (overuse) or a decreasing regeneration. The latter can result from an increasing surface runoff due to an increased rate of surface sealing or river regulation, or can be a result of climate change.

12.3 Exposure to Climatic Stimuli and Impacts up to Now

12.3.1 Past and Current Climatic Exposure and Physical Impacts

As mentioned before the sensitivity of WSS sector to changes, including climate changes, originates from many levels. Following a very coarse classification, one can distinguish between water resources; technical infrastructure; and consumer behaviour.

Water resources are generally sensitive to extreme events. Interference of normal service is following quite different mechanisms:

- Scarcity after long lasting dry and warm periods: Areas in Austria which are affected from aridity are mainly north eastern parts of the country. Nevertheless, scarcity can occur on a local or regional level according to absence of capable water resources (e.g. crystalline rock).
- Quality deterioration after long lasting dry and warm periods:
  - With regard to water supply the quality deterioration might result from new pollution pathways or less dilution of pollutants in the water body after a decreasing regeneration. These effects can occur on a local or regional level according to the soil structure and thickness of the covering layer.
  - With regard to wastewater the capacity of surface waters as receiving waters for wastewater discharge is decreasing with an increase of low water periods.
- Quality deterioration after heavy precipitation events or flooding:
  - With regard to water supply the quality deterioration is due to pollution of the water resources by massive infiltration of surface water, leading to a (at least) midterm deterioration of raw water quality in the water shed (mainly in Alpine, especially in Karst areas). Furthermore windthrow of timber (especially in forests with shallow-root-trees like spruce) can break up the covering layer of soil, enabling pollution to enter the water body.
  - With regard to wastewater heavy precipitation events will lead to combined sewer overflow (CSO) events with the discharge of untreated wastewater to the environment (every combined sewer system).

Technical infrastructures are sensitive to damages due to extreme events and natural disaster. Damage of infrastructure, including cleaning (removal of sediments from totally blocked sewer systems), disinfection of water supply assets, damage of machinery and electronic equipment etc., is following mainly two mechanisms:

- Flooding after extreme precipitation events
- Landslides after extreme precipitation events

To describe climatic stimuli to the consumer's behaviour the consumers have to be divided into three groups:

- Business, public, manufacturing and industrial use: The consumption of this group shows only a low dependency on the actual weather and will have only low dependency on climate changes as well. The only exceptions are water uses for cooling purposes.
- Agricultural use: In addition to the rain fed agriculture artificial irrigation is applied only in some parts of Austria and hardly uses water from the public water supply. However the demand for water resources will be affected if

irrigation increases due to climate changes. Further considerations concerning Agricultural issues see Chap. 8 (Agriculture).

- Private use: The consumption of this group shows a very strong dependency on the actual weather, especially to long lasting dry and warm periods and in combination with a high share of private gardens in the settlement structure (Neunteufel et al. 2013b).

12.3.2 Impact Chains up to Socioeconomic System

The relevant weather/climate parameters are mean and max values of temperature, drought duration, precipitation intensity and seasonal distribution as well as frequency of extreme events.

Short term (variation of the weather) and long term parameter (climate change—gradual shift of the mean values of temperature and precipitation) have impacts on water bodies (water resources for abstraction, and receiving waters); technical infrastructure (water supply, sewer system, and wastewater treatment); and consumer behaviour.

For example the forecasted rise of temperature will lead to a slight average increase of outdoor water consumption (+4 l per capita and day (l/Pd) in 2050 due to private garden irrigation and private swimming pools) but, very likely, a clear rise in peak consumption (Neunteufel et al. 2012a). On the other hand, the performance of wastewater treatment could be slightly enhanced by rising temperatures.

With increasing number or duration of droughts the rising temperatures will become more relevant to the water consumption as well as to the water resource availability. For the operation of combined sewer systems, longer duration of droughts will lead to an increased effort removing sediments.

An increase of intense precipitation events will influence all: water resources; water supply and sanitation infrastructure; as well as treatment performance (intensity) of wastewater due to an increase of sewer overflow events.

The synthesis of changing weather/climate parameters together with physical assets of the WSS sector and human behaviour results in an extensive matrix of quantifiable, semi-quantifiable and non-quantifiable impact chains. Tables 12.1 and 12.2 show a summary of the matrix for the water supply sector and the sanitation sector.

12.3.2.1 Water Supply Sector: Resource to Consumer

Table 12.1 Water supply related impact chains

Climate change parameter	Impact chain	Quantified in the model[a]
Precipitation: Lower precipitation or change of seasonal distribution, especially less precipitation during the summer	*Lower yield of springs (resource quantity) in the dry season* → Peak demand not covered by resource yield → enlargement of abstraction needed	Yes[b]
	Lower groundwater/surface water recharge → Lower dilution of pollutants (e.g. nitrate or pesticide) → enlargement of treatment needed	Yes[b]
	Less precipitation in the vegetation period → Increased outdoor demand (consumer behaviour) → enlargement of total capacity	No
Heavy precipitation events: Increase of frequency and/or intensity	*Increase of days with higher turbidity of spring water resources* → Possible microbiological contamination → discharge or treatment → adaptation of assets	Yes[b]
	Increase of land slide or mudflow events → Damage of infrastructure (abstraction, treatment, distribution system)	Yes[d]
	Increase of flood events → Damage of infrastructure → Infiltration into damaged pipe network → contamination →damage and restoration cost → Midterm deterioration of water resources (raw water) → damage and restoration cost	Yes[d] No No
Temperature: Increase of mean values and heat waves	*Change of resource quality (microbiological activity)* → Damage and restoration cost	No
	Change of withdrawal (resource quantity) → Increased outdoor and peak demand due to temperature rise (consumer behaviour)	Yes[c]

(continued)

Table 12.1 (continued)

Climate change parameter	Impact chain	Quantified in the model[a]
Drought: Increase of frequency and/or duration (consecutive dry days)	*Increase of depth of drying cracks in the soil* → Possible microbiological contamination → discharge or treatment → adaptation of assets	Yes[b]
	Increase of drought duration in the summer season → Increased outdoor demand (consumer behaviour) → enlargement of total capacity	No

[a]Costing method
[b]Change in production cost
[c]Change in final demand
[d]Replacement cost

12.3.2.2 Sanitation Sector: Consumer to Resource (Receiving Waters)

The data source of the impact chains displayed in Tables 12.1 and 12.2 are: Ertl et al. (2010), Kretschmer et al. (2010), Schöner et al. (2010), Lukas et al. (2011a, b), Mayr et al. (2009, 2010, 2011, 2012), Möderl et al. (2013), Neunteufel et al. (2009, 2011, 2012a, 2013b), Perfler et al. (2006, 2007), Plihal et al. (2013), Richard and Neunteufel (2012).

12.3.3 Economic Impacts up to Now

The focus of this chapter is laid on the vulnerability of WSS infrastructure including the vulnerability of water resources and the costs of past damage events.

12.3.3.1 WSS Sector: Past Damage Events and Their Costs

In recent years meteorological extreme events (e.g. 1991, 1997, 2002, 2003, 2005, 2007, 2009, and 2013) have caused emergency and disaster situations for the WSS sector in Austria. Especially weather situations related to the windstorm track V(5) b, deliver a vast amount of moist air from the Genoa low formation area along the pathway to the north of the Alps. Such Vb weather situations result in long lasting

Table 12.2 Sanitation related impact chains

Climate change parameter	Impact chain	Quantified in the model[a]
Precipitation: Change of seasonal distribution, less during summer, more during winter season	*Increase of wastewater volume during winter* → Increase of energy consumption at the treatment plant	 Yes[b]
	Lower surface water recharge (receiving waters chemical/biological/physical quality) → Lower dilution of pollutants (e.g. treatment plant runoff) → enhancement of treatment needed → Lower capacity for cooling (e.g. power plant or industry runoff) → enhancement of pre-cooling	 No No
Heavy precipitation events: Increase of frequency and/or intensity	*Increase of sewer flooding* → Increased volume of combined sewer overflow (CSO) → adaptation of CSO assets	 Yes[b]
	Increase of land slide or mudflow events → Damage of infrastructure (sewer and treatment plant)	 Yes[c]
	Increase of flood events → Damage of infrastructure (sewer and treatment plant) → Flooding of treatment plant → wash-out of biomass → decr. performance → increased pollution → Flooding of sewer system or treatment plant → sedimentation of mud → restoration cost	 Yes[c] No No
Temperature: Increase of mean values	*Lower oxygen solubility + higher microbial activity in surface waters* → Oxygen depletion/lower availability for higher organisms (e.g. fish) → change of biocenosis	 No
	Receiving waters temperature increase → Lower capacity for cooling (e.g. power plant or industry runoff) → enhancement of pre-cooling	 No
	Wastewater temperature increase → Higher microbial activity → increased treatment plant performance → decreased pollution	 No
Drought: Increase of duration (consecutive dry days)	*Increase of sedimentation in the sewer system during dry weather flow* → Cleaning by removal of sediments from combined sewer systems	 Yes[b]

[a]Costing method
[b]Change in production cost
[c]Replacement cost

Table 12.3 Austrian losses due to recent natural catastrophes

Year	Event	Austria total damage (M€)	Water supply sector damage (M€)	Affected areas
1991	Heavy precipitation event and flooding	73	n.a.	Danube region, Upper and Lower Austria
1997	Flooding	n.a.	n.a.	Lower Austria
1999	"Flooding of the century"	105	n.a.	Vorarlberg, Tyrol
2002	Long lasting heavy precipitation event and "flooding of the century" due to Vb weather situation	3,200	10	Vorarlberg, Tyrol, Salzburg, Upper and Lower Austria, Vienna
2005	Long lasting precipitation event and "flooding of the century" due to Vb similar weather situation	>70	3	Vorarlberg, Tyrol, Salzburg
2007	Long lasting precipitation event and minor flooding	n.a.	n.a.	Lower Austria
2009	Long lasting precipitation event, flooding, landslides due to Vb weather situation	About 100	n.a.	Upper and Lower Austria, Vienna, Burgenland, Styria
2013	Long lasting heavy precipitation event, flooding (not yet classified but similar to 2002) and landslides due to Vb weather situation	First estimates >2,000	No yet estimated	Tyrol, Salzburg, Upper and Lower Austria, Vienna

Source: Perfler et al. (2006); amended and supplemented for 2007, 2009 and 2013, Data: hydrographic service of Lower Austria; press release Kurier 05.06.2013

and very intense precipitation events, followed by floods and thus causing destruction of the infrastructure, heavy impact on the water quality in the distribution system, and midterm deterioration of raw water quality for example by contamination with mineral oil leaking from broken reservoirs. Table 12.3 shows Austrian total losses and exemplarily losses of the Austrian water supply sector.

Further details and a very basic extrapolation to estimate average current annual damage are provided as supplementary material (Economic Impacts Up to Now).

However the overall estimations result in average annual losses of 4 million euros for the water supply sector and 12 million euros for the sanitation sector but the estimations could be easily undervalued by far. With the future expansion of settlements and infrastructure it is most likely that damage will be at least equal or higher than with recent natural disaster events.

12.4 Future Exposure to and Impacts of Climate Change

12.4.1 Mid Range Climatic Scenario for Water Supply and Sanitation

The relevant climate indicators for Water Supply and Sanitation sector are temperature and duration of droughts during the summer season, seasonal distribution of precipitation and frequency and intensity of heavy precipitation events.

The mid-range COIN climatic scenario 2050 (changes relative to base period 1981–2010) relevant for Water Supply and Sanitation include:

- increase of mean temperatures (+2 °C)
- possible increase of heavy precipitation events
- decrease of summer precipitation (−9 %)
- decrease of total precipitation (−2 %)
- snow cover (mean day of snowmelt ends −13 days)
- increase of evapotranspiration
- change of water temperature
- increase of dry days (+3 %)

Of cause, the range of scenarios is much bigger and uncertainties have to be kept in mind.

12.4.2 High and Low Range Climatic Scenarios for Water Supply and Sanitation

Low-range climatic scenario: Temperature rise lower than mid-range scenario together with a more uniform distribution of precipitation and no increase of intense precipitation events.

High-range climatic scenario: Temperature rise higher than mid-range scenario together with a stronger shifting of precipitation towards the winter season, more/longer duration of drought events during the summer season and an increase of intense precipitation events.

12.4.3 Specific Method(s) of Valuation and Their Implementation Steps

An attempt was made to estimate the additional annual damages and benefits related to the impact chains for the periods 2016–2045 and 2036–2065, with and without climate change respectively.

First, the baseline scenario calculates the differences between today's WSS sector and the future WSS sector of the the periods 2016–2045 and 2036–2065 without the influences of the anticipated climate change. The current estimations of total assets, units (water consumption/wastewater) and on costs are based on existing empirical data (Neunteufel et al. 2009, 2011, 2012a, b, 2013a; Ertl et al. 2013). The future estimations of total assets and costs involved in the WSS sector are based on an extrapolation in accordance with the population growth, subsidy development (see shared socioeconomic pathway—Chap. 6) and anticipated change of asset rehabilitation (Neunteufel et al. 2012b) and consumer behaviour (Neunteufel et al. 2012a).

Second, the climate change scenario calculates the differences between the future WSS sector (baseline without climate change) and the future WSS sector of the periods 2016–2045 (calculation for 2030) and 2036–2065 (calculation for 2050) with the influences of climate change. The estimations for the future climate change situation are based on expert assumptions with regard to the impact chains listed in Tables 12.1 and 12.2.

As the impact chains and assumptions that had to be taken into account are manifold the procedure is pointed out with only one example: Based on the anticipation of a reduced groundwater recharge due to a seasonal/regional shift of precipitation the assumption was taken, that 10 % of the springs used for water supply could be temporary negatively affected by 2050. The anticipated shortage was estimated to be 10 % less yield than needed to cover the demand. This assumption was taken according to empirical sector performance indicators with regard to resource availability. Finally the demand for enlargement or enhancement of existing spring catchments due to quantitative seasonal/regional shift of precipitation was computed to be 10 % × 10 % = 1 %. Another 0.5 % were added due to anticipated qualitative changes (e.g. increased discharge due to turbidity), resulting in a total 1.5 % enlargement or enhancement of the existing catchments.

However, several non-quantifiable impact chains are not included within the calculation of the climate change scenario. Therefore the results of the quantification are undervalued for sure at a certain magnitude that can't be estimated.

These non-quantifiable impact chains are mainly related to:

- increases of environmental pollution;
- changes of biocenosis;
- changes of water temperature;
- compensation of exploitation;
- not quantifiable restoration of affected infrastructure.

The costing methods for quantifiable impact chains for both, baseline and climate change scenario are:

- Change in investment cost (for cases where enlargement or enhancement of existing assets is needed in order to provide a service similar to the service before socioeconomic, -demographic or climatic changes);

- Replacement cost (for cases where restoration, rehabilitation or renewal of damaged infrastructure)
- Economic value of change in final demand (when consumer demand changes due to changing socioeconomic, -demographic or climatic conditions).

Due to limited availability of input data the low reliability of extrapolations based on such data has to be considered with all results derived. Furthermore, detailed calculations were available for 2050 only. Exemplarily derivations and the compilation of estimated costs and benefits quantifications are therefore only included as supplementary material [Specific method(s) of valuation and their implementation steps].

12.4.4 Range of Sectoral Socio-Economic Pathway Parameters that Co-determine Climate Impact

For the water supply sector climate impact is predominately co-determined by three socio-economic parameters:

- Location of settlement development:

  With regard to quantitative and qualitative water resource availability the regional settlement development will determine which water resources are to be used to meet the increasing demand. As different resources in different regions can be more or less vulnerable to climate change the regional settlement development has a direct influence on the upcoming impact. Settlement development in regions with already scarce water resource availability, or due to climate change anticipated scarcity, could enhance the magnitude of the scarcity even more if there are no water transfer systems (regional bulk supply) which is usually the case in such regions. The same considerations are valid for qualitatively stressed water resources due to pollution of groundwater resources from agricultural fertilizer use.

  Especially with regard to pipe infrastructure settlement development in regions with higher risk of natural disaster will increase the magnitude of climate change impacts.
- Structure of settlement development:

  With regard to building and population density, the varying amount and type (e.g. private wells or central supply systems) of infrastructure involved directly determines the infrastructure at risk of damage caused by an increased number of extreme weather situations due to climate change. Thus, extensive settlement structures which require more pipe length will increase the magnitude of climate change impacts.
- Development of living environment:

  Usually the existence of private gardens is quite closely correlated to the settlement structure. However, the extent of private outdoor property directly influences the water demand for outdoor uses like irrigation or private swimming pools. Finally the outdoor uses directly correspond to rising temperatures or longer drought periods related to climate change.

For the waste water sector climate impact is predominately co-determined by the sewer system. As with the water supply sector similar considerations apply for:

- Location of settlement development:
  Settlement development in regions with higher risk of natural disaster will increase the magnitude of climate change impacts.
- Structure of settlement development:
  More extensive settlement structures will increase the magnitude of climate change impacts by increasing the amount of infrastructure at risk of damage.

As for the modelling socioeconomic assumptions are restricted to (i) population growth according to Chap. 6 (SSP) growth scenario (leading to stronger impact) and (ii) population growth according to Chap. 6 (SSP) low life expectancy scenario (leading to weaker impact). A differentiation according to development of living environment, location and structure of settlement was not realized due to high complexity.

12.4.5 *Monetary Evaluation of Impacts*

12.4.5.1 Direct Sector Impacts (Costs and Benefits) Without Feedback Effects from Other Sectors

Summing all estimations of quantified climate change impacts for the Water Supply Sector for the period 2010–2030 yields costs of 29 million euros, and 87 million euros up to 2050.

The total costs of all estimations of quantified climate change impacts for the Waste Water sector amount to 28 million euros for the period 2010–2030 and 83 million euros up to 2050.

On an annual basis, costs are dominated by the effect of higher investments while climate change and different assumptions on socioeconomic development contribute less to the costs.

For further details see supplementary material (Monetary evaluation of impacts)

12.4.5.2 Macroeconomic Effects

Those impact chains which are indicated as being quantified in Tables 12.1 and 12.2 are implemented in the macroeconomic (CGE) model, leading to changes of annual investment (and depreciation), changes of final demand and changes of subsidies (for more details on the implementation see supplementary material:

Table 12.4 Sectoral and total effects of quantified climate change impacts in sectors water supply and sanitation, average annual effects relative to baseline (for periods 2016–2045 and 2036–2065)

	Ø 2016–2045			Ø 2036–2065		
Changes in M€ p.a. relative to baseline	Gross output value	Intermediate demand	Gross value added	Gross output value	Intermediate demand	Gross value added
Gaining sectors	**+0.6**	**−0.0**	**+0.6**	**+1.9**	**+0.2**	**+1.7**
Water supply	+0.4	+0.0	+0.4	+1.1	+0.1	+0.9
Sewerage and waste collection	+0.2	−0.1	+0.2	+0.5	−0.1	+0.6
Construction	+0.0	+0.0	+0.0	+0.3	+0.2	+0.1
Losing sectors	**−4.8**	**−2.3**	**−2.5**	**−11.8**	**−5.7**	**−6.1**
Trade	−0.6	−0.3	−0.4	−1.6	−0.6	−1.0
Real estate	−0.4	−0.1	−0.3	−1.0	−0.2	−0.8
Rest of manufacturing	−0.3	−0.2	−0.1	−0.8	−0.5	−0.3
Rest of services	−0.3	−0.1	−0.2	−0.7	−0.2	−0.5
Energy	−0.3	−0.2	−0.1	−0.7	−0.5	−0.2
All other losing sectors	−2.9	−1.5	−1.4	−7.0	−3.7	−3.4
Total effect (all sectors)	**−4.2**	**−2.3**	**−1.9**	**−9.9**	**−5.4**	**−4.5**

Note: baseline scenario = reference socioeconomic development without climate change; climate change scenario = reference socioeconomic development and mid-range climate change; quantified impact chains: annual investments; demand changes for water supply and sanitation

Implementation of impact chains in the CGE model). Note that in the macroeconomic model there is a separate sector representing water supply (called "Water Supply"), whereas sanitation is part of the sector aggregate "Sewerage and Waste Collection". All macroeconomic effects are calculated for 2030 (representative for the period 2016–2045) and for 2050 (representative for the period 2036–2065) and effects are expressed for the climate change impact (CC mid-range) scenarios relative to the baseline without climate change in the same period (2030 and 2050).

The macroeconomic effects of the quantified impact chains in WSS are shown in Table 12.4. All effects are given as average changes of annual values in million euros relative to the respective baseline scenario for 2030 and 2050. Overall effects on the economy are rather small as the sectors Water Supply and Sewerage and Waste Collection contribute only about 1 % to economy wide value added (GDP). Regarding the two WSS sectors, gross output value in 2030 and 2050 is slightly higher in the climate change case compared to the baseline in the respective period (gross output value is determined by the sum of sectoral intermediate demand and gross value added). Concerning Water Supply, we see a positive effect because of higher demand in the climate change scenario. Furthermore, this effect on gross output value is influenced by the mentioned additional investment requirements. More precisely, the price for Water Supply is higher in the climate change scenario because of higher necessary annual investments, and therefore the demand induced quantity effect is reinforced by higher prices, leading to higher gross output value. This holds for 2030 as well as for 2050. Taking a closer look at sector Sewerage and Waste Collection we see also small positive effects regarding output value in both time periods. However, this result is driven mainly by price effects as in the macroeconomic model there is no additional demand for services from Sewerage and Waste Collection induced by climate change. Due to higher necessary annual investments in the climate change scenario, Sewerage and Waste Collection is getting more expensive. Thus, higher prices are leading to a higher output value.

To sum up, as the value of output is higher in the climate change scenario in the WSS sectors, gross value added is also rising and therefore WSS is contributing more to GDP than in the baseline scenario. However, due to economy wide feedback effects—mostly triggered by higher investment—other sectors are losing. As prices for Water Supply as well as Sewerage and Waste Collection are higher in the climate change scenario, consumption possibilities of households are lower, leading to less final demand for all other goods and services. Hence, all sectors except the WSS sectors as well as the Construction sector—which is gaining because of additional investments—are losing (in terms of output value). As shown in Table 12.4 the total effect on gross value added and therefore on GDP is slightly negative: Compared to the baseline scenario, due to the modelled impacts chains GDP is 2 million euros lower in the climate change scenario in the 2030s and 5 million euros lower in the 2050s on annual average. The same results hold for the change of welfare, which is measured as the quantity of consumed goods and services at prices of the baseline level. Unemployment remains unaffected.

Regarding public revenues and expenditures there are marginal negative effects arising from the modelled climate impact chains. The main driver of lower government revenues is attributable to lower revenues from the taxation of labour, as wages are decreasing slightly. Value added tax (VAT)—which is generating tax revenues from private household and government consumption—is also slightly lower, as final demand is below the baseline level. By assumption, government consumption expenditures remain unaffected by climate change impacts.

12.4.6 Qualitative Impacts (Non-monetised)

There are many impact chains whose costs have not been estimated. Tables 12.1 and 12.2 (impact chains) refer to whether a costing method can be applied or whether impact chains are not quantifiable.

These non-quantifiable impact chains are mainly related to:

- increases (or decreases) of environmental pollution—for example both, the increased pollution by combined sewer overflow (CSO) events and the increased waste water treatment performance with increasing temperatures are climate related but can't be easily evaluated in monetary terms.

- changes of water temperature will for example lead to changes of biocenosis which again can't be easily evaluated in monetary terms;
- compensation of exploitation; or
- not quantifiable restoration of affected infrastructure.

Finally, due to insufficient data, changes in investment and demand were taken account of in the macroeconomic assessment, but not changes in production cost (e.g. treatment effort, operation and maintenance, labour etc.).

12.4.7 Sector-Specific Uncertainties

The uncertainties with regard to all parameters involved are considerably high. Beside the general uncertainties with regard to socioeconomic, demographic, and climatic changes, there are sector-specific uncertainties including uncertain settlement development structures and locations, infrastructure specific uncertainties with regard to the extent of affected assets and damage and restoration cost estimations as well as uncertainties with regard to consumer behaviour.

However, major uncertainties are regarding the magnitude of the impacts on the WSS sector. The models of quantifiable impact chains are subject to many assumptions with regard to impact and cost. The causal nexuses are rather based on qualitative examples and extrapolations than on broad and quantifiable data. Besides that, the models of the impacts of climate change on the WSS sector are restricted to average values of the impacts as detailed estimates from a representative sample of utilities do not exist.

In addition, of the number of affected small and private water supply and wastewater facilities could not be estimated on a real data basis and thus may be easily undervalued by far.

Furthermore most uncertain assumptions can be seen in the non-quantifiable impact chains listed in Tables 12.1 and 12.2. In these cases either the costing is not quantifiable at all or, more often, the impact and amplitude of changes cannot be quantified with existing data. However, all non-quantifiable impact chains were at least evaluated with regard to their likely direction. The majority was found not to be beneficial for the WSS sector.

12.4.8 Relevance for Other Sectors

Changes in the WSS sector will influence every other sector and every private person using public water supply and sanitation services. Furthermore individual water supply and sanitation solutions will be effected the same way as they are facing the same changes. Moreover, climate change induced impacts in sectors like agriculture might lead to higher water demand and a competition for water resources.

12.5 Summary of Climate Costs for Water Supply and Sanitation

Besides the unquestioned influence of socioeconomic and demographic parameters the influence of weather and of climate change on the WSS sector is manifold.

The WSS sector is a complex system that involves all, very specific and geographically bound natural water resources; vast and diverse technical infrastructure; and a strong nexus to lifestyle and consumer behaviour. Therefore the sensitivity to changes, including climate changes, originates from many levels. The combination of changing weather/climate parameters and physical assets of the WSS sector as well as influences of human behaviour results in an extensive matrix of quantifiable, semi-quantifiable and non-quantifiable impact chains.

Climate change will result in a quantifiable change of production cost for cases where enlargement or enhancement of existing assets is needed in order to provide a service similar to the service before the change. Furthermore, a climate change related increase of floods or landslides will cause replacement cost for cases where restoration, rehabilitation or renewal of damaged infrastructure due to extreme events is needed. And finally, a change in consumer demand due to climatic conditions will change the economic value of the sector.

The macroeconomic consequences of climate change in WSS relative to the baseline scenario are overall slightly negative, both in terms of welfare and GDP. Unemployment is not affected by the modeled impact chains in WSS. In total, this translates to slightly lower government revenues. These negative macroeconomic effects are basically driven by two effects: First, higher investment, which on the one hand fosters the construction and building sector but on the other hand makes water supply and sewerage more expensive. Second, higher demand for water due to climate change will require a reallocation of consumption from other goods.

Beside the quantifiable impact chains there are several semi-quantifiable and non-quantifiable impact chains whose costs cannot be estimated. These include the increases of environmental pollution due to increased sewer floods, changes of water temperature and thus biocenosis or not yet quantifiable restoration of affected infrastructure.

However, the costs of climate change for the water supply and sanitation sector are substantially lower than the costs for enlargement of infrastructure due to the growing population. Fortunately the costs for enlargement will be financed by a growing number of people. On the other hand, unfortunately, the costs of climate change are additional costs for each customer.

Furthermore disturbing issues are that

(i) investigations on the effects of climate change revealed that there will be hardly any benefits but only a lot of different costs for the WSS sector
(ii) the uncertainties are considerably high, and include several non-quantifiable impacts.

In order to adapt the long-living assets of WSS sector in an efficient way, this is in combination with planned renewal, more and early information on the impacts and their magnitudes on the sector will be needed.

References

Ertl T, Kretschmer F, Mayr E, Plihal H, Perfler R (2010) Verknüpfung von Fehlernetz- und Risikoanalyse für das Risikomanagement in der Siedlungswasserwirtschaft, ÖWAV-Seminar: Risikomanagement und Vorsorge für kritische Wasserinfrastrukturen. ISBN 978-3-902084-67-5

Ertl T, Kretschmer F, Plihal H, Schwarz D (2013) INFOSAN – Strategische Informationsbeschaffung als Basis für die mittel- und langfristige ganzheitliche generelle Sanierungsplanung von Kanalisationsanlagen in Österreich, current research project at the University of Natural Resources and Life Sciences, Vienna. https://forschung.boku.ac.at/fis/suchen.projekt_uebersicht?sprache_in=de&menue_id_in=300&id_in=7965. Accessed 12 Aug 2013

Kretschmer K, Plihal H, Mayr E, Perfler R, Ertl T (2010) Failure experience and risk analysis for strategic sewer rehabilitation planning. In: 1st IWA Austrian national young water professionals conference, Wien, 09–11 Juni 2010

Lukas A, Mayr E, Katzmaier H, Perfler R (2011a) Failure experience improvement system (FEIS) for water supply systems. J Hydroinf. Accessed online 7 Nov 2011. doi:10.2166/hydro.2011.125

Lukas A, Mayr E, Richard L, Perfler R (2011b) Supporting the water safety plan (WSP) approach with the failure experience improvement system (FEIS). Water Sci Technol Water Supply 11 (3):288–296, IWA Publishing 2011

Mayr E, Lukas A, Ruhri M, Neunteufel R, Perfler R (2009) FEIS – failure experience improvement system; IWA international water association, 3rd leading edge conference on strategic asset management, conference proceedings, Miami, FL, 11–13 Nov 2009

Mayr E, Lukas A, Aichlseder W, Perfler R (2010) Experiences and lessons learned from practical implementation of a software-supported water safety plan (WSP) approach. Water Sci Technol Water Supply 12(1):101–108

Mayr E, Lukas A, Möderl M, Rauch W, Perfler R (2011) Integrales Risikomanagement in der Trinkwasserversorgung in Österreich, Österreichische Wasser- und Abfallwirtschaft, Band 63/Heft 3–4

Mayr E, Lukas A, Aichlseder W, Perfler R (2012) Das wasserLOGbuch: Ein softwaregestützter Wassersicherheitsplan auf Grundlage des Betriebs- und Wartungsbuchs. Österreichische Wasser- und Abfallwirtschaft 9–10(12):447–452

Möderl M, Rauch W, Lukas A, Mayr E, Neunteufel R, Perfler R, Neuhold C, Friedl F, Fuchs-Hanusch D, Lammel J, Hohenblum P, Skopik F, Bleier T, Weber K, Eder F, Brugger M (2013) Austrian activities in protecting critical water infrastructure. Springer, New York, in press. ISBN 978-3-319-01091-5

Neunteufel R, Bock R, Perfler R (2009) ÖVGW Statistikprojekt: Langjährige Datenreihe aus DW 1 Daten – Untersuchung möglicher Einflussfaktoren – Endbericht. Wien, im Dezember 2009, Hrsg. ÖVGW. www.ovgw.at

Neunteufel R, Richard L, Perfler R, Tuschel S, Mader K, Haas E (2011) Studie Wasserverbrauch und Wasserbedarf – Teil 1: Literaturstudie zum Wasserverbrauch – Einflussfaktoren, Entwicklung und Prognosen; Bundesministerium für Land- und Forstwirtschaft, Umwelt und Wasserwirtschaft, 2010, Wien. http://publikationen.lebensministerium.at/publication/publication/view/3416/19884

Neunteufel R, Richard L, Perfler R, Tuschel S, Mader K, Haas E (2012a) Wasserverbrauch und Wasserbedarf – Auswertung empirischer Daten zum Wasserverbrauch (Teil 2); Bundesministerium für Land- und Forstwirtschaft, Umwelt und Wasserwirtschaft, Wien. http://www.lebensministerium.at/publikationen/wasser.html

Neunteufel R, Ertl T, Spindler A, Lukas A, Perfler R, Schwarz D, Zessner M, Haberl R (2012b) Technische Herausforderungen in der Siedlungswasserwirtschaft Wien: Bundesministerium für Land- und Forstwirtschaft, Umwelt und Wasserwirtschaft, Wien. http://www.lebensministerium.at/publikationen/wasser/Technische-Herausforderungen-in-der-Siedlungswasserwirtschaft.html

Neunteufel R, Friedl F, Perfler R, Schrotter S, Mayr E, Fuchs-Hanusch D (2013a) Abschlussbericht zum ÖVGW Benchmarking 2012, Wien/Graz, Februar 2013, Hrsg.: ÖVGW – Österreichische Vereinigung für das Gas- und Wasserfach. www.trinkwasserbenchmarking.at

Neunteufel R, Richard L, Perfler R (2013b) Water demand – the Austrian end-use-study and conclusions for the future. IWA Publishing 2013: Water Sci Technol Water (in press: WS-EM13124 accepted on 29 July 2013)

Perfler R, Unterwainig M, Formayer H (2006) StartClim – Auswirkungen von Extremereignissen auf die Sicherheit der Trinkwasserversorgung in Österreich; StartClim2005.A4, Teilprojekt von StartClim2005 „Klimawandel und Gesundheit". http://www.austroclim.at/startclim/

Perfler R, Unterwainig M, Mayr E und Neunteufel R (2007) Sicherheit und Qualität in der Trinkwasserversorgung in Österreich – Einflussfaktoren, aktuelle Anforderungen und Initiativen. Österreichische Wasser- und Abfallwirtschaft, 59. Jrg, Heft 9–10, 125

Plihal H, Münch Ph, Kretschmer F, Ertl Th (2013) Innovative Kanalinspektion als Grundlage für die bedarfsorientierte Kanalreinigung. Österreichische Wasser- und Abfallwirtschaft 65:89–93. ISSN 0945-358X

Richard L, Neunteufel R (2012) An analysis of water consumption and usages at the household level in Austria. In: International water association (IWA) European utility conference, Vienna, Austria, 18–20 April 2012

Schöner W, Böhm R, Haslinger K, Blöschl G, Merz R, Blaschke AP, Viglione A, Parajka J, Kroiß H, Kreuzinger N (2010) Anpassungsstrategien an den Klimawandel für Österreichs Wasserwirtschaft. Im Auftrag des BMLFUW und der Länder. ZAMG, TU-Wien, Wien

Chapter 13
Buildings: Heating and Cooling

Lukas Kranzl, Marcus Hummel, Wolfgang Loibl, Andreas Müller, Irene Schicker, Agne Toleikyte, Gabriel Bachner, and Birgit Bednar-Friedl

Abstract While energy savings in buildings is among the key prerequisites for a low-carbon future, our ability to maintain temperatures in buildings within a specific comfort range, and thus our demand for heating and cooling energy, are also highly sensitive to climate change. We quantify two main impact chains: (1) a higher temperature in winter leads to a reduction of heating energy demand and (2) a higher temperature in summer leads to an increase in demand for cooling. The demand for cooling energy depends largely on the future uptake of air conditioning in the building sector and is subject to considerable uncertainty. On quantifying these two impacts for the example of Austria for the period around 2050 a net saving of about 230 million euros per year is found, triggering slightly positive effects on welfare and GDP. The result is depending on the development of energy prices and in particular by the ratio of electricity to fuel price in the heating sector. The results show that, in absolute terms, the energy reduction in heating is much higher than the increased energy demand for cooling for the time horizon and the geographical location investigated. This stems from the fact that energy demand for

L. Kranzl (✉) • M. Hummel • A. Müller • A. Toleikyte
Institute of Energy Systems and Electrical Drives, Vienna University of Technology, Vienna, Austria
e-mail: kranzl@eeg.tuwien.ac.at; hummel@eeg.tuwien.ac.at; mueller@eeg.tuwien.ac.at; toleikyte@eeg.tuwien.ac.at

W. Loibl
Energy Department, Austrian Institute of Technology, Vienna, Austria
e-mail: wolfgang.loibl@ait.ac.at

I. Schicker
Climate Research Section, Division Data, Methods, Models, ZAMG—Central Institute for Meteorology and Geodynamics, Vienna, Austria
e-mail: irene.schicker@zamg.ac.at

G. Bachner
Wegener Center for Climate and Global Change, University of Graz, Graz, Austria
e-mail: gabriel.bachner@uni-graz.at

B. Bednar-Friedl
Wegener Center for Climate and Global Change, University of Graz, Graz, Austria

Department of Economics, University of Graz, Graz, Austria
e-mail: birgit.friedl@uni-graz.at

K.W. Steininger et al. (eds.), *Economic Evaluation of Climate Change Impacts*, Springer Climate, DOI 10.1007/978-3-319-12457-5_13

air conditioning in Austria in 2008 was only 0.4–0.5 % of the final energy demand for heating. The impacts and costs resulting from a strong increase in electricity peak loads in summer are investigated in Chap. 14 (Electricity).

13.1 Introduction

One of the crucial purposes of buildings is to protect people against weather conditions and ensure a comfortable indoor climate. Construction systems and technologies are designed to meet specific climatic conditions and indoor requirements. Ceteris paribus, a change in climatic conditions not only affects indoor climate, it also has an impact on the suitability of prevailing building configurations. While some building occupants adapt autonomously, e.g. by changing heating mode or behaviour or by investing in additional cooling devices, several individuals may not have the means to do so.[1] This can result in loss of comfort, productivity, or even worse, loss of health. The latter possibility is discussed in Chap. 11 (Human Health). Climate change may affect the functionality of buildings in several ways due to a higher frequency of extreme events and natural disasters. In Austria, singular and local storms follow no specific pattern or frequency. Usually they are short events concentrated in the eastern lowlands and in valleys and typically damage only a few objects. Only hurricanes (wind speed >118 km/h) are large to continental-scale events which may result in large damage to settlements or forests. During the last 10 years, seven of such continental scale events with wind speeds between 100 and 200 km/h have been observed. However, the impact on roofs due to the higher magnitude and frequency of storms is not quantified in this work since the respective climate scenarios do not provide reliable predictions. How climate change affects buildings in particular the direct impact of floods and storm damage, is addressed in Chap. 18 (Catastrophe).

This chapter concentrates on climate induced effects on heating and cooling. Owing to the volume of buildings exposed and the expected regularity of events, the general annual costs associated with the resulting changes in heating are significant. This is why climate change impact has the potential for releasing large changes in this impact field and making it highly relevant for our analysis.

Section 13.2 describes climatic and non-climatic factors which have an impact on heating and cooling of buildings in the next decades. Section 13.3 starts with a discussion on past and current climate exposure and describes the impact chains in the sector. Section 13.4 provides the approach and results of our evaluation. Finally, in Sect. 13.5 we derive conclusions.

[1] On the relation between indoor comfort and control strategies see e.g. Roberts (2008).

13.2 Dimensions of Sensitivity to Climate Change

The energy demand of buildings for space heating and cooling is determined by the nature of building components, solar and internal appliance gains, and the difference between indoor and outdoor temperature. A change in outdoor temperature may lead to a change in energy demand for heating and cooling.

13.2.1 Climatic Factors

We have divided the main factors into two impact chains:

(1) Impact on energy demand for heating: temperature change (heating degree days) and solar radiation. Other factors such as wind speed have some relevance, but are omitted here.
(2) Impact on energy demand for cooling and ventilation: temperature change (cooling degree days). While changes in solar radiation (due to a possible change in cloud cover), and changes in wind and in humidity would play a role (for heating and for cooling), the impact of climate change impact is rather small and uncertain (Bednar et al. 2013). In this present analysis they have thus been ignored. In contrast, the expected length, severity and frequency of heat waves are likely to have a distinct impact on the market penetration of cooling devices and thus have to be considered.

13.2.2 Non-Climatic Factors

Buildings are among the most durable goods in our society. As a result, changes within the building sector tend to occur at a slower pace than changes taking place in other economic sectors (e.g. tourism, industry etc.). Given that the building sector accounts for about 40 % of European greenhouse gas emissions (EPBD recast 2010), it is of no surprise that this has become a focus in the energy and climate policy. Improving the energy efficiency of building envelopes, heating and cooling systems etc. offers high potential for reducing GHG emissions if fossil energy sources are used. Efficiency targets for new and old buildings have been established all over the world at both the national and regional level. Apart from the demands of climate policy, other reasons for increasing building energy efficiency include the need to reduce fuel poverty and the desire to increase energy security (as a result of rising energy prices). All of these factors drive the change towards improving the thermal quality of the building stock.

Other major drivers behind changes in the structure of the building stock include: population growth, changes in family structures and related household size, GDP growth, comfort requirements in terms of dwelling size per capita and indoor temperature levels, the impact of spatial planning on building structure, and the mix of different type of buildings. Particularly economic development plays a crucial role in the overall development of residential and non-residential areas and thus in the market penetration of air conditioning (AC) devices (Isaac and van Vuuren 2009).

The Austrian population in 2010 was 8.38 million. The baseline scenario (based on Statistik Austria projections, see Chap. 6—SSP) let expect for 2030 around 9.0 million, while in 2050 the number is expected to reach 9.5 million people. The relative increase in the number of households is higher, rising from a current figure of 3.6 million households (in 2010) to 4.05 million in 2030 and to 4.31 million in 2050. Currently, no specific figures are available on the expected future number of buildings. However, figures do exist with respect to the gross floor area of residential and non-residential structures (excluding industrial facilities). The figures for our reference scenario are: 570 km^2 gross floor area in 2010, expected to grow to 690 km^2 by 2030, and to 730 km^2 by 2050,[2] which is an increase of 120 km^2 and 160 km^2 respectively.

A more wealthy society also increases the demand for additional floor space leading to higher personal comfort, by raising individual flat size. All this increases the demand for energy- first for heating, and perhaps in future, more and more for cooling.

Currently, in Austria space heating is much more relevant than cooling: At the moment energy consumption for space cooling amounts to only 0.4–0.5 % of the energy consumption for space heating (see Müller and Kranzl 2013). Thus, in the case of Austria, cooling is a relatively new subject as in former years it was not a distinct requirement for securing indoor thermal comfort. However, increasing temperatures and growing heat island effects in urban environments as a result of densification and less nocturnal cooling (see Chap. 17—Cities) are expected to increase future demand for the active cooling of flats. Moreover, under typical climatic conditions in Austria thermal building insulation may lead to higher cooling loads unless specific measures for the reduction of cooling energy needs are taken (e.g. shading devices).

In Austria, a considerable effort is currently being made to further improve thermal building quality in new and renovated buildings and to increase the extent of renovation activities. National targets for zero-energy-buildings (for the period up to 2020) were established in 2012 (OIB 2012). However, the current rate of renovation is in the range of about 1 %, and thus still remains far below expectations and official targets (Müller and Kranzl 2013; Bundesministerium für Wirtschaft 2010).

[2] Based on extrapolation of "Energieszenarien bis 2050: Wärmebedarf der Kleinverbraucher" on the reference scenario.

13.2.3 Identification of Potential Large-Damage Combinations

Factor interplay in the assessment of potential climate change damage in the building sector has both positive and negative effects. On the one hand, a benefit may arise due to an expected reduction in the demand for heating energy. On the other hand, an adverse impact may be expected due to increasing demand for cooling energy or due to rising costs associated with storm damage to roofing (where the latter is not addressed specifically in this chapter). These effects have been discussed e.g. in Aguiar et al. (2002), Cartalis et al. (2001), Olonscheck et al. (2011) or Kranzl et al. (2010), however only partly with respect to related costs.

In terms of absolute energy levels, the extent to which benefits materialize depends on the strength of the following: (1) uptake of energy efficiency measures in the building sector, (2) whether temperature increases in summer are stronger than those in winter. Regarding the first point, one must remember that in more efficient buildings the heating period is shorter than in less efficient buildings (see e.g. Zangheri et al. 2014). Thus, more efficient buildings are better able to make use of higher winter temperatures (Kranzl et al. 2010; Bednar et al. 2013). On the other hand, the lower the efficiency of buildings, the higher the difference in benefits between the baseline and mid-range climate scenario, since the absolute energy demand level is higher in the case of a low efficient building stock. Of course, this should only be understood as a ceteris paribus condition and must not be understood as an argument for lower efficiency standards in the building sector.

Regarding the adverse impact of climate change in the form of increased cooling loads, the following factors lead to increased costs (when acting in combination, due the high degree of non-linearity, the impact of such factors is particularly problematic): (1) Higher temperature increases in summer than in winter, (2) longer, more frequent and more severe heat waves, leading to a higher uptake of air conditioning, (3) greater urban heat island effects due to expansion and densification of urban areas (see Chap. 17—Cities), (4) limitations on reductions in appliance internal loads (e.g. no improvement in appliance efficiencies) and higher demand for electric appliances in buildings and (5) lack of measures for combatting gains in solar radiation e.g. by shading. Other factors relating to damage caused by adverse changes in cooling peak loads in the electricity system are discussed in the Chap. 14 (Electricity).

13.3 Exposure to Climatic Stimuli and Impacts to Date

13.3.1 Past and Current Climatic Exposure and Physical Impacts

Both for heating and cooling the past years have already shown increasing cooling degree days and decreasing heating degree days. Regarding the impact on heating, corresponding data is given in Sect. 13.3.3. The impact of climate change on cooling energy demand in recent years in particular is under discussion. Only rather few estimates are available for Austria concerning cooling energy demand in buildings (Haas et al. 2007; Prettenthaler and Gobiet 2008; Zoll 2010; Müller and Kranzl 2013) which show slightly different results in the range of about 250–500 GWh/yr for the time frame around 2005–2010. Official energy statistics do not provide separate data for cooling energy demand (Statistik Austria 2011). Thus, there is no clear evidence on how energy demand for cooling in Austria has developed in recent years. More information in this respect is available for other countries, in particular for more southern countries with higher cooling load requirements (Toleikyte et al. 2012). E.g. Giannakopoulos and Psiloglou (2006) show the historical data concerning variations in energy consumption, temperature and the gross national product (GNP) for the case of Greece. They argue that all these parameters show a clear upward trend in the period 1993–2001. The maximum daily energy consumption was 38 GWh in 1993 while in summer 2001 it had reached 58 GWh. While it is not clear which factors exactly triggered this development, the study shows that there is a declining trend for HDD (heating degree days) and an increasing trend for CDD (cooling degree days) in the investigated period. Beccali et al. (2007) indicate that summer electricity consumption in the building sector in Italy has grown steadily due to the growing demand for cooling. Moral-Carcedo and Vicéns-Otero (2005) get similar results for the case of Spain from a correlation analysis of electricity peak loads, temperature and cooling energy demand.

Summing up, based on the literature one can say that it is not always possible to clearly separate the effects of increased demand for cooling energy (i) due to climate change and the impact of corresponding autonomous adaptation, and (ii) due to changes in behaviour reflecting a higher demand for personal comfort and thus changes in lifestyle. Notwithstanding this, there is still sufficient evidence that cooling energy demand has increased at least in some southern countries. One indication of the increase in demand for cooling energy in Austria—or at least for an increase in the amount of attention being paid to cooling energy demand—can be witnessed in the development of the respective official standards. The relevant 2011 standard (OIB 2011) requires that the overheating of residential buildings during summer has to be avoided (see ÖNORM B 8110-3).

13.3.2 Impact Chains in the Socioeconomic System

Table 13.1 lists the identified impact chains for the impact field Buildings: heating and cooling which are triggered by temperature increases.

In order to determine the quantitative impact of changing temperature levels in climate scenarios on heating and cooling energy demand, we applied the model Invert/EE-Lab according to the work done in the ACRP project PRESENCE (Kranzl et al. 2013b). Invert/EE-Lab is a dynamic bottom-up simulation tool used

Table 13.1 Impact chains "buildings: heating and cooling"

Climate change parameter	Impact chain	Quantified in the model
Increase in temperature in winter[a]	→ Reduced heating energy demand → [change in final demand for energy]	Yes
Increase in temperature in summer	→ Higher cooling energy demand and stronger growth of air conditioning in buildings → [change in final demand for energy and for AC units]	Yes
	→ Higher temperature levels in buildings (in case that there is no air conditioning and no passive adaptation measures at the building level) → lower comfort of occupants	No

[a]According to Petoukhov and Semenov (2010) climate change could also lead to lower temperature in winter. However, this is not the case in the climate scenario taken into account in our research

to develop scenarios (price scenarios, insulation scenarios, different consumer behaviours, climate change impact, etc.) and their respective impact on future trends of renewable as well as conventional energy sources at a national and regional level. The basic idea is to model building stock, heating, cooling and hot water systems at a highly disaggregated level in order to calculate related energy needs and energy supplies, to determine reinvestment cycles and new investment in building components and technologies, and to simulate the decisions of various agents (i.e. owner types) with respect to investment decisions in a specific building segment. Rebound effects of renovation activities are covered in terms of higher effective indoor temperature after building renovation. More details are available e.g. in Müller et al. (2010), Kranzl et al. (2010), Müller (2012), Kranzl et al. (2013a).

The building stock has been subdivided according to climatic regions. Energy demand was calculated on a static, monthly basis to derive hourly load profiles based on COIN climate scenarios (Kranzl et al. 2013b). Based on the Chap. 6 (SSP), a reference scenario has been developed and applied in the model Invert/EE-Lab to estimate the uptake of renovation measures and investments on an annual basis.

13.3.3 Economic Impacts Up to Now

For the evaluation of economic impacts of climate change up to now, we compared the heating and cooling expenses in a climate base period (1980–2010) with the expenses under current climate (HDD bias corrected value of 2010). For this purpose we considered the building stock in the structure of the year 2010 including the stock of heating systems of this year.

In 2010, the energy consumption for space heating and hot water in Austria was 380 PJ, adjusted to mean climate[3] (Müller and Kranzl 2013; Statistik Austria 2011). The corresponding expenses for heating energy amounted to about 7.2 billion Euros (retail prices including taxes). Cooling energy demand is estimated to be around 400 GWh_{el} ((Müller and Kranzl 2013) with related corresponding energy expenses of 70 million euros. Assuming a constant climate for the reference period 1980–2010 and with energy prices according to the assumptions in this study (Chap. 6—SSP), heating energy expenses would have been about 11 million euros higher and cooling energy expenses 5 million euros lower. These data relate to an overall GDP of about 300 billion Euros in the year 2011. In the past few years, the market penetration of cooling devices has risen strongly. However, it remains unclear to what extent this is due to climate change and to what extent it is due to a general trend towards higher levels of comfort.

13.4 Future Exposure to and Impacts of Climate Change

13.4.1 Mid-Range Climatic Scenario for Heating and Cooling

In order to consider regional differences in climate and climate change, a set of different climatic regions was defined, based on Schicker and Formayer (2012). Semi-synthetic climate data (SSCD) sets based on observations and regional climate model (RCM) simulations of the A1B scenario were created. For the analyses in this chapter, MPI-REMO A1B climate data were used as proxy for the COIN A1B data in order to allow the use of substantial previous modelling results (Kranzl et al. 2014).[4] Compared to other RCM projections (RCM-ALADIN driven by the ARPEGE GCM), these results show a rather large increase in winter temperatures and small increase in summer temperatures. These data serve as input for a building energy model. Due to topography and other differences in climatic conditions in Austria it was decided that a set of various climatic clusters needed to be defined. On the one hand, these clusters are based on the INCA climatology (Haiden et al. 2011) of temperature and radiation conditions in January and July. On the other hand, a more robust clustering was applied using a 30 year data set (1971–2000) from the Austrian Central Institute for Meteorology and Geodynamics (ZAMG).

For the two months January and June, temperature and radiation classes were defined:

[3] Climate adjustment has been carried out for the year 2010 according to the mid-range climate scenario of Chap. 5 (Climate).

[4] We are aware that the RCP scenarios derived for the IPCC AR5 would be more up-to date. However, at the time when the analyses in this chapter started, these results were not yet available.

- June: temperature < 18 °C, 18 °C < temperature < 22 °C, and temperature > 22 °C, and radiation < 230 W/m^2 and radiation > 230 W/m^2.
- January: temperature < 15 °C and temperature > 15 °C, radiation < 50 W/m^2 and radiation > 50 W/m^2.

Not all possible cluster combinations are present in Austria. In total, 19 climatic clusters are available.[5] These were taken into account for the further analysis. The SSCD program (Heindl et al. 1990) was used to calculate the semi-synthetic hourly conditions for a representative year. It requires data on temperature, global radiation, diffuse radiation, wind speed, and relative humidity on an hourly basis as input. Additionally, mean monthly values of each parameter are needed to generate a SSCD year.

For future climatic conditions based on the three bias-corrected and localised RCM simulations and on three time slices (1981–2010 E-OBS/past, 2011–2040 present and near future, 2036–2065 future), grid cells corresponding to each of the 19 climatic cluster were used. Differences between the simulated and observed mean monthly values of the parameters were calculated for every time slice and added to the observed monthly values which could then be used as data in the future SSCD.

We used hourly data for temperature and solar radiation in order to determine heating and cooling demand load profile. The annual energy demand was calculated by a static monthly approach implemented in the model Invert/EE-Lab, as described in Bednar et al. (2013), Müller et al. (2010) and Kranzl et al. (2013b).

13.4.2 High and Low Range Climatic scenarios for Heating and Cooling

According to the specification in Chap. 5 (Climate), high- and low-range climatic scenarios were taken into account to assess the cost range of inaction for heating and cooling. Table 13.2 shows the heating and cooling degree days as exemplary indicators. The values indicate a substantial increase of cooling degree days and decrease of heating degree days in the hottest climate scenario, whereas the coldest climate scenario is relatively near to the mid-range climate scenario. For the quantitative assessment of the range of cost of inaction, the data described above for the mid-range scenario were up- and downscaled with HDD and CDD development for the different climate regions described above.

Besides these effects of increasing CDD, heat waves and extreme heat periods could also lead to a substantial increase in air conditioning unit sales. Thus, such extreme periods can have an impact on the buildings' energy demand which goes beyond the pure cooling energy need during this period.

---

[5] A map of these clusters is presented in the supplementary materials (Supplementary Material Fig. 13.1).

Table 13.2 High- and low-range climate scenarios for the example of cooling degree days (CDD) and heating degree days (HDD), Chap. 5 (Climate)

		CDD		HDD	
		Absolute increase (Kd)	Relative increase (%)	Absolute increase (Kd)	Relative increase (%)
	Ø 1981–2010	89	0	4,338	0
Hottest	2030	233	262	−1,235	−28
	2050	583	655	−1,879	−43
Coldest	2030	38	43	−391	−9
	2050	89	100	−554	−13

13.4.3 Specific Method(s) of Valuation and Their Implementation

The following steps were carried out to assess the costs of the two impact chains space heating and space cooling:

- Development of a reference scenario as well as scenarios with diminishing and enhancing heating and cooling energy demand. First, these scenarios are calculated ignoring the impact of climate change, and then taking climate change into account. This includes the uptake of renovation measures and changes in the mix of heating and cooling technologies. This step was carried out using the model Invert/EE-Lab (see above). Input data regarding energy prices,[6] growth and the regional distribution of population and building stock was based on Chap. 6 (SSP). Factors such as building codes and support instruments were determined according to the reference scenario assumptions, i.e. slow progress in building codes etc. was assumed, reflecting the requirements of the European Energy Performance of Buildings Directive (recast) but not beyond. The scenarios are based on Kranzl et al. (2013b).
- For the cost evaluation, the costing method "change in final demand" was selected (Chap. 7—Economic Framework). This method is primarily based on private, residential buildings. For private, non-residential buildings the change of heating and cooling systems and related energy demand may be understood as a change in the system of production. In the case of public buildings, the change in heating and cooling systems and related energy demand would reflect a change in public final demand. Due to constraints in data availability and uncertainty, we decided to use the costing method "change in final demand" for all building types. A change in costs occurs for heating and cooling energy demand as well as for investment in heating and cooling systems. However, we decided not to take into account the change in the investment in heating systems,

[6] In fact, energy prices may also be affected by climate change. This is discussed in Chap. 14 (Electricity).

assuming that installers would continue to design heating systems for the same design outdoor temperature, although this temperature occurs less often in a warmer climate than in a climate where no climate change occurs. Thus, the scenarios were evaluated in terms of costs of energy carriers for heating and cooling (C_{en}), calculated on the basis of energy prices (p_{en}) and energy demand (Q_{en}) for all energy carriers (en) as well as regarding required investment costs for cooling and ventilation units ($C_{inv,ac}$), derived from the specific costs of cooling systems (c_{ac}) and the installed capacity (P_{ac}) in the different building categories (bca).

$$C_{en} = \sum_{en} p_{en} \cdot Q_{en}$$

$$C_{inv,ac} = \sum_{bca} c_{ac,\ bca} \cdot P_{ac,bca}$$

- The costs of inaction are calculated by taking the difference between the costs in the climate change scenario and those in the baseline scenario (i.e. no climate change).
- To gain input for the macro-model and to assess feedback from other sectors the effects have been divided into the following sectors, corresponding to the related macro-economic sectors:
  - Costs for biomass fuels
  - Costs for heating oil and coal
  - Costs for natural gas, electricity and district heating
  - Costs for air conditioning and ventilation devices

13.4.4 Range of Sectoral Socio-Economic Pathway Parameters Co-Determining Climate Impact

The possible pathways involved have an impact on the building sector's future exposure and sensitivity with respect to climate change. Sectoral exposure is mainly linked to the growth of the building stock and the location of buildings in various regions with different climate change signals. Sensitivity is mainly a function of the thermal quality of buildings, the energy efficiency of heating and cooling systems and their energy characteristics in summer, and also of required comfort levels and behaviour, technology, energy carrier mix, and energy price levels.

In the cost assessment, some—but not all—of the socio-economic factors described above were taken into account. Relevant factors for these socio-economic ranges are documented in more detail in the supplementary material to this chapter. Some of these factors, which are listed there were taken into account in the quantitative assessment, like summer indoor temperature, improving thermal quality by building renovation or energy carrier mix for heating. Other factors, like growth and regional distribution of building stock, development of electric appliances and related internal loads, number of buildings with AC and on-site PV were not considered in the assessment of cost ranges in socio-economic scenarios.

The methodological approach used in considering the various scenarios is documented in the description of the monetary evaluation.

13.4.5 Monetary Evaluation of Impacts

13.4.5.1 Direct Sector Impacts (Costs and Benefits) in the Absence of Feedback Effects from Other Sectors

The baseline development of energy demand for space heating and hot water preparation (reference scenario) results in a decline by about 40 % till the 2050s. The reduction in practice will of course depend on how strictly policies are applied. For example, there is considerable room for leeway with respect to the thermal insulation of buildings. In general, this scenario seems roughly in line with the renewable energy targets for 2020 described in the European renewable energy directive. Some doubt remains since the national renewable energy action plans (BMWFJ 2010) do not distinguish between space heating and process heat. The same holds for the climate mitigation and energy efficiency targets (Table 13.3).

This is related to energy expenses for biomass in the range of about 0.6 billion Euros, for coal and oil of 2 billion Euros and for natural gas, electricity and district heating of about 3.7 billion Euros in the base year and 2050, respectively.

The climate change signal reduces the final energy demand for space heating by about 5.8 TWh/yr in 2050. In order to calculate the economic benefits of this reduction in energy demand two steps were carried out: first the energy demand reduction was derived by energy carrier, and second, for each energy carrier the prices prevailing in 2050 (based on reference SSP assumptions) were applied.

Table 13.3 Baseline scenario final energy demand by energy carrier for space heating and hot water preparation (reference scenario with constant climate, GWh)

GWh	2010	2020	2030	2040	2050
Coal	930	391	78	75	136
Oil	22,121	16,256	8,218	3,726	2,351
Natural gas	27,869	26,921	24,008	20,060	16,922
District heating	17,417	19,233	20,909	20,315	18,608
Electricity	8,118	5,603	3,852	3,512	3,382
Biomass	19,479	20,890	21,346	20,678	20,004
Ambient Energy	1,313	2,493	3,819	5,003	5,714
Solar thermal Energy	1,276	2,080	3,486	4,499	5,182

Without any discounting of the costs of inaction, this results in about 383 million euros/year in the 2050s.

Comparing baseline (no climate change) and mid-range climate change final energy demand for space cooling reveals an additional electricity demand for cooling of 470 GWh_{el}/yr, and additional 120 million euros/year in 2050 (taking into account retail electricity prices from above and assuming no discounting of costs of inaction). Moreover, additional investments in air conditioning of about 25 million euros/year are also expected in 2050. One of the key uncertainties in this field relates to the market penetration of air conditioning in the building stock. Our approach here was to link the penetration of air conditioning to indoor temperature assuming the absence of an active cooling system: the higher the indoor temperature and the more frequent high temperature levels occur in certain building types and certain regions, the more likely becomes the installation of an active AC system (Müller et al. 2014).

While the low-range climate scenario does not deviate strongly from the mid-range scenarios, it becomes clear that the high-range scenario leads to a strong increase of the effects—both benefits in terms of reduced heating energy demand and costs in terms of increased cooling energy demand. A few aspects of diminishing and enhancing effects of socio-economic development (summer comfort requirements, energy efficiency measures)[7] were also taken into account. For cooling, the consideration of high-range climate scenarios and enhancing socio-economic development leads to more than doubling of costs. However, it should be taken into account that there are still a lot of factors which have not been taken into account in this analysis (see discussion of uncertainties in Sect. 13.4.7; Figs. 13.1 and 13.2; Tables 13.4 and 13.5).

13.4.5.2 Macroeconomic Effects

Concerning heating and cooling those impact chains which are triggering changes of energy demand[8] of private households, the government as well as industry (intermediate demand) are implemented in the macroeconomic model (see Sect. 13.3.2 for a detailed description of the impact chains and Supplementary Material Table 13.4 for a summary of how the effects are implemented into the macroeconomic model). All macroeconomic effects are calculated for 2030 (representative for the period 2016–2045) and for 2050 (representative for the period 2036–2065) and effects are expressed for the climate change impact (CC mid-range) scenarios relative to the baseline without climate change in the same respective period (2030 and 2050).

[7] The full list of socio-economic factors is shown in the supplementary materials (Supplementary Material Tables 13.1 and 13.2.)

[8] The "rebound effect" is neglected in the macroeconomic assessment. Some aspects of the rebound-effect are covered implicitly in Invert/EE-Lab (increased effective indoor temperature after building renovation).

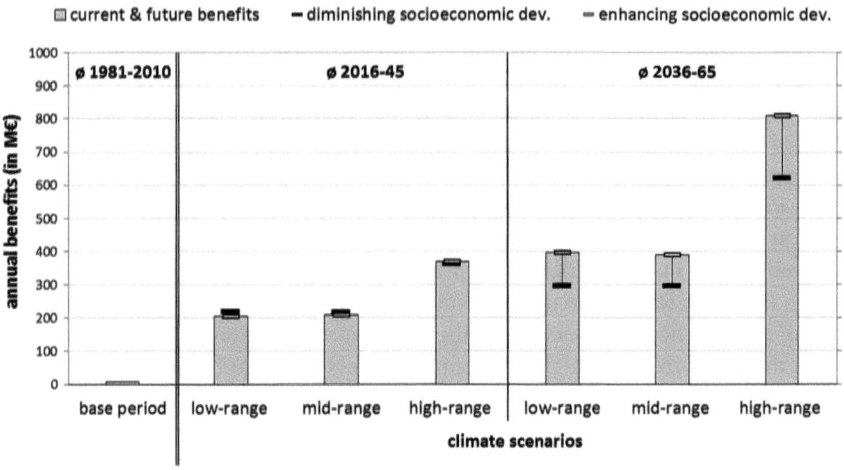

Fig. 13.1 Average annual economic impacts for heating energy in climate change and socio-economic scenarios

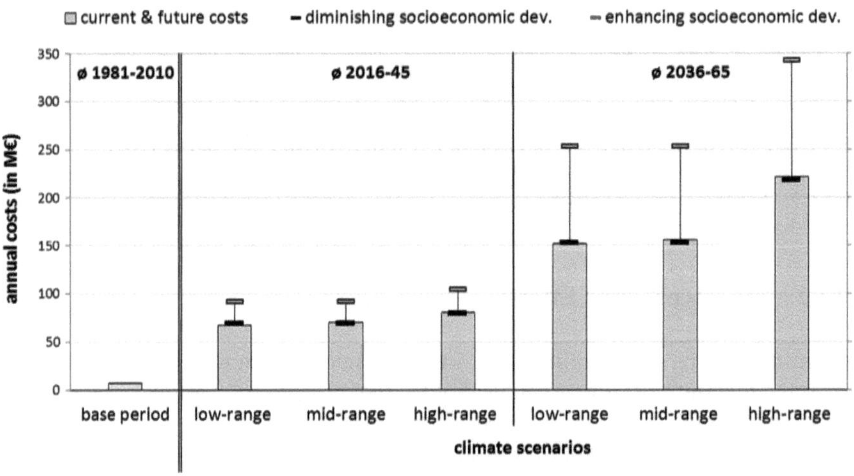

Fig. 13.2 Average annual economic impacts for cooling in climate change and socio-economic scenarios

When combining heating and cooling effects which are triggered by climate change, the absolute reduction in heating is stronger than the increase for cooling (measured in expenditures). Hence, expenditures for energy (i.e. demand) is lower in the climate change scenario compared to the baseline. Table 13.6 gives an overview of selected winners and losers after implementation of the quantified impact chains regarding final and intermediate demand changes. All numbers show absolute changes between the climate change and the baseline scenario, given in million euros. The sectoral effect on gross output value is decomposed into two

Table 13.4 Average annual economic impacts for heating in Austria

Projected future benefits (M€ p.a.)			Climate change		
			Low-range	Mid-range	High-range
Ø 2016–2045	Socioeconomic development	Diminishing	217	216	360
		Reference	205	208	368
Ø 2036–2065		Diminishing	293	294	619
		Reference	395	390	809

Table 13.5 Average annual economic impacts for cooling in Austria

Projected future costs (M€ p.a.)			Climate change		
			Low-range	Mid-range	High-range
Ø 2016–2045	Socioeconomic development	Diminishing	69	69	79
		Reference	67	70	80
		Enhancing	92	92	104
Ø 2036–2065		Diminishing	153	153	218
		Reference	152	156	222
		Enhancing	253	253	343

additive components, namely intermediate demand and value added (equivalently value added is obtained by subtracting all intermediate inputs from gross output value). The change of value added is giving information on how much the sectoral contribution to GDP is changing.

Starting with the sectoral losers (in terms of value added), we see that those sectors which are supplying energy carriers[9] have a lower value added in the climate change scenario, as demand is lower. The Energy sector is hit hardest (−39 million euros on average per year in 2030 and −79 million euros in 2050), followed by Forestry (−14 million euros and −33 million euros) and Coke and Petroleum products (−1 million euros and −1 million euros, respectively). As demand is lower, less output is necessary and therefore also intermediate demand as well as output value is lower in the climate change scenario for those sectors.

Regarding sectoral winners (in terms of value added), we see that the Trade sector is on top, as there more demand for air conditioners is leading to higher output (as well as price). Next to that, there are also positive effects on other sectors, as private households can expand their consumption for other goods and services than heating and cooling. Therefore consumption for Real Estate, Accommodation (including a part of tourism) as well as Rest of Services is higher in the climate

[9] Sector "Energy" is providing electricity, gas and district heat; sector "Coke and Petroleum Products" is providing coke and fuel oil; sectors "Forestry" and "Trade" are providing biomass; sector "Trade" is providing air conditioners.

Table 13.6 Sectoral and total effects of quantified climate change impacts for heating and cooling, average annual effects relative to baseline (for periods 2016–2045 and 2036–2065)

Changes in M€ p.a. relative to baseline	Ø 2016–2045 Gross output value	Inter-mediate demand	Gross value added	Ø 2036–2065 Gross output value	Inter-mediate	Gross value added
Gaining sectors	**+139**	**+57**	**+81**	**+285**	**+119**	**+166**
Trade	+33	+13	+20	+55	+22	+33
Real estate	+26	+8	+18	+55	+17	+38
Accommodation	+14	+5	+9	+30	+11	+19
Rest of services	+10	+2	+8	+21	+5	+16
All other gaining sectors	+56	+29	+27	+123	+64	+59
Losing sectors	**−260**	**−203**	**−58**	**−532**	**−414**	**−118**
Energy	−209	−170	−39	−435	−356	−79
Forestry	−30	−16	−14	−71	−38	−33
Coke and petro-leum products	−7	−7	−1	−9	−9	−1
All other losing sectors	−14	−9	−4	−16	−11	−5
Total effect (all sectors)	**−122**	**−146**	**+24**	**−246**	**−295**	**+48**
GDP at producer price			+0.01 %			+0.01 %

Note: Baseline scenario = reference socioeconomic development without climate change; climate change scenario = reference socioeconomic development and mid-range climate change; quantified climate impact chain: final and intermediate demand changes

change scenario, leading to more value added in those sectors. Compared to the baseline scenario, the overall effect concerning sectoral winners is a higher value added in the amount of +81 million euros in 2030 and of +166 million euros in 2050 (on average per year). Note that the positive effects are accompanied by higher prices, which amplifies the demand driven quantity effect.

Summing up across all sectors, value added in 2030 is by 24 million euros higher in the climate change scenario due to the implemented climate impact chains regarding heating and cooling effects. The effect is stronger in 2050, where gross value added is higher by 48 million euros (compared to the baseline scenario in 2030 and 2050 respectively).

By summing up sectoral effects on value added and correcting for indirect taxes minus subsidies, we obtain the effect on GDP. The impact chains regarding final and intermediate demand changes lead to effects on GDP of +27 million euros (+0.01 %) on average per year in 2030 and of +54 million euros in 2050 (+0.01 %)

(relative to the baseline scenario in 2030 and 2050). Regarding welfare—measured as the quantity of consumed goods and services in the climate change scenario at baseline prices—the effects are stronger (+86 million euros in 2030 and +125 million euros in 2050) because the climate change induced decrease in demand for energy does not decrease welfare since in the climate change scenario the same utility out of heating and cooling can be achieved but less expenditure is needed. Unemployment is on average slightly lower in the climate changes scenario. This is triggered by the overall positive trend and expansion of production.

In the climate change scenario government revenues are slightly higher compared to the baseline scenario; namely by +12 million euros on average per year in 2030 and by +24 million euros in 2050. Compared to the baseline which lies at 149,044 million euros in 2030 and at 206,390 million euros the climate change impacts are about +0.01 % in 2030 and 2050. The higher revenues are mainly covered by higher labour tax revenues as well as value added tax (triggered by more consumption). As revenues are rising, also expenditures do (as we assume equality between revenues and expenditures): Unemployment benefits are lower as employment is rising, but other transfers to private households are rising, as government does not expand its consumption due to climate change, but is giving additional tax revenues back to the households as transfers (see Supplementary Material Table 13.5 for more details about the effects on public budgets).

13.4.6 Qualitative Impacts (Non-monetised)

Although we assumed a rising share of air conditioning with increased temperature levels, in our scenario there is still a substantial part of the building stock without AC. Occupants of such buildings can expect a significant loss of comfort during periods of high temperatures. While this loss of comfort implies a significant additional welfare loss, it could not be quantified. Therefore, both the data on air-conditioning and those on the overall health implications were cross-checked and confirmed by the authors of the Chap. 11 (Human Health). However, there is still the additional aspect of comfort loss which has no direct health implication but definitely leads to welfare loss. Dealing with them was not subject of the present study. So, we explicitly want to emphasize that these costs are not included in the quantitative data presented.

13.4.7 Sector-Specific Uncertainties

The estimations of costs and benefits presented here involve substantial uncertainties:

- Poor data availability regarding the current diffusion of AC. In addition, the development trends concerning the diffusion of air conditioning units in residential and non-residential buildings are far from clear. Factors such as comfort requirements, economic development, length, frequency and severity of heat waves are all subject to considerable uncertainty. The approach taken here links indoor temperature levels in buildings directly to the diffusion of air conditioning. However, the empirical evidence for this link remains rather poor.
- Autonomous adaptation in the form of shading devices and reduction of cooling loads. We assumed that autonomous adaptation mainly refers to maladaptation in the form of a higher share and operation of air conditioning. However, the uptake of shading devices and other efficiency measures for reducing cooling loads is also a possible form of an autonomous, uncoordinated adaptation response.
- The heat island effect was not quantified in our study. This indicates that we are probably underestimating the future cooling energy demand and load. Further related aspects are investigated in Chap. 17 (Cities).
- The overall net monetary result is strongly influenced by the level of energy prices. Since the reduced costs for heating energy demand (which is mainly non-electrical energy) are offset to some extent by the increased costs for cooling energy demand, the ratio of electricity price to fuel prices has a significant impact on magnitudes when assessing the net effect.
- When interpreting the results we should be aware that the COIN mid-range climate scenario is among those scenarios in the A1B family with relatively low summer temperature levels. So, the results of the mid-range climate scenario are probably underestimating the effects of cooling energy need, related costs and comfort losses.

13.4.8 Relevance for Other Sectors

Energy demand for cooling is mainly covered by electricity. The results in Chap. 14 (Electricity) indicate that the impact on peak electricity loads in summer could become highly significant. The feedback loop from potential higher electricity peak prices in summer on costs for cooling energy demand was not considered since real time pricing is not very common up to now.

As far as cooling energy demand is not covered by passive or active technologies, higher indoor temperature results and may impact human health. This is covered in Chap. 11 (Human Health) of this book.

13.5 Summary and Conclusions Regarding Climate Costs for Heating and Cooling

Based on the analyses performed for this study we conclude that the change in heating energy demand and cooling energy demand, together with the additional investments required in cooling devices, represent the main areas through which the impact of climate change on the investigated sector will be felt. We quantified these effects in our analysis. The overall final energy demand for space cooling in Austria in 2008 was only about 0.4–0.5 % of the energy demand for space heating. Thus, the climate-induced decrease in energy demand for heating strongly outweighs the increase in energy demand for cooling and related investment requirements. This is true even though the price of electricity is much more significant in the cooling sector than in the fuel mix applied in the heating sector. There is thus a climate-induced net benefit (i.e. lower net costs for heating and cooling) of about 120 million euros/year in 2030, and of 226 million euros/year in 2050. This does not take into account the additional costs which may result from the need to increase plant capacity to meet higher peak electricity demand for cooling in summer. This is further investigated in Chap. 14 (Electricity). Regarding the macroeconomic consequences of climate change concerning heating and cooling positive effects on welfare are emerging as the same level of utility can be achieved with less expenditures. The effects on GDP are also slightly positive and unemployment is marginally lower.

The analysis of climate and socio-economic ranges indicate that both effects (reduced heating energy expenses and increased cooling energy expenses) could strongly increase in hotter climate scenarios.

Several impacts could not be quantified in the present study. This includes for example changes in comfort levels.

High efforts in energy efficiency improvement in the building sector are one of the key prerequisites for ambitious climate mitigation targets. Due to the very long lead times in the building sector, there is an urgent need to adopt effective policies creating the regulatory and economic framework for a low-carbon building stock in 2050. This has to be accompanied by adaptation measures in order to reduce not only heating energy demand but also address cooling energy demand. In particular, considering passive measures to reduce cooling energy demand (e.g. shading, night cooling) in building codes is of high relevance.

The results in this book are derived for the case of Austria. To which extent the results can be transferred to other regions, mainly depends on the following conditions: (1) The relation and absolute level of heating and cooling energy demand should be comparable. At least, this is the case in Western and central EU countries. (2) The results are strongly driven by the energy price level. Thus, in regions with strongly different energy prices and in particular with different relation of fuel and electricity prices, the results would deviate correspondingly. (3) The results depend also on the energy policy targets and framework which can be assumed for the development in the next decades. Thus, these conditions should be considered as comparable. Last but not least, in countries with a significantly higher current share of air conditioning devices in the building stock, the uncertainty regarding the future market penetration of these units would be much lower. Thus, the corresponding methodological approach could and should be adapted.

References

Aguiar R, Oliveira M, Gonccedilalves H (2002) Climate change impacts on the thermal performance of Portuguese buildings. Results of the SIAM study. Build Serv Eng Res Technol 23:223–231. doi:10.1191/0143624402bt045oa

Beccali M, Cellura M, Brano VL, Marvuglia A (2007) Short-term prediction of household electricity consumption: Assessing weather sensitivity in a Mediterranean area. Renew Sust Energ Rev 12(8):2040–2065

Bednar T, Neusser M, Gladt M (2013) Modelling heating and cooling energy demand of buildings under climate change. Working paper in the frame of the project PRESENCE, Wien

Bundesministerium für Wirtschaft, Familie und Jugend (2010) Nationaler Aktionsplan 2010 für erneuerbare Energie für Österreich (NREAP-AT) gemäß der Richtlinie 2009/28/EG des Europäischen Parlaments und des Rates

Cartalis C, Synodinou A, Proedrou M, Tsangrassoulis A, Santamouris M (2001) Modifications in energy demand in urban areas as a result of climate changes: an assessment for the southeast Mediterranean region. Energy Convers Manag 42:1647–1656

EPBD Recast (2010) European Parliament and the council, 2010. Directive 2010/31/EU of the European Parliament and of the council on the energy performance of buildings (recast)

Giannakopoulos C, Psiloglou BE (2006) Trends in energy load demand for Athens, Greece: weather and non-weather related factors. Climate Res 31:97

Haas R, Biermayr P, Müller A, Kranzl L, Schriefl E (2007) Wärme und Kälte aus Erneuerbaren 2030. Studie im Auftrag der Wirtschaftskammer Österreich. Dachverband Energie-Klima, Wien

Haiden T, Kann A, Wittmann C, Pistotnik G, Bica B, Gruber C (2011a) The integrated nowcasting through comprehensive analysis(INCA) system and its validation over the eastern Alpine region. Weather Forecast 26:166–183

Heindl W, Kornicki T, Sigmund A (1990) Erstellung halbsynthetischer Klimadatensätze für meteorologische Messstationen, Project report

Isaac M, van Vuuren DP (2009) Modeling global residential sector energy demand for heating and air conditioning in the context of climate change. Energy Policy 37:507–521. doi:10.1016/j.enpol.2008.09.051

Kranzl L, Formayer H, Haas R, Kalt G, Manfred L, Müller A, Nachtnebel HP, Redl C, Schörghuber S, Seidl R, Stanzel P (2010) Ableitung von prioritären Maßnahmen zur Adaption des Energiesystems an den Klimawandel.Endbericht im Rahmen der Programmlinie "Energie der Zukunft", Wien

Kranzl L, Hummel M, Müller A, Steinbach J (2013a) Renewable heating: perspectives and the impact of policy instruments. Energy Policy 59:44–58. doi:10.1016/j.enpol.2013.03.050

Kranzl L, Matzenberger J, Totschnig G, Toleikyte A, Schicker I, Formayer H, Gorgas T, Stanzel P, Nachtnebel HP, Bednar T, Gladt M, Neusser M (2013b) Modelling climate change impact on energy systems. Working Paper for the Second Review-Workshop in the project PRESENCE, Vienna

Kranzl L, Matzenberger J, Totschnig G, Toleikyte A, Schicker I, Formayer H, Gorgas T, Stanzel P, Nachtnebel HP, Bednar T, Gladt M, Neusser M (2014) Power through resilience of energy system. Final report of the project PRESENCE. Project in the frame of the Austrian climate research program, Vienna

Moral-Carcedo J, Vicéns-Otero J (2005) Modelling the non-linear response of Spanish electricity demand to temperature variations. Energy Econ 27:477–494. doi:10.1016/j.eneco.2005.01.003

Müller A (2012) Stochastic building simulation, working paper, Berkeley. http://www.marshallplan.at/images/papers_scholarship/2012/Mueller.pdf

Müller A, Kranzl L (2013) Energieszenarien bis 2030: Wärmebedarf der Kleinverbraucher. Ein Projekt im Rahmen der Erstellung von energiewirtschaftlichen Inputparametern und Szenarien zur Erfüllung der Berichtspflichten des Monitoring Mechanisms, Wien

Müller A, Biermayr P, Kranzl L, Haas R, Altenburger F, Weiss W, Bergmann I, Friedl G, Haslinger W, Heimrath R, Ohnmacht R (2010) Heizen 2050: Systeme zur Wärmebereitstellung und Raumklimatisierung im österreichischen Gebäudebestand: Technologische Anforderungen bis zum Jahr 2050. Gefördert vom Klima- und Energiefonds

Müller A, Hummel M, Kranzl L, Bednar T, Neusser M (2014, in press) Climate change impact on heating and cooling: the example of Austria. Energy Build

OIB (2011) Leitfaden Energietechnisches Verhalten von Gebäuden. OIB-330.6-111/11-010. Österreichisches Institut für Bauordnung, Wien

OIB (2012) OIB-Dokument zur Definition des Niedrigstenergiegebäudes und zur Festlegung von Zwischenzielen in einem „Nationalen Plan"gemäß Artikel 9 (3) zu 2010/31/EU. Österreichisches Institut für Bauordnung, Wien

Olonscheck M, Holsten A, Kropp JP (2011) Heating and cooling energy demand and related emissions of the German residential building stock under climate change. Energy Policy 39:4795–4806. doi:10.1016/j.enpol.2011.06.041

Petoukhov V, Semenov VA (2010) A link between reduced Barents-Kara sea ice and cold winter extremes over northern continents. J Geophys Res 115. doi: 10.1029/2009JD013568

Prettenthaler F, Gobiet A (2008) Heizen und Kühlen im Klimawandel—Teil 1, Studien zum Klimawandel in Österreich. Verlag der Österreichischen Akademie der Wissenschaften, Wien

Roberts S (2008) Altering existing buildings in the UK. Energy Policy 36:4482–4486. doi:10.1016/j.enpol.2008.09.023

Schicker I, Formayer H (2012) Working paper on climate change scenarios. In the frame of the project PRESENCE, Wien

Statistik Austria (2011) Energiebilanzen Österreich 1970–2010. Statistik Austria, Wien

Toleikyte A, Hummel M, Kranzl L (2012) Discussion of climate sensitive scenarios for the heating and cooling sectors in selected European countries. Working paper of the project PRESENCE, Wien

Zangheri P, Armani R, Pietrobon M, Pagliano L, Fernandez-Boneta M, Müller A (2014) Heating and cooling energy demand and loads for building types in different countries of the EU. Report in the frame of the IEE project ENTRANZE

Zoll R (2010) Szenarien zur Entwicklung der Gebäudeklimatisierung in Österreich. Diplomarbeit an der Technischen Universität Wien, Institut für elektrische Anlagen und Energiewirtschaft

Chapter 14
Electricity

Lukas Kranzl, Gerhard Totschnig, Andreas Müller, Gabriel Bachner, and Birgit Bednar-Friedl

Abstract This chapter investigates the impact of climate change on the electricity sector. We quantified two main impact chains: (1) impact of climate change on electricity supply, in particular on hydropower and (2) impact of climate change on electricity demand, in particular for heating and cooling. The combined effects of these two impact chains were investigated using the optimization model HiREPS. This takes the hourly resolution of the electricity system into account and considers, in particular, the interaction of the Austrian and German electricity markets. The results show that by 2050 there is a robust shift in the generation of hydroelectric power from summer to winter periods and a slight overall reduction in hydropower generation. The absolute increase in electricity demand is moderate. However, the electricity peak for cooling approximately reaches the level of the overall electricity load in 2010. These two effects—decreasing hydropower supply and increasing cooling electricity peak load (cf. Chap. 13)—lead to moderate sectoral climate change costs in 2050 compared to the baseline scenario without climate change. Regarding macroeconomic effects coming from climate change impacts on the electricity sector we see negative impacts on welfare as well as GDP. However, significant uncertainties remain and the effect of extreme events and natural hazards on electricity supply and transmission infrastructure also needs further examination. The costs of a potential increase in black out risk may be orders of magnitude higher than the costs indicated in our mid-range scenario.

L. Kranzl (✉) • G. Totschnig • A. Müller
Institute of Energy Systems and Electrical Drives, Vienna University of Technology, Vienna, Austria
e-mail: kranzl@eeg.tuwien.ac.at; totschnig@eeg.tuwien.ac.at; mueller@eeg.tuwien.ac.at

G. Bachner
Wegener Center for Climate and Global Change, University of Graz, Graz, Austria
e-mail: gabriel.bachner@uni-graz.at

B. Bednar-Friedl
Wegener Center for Climate and Global Change, University of Graz, Graz, Austria

Department of Economics, University of Graz, Graz, Austria
e-mail: birgit.friedl@uni-graz.at

K.W. Steininger et al. (eds.), *Economic Evaluation of Climate Change Impacts*,
Springer Climate, DOI 10.1007/978-3-319-12457-5_14

14.1 Introduction

As the highest emission of greenhouse gases occurs in the energy sector it is clearly one of the key drivers of climate change. At the same time, however, climate change itself has an effect on energy provision and consumption: e.g. the demand for cooling energy increases, the demand for heating energy decreases, hydropower generation is affected by changes in levels of precipitation and evaporation, power plants may be affected by rising sea level, decreasing cooling water supply (quantity and temperature) and the risk of damage to energy grid infrastructure as a result of an increase in natural hazards is also likely to increase [on a global scale, these aspects are discussed in detail in Arent and Tol (2014)].

The energy sector has been subject to radical change in the last few years and decades (see e.g. Haas et al. 2008; Grübler et al. 2011). In the future, the sector clearly will have a major impact on global sustainability indicators and the level of greenhouse gas emissions. The objective of this chapter is to assess the cost of climate change in the electricity sector. Despite of the fact that the costs of climate change in the electricity sector are also driven by non-climatic factors (e.g. population growth) and by the whole development of the sector itself (e.g. technological development) which is highly uncertain, the focus of this chapter is not to provide a detailed assessment of future scenarios of the electricity system.

The relevant impact of climate change on electricity demand was specifically taken into account in the field of activity "buildings: heating and cooling" but also with regard to demand of other energy carriers than electricity (see Chap. 13). In this chapter, we focus on relevant aspects of climate change costs in the electricity sector for the case of Austria. We are aware that a comprehensive study of climate change impact across the whole energy sector would have to include also other sectors, aspects and effects than those considered and quantified here (see also Table 14.1 and the list of considered impact chains).

14.2 Dimensions of Climate Sensitivity to Climate Change

The electricity sector exhibits several dimensions of sensitivity to climate change. However, for the purposes of the present study, only a sub-set of these dimensions is investigated in any detail (see Table 14.1).

Relevant topics are:

- Sensitivity of infrastructure to natural hazards and on the international scale to sea level rise (as a landlocked country the latter is not relevant for Austria): This includes impact on transmission infrastructure and impact on supply infrastructure (e.g. refineries, power plants, mining). In the present work, this aspect is not quantified.
- Sensitivity of electricity demand: Climate change has an impact not only on electricity consumption for heating and cooling but also on the related load

Table 14.1 Impact chains sector electricity

Climate change parameter	Impact chain	Quantified in the model
Precipitation and temperature: Change in precipitation or change of seasonal distribution, increase in temperature	*Change in river run-off levels for hydropower catchments* → Change in overall annual hydropower generation and change in seasonal hydropower generation profile → change in electricity generation mix [change in production cost]	Yes
	Change in river run-off and water temperature in summer → Lower availability of cooling water for thermal and nuclear power plants → change in electricity generation mix and/or reduction in reliability of the electricity system	No
Wind speed and solar radiation: Change of wind speed (including frequency of storms) and solar radiation	*Change in wind and PV power generation* → Change in electricity generation mix [change in production cost]	Yes
Temperature: Increase of mean values and heat waves	*Increased cooling energy demand in summer* → Increased electricity demand, change in load profile and increase in summer electricity peak load [change in final demand] → change in electricity generation mix [change in production cost]	Yes
	Decreased heating energy demand in winter → Decreased electricity demand, change in load profile and decrease in winter electricity peak load [change in final demand] → change in electricity generation mix [change in production cost]	Yes
Precipitation and temperature, wind speed and solar radiation: see impact chains above	*Change in supply and demand profiles and resulting residual loads* → Change in reliability of electricity supply and change in probability of blackouts (if no corrective actions are taken)	No
Storms, temperature increase, floods, drought and other extreme events: Increase of extreme events	*Natural hazards, sea level rise etc.* → Electricity infrastructure at risk (supply and transmission) → change in reliability of electricity supply and change in probability of blackouts (if no corrective actions are taken)	No

profiles. Moreover, climate change impact in other sectors such as transport and mobility, or manufacturing, might also have an effect on electricity demand and related load profiles. In our work, electricity demand for heating and cooling as well as related load profiles are based on a detailed techno-economic bottom-up model Invert/EE-Lab applied in Chap. 13.

- Sensitivity of energy supply: Electricity supply is affected in various ways by climate change: First, cooling water temperature and quantity has an effect on the availability and efficiency of thermal power plants or on the corresponding additional costs for cooling towers (see e.g. Förster and Lilliestam 2010; Klein et al. 2013). This is driven in particular by the maximum permitted temperature increase of cooling water in rivers. Second, ambient air temperature levels to some extent have an impact on the electrical efficiency levels of thermal power plants. Third, all renewable power plants are dependent on parameters such as precipitation/evaporation, wind velocity or radiation (cloudiness) and thus sensitive to shifts induced by climate change. In particular, run-off in river basins and hydropower availability will change as precipitation and evaporation levels adapt to new temperatures. In our work, the third aspect has been quantified by taking into account results from a hydrological model of Austrian river run-off and dependence on precipitation and temperature levels (Nachtnebel et al. 2013).

14.2.1 Climatic Factors

The following climatic factors are relevant for the electricity sector: (1) Temperature increases in winter and summer, in particular heat waves, have an impact on heating and cooling (see e.g. Aebischer et al. 2007; De Cian et al. 2007; Isaac and van Vuuren 2009; Olonscheck et al. 2011). (2) Temperature and precipitation are relevant for river run-off and affect hydropower and cooling water availability (see e.g. Koch and Vögele 2009; Nachtnebel et al. 2013; Felberbauer et al. 2010). (3) Solar radiation is relevant for PV and solar thermal generation. (4) Wind speed and the frequency of storms have an impact on wind power generation (see e.g. Pryor and Barthelmie 2010). (5) A greater prevalence of storms, flooding, avalanches and other natural hazards may lead to increased damage to electricity transmission infrastructure (see e.g. Altvater et al. 2011; Francis et al. 2011; Kirkinen et al. 2005; Mima et al. 2012).[1]

With respect to all these factors, particular attention has to be paid to potential changes in seasonal patterns since these determine the required additional fossil power plant capacity.

[1] Section 14.4.1 documents in a more detailed way the climate input data and Sect. 14.4.3 explains the methods how the related impact chains are assessed.

14.2.2 Non-climatic Factors

Within the last few years and decades, the electricity sector has developed in a highly dynamic fashion. Technology and energy carrier mix change continuously, as does the political framework. The implementation and compliance with policy targets will be a highly relevant factor for the future development of the electricity system. On the one hand, the state of the electricity system in 2050 and beyond will depend on how these targets are balanced. On the other hand, due to the complexities of global interaction within the energy sector, numerous exogenous non-climate factors such as prevailing geopolitical constellations, the overall development of energy demand and supply around the globe can all exert a major influence. Bearing this in mind, the following non-climatic factors will be particularly relevant in the sector's development up to 2050 and beyond:

- Transformation towards low-carbon electricity: the expected intensity of such a transformation is highly relevant to assess climate change impacts since renewable energy sources have other vulnerabilities than fossil sources.
- Development of a decentralized electricity system: The level of decentralization has consequences for the vulnerability of the electricity grid infrastructure.
- Development of electricity demand: The absolute level of electricity demand is a strong driver of the vulnerability of the electricity system towards climate change since it drives the need for additional infrastructure and capacities (for general drivers of the energy demand and the socio-economic framework in the reference scenario see Chap. 6).

Despite these challenges, in Austria, currently no official long term target, strategy or vision for the future of the electricity sector exists at national energy policy level, which could directly serve as a socio-economic reference scenario. Thus, for the future relevance of non-economic factors we rely on studies like Totschnig et al. (2013), Reichl et al. (2010), Schleicher and Köppl (2013), Streicher et al. (2010) or Köppl et al. (2011). Chapter 3 outlines how socio-economic scenario-assumptions have been considered in the quantitative assessment.

14.2.3 Identifying Combinations Causing Greatest Potential Damage

In general, periods of crisis in the electricity system are associated with increases in periods of high electricity demand, low electricity supply and interruptions in the transmission infrastructure (see e.g. Totschnig et al. 2013; Reichl et al. 2013; Kranzl et al. 2014). The increase in the frequency of such situations is related to the following: increasing demand for cooling energy (i.e. given that no passive counter-measures for reducing cooling loads are taken, and that the use of air conditioning rises to reflect heat wave periods and changes in personal comfort levels); low generation of hydropower and thermal power plants; low share of PV to cover this demand; low storage and grid capacities; large scale regional occurrence of such conditions; low flexibility of loads (cooling loads as well as other electric loads); high energy prices.

As the Austrian electricity sector is closely linked to that of neighbouring regions, conditions prevailing outside the country also impact on all the above factors.

14.3 Exposure to Climatic Stimuli and Impacts to Date

14.3.1 Past and Current Climatic Exposure and Physical Impacts

Exposure of the electricity system to climate change is mainly driven by the type and location of infrastructure, the number of thermal and nuclear power plants relying on cooling water availability and the number of hydropower plants. Moreover, the relevance of air conditioning is also a major driver of summer peak loads and related exposure. The increasing trend towards air conditioning is discussed in the Chap. 13.

Several authors have discussed the issue of growing electricity peaks in summer periods, in particular in countries with higher cooling loads (some of them also discussed in Chap. 13). Beccali et al. (2007) point out that summer electricity consumption in the building sector in Italy has grown steadily. According to the annual reports published by the Italian National Grid Operator, summer peak load for 2000–2005 showed a rise of 25 %, or 8.38 GW. Temperature and corresponding adjusted electricity demand for Spain have been discussed by Moral-Carcedo and Vicéns-Otero (2005). Pechan and Eisenack (2013) discuss the impact of the 2006 heat wave on electricity spot markets. They found that over a two week period in Germany, the heat wave and the resulting reduction in the availability of cooling water led to an average price increase of 11 % and to additional costs of 15.9 million euros.

14.3.2 Impact Chains

Based on the analysis of sensitivity and exposure of the electricity system to climate change, we identified the main impact chains as described in Table 14.1. Section 14.4 describes in more detail the approaches and data how these impact chains have been assessed.

14.4 Future Exposure to and Impacts of Climate Change

14.4.1 Mid Range Climatic Scenario for Electricity

The results for the sector electricity are based on the project PRESENCE (Nachtnebel et al. 2013; Totschnig et al. 2014; Kranzl et al. 2013). We selected the climate change scenario A1B from the model REMO (driven by ECHAM5) in order to allow the use of comprehensive model results based on Kranzl et al. (2014). This shows very similar climate change signals to those found in the COIN climate change scenario. For modelling the impact of climate change on river run-off and hydropower generation (Nachtnebel et al. 2013), we used bias-corrected and localized climate data. The bias-corrected RCMs and the observed gridded data (E-OBS data) had to be spatially downscaled from the 25 × 25 km grid to a 1 × 1 km grid. This was done using the high-resolution Austrian INCA data set (Haiden et al. 2011). Thus, it was possible to capture the major Austrian valleys and mountain regions. As the INCA data set only applies to the period from 2003 onwards it could not be used directly for bias correction, but it was possible to use it to estimate spatial variability e.g. of temperature and precipitation on a monthly basis. This information was then included in the localized RCM scenario data (Pospichal et al. 2010).

For the localization of the parameters, monthly means for the corresponding time period in the RCM data, the hydrological data, and the INCA data were calculated. Corresponding grid cells (1 × 1 km) in the INCA/hydrological model are assigned to RCM grid cells (25 × 25 km). For each month correction factors for each parameter were calculated for every INCA grid cell. These correction factors were subsequently applied to the daily values of the RCM data, thus inserting the spatial variability of the high-resolution data set into the model. This method assumes that the differences between the RCM data and the INCA data are the result of altitude and orographical effects, i.e. are constant over time. While this is likely to be true with respect to temperature and shortwave radiation, it is not likely to hold for precipitation. Nonetheless, it is still the best available method. Calculation of parameter correction factors is now described below.

Climate scenario data applied to the sector heating and cooling are further documented in Chap. 13.

The electricity sector was modeled using HiREPS. Three meteorological parameters were needed for the analysis: temperature, radiation, and wind speed (in addition to the input from the hydrological model regarding hydropower). The highest possible temporal resolution available was used (daily for temperature and radiation, 12-hourly for wind speed).

For wind speed and radiation monthly percentiles of the hindcast were calculated and then assigned to the control and scenario data to generate look-up tables for the HiREPs model. As radiation input, the global radiation calculated was used. Wind speed at a height of 850 hPa were considered for Austria and Germany for each grid cell. Then it was averaged over the entire domain and then the cubic root was calculated to provide input for the HiREPs model. For temperature, a similar approach to that used for radiation was employed. However, here an additional weight using the population density of the lspop (1 × 1 km, Dobson et al. 2000) was applied.

14.4.2 High and Low Range Climatic Scenarios for Electricity

The sensitivities for high- and low-range climatic scenarios were carried out for the aspect of temperature impact on heating and cooling. The evaluation of direct monetary effects was carried out for the change in electricity demand and electricity load for heating and cooling. Where the climate change scenario "low" results in only slightly reduced cooling electricity demand and loads, the impact of the "high" scenario is significant: cooling load increases by more than 45 %. However, as pointed out above, in more extreme climate scenarios the uncertainty regarding the market penetration of air conditioning units increases and might strongly affect this result, compare also the discussion in Chap. 13.

14.4.3 Specific Method(s) of Valuation and Their Implementation

The impact of climate change on supply and demand shifts in electricity was investigated using an integrated modelling approach. Cost evaluation of electricity generation costs needs to be applied to the impact chains described above. The decomposition of effects is not straightforward, since both demand and supply lead to a new electricity price level.

Figure 14.1 shows the documentation of the model cluster which has been applied in the project PRESENCE (Kranzl et al. 2014) and on which the results in the present study are based. HiREPS builds on data directly from the climate scenarios, from the hydrological modelling and from building stock model Invert/EE-Lab, see Chap. 13.

All impact chains which we labeled as "quantified" in Table 14.1 are covered in this approach (Fig. 14.1): change in river run-off levels, wind and PV power generation are taken into account in HiREPS via the hourly climate data described in Sect. 14.4.1 and on a monthly basis the derived results for river run-off in Austrian water basins. Increased cooling energy demand in summer and decreased heating energy demand in winter are considered in HiREPS via the total change of annual final energy demand as well as the change in hourly load profiles and thus also the changes in peak loads.

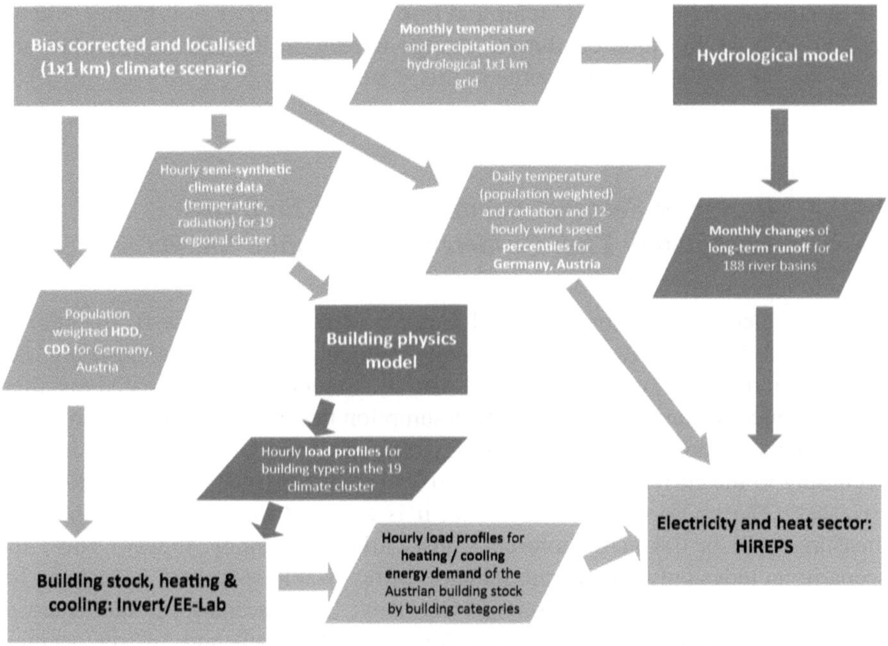

Fig. 14.1 Documentation of flow diagram of data, analytical steps and input of climate data for deriving the impact of climate change on the electricity sector in Austria

Climate change leads to a shift in both the supply and demand curves. The area below the supply curve corresponds to the overall electricity generation costs. The main indicator for monetary evaluation is the difference between the respective electricity generation costs in the baseline case and those in the mid-range scenario. This difference is derived on an hourly basis for the simulation year 2050.

In addition to the above effect, there is also a change in the final demand for electricity. As this is already covered in Chap. 13 and is not considered here.

The change in the electricity generation mix is modelled in the optimisation model HiREPS (Totschnig et al. 2013; Kranzl et al. 2013; Totschnig et al. 2014). The HiREPS model is a dynamical simulation and optimization model of the electricity and heating system. The model focuses on analyzing the integration of fluctuating renewable electricity generation into the power system, and specifically uses an approach whereby important system constraints are treated endogenously. For the investigation in the project COIN the model was applied to the electricity system in Austria and Germany. After optimising the model for both countries, the individual effects for Austria were then separated out for use as input in the HiREPS model. The latter addresses these aspects endogenously by using spatially and temporally highly resolved wind, solar and hydro inflow data, and by including a detailed model of hydropower and pumped storage, thermal power plants (including startup costs and efficiency losses during part-load operation), interaction of the electricity and heating systems, load flow calculation (including thermal limits of the electricity grid), and hourly temporal resolution. Therefore, it is highly suitable to deal with the question of climate change impact on the electricity sector and assessing the cost of climate change in this sector.

14.4.4 Range of Sectoral Socio-economic Pathway Parameters That Co-determine Climate Impact

Since the electricity sector in central Europe is closely interlinked, in order to undertake a dynamic investigation a purely national analysis is not sufficient. For this reason, the Austrian and German electricity sectors were investigated together. The reference scenario is based on the assumption that Austria and Germany meet their targets for renewable energy, energy efficiency and GHG-emissions in 2020 according to the corresponding EU directives (in particular 2009/28/EU, 2010/31/EU, 2012/27/EU). However, after 2020, it is assumed that no further ambitious measures will be taken to enforce a low-carbon electricity supply. Thus, the scenario includes only a moderate increase of renewable electricity generation for the investigated region of Austria and Germany. In the scenario, a total share of about 30 % renewable electricity generation (i.e. for Germany and Austria together) is assumed, with wind generation accounting for almost 12 and PV for 6 % (Totschnig et al. 2013).

Electricity consumption growth is based on Capros et al. (2013). According to this source, total electricity consumption increases by about 48 % for Austria from 2005 until 2050. The evolution of electricity demand for heating and cooling (the reference scenario) has been described in more detail in Chap. 13.

14.4.5 Monetary Evaluation of Impacts

14.4.5.1 Direct Sector Impacts (Costs and Benefits) Excluding Feedback Effects from Other Sectors

As pointed out above, there are mainly two aspects driving the costs of climate change in the electricity sector: (1) impact on electricity supply and (2) impact on electricity demand. While the impact on wind and PV generation is almost negligible, the shift of hydropower generation from summer to winter season is quite significant. Also, results from other studies show that this result is robust over a wide number of studies, see e.g. Bachner et al. (2013), Felberbauer et al. (2010), Nachtnebel et al. (2013), Kranzl et al. (2010).

The second major impact is due to the increase in cooling energy demand, see Chap. 13. However, not only the total increase in electricity demand (relatively moderate) is relevant, but also the strong increase in the cooling peak load.

As described in Chap. 13, climate change also leads to reduced heating loads during winter. However, due to the higher simultaneity in the cooling energy demand and the lower full load hours of cooling devices, the impact on electricity peak load is expected to be higher for cooling as for heating. For this reason we decided to take into account only the costs for increasing cooling peak loads.

In total, both in the baseline and in the mid-range climate change scenario electricity demand for heating and cooling declines from the base year until 2030 and 2050. Thus, the additional electricity consumption for cooling which occurs in both scenarios until 2050 is compensated by increasing energy performance for heating, because the latter is also partly covered by electricity for heat pumps or by direct electric heating. Until 2050, the baseline scenario results in a reduced electricity consumption for heating and cooling by 20 %, whereas in the mid-range scenario the reduction is only 18 % due to the increased relevance of cooling energy.

Both effects (supply and demand) lead to a change in the electricity generation mix and an increase in electricity generation costs. In particular the increase in summer peak loads as a result of greater cooling demand lead to a high electricity price, though this holds only for a very limited time over the whole year. Overall, the effects result in a slight increase of fuel costs for natural gas from 1,000 million euros in the baseline to about 1,040 million euros in the mid-range climate change scenario in 2030 and from 2,300 million euros to 2,420 million euros in 2050. The fuel expenditures for coal and biomass power plants are not affected by climate change according to the model results.

In order to deal with the higher peak load in summer, additional investments in power generation plants are required (assuming no countermeasures are taken to reduce this peak load) and higher electricity generation costs occur. The sums of these effects are shown in Table 14.2 and Fig. 14.2. The cost data are based on the assumptions regarding energy prices of the SSP (Shared Socio-economic pathways, see Chap. 6) and without discounting of cost of inaction.

In addition to the reference socio-economic scenario and the mid-range climate change scenario, results for low and high climate scenarios are included in Figure 14.2 and Table 14.2. They are driven by the additional electricity demand and related peak loads for cooling.

Table 14.2 Selected economic impacts of climate change on the electricity sector[a]

Projected future costs relative to Ø 1981–2010 (M€)			Climate change		
			Low-range	Mid-range	High-range
Ø 2036–2065	Socioeconomic development	Diminishing	207	207	355
		Reference	205	227	363
		Enhancing	435	435	638

[a]Results for low and high climate change scenarios have been calculated only for the change in cooling load. Effects of hydropower generation have not been evaluated in the low and high climate change scenarios and in the diminishing and enhancing socio-economic scenarios

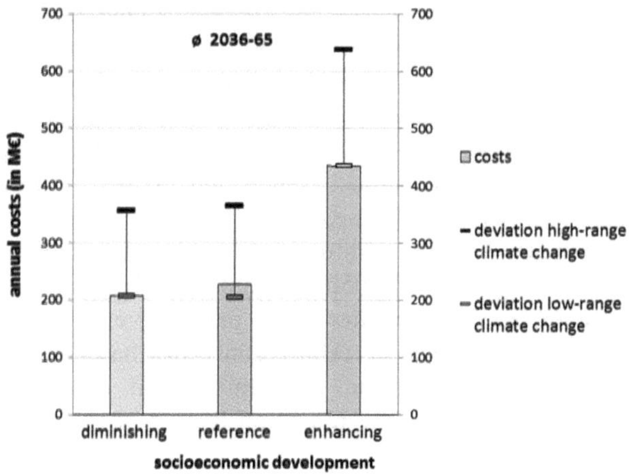

Fig. 14.2 Cost of inaction in the sector electricity

14.4.5.2 Macroeconomic Effects

In the macroeconomic model we implemented those impact chains, which trigger a change in electricity production mix (i.e. change in production costs) as well as a change in final demand, including investments requirements to meet higher peak loads in summer (see Table 14.1 for a detailed description of the individual impact chains).

Compared to the model base year (2008) production costs for the generation of electricity are changing: In the baseline scenario annual expenditures for gas and coal are rising until the 2030s (2016–2045) and the 2050s (2036–2065), whereas expenditures for biomass and biogas are decreasing (real price effects are included). In the climate change scenario the requirements to meet higher peak loads are met by additional gas turbines. Thus, there are higher expenditures for gas as more input is needed relative to the baseline. The change regarding the electricity production mix is implemented in relative terms. For absolute numbers see Sect. 14.4.5.1.

Furthermore, final demand is changing[2]: In the baseline scenario private households as well as the government are decreasing their consumption of electricity by −21.3 % in the 2030s and by −20.4 % in the 2050s (relative to the model base year).[3] This decrease in future electricity demand is driven by the underlying socio-economic development (less electricity demand for heating and a slight increase for cooling with a negative net effect). In the climate change scenario cooling demand is rising, relative to the baseline, as more air conditioning is assumed. Therefore, the before mentioned socio-economic driven decrease is less strong: Private

[2] Note that only changes of electricity demand are analysed in this chapter. Other energy carriers are not relevant for final demand changes.

[3] By assumption and due to lack of more detailed data the government final demand is showing the same relative demand changes as private households.

households as well as the government are reducing their demand for electricity by −20.5 % in the 2030s and by −18.3 % in the 2050s; relative to the model base year. The impacts regarding electricity demand is implemented in relative terms. For more information on that see Sect. 14.4.5 and Chap. 13.

Finally, the Electricity sector faces higher annual investments. Due to the assumed socio-economic development, additional investments are 99 million euros per year in the 2030s and 298 million euros in the 2050s. Due to climate change peak loads are assumed to be higher because of air conditioning demand. By assumption the higher peak load for cooling is provided by additional gas turbines. Therefore, in the climate change scenario annual investments are higher compared to the baseline scenario: Investments are rising by 130 million euros in the 2030s and by 390 million euros in the 2050s (see Sect. 14.4.5). Table 14.3 summarises the implementation of the stated impacts into the CGE model.

Table 14.4 gives an overview of sectoral effects of climate change impacts relative to the baseline scenario. All effects are given as average changes of annual values in million euros (m€) relative to the respective baseline scenario (price changes by feedback effects are included). Concerning the Energy sector (which is an aggregate including the Electricity sector) we see negative impacts on gross output value as additional investment requirements are leading to higher prices for electricity and therefore, despite the additional demand for cooling by private households and the government, overall demand is lower in the climate change scenario (industry demands less electricity and shifts to other energy sources instead). However, due to higher investments annual depreciation is higher, leading to a higher gross value added. In the climate change scenario annual gross value added is on average +5 million euros above the baseline level in the 2030s and

Table 14.3 Implementation of baseline and climate change scenario for electricity in the macroeconomic model, average annual effects for periods 2016–2045 and 2036–2065

	Ø 2016–2045		Ø 2036–2065	
Change relative to base year (2008)	Baseline	Climate change	Baseline	Climate change
Change of electricity production mix (change of average additional expenditure p.a.)				
Gas	+4.8 %	+5.0 %	+4.6 %	+4.7 %
Coal	+3.8 %	+3.8 %	+3.8 %	+3.8 %
Biomass	−0.3 %	−0.3 %	−0.5 %	−0.5 %
Biogas	−1.3 %	−1.3 %	−3.1 %	−3.1 %
Final demand				
Private households	−21.3 %	−20.5 %	−20.4 %	−18.3 %
Government	−21.3 %	−20.5 %	−20.4 %	−18.3 %
Additional annual investments (in M€ p.a.)	+99	+130	+298	+390

Note: baseline scenario = reference socioeconomic development without climate change; climate change scenario = reference socioeconomic development and mid-range climate change; quantified climate impact chains: change of electricity production mix, final demand changes, additional investments

Table 14.4 Sectoral and total effects of quantified climate change impacts in sector Electricity, average annual effects relative to baseline (for periods 2016–2045 and 2036–2065)

	Ø 2016–2045			Ø 2036–2065		
Changes in M€ p.a. relative to baseline	Gross output value	Inter-mediate demand	Gross value added	Gross output value	Inter-mediate demand	Gross value added
Gaining sectors	**+8**	**−6**	**+13**	**+10**	**−28**	**+39**
Energy (incl. Electricity)	−13	−18	+5	−55	−67	+13
Construction	+19	+11	+7	+59	+36	+23
Rest of extraction (incl. Gas)	+2	+1	+1	+6	+3	+3
Losing sectors	**−263**	**−105**	**−157**	**−748**	**−301**	**−448**
Trade	−39	−15	−24	−110	−42	−68
Real estate	−39	−11	−27	−108	−32	−77
Accommodation	−20	−7	−13	−56	−19	−37
All other losing sectors	−165	−72	−93	−474	−208	−266
Total effect (all sectors)	**−255**	**−111**	**−144**	**−738**	**−329**	**−408**
GDP at producer price			−0.04 %			−0.08 %
...thereof price effect			−0.01 %			−0.02 %
...thereof quantity effect			−0.03 %			−0.06 %

Note: baseline scenario = reference socioeconomic development without climate change; climate change scenario = reference socioeconomic development and mid-range climate change; quantified climate impact chains: change of electricity production mix, final demand changes, additional investments

+13 million euros above the baseline in the 2050s.[4] Intermediate demand of the Energy sector is also lower in the climate change case, because less output is produced.

In terms of value added, many other sectors are losing because of climate change; especially consumption goods coming from the sectors Trade, Real Estate or Accommodation (which covers a part of Tourism). This reflects the fact that private households face higher prices (at a higher demand) for electricity and therefore have lower consumption possibilities for other goods and services. The only two sectors which are gaining are Construction and Rest of extraction

[4] The interaction of price and quantity effect is leading to a negative net effect concerning gross output value of the Energy sector, reflecting that the negative demand (quantity) effect is stronger than the positive price effect.

(including the extraction of gas). The former is gaining because of additional power plant investments which are carried out by the construction sector; the latter is gaining because of higher demand for gas to meet the demand for air conditioning. Summing up, due to the implemented climate impact chains gross value added in the 2030s is by −144 million euros lower in the climate change scenario (compared to the baseline scenario). The effect is much stronger in the 2050s, where gross value added is lower by −408 million euros.

After adding indirect taxes less subsidies to the sum of sectoral value added, we obtain effects on GDP, which is shown in Table 14.5, together with effects on welfare and unemployment. After modelling the climate impacts chains with effects on electricity production, final demand as well as investments GDP is decreasing on average by −165 million euros p.a. in the 2030s and by −467 million euros in the 2050s. Note, that about three quarters of the GDP effect is induced by quantity effects and only one quarter by price effects. Regarding welfare—measured as the quantity of consumed goods and services at prices of the baseline level—we see similar effects as of GDP. The effect on welfare is less strong as on GDP, as a part of the losses is carried by the industry. As economy wide output is lower in the climate change scenario employment is lower as well; in the 2030s by −0.02 %–points and by −0.04 %–points in the 2050s. Together with less consumption this leads to lower government revenues which are shown in Table 14.6. Compared to the baseline scenario government revenues are lower by −61 million euros in the 2030s and by −173 million euros in the 2050s. As we assume equality between revenues and expenditures, expenditures decrease by the same amount (more unemployment benefits but a reduction of other transfers to households). By assumption, government consumption expenditures remain unaffected by climate change impacts (therefore not shown in the table).

14.4.6 Qualitative Impacts (Non-monetised)

We need to bear in mind that the monetary evaluation indicated above is far from complete. A considerable number of aspects were not quantified, e.g.:

1. Infrastructure at risk through more frequent natural hazards: In order to guarantee the same level of system reliability in the future as is existing today, higher levels of systemic redundancy and back up are required, both with respect to transmission and to power generating capacity.
2. An increased frequency of natural hazards and other extreme events, together with a growing frequency of adverse constellations in the power system (i.e. high demand peaks with lack of supply capacity, lack of storage and transmission capacity) could lead to a higher probability of black outs (for cost estimation of black outs see the following section uncertainties).

Table 14.5 Effects of quantified climate change impacts in sector Electricity on GDP, welfare and unemployment across different climate and socioeconomic scenarios, average annual effects relative to baseline (for periods 2016–2045 and 2036–2065)

		Ø 2016–2045			Ø 2036–2065		
	Socioeconomic scenarios						
Changes in M€ p.a. relative to baseline	Climate scenarios	Diminishing	**Reference**	Enhancing	Diminishing	**Reference**	Enhancing
GDP (changes in M€)	High						
	Mid		**−165**			**−467**	
	Low						
Welfare (changes in M€)	High						
	Mid		**−159**			**−443**	
	Low						
Unemployment rate (in % points)	High						
	Mid		**+0.02**			**+0.04**	
	Low						

Note: Empty cells not quantified

Table 14.6 Effects of quantified climate change impacts in sector Electricity on government budget, average annual effects relative to baseline (for periods 2016–2045 and 2036–2065)

Changes in M€ p.a. relative to baseline	Ø 2016–2045	Ø 2036–2065
Revenues	**−61**	**−173**
Production tax	−5	−13
Labour tax	−25	−69
Capital tax	−10	−29
Value added tax	−20	−57
Other taxes	−1	−4
Expenditures	**−61**	**−173**
Unemployment benefits	+22	+62
Transfers to households net of other taxes	−83	−234
Government budget in baseline (p.a.)	**148,480**	**204,500**
Climate change impact on government budget	−0.04 %	−0.08 %

Note: baseline scenario = reference socioeconomic development without climate change; climate change scenario = reference socioeconomic development and mid-range climate change; quantified climate impact chains: change of electricity production mix, final demand changes, additional investments

14.4.7 Specific Uncertainties for the Electricity Sector

Apart from those listed above several other factors could also have a substantial impact:

- **Autonomous adaptation by utilities**. Some of them were explicitly taken into account in our modelling approach (e.g. investment in additional power plant capacities) some others were not e.g. restricting maximum loads during peak times. This would result in some part of the demand for cooling having to remain unsatisfied.
- There is uncertainty regarding the need for current **electricity consumption for cooling**. In addition, the possible development and further uptake of AC-units in the course of heat waves is also subject to a considerable degree of uncertainty.
- Assumptions regarding **technological development**, in particular regarding the cost-evolution of PV and storage capacities have a strong impact on the future characteristics of the electricity system. Lower than expected cost reductions in PV and storage capacities would increase the cost of inaction in the sector.
- **Social acceptability and perceptions** with respect to new technologies and electricity grid expansion: we assume that there are almost no barriers to an expansion of the electric grid and thus to the provision of greater cross-regional flexibility. In reality, we know that there are acceptability barriers of grid expansion. The higher these barriers, the higher the cost of inaction, and the higher the risk of black outs.
- The standard assumption regarding **growth of electricity consumption** was taken from PRIMES (Capros et al. 2013). However, we need to be aware that strong growth in electricity consumption is not in line with current policy targets.

A change in electricity consumption would also change the required additional power plant capacities and related costs.

- **Fluctuation of hydropower generation**: It remains an open question whether increasing fluctuations will lead to some threshold level being surpassed such that there will be a discontinuous jump in costs for the whole electricity system.
- We did not deal at all with potential interruption to supply or transmission **infrastructure** as a result of higher frequency of extreme events such as storms, floods or avalanches.
- There are other extreme events like heat waves and droughts which could—in combination—have a multiplier effect resulting in periods in which high cooling loads, low hydro power generation, and low output from thermal power plants, all occur simultaneously.
- Probability of black outs: All the aspects listed above could not only lead to higher costs in the electricity sector but also to an increased probability of electricity black outs.

Costs of Power Outages in Austria

Johannes Reichl[*]

[*]Energieinstitut at Johannes Keppler University Linz

Extreme meteorological events already cause power supply interruptions under current climate conditions every now and then, and a further increase of such events as a consequence of climate change is expected to result in a coeval increase of power outages. While science still lacks of quantifying the risks of the power system under climate change, simulation studies can provide estimates of the socio-economic damage of such disruptive events. A view on the simulation outcomes reveals the economic relevance of blackouts, and thus the importance of knowledge to prevent and deal with the threat of power outages in the light of climate change. As a first example it is assumed a power outage hits the whole of Austria on a weekday morning in summer and lasts for 6 h. The total expected damage of power outage summarises to about 350–400 million euros for the Austrian economy, of which the energy-intensive manufacturing sector is most vulnerable and thus bares the largest share of these outage costs. Considering the same outage scenario but assuming a duration of 24 h until electricity supply can be restored sums up to damage costs between 750 and 1,100 million euros (Reichl et al. 2013).

The damage costs presented in the last paragraph contain those values sustainably lost during and after the blackout event. This means that while some businesses can catch up with work once the electricity supply is restored, this will usually result in higher production costs and for many goods and services such an option does not exist at all. As a consequence, it is required to better understand the consequences of climate change in this

(continued)

particularly vulnerable field and to learn about its human dimension. Consequently, supporting society and policy makers in adapting to the increased risk of power outages, based on scenarios for future extreme meteorological events, should be much more considered.

References

Reichl J, Schmidthaler M, Schneider F (2013) The value of supply security: the costs of power outages to Austrian households, firms and the public sector. Energy Econ 36:256–261

Lower summer river flows can reduce the operating efficiency and output of hydro power plants. For a closer assessment for this effect, daily hydrological modelling would be required, which was not possible in the present study.

14.5 Summary of Climate Costs for Electricity and Conclusions

For Austria, our scenario results show that climate change leads to a slight reduction in overall hydropower generation. There is also a shift in hydropower generation to the winter period, where, normally, electricity prices are higher (see e.g. Nachtnebel et al. 2013). Reduced electricity demand for heating over the year compensates for the increase in electricity needed for cooling. However, cooling potentially leads to substantially higher electricity peak loads in summer, resulting in high electricity prices and higher requirements for back-up capacity, particularly in cases where the additional demand cannot be covered by PV/wind generation.

For the two impact chains described above, the mid-range climate change scenario reveals a moderate increase in costs for the sector for 2050 of about 230 million euros per year. Including spillover effects to and from other sectors, these costs translate into reductions in annual GDP of 467 million euros in the climate change scenario for 2050 (compared to the baseline scenario). Due to these quantified impact chains, prices for electricity are on average increasing leading to lower consumption possibilities for private households, and so the sectors trade, real estate, and accommodation are affected negatively as well. In contrast, construction and the extraction sector (which includes gas) are gaining due to higher investments into back-up capacity.

Taking into account the uncertain impact of cooling energy demand leads to an estimated increase from 230 million euros to about 640 million euros of direct costs in the case of enhancing socio-economic conditions and high-range climate change scenario. However, several additional impact chains were not quantified in the present study. These include the impact of extreme events on electricity supply and transmission infrastructure and the risk of black outs arising from some combination of effects. Lack of resources meant that such effects could not be dealt with sufficiently here.

One must bear in mind that huge uncertainties remain (see discussion on uncertainties above). Probably the most relevant uncertainty relates to the question of an increased probability of black outs. All the aspects listed above could lead not only to higher costs in the electricity sector but also to an increase in the probability of electricity black outs. The costs associated with such black outs are likely to be orders of magnitude higher than those associated with electricity generation. Given the high importance of the electricity sector to society, this is obviously a question which requires considerable attention. In particular, future research needs to focus on two related aspects, i.e. the potential impact of extreme events, and how such events might affect the probability of electricity black outs.

References

Aebischer B, Catenazzi G, Henderson G, Jakob M (2007) Impact of climate change on thermal comfort, heating and cooling energy demand in Europe. ECEEE 2007 Summer study: saving energy – Just do it!, pp 859–870

Altvater S, van de Sandt K, Marinova N, de Block D, Klostermann J, Swart R, Bouwma I, McCallum S, Dworak T, Osberghaus D (2011) Assessment of the most significant threats to the EU posed by the changing climate in the short, medium and long term – Task 1 report. Ecologic, Berlin

Arent D, Tol RSJ (2014). IPCC AR5. Climate change 2014: impacts, adaptation, and vulnerability. Key economic sectors and services, Chap 10. IPCC

Bachner G, Bednar-Friedl B, Birk S, Feichtinger G, Gobiet A, Gutschi C, Heinrich G, Kulmer V, Leuprecht A, Prettenthaler F, Rogler N, Schinko T, Schüppel A, Stigler H, Themessl M, Töglhofer C, Wagner T (2013) Impacts of climate change and adaptation in the electricity sector – the case of Austria in a continental european context (EL.ADAPT), Wegener Center Scientific Report 51-2013. Wegener Center Verlag, Graz

Beccali M, Cellura M, Brano VL, Marvuglia A (2007) Short-term prediction of household electricity consumption: assessing weather sensitivity in a Mediterranean area. Renew Sustain Energy Rev 12(8):2040–2065

Capros P, De Vita A, Tasios N, Papadopoulos D, Siskos P, Apostolaki E, Zampara M, Paroussos L, Fragiadakis K, Kouvaritakis N et al (2013) EU energy, transport and GHG emissions: trends to 2050. Reference scenario 2013

De Cian E, Lanzi E, Roson R (2007) The impact of temperature change on energy demand: a dynamic panel analysis. http://ssrn.com/abstract=984237

Dobson JE, Bright EA, Coleman PR, Durfee RC, Worley BA (2000) A global population database for estimating populations at risk. Photogramm Eng Remote Sens 66(7):849–857

Felberbauer K, Muster M, Schloffer M, Theißing M, Tragner M, Wanek M, Zisler M (2010) KRAKE – Kraftwerke im Klimawandel Auswirkungen auf die Erzeugung von Elektrizität. Endbericht im Rahmen des Programms "Neue Energie 2020

Förster H, Lilliestam J (2010) Modeling thermoelectric power generation in view of climate change. Reg Environ Change 10(4):327–338

Francis RA, Falconi SM, Nateghi R, Guikema SD (2011) Probabilistic life cycle analysis model for evaluating electric power infrastructure risk mitigation investments. Clim Change 106 (1):31–55

Grübler A, Johansson TB, Mundaca L, Nakicenovic N, Pachauri S, Riahi K, Rogner H-H, Strupeit L (2011) Energy primer. In: Gomez-Echeverri L, Johansson TB, Nakicenovic N, Patwardhan A (eds) Global energy assessment: toward a sustainable future. IIASA, Cambridge University Press, Laxenburg

Haas R, Nakicenovic N, Ajanovic A, Faber T, Kranzl L, Müller A, Resch G (2008) Towards sustainability of energy systems: a primer on how to apply the concept of energy services to identify necessary trends and policies. Energy Policy 36(2008):4012–4021. doi:10.1016/j.enpol.2008.06.028

Haiden T, Kann A, Wittmann C, Pistotnik G, Bica B, Gruber C (2011) The integrated nowcasting through comprehensive analysis(INCA) system and its validation over the eastern Alpine region. Weather Forecast 26:166–183

Isaac M, van Vuuren DP (2009) Modeling global residential sector energy demand for heating and air conditioning in the context of climate change. Energy Policy 37:507–521. doi:10.1016/j.enpol.2008.09.051

Kirkinen J, Martikainen A, Holttinen H, Savolainen I, Auvinen O, Syri S (2005) Impacts on the energy sector and adaptation of the electricity network business under a changing climate in Finland. FINADAPT Working Paper 10, Finnish Environment Institute Mimeographs 340, Helsinki, p 43

Klein DR, Olonscheck M, Walther C, Kropp JP (2013) Susceptibility of the European electricity sector to climate change. Energy 59:183–193

Koch H, Vögele S (2009) Dynamic modeling of water demand, water availability and adaptation strategies for power plants to global change. Ecol Econ 68(7):2031–2039

Köppl A, Kettner C, Kletzan-Slamanig D, Schleicher S, Schnitzer H, Titz M, Damm A, Steininger KW, Wolkinger B, Lang R, Wallner, G, Artner H, Karner A (2011) Energy transition 2012/2020/2050. Strategies for the transition to low energy and low emission structures. Final report (No. 2011/019-2/S/WIFO Project No. 7407). Wien

Kranzl L, Formayer H, Haas R, Kalt G, Manfred L, Müller A, Nachtnebel HP, Redl C, Schörghuber S, Seidl R, Stanzel P (2010) Ableitung von prioritären Maßnahmen zur Adaption des Energiesystems an den Klimawandel.Endbericht im Rahmen der Programmlinie "Energie der Zukunft", Wien

Kranzl L, Matzenberger J, Totschnig G, Toleikyte A, Schicker I, Formayer H, Gorgas T, Stanzel P, Nachtnebel HP, Bednar T, Gladt M, Neusser M (2013) Modelling climate change impact on energy systems. Working Paper for the Second Review-Workshop in the project PRESENCE, Vienna

Kranzl L, Matzenberger J, Totschnig G, Toleikyte A, Schicker I, Formayer H, Gorgas T, Stanzel P, Nachtnebel HP, Bednar T, Gladt M, Neusser M (2014) Power through resilience of energy system. Final report of the project PRESENCE, Project in the frame of the Austrian climate research program, Vienna

Mima S, Criqui P, Watkiss P (2012) Technical policy briefing note no. 4. The impacts and economic costs of climate change on energy in the European Union. Final draft report. http://www.climatecost.cc/reportsandpublications.html

Moral-Carcedo J, Vicéns-Otero J (2005) Modelling the non-linear response of Spanish electricity demand to temperature variations. Energy Econ 27:477–494. doi:10.1016/j.eneco.2005.01.003

Nachtnebel HP, Stanzel P, Hernegger M (2013) Power through resilience of energy systems: energy crises, trends and climate change (PRESENCE). Contributions to work packages 4 – hydrology and hydropower and 5 – availability of cooling water for thermal power plants and the industry, Vienna

Olonscheck M, Holsten A, Kropp JP (2011) Heating and cooling energy demand and related emissions of the German residential building stock under climate change. Energy Policy 39:4795–4806. doi:10.1016/j.enpol.2011.06.041

Pechan A, Eisenack K (2013) The impact of heat waves on electricity spot markets. Oldenburg discussion papers in economics

Pospichal B, Formayer H, Haas P, Nadeem I (2010) 10th EMS annual meeting, 10th European conference on applications of meteorology (ECAM) Abstracts, Zürich, Switzerland, 13–17 Sept. http://meetingorganizer.copernicus.org/EMS2010/EMS2010-792.pdf. Accessed 14 Jan 2014

Pryor SC, Barthelmie RJ (2010) Climate change impacts on wind energy: a review. Renew Sustain Energy Rev 14: 430–437. (http://www.sciencedaily.com/releases/2011/05/110502151355.htm and http://www.mendeley.com/research/climate-change-impacts-wind-energy-review-1/ for abstracts)

Reichl J, Kollmann A, Tichler R, Pakhomova N, Moser S, Goers S, Lindorfer J, Greibl E, Fazeni K, Haas R, Müller A, Sagbauer N, Kloess M (2010) AWEEMSS – Analyse der Wirkungsmechanismen von Endenergieeffizienz-Maßnahmen und Entwicklung geeigneter Strategien für die Selektion ökonomisch-effizienter Maßnahmenpakete. Endbericht, Energieinstitut an der JKU Linz, Energy Economics Group, February 2010

Reichl J, Schmidthaler M, Schneider F (2013) The value of supply security: the costs of power outages to Austrian households, firms and the public sector. Energy Econ 36:256–261

Schleicher S, Köppl A (2013) Energieperspektiven für Österreich Zielorientierte Strukturen und Strategien für 2020. Im Auftrag des BMWFJ. Wien

Streicher W, Schnitzer H, Titz M, Tatzber F, Heimrath R, Wetz I, Hausberger S, Haas R, Kalt G, Damm A, Steininger K, Oblasser S (2010) Energieautarkie für Österreich 2050. Feasibility Study. Endbericht, Energie der Zukunft

Totschnig G, Kann A, Truhetz H, Pfleger M, Ottendörfer W, Schauer G (2013) AutRES100 – Hochauflösende Modellierung des Stromsystems bei hohem erneuerbaren Anteil – Richtung 100% Erneuerbare in Österreich. Endbericht im Rahmen des Programms Neue Energie 2020, Wien

Totschnig G, Müller A, Hummel M, Hirner R, Nachtnebel HP, Stanzel P, Schicker I, Formayer H (2014) Climate change impact on the electricity sector: the example of Austria. Paper submitted to the journal Energy

Chapter 15
Transport

Birgit Bednar-Friedl, Brigitte Wolkinger, Martin König, Gabriel Bachner, Herbert Formayer, Ivo Offenthaler, and Markus Leitner

Abstract 30 to 50 % of road maintenance costs in Europe are weather-related, with precipitation triggered events, like flooding and mass movement, contributing most. As most transport occurs on roads, damage implications of road transport infrastructure are explicitly relevant. In this chapter, we focus therefore on damages to road transport infrastructure and assess the costs of climate change induced repair and investment for the Austrian road network until mid-century. In addition to changed precipitation patterns, we also take road network expansion into account. We find that precipitation triggered damage costs to the Austrian road network are 18 million euros per year in the period 1981–2010. These damages increase to 27 million euros per year in the period 2016–2045 and 38 million euros in the period 2036–2065. For Austria in total, the lion's share of this cost increase is caused by an increase in exposed values (road network expansion), not climate change. While some regions are characterised by increases in precipitation, precipitation is decreasing in others, and there is also a seasonal shift. As a consequence, the overall effect of changes in precipitation is modest for Austria in total. The induced additional investment needed for road maintenance due primarily to road network extension and only secondarily to climate change is beneficial for the construction sector, but affects other sectors negatively due to higher prices. As a

B. Bednar-Friedl (✉)
Department of Economics, University of Graz, Graz, Austria

Wegener Center for Climate and Global Change, University of Graz, Graz, Austria
e-mail: birgit.friedl@uni-graz.at

B. Wolkinger • G. Bachner
Wegener Center for Climate and Global Change, University of Graz, Graz, Austria
e-mail: brigitte.wolkinger@uni-graz.at; gabriel.bachner@uni-graz.at

M. König • I. Offenthaler • M. Leitner
Environmental Impact Assessment & Climate Change, Environment Agency Austria, Vienna, Austria
e-mail: markus.koenig@umweltbundesamt.at; ivo.offenthaler@umweltbundesamt.at; markus.leitner@umweltbundesamt.at

H. Formayer
Institute of Meteorology, BOKU—University of Natural Resources and Life Sciences, Vienna, Austria
e-mail: herbert.formayer@boku.ac.at

K.W. Steininger et al. (eds.), *Economic Evaluation of Climate Change Impacts*, Springer Climate, DOI 10.1007/978-3-319-12457-5_15

consequence, the decline in welfare and GDP is about three times larger than the additional investment cost for both periods (2016–2045 and 2036–2065).

15.1 Introduction

The link between climate change and the transport sector has been examined extensively in the past in the context of the sector's contribution to climate change as well as the changes needed in transport for a decarbonisation of the economy (Tapio et al. 2007; Uherek et al. 2010; Pietzcker et al. 2014).[1] Only more recently, the recognition has risen of the vulnerability of the transport sector to extreme weather events and other climatic impacts and the implied needs for adaptation of the transport sector (European Commission 2013a, b).

According to Nemry and Demirel (2012), 30–50 % of current road maintenance costs in Europe are weather-related, equal to 8–13 billion euros per year. Another study on weather induced costs on the total transport network (Enei et al. 2011) finds costs of extreme weather events equal to 2.25 billion euros per year for Europe, with 80 % of these damages affecting roads, 16 % air, and 3 % rail. While higher and more extreme precipitation leads to additional costs for road transport in Europe in the future (0–192 million euros per year for the period 2040–2100), milder winters result in a reduction of road infrastructure costs of about 170–500 million euros per year and thus the overall effect is unclear (Nemry and Demirel 2012). In the Alps, however, the costs of more extreme events (mainly heavy precipitation) are found to outweigh the savings from reduced frost weathering on roads, rails, and bridges and hence overall costs are found to increase there.[2]

Disruption of transport infrastructure and transport service due to changing climate conditions has far reaching (economic) consequences since transport represents a great share of economic activity and contributes to the functioning of other sectors (Enei et al. 2011). Moreover, transport systems perform worse under extreme weather conditions (see Koetse and Rietveld 2009, for an overview). Thus, besides the direct effects of climate change on transport infrastructure, indirect costs due to destruction of assets arise (Larsen et al. 2008) as well as network effects through e.g. delays, detours, and trip cancellations (see Suarez

[1] The contribution of transport to climate change is both from transport operation (greenhouse gas emissions) and from transport infrastructure (paving of natural surface), and the latter may enhance susceptibility to e.g. flooding and aggravate climate change impacts.

[2] Vulnerability and impact assessments for road infrastructure are also available for individual countries such as the United Kingdom (Defra 2012), the Netherlands (Kwiatkowski et al. 2013), Sweden (Swedish Commission on Climate and Vulnerability 2007) or the United States (TRB 2008).

et al. 2005, for an estimation of Boston). Delays are especially critical when production gets more and more fragmented and specialised and some branches are dependent on just-in-time production (TRB 2008).

The aim of this chapter is to provide a discussion of possible climate change impacts for transport infrastructure and operation and to quantify the economic costs of these impacts for Austria up to the middle of the century. The economic assessment is limited to damages to the road network as damage cost data is not available for other modes of transport and for transport operation. Yet, as shown in another previous study for the European Alps (Enei et al. 2011), road damage covers 80 % of damage costs to transport infrastructure.[3] In our assessment, we consider both changes in daily precipitation according to a mid-range climate change scenario (see Chap. 5—Climate) and three different socioeconomic scenarios (i.e. different assumptions on network expansion). As a final step, we assess the total economic consequences, i.e. the direct and indirect effects triggered by the induced additional investments in road maintenance. The next section reviews the sensitivity to climate change for the transport sector both in general and for Austria in particular. Section 15.3 provides a detailed description of possible impact chains for all modes of transport. Moreover, past and current climatic exposure and observed impacts for Austria are discussed (road damage events). Section 15.4 provides an evaluation of future economic impacts for Austria. This involves a description of the costing methodology, as well as presentation and discussion of results. A final section summarises key findings.

15.2 Dimensions of Sensitivity to Climate Change

For identifying the sensitivity of the transport sector i.e. the likelihood and degree of effects and thus the damage potential due to climate change, it is important to consider key factors determining this sensitivity (Lapp 2010). Sensitivity may change over time due to socioeconomic change (Füssel and Klein 2006). In the transport system, this socioeconomic change is visible as growth in transport volume and expansion of the transport network. The role of climatic factors which determine sensitivity of transport to climate change is given in Sect. 15.2.1 whereas the non-climatic factors are given in Sect. 15.2.2.

[3] Nevertheless, the resulting estimates should be understood as lower bound estimates for the climate change induced costs which may arise to transport and mobility in Austria.

15.2.1 Climatic Factors

The transport system is influenced by heat and drought, ice and snow, rainfall and flooding (for waterway transport also droughts with low flow), and storms (Nemry and Demirel 2012). Sensitivity varies across modes of transport and among transport infrastructure (roads, rails, bridges, ports, airports) and transport operation (vehicles, rolling equipment and users). For more details on the impact chains, see Sect. 15.3.1.

It is important to note that the climatic sensitivity of transport infrastructures is site-specific and influenced by (almost) static factors such as geologic bedrock, sediments and geomorphologic conditions. As a consequence, the same amount of e.g. rainfall may cause large damages in some regions but no damages elsewhere.

15.2.2 Non-climatic Factors

The degree of potential impact is also dependent on the state of the roads (the pavement structure with its different layers, its age or the frequency of maintenance), repair and replacement costs, and traffic intensity (more variable factors). There is also a feedback between higher transport volumes and road network expansion and transport volumes are influenced in turn by the availability and price of energy (fuels) and other resources (for the construction of vehicles).

For transport operation, sensitivity is higher for arterial, connecting or access roads and if there are no alternative routes available. Furthermore spatial patterns of settlement and production influence the degree of impact affecting e.g. commuters and supply chains.

15.3 Exposure to Climatic Stimuli and Impacts Up to Now

15.3.1 Impact Chains for the Transport System

Impacts are described in Table 15.1 by climatic stimulus for all modes of transport, concentrating on land transport. Impact chains comprise direct as well as indirect impacts: Direct impacts mean impacts on the physical infrastructure and the operating performance of the system, while indirect means higher-order effects such as interruptions in production chains or loss-of-time for commuters due to interrupted transport infrastructure service provisions. Macroeconomic effects in our chapter in turn refer to the effects of additional costs for road maintenance on other sectors and the public budget (cf. 15.4.4.2).

1. Increased temperature and more heat waves
   Changes in average temperature and extremes, especially over periods of several days may cause damages to the material, structural damages and traffic-related

Table 15.1 Transport related impact chains

Climate change parameter	Impact chain	Quantified in the model
Precipitation: Increase in intense precipitation events, changes in snowfall occurrence and freezing precipitation	*Increase in floods, landslides and mudflows*	
	→ Damages or disruption to roads (e.g. concrete deterioration), bridges (e.g. scouring), rail embankments, earthworks, culvert washout → restoration costs (possible long term costs if shift from rail freight transport to road occurs not temporarily but permanently)	Yes
	→ Transport service interruption → interruptions in production chains (freight transport), loss-of time due to detours or traffic interruption (passenger transport) → increase in production costs and compensation costs	No
	→ Flight delays → compensation costs	No
	Decrease in snowfall occurrence and freezing precipitation → Reduction of accidents and delays due to equipment failure (e.g. blocking of breaks) → reduction of compensation costs	No
Increase in extended drought periods with low water flows and drying out of forests	*Reduction of depth of inland waterways* → Transport service interruption → interruptions in production chains, loss-of time → increase in production costs and compensation costs	No
	Increase of susceptibility to wildfires → Transport service interruption → increase in operation/production costs and compensation costs	No
Storm events: Increase of frequency and/or intensity	*Increase of debris and broken trees*	
	→ Damage to roads and rail tracks, installations (cables, signs, lighting)	No
	→ Transport service interruption due to fallen trees, loss of electricity or reduced visibility → increase in operation/production costs and compensation costs	No

(continued)

Table 15.1 (continued)

Climate change parameter	Impact chain	Quantified in the model
Temperature: Increase in very hot days and heat waves, higher temperatures in winter, more freeze – thaw conditions	*Road: deformation of road surfaces (ruts), migration of liquid asphalt, bridge material degradation, cracking and potholes* →Damage of infrastructure → restoration/replacement costs →Degradation of road safety → accident costs →Road service interruption → interruptions in production chains, loss of time → increase in production costs and compensation costs	No
	Railway: track buckling, sag of overhead line equipment, failures of pneumatic appliances, brakes, sliding doors (decrease when winter temperatures increase)	
	→Damage of infrastructure → restoration/replacement costs	No
	→Rail service interruption → interruptions in production chains, loss of time → increase in production costs and compensation costs	No
	Air Transportation: buckling of pavements and concrete facilities, reduction of snow and ice removal, airplane de-icing	
	→Damage of infrastructure → restoration/replacement costs	No
	→Reduction of snow removal and de-icing costs	No
	Passenger discomfort and higher energy demand due to air conditioning in vehicles → Higher ambient temperatures in all vehicles → more cooling demand → higher energy demand and costs	No
	Waterways: longer ice-free shipping season, less ice accumulation →Decrease of operation costs	No
Milder winters	*Reduced snow fall and ice conditions*	
	→ Decrease of maintenance costs	No
	→ Improvement of transport times in winter season → less loss of time due to heavy winter conditions	No

Sources: Enei et al. (2011), Nemry and Demirel (2012), Network Rail (2011), Nilson et al. (2012), Peterson et al. (2008), Thornes et al. (2012), Toivonen et al. (2011), TRB (2008) deformation to road surfaces (ruts) as well as migration of liquid asphalt (TRB 2008). Heat induced thermal expansion may also harm bridges which are often important traffic nodes. Subsequently material damages have substantial consequences for road safety (e.g. higher accident risk on road traffic). Increasing drought can also lead to a degradation of road foundation. For rail tracks, not only high temperatures but also fast changing temperatures are considered to be very destructive leading to track buckling, overheating of electronic equipment such as rail signalling installations or sag of overhead line equipment (Nolte et al. 2011; Dora 2011). In winter, higher temperatures reduce the numbers of cold days and thus directly the snow and ice removal costs (Jaroszweski et al. 2010) but more frequent freeze-thaw conditions may cause cracking and potholes, and soil erosion may damage embankments and earthwork along rail tracks. Also for roads, snow can erode construction material and block culverts and drainage as well as damage or disrupt bridges and embankments and slopes. Consequently, snowmelt can increase catchment runoff by releasing volumes of surface water previously held in a frozen state (Galbraith et al. 2005).

2. A higher frequency of droughts and less soil moisture in combination with heavy precipitation and extreme rainfall events and flooding
   Heavy rain events may overload the drainage system and increase the risk of flooding of road tracks in flat areas and valleys. In addition, there is a substantial risk of landslides, especially caused by heavy rain events. Road scouring and washout increases affect the structural integrity of roads, bridges and tunnels as well as blockings of rail tracks or derailments. Furthermore, heavy rain events lead to deterioration of traffic safety due to reduced visibility or flooded underpasses and influence the operation of transportation (TRB 2008). In combination with very low temperature, the impacts are similar to those for low temperature only. For example, snowfall can cause inefficient acceleration or breaking, accidents and damages to rail and road infrastructure (Bläsche et al. 2012).
3. More intense and frequent storms cause damages to infrastructure itself but also lead to blocked roads by fallen trees.
   Heavy wind damages destructs road and rail infrastructure, vehicles and cables resulting in loss of electricity along rail tracks. It further impedes the transport operation e.g. by hazardous vehicle behaviour especially for high-sided vehicles. Roadside furniture like signs, streetlights, gantries etc. can be affected by strong wind if they are not designed for certain wind gusts (Galbraith et al. 2005). Falling trees may lead to blocking of roads and rail tracks, may cause accidents and over power lines lead to sag or tensional failures. Depending on the function of the road (arterial or connector) or rail track this may lead to far reaching disturbances of the traffic flow. For countries with connection to the sea, storm surges may become more important.
4. Sea level rise affects transportation (infrastructure) in coastal regions
   Roads and rail tracks in coastal regions are more likely to be inundated with interruption of transport service, damages to infrastructure and (in the long run) the possibility of closures rather than reconstruction of old routes in more sparsely populated regions.

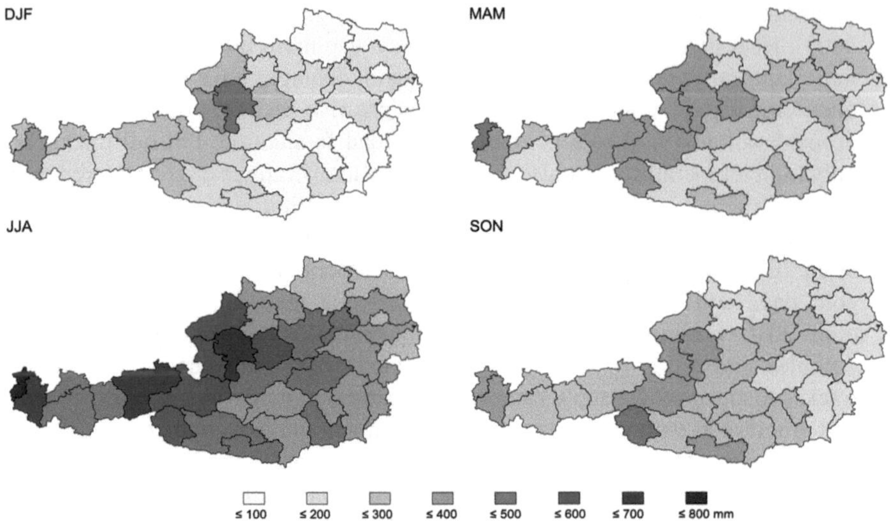

Fig. 15.1 Average seasonal precipitation across Austrian NUTS-3 regions (2003–2012). *DJF* winter, *MAM* spring, *JJA* summer, *SON* autumn. *Data*: ZAMG (2012) (Color figure online)

15.3.2 Past and Current Climatic Exposure

As precipitation is found as the main trigger for weather-induced damages to road transport infrastructure according to European studies (Enei et al. 2011; Nemry and Demirel 2012), we focus on this climate stimulus in the estimation of current and future economic impacts. Recent seasonal precipitation patterns for Austria are depicted in Fig. 15.1. The potential of damage due to precipitation varies both across Austrian NUTS-3 regions[4] as well as across seasons, with the western and central parts (Central Alps) characterised by higher precipitation, especially during summer, than the eastern parts.

15.3.3 Impacts Up to Now

Regarding impacts, data on damages to the Austrian transport network is very limited, with the exception of the secondary road network (federal and provincial roads; i.e. higher road network excluding highways) for which time series data is available for the provinces of Salzburg (which is typical for the alpine landscapes found in Western and Central parts of Austria) and Styria (which represents a

[4] NUTS (Nomenclature of Territorial Units for Statistics) is an Eurostat classification for different levels of a country. Austria (NUTS-0 level) is divided in nine provinces (NUTS-2 level) and 35 groups of districts (NUTS-3 level).

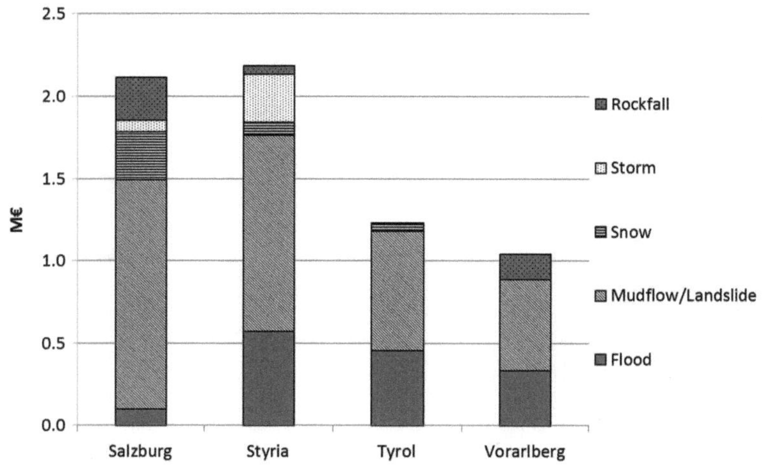

Fig. 15.2 Average annual damage costs to secondary road network in four Austrian provinces (in M€): Salzburg (ø 2007–2010), Styria (ø 2008–2011), Tyrol (ø 2006–2011), and Vorarlberg (ø 2006–2010). *Data source*: Amt der Salzburger Landesregierung (2012), Amt der Steiermärkischen Landesregierung (2012), Amt der Tiroler Landesregierung (2011), Amt der Vorarlberger Landesregierung (2011). Note: Styria and Salzburg have a comprehensive database for the survey periods 2008–2011 and 2007–2010 respectively while Tyrol and Vorarlberg lack an unbroken data base although survey periods are slightly longer. For Vorarlberg, a hundred year flood occurred in 2005 (Pfurtscheller and Kleewein 2010) and hence these damage costs were not used for calculating average annual damage costs

gradient from pre-Alpine areas to the central Alps). For both provinces in total, 70 % of road damage events in the period 2007–2011 can be attributed to precipitation triggered events like floods, landslides and mudflows.

The average annual road damage costs to the secondary road network (federal and provincial roads) in four Austrian provinces are given in Fig. 15.2. In all four provinces, floods, mudflows and landslides contribute about 80 % of all road damage costs. In comparison, damages due to avalanches and snow pressure play a minor role for road damage costs, mostly due to installed avalanche control measures like galleries. Rockfall is of importance in high alpine areas with steep valleys (see e.g. Salzburg and Vorarlberg), but is overall less important when a large part of the secondary road network is in pre-alpine areas (e.g. Styria). In contrast, storm has been of more importance in Styria.

15.4 Future Exposure to and Impacts of Climate Change

15.4.1 Mid-range Climatic Scenario for Transport

For assessing future damage events to road infrastructure, we use the mid-range climate scenario (see Chap. 5—Climate). As climate indicator, we use daily

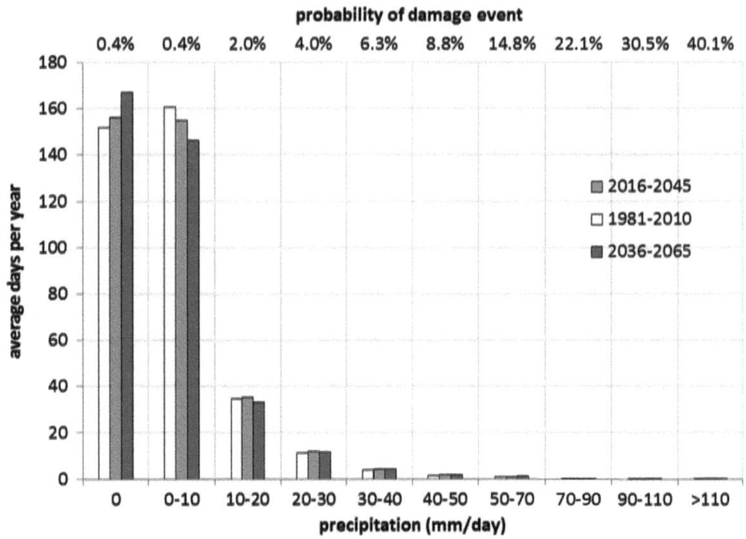

Fig. 15.3 Average number of days per year in each precipitation class for base period 1981–2010, mid-range climate change 2016–2045 and 2036–2065 (for Austria on average)

precipitation data because according to Enei et al. (2011) highest damages are precipitation triggered.[5] Precipitation is represented by a gamma distribution which gives the daily distribution of precipitation for each month for each of the 35 Austrian NUTS-3 regions. This distribution indicates that there are many days with no or only a few millimetres (mm) precipitation but only few days with precipitation of 60 mm or higher (see Chap. 5—Climate for details). For some NUTS-3 regions and some months, there is a shift towards a lower occurrence of days with low precipitation and a slight increase in days with high precipitation in the two future periods 2016–2045 and 2036–2065 compared to the base period 1981–2010. While this pattern can be detected for selected NUTS-3 regions and selected months, the effect on the national scale and over the year is less visible (see Fig. 15.3) as the effects cancel each other out across regions and seasons/months.

15.4.2 *Specific Method(s) of Valuation and Their Implementation Steps*

The evaluation of damages comprised five steps: the estimation of an impact function based on past damage cost data. This was done for the damage data

[5] We tested also for the influence of different climatic factors on the occurrence of damage events and costs in the dataset for Styria and Salzburg and found that daily precipitation has a significant and positive influence on damages but temperature does not.

described in Sect. 15.3.3 above for the secondary road network in the province of Styria. The impact function is a quadratic function which relates daily precipitation to the probability for a damage event of 30,000 € or higher (per day and NUTS-3 region).[6] The second step required the application of this impact function to precipitation data for all NUTS-3 regions in Austria, both for past precipitation data (1980–2010) and for the mid-range climate scenario (periods 2016–2045 and 2036–2065). See Fig. 15.3 for results on the annual distribution of precipitation for Austria in total in all three periods and associated probability of damage events. In a third step, damage estimates were scaled up from the secondary to the total road network. In a final step, physical impacts (events with damage potential of 30,000 € or higher per day and NUTS-3 region) were translated into costs and aggregated to province, then scaled up to national level. For more details, see the supplementary material.

15.4.3 Range of Sectoral Socioeconomic Pathway Parameters that Co-determine Climate Impact

In order to elicit potential impacts of climate change up to 2050, it is necessary to distinguish developments due to socioeconomic change from those due to climate change. Regarding socioeconomic change, the main driver is change in the length of the road network which has, in turn, up to now been triggered by transport performance.

Both freight and passenger transport performance in Austria have grown steadily in the past and are expected to increase in the future, particularly on roads. For freight transport, transport performance (tkm) is expected to be 54 % higher in 2030 relative to 2005. For passenger transport, an increase of transport performance of 24 % is expected (23 % for motorised individual transport and 28 % for public transport) between 2005 and 2030 (UBA 2012 and BMVIT 2012a).

Due to the increase of passenger as well as freight transport performance, the Austrian road network has been and will be further expanding. Both the priority network (highways and speedways) and local road network increased markedly over the last two decades, while the intermediate network (federal and regional roads) remained fairly constant (BMVIT 2012b). Network growth rates (differentiated by the three hierarchy levels) over the period 1990–2010 were used to set up three different scenarios for socioeconomic developments up to 2030 and 2050. For the reference socioeconomic development, an expansion of 340 km per year is

[6] As damage events with small damage cost are found frequently in the data also for days with very little precipitation, we compared distribution functions for different threshold levels (€ 20,000, € 30,000, and € 40,000). While all three regressions had a good fit, € 30,000 was chosen as the threshold for further analysis because a threshold of € 20,000 was well below mean damages and a threshold of € 40,000 reduced the number of included events by 50% relative to the a threshold of € 30,000.

assumed, which is half the annual increase in the period 1990–2010. For the impact diminishing socioeconomic development, it is assumed that there is no expansion of road infrastructure while in the impact enhancing socioeconomic scenario, it is assumed that the expansion of road infrastructure follows the expansion in the period 1990–2010 with an increase of about 80 km per year.

15.4.4 *Monetary Evaluation of Impacts*

15.4.4.1 Direct Sector Impacts Without Feedback Effects from Other Sectors

Table 15.2 provides the estimated average annual number of events with damages equal to 30,000 € or higher for the base period (1981–2010), as well as the average annual damage costs. It shows that both the number of events and total damage costs are highest for Upper Austria and Lower Austria, the provinces with the longest road network. Tyrol, Salzburg and Vorarlberg are the most alpine provinces and have therefore the highest number of events per 1,000 km of road network and therefore also the highest damage cost per km of road network. The remaining provinces fall in between.

In addition to counteracting regional effects, Fig. 15.3 helps in understanding why the increase in the number of damage events due to climate change is small: by comparing precipitation patterns for Austria in 2016–2045 to those in 1981–2010,

Table 15.2 Average annual weather and climate-triggered economic impacts in transport (damage costs for the total Austrian road network) across Austrian provinces in base period 1981–2010 (prices of 2008)

Province	Length of road network (km)[a]	Number of events ≥ 30,000€	Events per 1,000 km of road network	Total damage costs (€/a)	Costs per km of road network (€/km/a)
Burgenland	5,826	13.2	2.27	666,343	114
Lower Austria	31,108	76.5	2.46	3,854,550	124
Vienna	2,811	8.0	2.86	405,587	144
Carinthia	9,504	29.8	3.13	1,500,614	158
Styria	18,295	50.9	2.78	2,567,845	140
Upper Austria	26,836	88.6	3.30	4,465,326	166
Salzburg	5,202	24.8	4.76	1,248,794	240
Tyrol	11,116	51.1	4.60	2,576,794	232
Vorarlberg	3,891	21.3	5.47	1,072,164	276
Austria (total)	**114,589**	**364.1**	**3.18**	**18,358,019**	**160**

[a]For the road network, values of 2010 are given

Table 15.3 Change in estimated average annual damage costs across Austrian provinces for periods 2016–2045 and 2036–2065, reference socioeconomic development and mid-range climate change compared to the base period 1981–2010 (prices of 2008)

	ø 2016–2045			ø 2036–2065		
Change relative to ø 1981–2010	Road network (km)[a]	Number of events ≥ 30,000 €	Total damage costs (€/a)	Road network (km)[a]	Number of events ≥ € 30,000	Total damage costs (€/a)
Burgenland	+349	+0.9	+304,211	+700	+1.6	+717,280
Lower Austria	+1,526	+4.2	+1,687,350	+3,063	+7.6	+4,013,760
Vienna	+214	+0.6	+191,623	+430	+1.3	+465,402
Carinthia	+566	+2.5	+713,346	+1,136	+4.6	+1,719,198
Styria	+1,134	+3.1	+1,143,564	+2,273	+6.6	+2,822,057
Upper Austria	+1,740	+7.6	+2,142,302	+3,486	+12.1	+4,961,719
Salzburg	+318	+1.8	+579,912	+640	+2.9	+1,346,038
Tyrol	+738	+3.7	+1,188,152	+1,481	+5.9	+2,758,373
Vorarlberg	+259	+1.6	+497,759	+520	+2.3	+1,134,070
Austria (total)	**+6,844**	**+26.0**	**+8,448,219**	**+13,730**	**+44.8**	**+19,937,899**

[a]For the road network, changes relative to 2010 are stated

we see a shift from days with a damage probability below 5 % (corresponding to daily precipitation up to 30 mm) to days with a damage probability up to 40 % (corresponding to daily precipitation between 30 and 110 mm). But there is also an increase in days without precipitation for which the probability of precipitation events is lowest. Due to these two counteracting effects, the total number of damage events is only slightly higher in 2016–2045. As the increase in days without precipitation is even stronger in the period 2036–2045, the number of damage events is even smaller than in 2016–2045.

Table 15.3 provides the estimated average annual damage costs for the mid-range climate change and reference socioeconomic development in the periods 2016–2045 (left) and 2036–2065 (right). For Austria as a whole, 364 damage days are estimated for the base period 1981–2010. This number is estimated to increase to 390 days in 2016–2045 and to 409 days in 2036–2065. Again, differences across provinces are due to different sizes of the road network and the corresponding expansion as well as due to changes in precipitation (which is visible in the increase of event days).

The average annual damage costs for Austria increase from 18.4 million euros in the base period (1981–2010) to 26.8 million euros in 2016–2045 and to 38.3 million euros in 2036–2065. As costs per damage event are assumed to increase with the average economic growth rate, there is thus a much larger difference in damage costs than there is in damage days between period 2016–2045 and 2036–2065. Comparing changes in events to those in damage costs shows the importance of looking at both changes in frequencies of events as well as exposed values. While

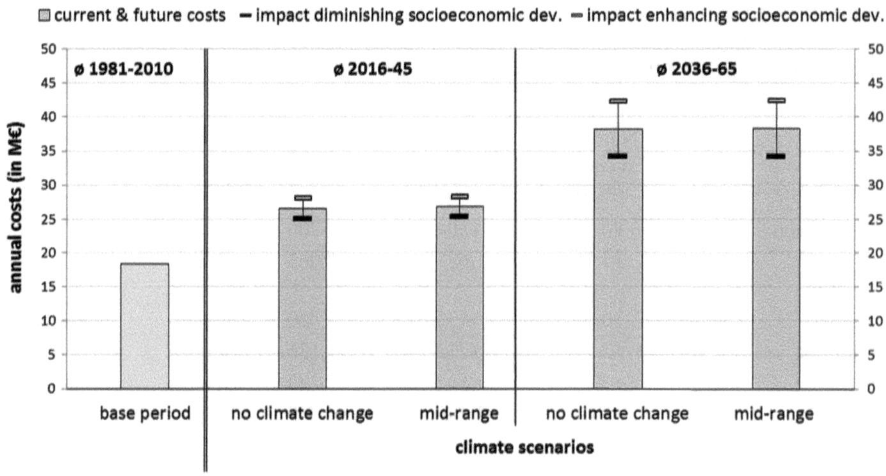

Fig. 15.4 Average annual weather and climate triggered damage costs for the total Austrian road network arising from socioeconomic development and climate change (in M€). Notes: "impact enhancing": Annual growth in road network as in period 1981–2010; "impact diminishing": no growth in road network (fixed at size of 2010)

the number of damage events increases by 7 % in the first period, costs increase by 46 % because both exposed values and maintenance costs per km of destroyed road are assumed to be higher in the future.[7] Moreover, while the increase in damage costs in the second period is more than twice as strong as the increase in the first period, the increase in damage days is smaller.

Figure 15.4 compares damage costs for the total Austrian road network for different socio-economic developments (reference, diminishing and enhancing assumption on network expansion) and due to climate change (baseline without climate change and mid-range climate change). For each future period, costs for both—the baseline without climate change and with mid-range climate change—are very similar. When road expansion is doubled as compared to the reference socioeconomic assumption ("enhancing" socioeconomic assumption), direct sector impact costs increase to 28 million euros in the first period and to 42 million euros in the second. When instead road expansion is completely stopped ("diminishing" socioeconomic assumption), damages are smaller than in the reference specification. Thus, according to this analysis, the main damage trigger is road network expansion and hence the increase in the exposed values.

Finally, it is important to note that estimated damage events and costs are lower bound estimates of road damage costs because damage events are evaluated with

[7] As suggested by one of the reviewers, an alternative assumption on the development of costs per event could be that improved technology might reduce vulnerability of roads and hence real costs could also decrease over time. Due to lack of information on such cost developments, we do not explore this alternative option in our assessment.

today's mean damage costs at or above the 30,000 € threshold level. Thus, smaller damages are ignored and it is also ignored that average damage costs above the threshold might change due to increasing severity of events. Moreover, according to the overview of impact chains given in Table 15.1, these projections are only reporting the direct damages to the Austrian road network and neither any other impact chain nor the indirect effects arising from road service disruptions.

15.4.4.2 Macroeconomic Effects

For road transport infrastructure those impact chains which are leading to additional investments for road damages are implemented into the macroeconomic model which is then used to assess the difference of direct and indirect effects between the climate change scenario (mid-range climate change and reference socioeconomic development) and a baseline scenario (reference socioeconomic development without climate change) in the same period. The CGE model which was used to compute these tables is described in Chap. 7 (Economic Framework). As only the impacts on road transport infrastructure are implemented in the model, the results of this section can be regarded as a lower bound. Other possible impacts like forced detours, loss of time and production losses are not included. Direct impacts on other transport modes are also excluded.

Higher damages to road infrastructure lead to additional investments (modelled as a shift within the investment expenditures). In the baseline scenario, additional investments for weather related damages on road infrastructure emerge because of a denser road network (i.e. more roads in km). In the climate change scenario, this effect is slightly amplified, leading to higher necessary investments for road damage reconstruction. In addition, capital becomes less effective as it is used to rebuild damaged infrastructure and by this not increasing the capital stock of the economy.

The macroeconomic assessment was carried out for three different socioeconomic scenarios (enhancing, reference and diminishing) and the respective model parameters are shown in Table 15.4.

All macroeconomic effects are calculated for 2030 (representative for the period 2016–2045) and for 2050 (representative for the period 2036–2065). All effects are shown for the climate change scenario relative to the baseline scenario in the same period (2030 and 2050). Table 15.5 shows the effects of the implemented impact chain "additional investments for road damages" for selected economic sectors (socioeconomic reference scenario and mid-range climate change scenario). The only gaining sector (in terms of value added) is the construction sector. Its gross value added (and therefore its contribution to GDP) is slightly higher in the climate change scenario compared to the baseline in 2030 as well as in 2050. Note that the effect in the second period is weaker than in the first period. This is due to the underlying climate change scenario in which the difference to the baseline scenario is smaller in the second period. As there is more demand for construction in the climate change scenario, its output (as well as price) is higher as well, leading

Table 15.4 Implementation of baseline and climate change scenario for road transport infrastructure in the macroeconomic model, average annual effects for periods 2016–2045 and 2036–2065

	ø 2016–2045		ø 2036–2065	
Annual additional investments for road damage reconstruction (in M€), relative to base year (2008)	Baseline	Climate change	Baseline	Climate change
Socioeconomic pathway				
Enhancing	+9.73	+10.00	+24.01	+24.08
Reference	+8.19	+8.45	+19.87	+19.94
Diminishing	+6.66	+6.90	+15.73	+15.79

Note: baseline scenario = socioeconomic development (impact enhancing, reference, impact diminishing) without climate change; climate change scenario = socioeconomic development (impact enhancing, reference, impact diminishing) and mid-range climate change; quantified impact chain: additional investments for road damages

Table 15.5 Sectoral and total effects of quantified climate change impacts for road transport infrastructure, average annual effects relative to baseline (for periods 2016–2045 and 2036–2065)

	ø 2016–2045			ø 2036–2065		
Changes in M€ p.a. relative to baseline	Gross output value	Intermediate demand	Gross value added	Gross output value	Intermediate demand	Gross value added
Gaining sectors	**+0.3**	**+0.2**	**+0.1**	**+0.1**	**+0.0**	**+0.0**
Construction	+0.3	+0.2	+0.1	+0.1	+0.0	+0.0
Losing sectors	**−1.4**	**−0.8**	**−0.7**	**−0.4**	**−0.2**	**−0.2**
Trade	−0.2	−0.1	−0.1	−0.0	−0.0	−0.0
Real Estate	−0.1	−0.0	−0.1	−0.0	−0.0	−0.0
Rest of Manufacturing	−0.1	−0.1	−0.0	−0.0	−0.0	−0.0
All other losing sectors	−1.0	−0.6	−0.4	−0.3	−0.2	−0.1
Total effect (all sectors)	**−1.2**	**−0.6**	**−0.6**	**−0.3**	**−0.2**	**−0.1**

Note: baseline scenario = reference socioeconomic development without climate change; climate change scenario = reference socioeconomic development and mid-range climate change; quantified impact chains: productivity gains due to prolonged growing seasons; change in cost structure

in turn to more intermediate demand and also gross value added (which is obtained by subtracting intermediate demand from gross output value). Regarding losses we see a total output value loss due to climate change of 1.4 million euros in 2030 and 0.4 million euros in 2050 (again, the effects in the second period are smaller). Those losses are emerging as the additional capital which is needed for the reconstruction of climate change induced road damages cannot be used in a more efficient way. This translates to higher prices and therefore to less final demand and welfare of private households. Hence, we see demand driven output reductions of the trade sector, real estate as well as manufacturing (i.e. typical consumption goods). These output reductions lead to lower net value added, compared to the baseline scenario.

Table 15.6 Effects of quantified climate change impacts for road transport infrastructure on government budget, average annual effects relative to baseline (for periods 2016–2045 and 2036–2065)

Changes in M€ p.a. relative to baseline	ø 2016–2045	ø 2036–2065
Revenues	**−0.2**	**−0.1**
Production tax	−0.0	−0.0
Labour tax	−0.1	−0.0
Capital tax	−0.0	−0.0
Value added tax	−0.1	−0.0
Other taxes	−0.0	−0.0
Expenditures	**−0.2**	**−0.1**
Unemployment benefits	+0.2	+0.0
Transfers to households net of other taxes	−0.4	−0.1

Note: baseline scenario = reference socioeconomic development without climate change; climate change scenario = reference socioeconomic development and mid-range climate change; quantified impact chains: productivity gains due to prolonged growing seasons; change in cost structure

Summing up the effects on gross value added for all sectors, we obtain the effect on GDP (after correcting for indirect taxes less subsidies). The climate change effect on GDP[8] is a reduction of 0.64 million euros in 2030 and of 0.17 million euros in 2050 with reference socioeconomic development, which is in both periods about three times larger than the direct costs (see Table 15.2) in terms of higher investments in road infrastructure. With diminishing (enhancing) socioeconomic development, negative effects are slightly weaker (stronger): 0.60 million euros (0.68 million euros) in 2030 and 0.16 million euros (0.18 million euros) in 2050. Thus, economic feedback effects from redirected investment, higher prices and hence lower purchasing power of households amplify the direct effect considerably.

From the perspective of households, welfare[9] is a better indicator than GDP as it reflects the purchasing power of private and public households. But the effect is of equal magnitude as that on GDP.

Table 15.6 shows the effects on government budgets. In 2030, government revenues in the climate change scenario are lower than in the baseline scenario by 0.2 million euros whereas it is 0.1 million euros in 2050. As we assume equality in revenues and expenditures in the CGE model, expenditures are also lower in the

[8] Gross domestic product (GDP) of a country is the value of all produced goods and services within a year. GDP can be determined by the sum of all sectoral net value added. Note that GDP only measures flows and therefore gives no information about the development of (natural) stocks,

[9] Welfare is measured as the quantity of private and government final demand (consumption) priced with baseline prices. The standard welfare measure is corrected by forced consumption which is not welfare enhancing (e.g. additional water consumption which is necessary to provide the same service level as without climate change).

climate change scenario by the same amounts, which implies that net transfers on households need to be decreased or taxes increased.

15.4.5 Qualitative Impacts (Non-monetised)

One important limitation of the current analysis is the limitation to damages of road infrastructure. Damages on other transport modes could not be assessed quantitatively, but additional costs are likely to occur.

Moreover, disruption of transport services (delays, detours, trip cancelations) is likely to be cost intensive as well, but time losses are difficult to estimate and monetise. Since transport and mobility have a highly important service function for production as well as for all other sectors by transporting people and goods, such additional indirect damages may exceed direct damage costs.

Scale of loss-of-time—as one possibility to assess indirect effects of transport infrastructure interruptions to mobility of freight and people—is dependent on economic growth and wages. Therefore indirect damages due to loss of time tend to be higher in a prosperous economy, while in an economic downturn direct damages (in the form of a comparatively higher burden for repairing and reconstructing) tend to dominate.

15.4.6 Sector-Specific Uncertainties

The main uncertainties are:

- Development of freight and transport volumes, congestion, and induced expansion of the transport network (this is also dependent on the development of the policy paradigm that more traffic triggers additional road construction)
- Development of the transport network and the respective shares of highways, secondary and local road network. Higher growth rates of the local road network might raise future damage costs due to its higher susceptibility towards landslides, mudflows and flooding
- Development of settlement and production location structures, which reinforce development of transport volumes
- Damage costs are approximated by expenditures for maintenance, yet budgetary limitations might lead to partial replacement of damages only. Damage estimates can therefore be understood as lower bound estimates for the true costs
- Future development of damage costs, with either increasing exposed values or improved technology and hence lower costs

15.5 Summary of Climate Costs for Transport and Conclusions

Around 70 % of all registered damage costs to Austrian road network are related to precipitation-triggered mass movements (mudflows and landslides) and floods. Rock fall and avalanche damages are considerable in western provinces, while damages from storms, snow pressure and hail remain below 10 %. Not included in damage data bases, but nevertheless important are frost weathering (as this has severe impacts on road maintenance) and heat (as this has impacts mainly on lane grooves along asphalt tracks and buckling of concrete slab sections).

Estimating the costs of these damages in the base period 1981–2010 leads to average annual costs equal to 18 million euros for the total Austrian road network. These costs are estimated to increase to 27 million euros per year in the period 2016–2045 and to 38 million euros in the period 2036–2065. About 80 % of this cost increase relative to the base period 1981–2010 is due to socioeconomic development (expansion in the road network and increase in exposed values) and only about 20 % is influenced by changes in precipitation. The relatively small role of climatic changes is due to three counteracting effects: in almost all Austrian regions, there is an increase in intensity, i.e. a decline in days with low precipitation and an increase in days with high precipitation, which leads to an increase in damage events. However, over the year there is also an increase in days with no precipitation which reduces the number of damage events. Finally, there is also a shift in precipitation between regions, such that some regions are characterised by higher and others by lower precipitation. As a consequence of all three effects, the net increase in damage events and costs due to climate change is small compared to the contribution of socioeconomic development.

Future damage costs will be determined considerably by road network developments, both regarding network expansion but also regarding costs of construction, maintenance and repair. In the past, growth rates for the road network have been influenced by growing transport volume and urban sprawl. When applying past rates of growth in the road network, direct sector impact costs increase to 28 million euros in the first period and to 42 million euros in the second.[10] Regarding costs of repair and maintenance, we have assumed that they increase with the average economic growth rate reflecting that in the future higher quality road surfaces are applied. The consequence of this assumption is that future damage costs increases sharply compared to the base period. For keeping damage costs low, i.e. reducing the vulnerability of road transport, there are therefore two broad options available: limiting the expansion of the road network and increasing efficiency in road

[10] But on the other hand, additional bypass/detour routes might lead to less loss-of-time and thus lower indirect /costs of service interruptions, if this effect is not offset by higher traffic volumes. While the public budget is mostly affected by infrastructure damages (considered in this chapter), damages due to service interruptions will materialise mainly in the private sector (not considered in this chapter).

maintenance (by improving technology for pavement etc.) and thus decreasing the cost of each damage event.

Climate change poses additional regional risks and add-on costs on transport infrastructures and mobility. Core climatic damage triggers are strong and extreme precipitation events throughout the year. Yet, compared to gradual changes in temperature which show clearly positive climate change signals, these high precipitation events are far more difficult to detect in past data and projections of future changes are less certain. For all future transport infrastructure planning processes, a risk assessment for mass movement and flood-related damages is therefore highly recommended.[11]

Acknowledgements This chapter presents research undertaken as part of the adapt2to4 project. This project was funded by the Austrian Climate and Energy Fund and carried out under the Austrian Climate Research Programme (ACRP).

References

Amt der Salzburger Landesregierung (2012) Schadensereignisse und -kosten an Landesstraßen nach Schadkategorie und Straßenabschnitt, 2004–2010, Salzburg

Amt der Steiermärkischen Landesregierung (2012) Schadensereignisse und -kosten an Landesstraßen nach Schadkategorie und Straßenabschnitt 2008-2011, Graz

Amt der Tiroler Landesregierung (2011) Schadensereignisse und -kosten an Landesstraßen nach Schadkategorie, Bezirk und Straße 2005-2011, Innsbruck

Amt der Vorarlberger Landesregierung (2011) Katastrophenschäden an Landesstraßen (Aufwendungen Oktober 2005 bis Dezember 2010), Bregenz

Bläsche J, Kreuz M, Mühlhausen T, Schweighofer J, Leviäkangas P, Molarius R, Nokkala M, Athanasatos S, Michaelides S, Papadakis M, Ludvigsen J (2012) Consequences of extreme weather, EWENT Project, Deliverable D3.4

BMVIT (2012a) Ausbauplan Bundesverkehrsinfrastruktur—Klug investieren, verantwortlich sparen. http://www.bmvit.gv.at/verkehr/gesamtverkehr/ausbauplan/downloads/prognose_2012_8.pdf

BMVIT (2012b) Verkehr in Zahlen, Ausgabe 2011, Wien. http://www.bmvit.gv.at/verkehr/gesamtverkehr/statistik/downloads/viz_2011_gesamtbericht_270613.pdf

Defra (2012) The UK climate change risk assessment 2012 evidence report project deliverable number D.4.2.1, Release 8, Defra Project Code GA0204, London

Dora J (2011) Adapting to extreme climate change (TRaCCA), Phase 3 report, Tomorrow's railway and climate change adaptation, Rail Safety and Standards Board, London

Enei R, Doll C, Klug S, Partzsch I, Sedlacek N, Kiel J, Nesterova N, Rudzikaite L, Papanikolaou A, Mitsakis V (2011) Vulnerability of transport systems—main report. Transport sector vulnerabilities within the research project WEATHER (Weather Extremes: Impacts on

[11] Some alpine access roads might need either significant adaptation investment or deconstruction/resettlements should be considered if investments do not pay off. In fact, the regional distribution of costs and benefits is diverse: road sections on or along slippery terrain (or in vicinity of melting permafrost rocky terrain, accelerating rock decomposition/rock fall, which has not been investigated here) would produce significant damages further on and—at least seasonally—face additional damage costs and interruptions of services.

Transport Systems and Hazards for European Regions) funded under the 7th framework program of the European Commission. Project coordinator: Fraunhofer - ISI. Karlsruhe, 30.9.2010

European Commission (2013a) An EU Strategy on adaptation to climate change, COM(2013) 216 final, Brussels

European Commission (2013b) Adapting infrastructure to climate change, Commission Staff Working Document, Accompanying the EU Strategy on adaptation to climate change, SWD (2013) 137 final, Brussels

Füssel H-M, Klein RJT (2006) Climate change vulnerability assessments: an evolution of conceptual thinking. Clim Change 75(3):301–329

Galbraith RM, Price DJ, Shackman L (eds) (2005) Scottish road network climate change study. Scottish Executive, Edinburgh

Jaroszweski D, Chapman L, Petts J (2010) Assessing the potential impact of climate change on transportation: the need for an interdisciplinary approach. J Transp Geogr 18(2):331–335

Koetse MJ, Rietveld P (2009) The impact of climate change and weather on transport: an overview of empirical findings. Transp Res Part D 14:205–221

Kwiatkowski KP, Oslakovic IS, ter Maat HW, Hartmann A, Chinsowsky P, Dewulf GP (2013) Climate change adaptation and roads: Dutch case study of costs impacts at the organization level, Working Paper Proceedings at the Engineering Project Ogranization Conference, July 9–11, Colorado

Lapp D (2010) National assessment of the engineering vulnerability of infrastructure to climate change: tools to engage and enable engineers. http://www.pws.gov.nt.ca/asset/conference_files/PMC%20Presentations/National%20Assessment%20Vulnerability%20of%20Infrastructure%20to%20Climate%20Change%20-%20David%20Lapp.pdf

Larsen PH, Goldsmith S, Smith O, Wilson ML, Strzepek K, Chinowsky P, Saylor B (2008) Estimating future costs for Alaska public infrastructure at risk from climate change. Global Env Change. http://www.climatechange.alaska.gov/aag/docs/O97F18069.pdf

Nemry F, Demirel H (2012) Impact of climate change on transport: a focus on road and rail transport infrastructure. JRC Scientific and Policy Reports, Sevilla

Network Rail (2011) Network rail climate change adaptation report, In Response to the UK Government's Adaptation Reporting Power, 30 April 2011

Nilson E, Lingemann I, Klein B, Krahe P (2012) Impact of hydrological change on navigation conditions. ECCONET report, Brussels

Nolte R, Kamburow Ch, Rupp J (2011) ARISCC—adaptation of railway infrastructure to climate change, Final Report (6th draft version), Berlin

Peterson TC, McGuirk M, Houston TG, Horvitz AH, Wehner MF (2008) Climate variability and change with implications for transportation, Transportation Research Board

Pfurtscheller C, Kleewein K (2010) Are alpine regions especially affected by natural hazards? Meso-scale assessment of economic resilience and vulnerability based on the 2005 floods in the Austrian federal states of Tyrol and Vorarlberg. In: Borsdorf A, Grabherr G, Heinrich K, Scott B, Stötter J (eds) Challenges for mountain regions—TACKLING complexity. Böhlau, Vienna, pp 94–101

Pietzcker RC, Longden T, Chen W, Fu S, Kriegler E, Kyle P, Luderer G (2014) Long-term transport energy demand and climate policy: alternative visions on transport decarbonization in energy-economy models. Energy 64:95–108

Suarez P, Anderson W, Mahal V, Lakshmanan TR (2005) Impacts of flooding and climate change on urban transportation: a system wide performance assessment of the Boston metro area. Transp Res Part D 10:231–244

Swedish Commission on Climate and Vulnerability (2007) Sweden facing climate change—threats and opportunities, SOU 2007:60, Final Report, Stockholm

Tapio P, Banister D, Luukkanen J, Vehmas J, Willamo R (2007) Energy and transport in comparison: Immaterialisation, dematerialisation and decarbonisation in the EU15 between 1970 and 2000. Energy Policy 35(1):433–451

Thornes J, Rennie M, Marsden H, Chapman L (2012) Climate change risk assessment for the transport sector, UK 2012 CCRA, Defra Project Code GA0204, London

Toivonen S, Keränen J, Ludvigsen J, Vajda A, Tuomenvirta H, Juga I, Nurmi P, Rauhala J, Rehm F, Gerz T, Muehlhausen T, Schweighofer J, Michaelides S, Papadakis M, Dotzek N, Groenemeijer P (2011) Extreme weather impacts on transport systems, EWENT Project, D.1, VTT Working Papers, Finland

TRB (2008) Potential impacts of climate change on U.S. transportation, Special Report, Transportation Research Board, National Research Council, 290

UBA (2012) Austria's National Inventory Report 2012, REP-0381, Wien.

Uherek E, Halenka T, Borken-Kleefeld J, Balkanski Y, Berntsen T, Borrego C, Gauss M, Hoor P, Juda-Rezler K, Lelieveld J, Melas D, Rypdal K, Schmid S (2010) Transport impacts on atmosphere and climate: Land transport. Atmos Environ 44(37):4772–4816

Zentralanstalt für Meteorologie und Geodynamik (ZAMG) (2012) Integrated Nowcasting through Comprehensive Analysis (INCA) database 2003-2012, Vienna

Chapter 16
Manufacturing and Trade: Labour Productivity Losses

Herwig Urban and Karl W. Steininger

Abstract The sector "Manufacturing and Trade" exhibits relatively high climate sensitivity as it depends on climate sensitive raw materials and intermediary inputs (such as agricultural products, timber and energy). In addition, changes in climatic stimuli (such as in temperature and relative humidity) may also influence production processes and/or the productivity of workers.

In the present chapter all these effects are discussed qualitatively. The productivity losses of workers, however, are also estimated on the basis of a quantitative model using a relationship between the Wet Bulb Globe Temperature (WBGT) index and the productivity of workers. The Human Capital Approach (HCA) and a GDP per employee approach are used for monetising the direct productivity losses.

Changing working conditions can have serious effects on the productivity of workers and thus on companies. Depending on the climatic development and the degree of adaptation the degree of damage caused can vary significantly. The direct climate impacts observed in the sector "Manufacturing and Trade" are magnified fourfold by associated macroeconomic feedback effects. For the mid-range climate scenario, there is a decline in economic welfare of 6 million euros per year for the period 2016–2045 (and 54 million euros for 2036–2065). For the high-range climate scenario respective welfare losses amount to 58 million euros (296 million euros). As declining demand also triggers price declines, losses in GDP are thus stronger, about 1.5 times the welfare losses. Note, however, that we only estimate the effects of productivity changes within the sector "Manufacturing and Trade". Similar productivity changes could affect the remaining sectors of the whole economy as well, and thus could increase the economy-wide effects of climate-induced productivity changes above those quantified in the present chapter.

H. Urban (✉)
Wegener Center for Climate and Global Change, University of Graz, Graz, Austria
e-mail: herwig.urban@uni-graz.at

K.W. Steininger
Wegener Center for Climate and Global Change, University of Graz, Graz, Austria

Institute of Economics, University of Graz, Graz, Austria
e-mail: karl.steininger@uni-graz.at

K.W. Steininger et al. (eds.), *Economic Evaluation of Climate Change Impacts*, Springer Climate, DOI 10.1007/978-3-319-12457-5_16

16.1 Introduction

The economic activities covered in this chapter include those activities occurring in the subsector "manufacturing", i.e. all those activities related to production facilities. Such activities entail the transformation of materials or parts of products, either mechanically, physically or chemically, into final goods and products. Raw and basic materials originate from sectors such as agriculture, forestry or mining. The crucial criterion for inclusion in this sector is that there be substantial change or reshaping of inputs. The outputs of production facilities are either products for use and consumption, or for intermediary products required in further processing. The assembling of product parts (originally produced and/or bought as prefabricated intermediate inputs) is also part of this sector, as is the branch "repair and installation of machinery and equipment" (Statistics Austria n.y.a).

Secondly, the subsector "trade", which is also dealt with here, comprises wholesale and retail activities that do not entail processing of the goods sold. Standard activities in this respect are sorting, classifying, mixing and filling, and companies engaging in such activities are therefore classified as wholesale or retail traders. Wholesale and retail traders can be seen as the last link in the goods distribution chain. While the main customers of wholesale traders are companies, agencies, professional users and retail traders, those of retail traders are private consumers and users. In addition, the branch "repair of motor vehicles and motorcycles" is also included in this subsector (Statistics Austria n.y.a).

The combination of the above two subsectors makes up the full sector "Manufacturing and Trade". It consists of all branches producing new products from either raw materials or intermediary inputs, together with all wholesale and retail companies engaged in distributing and selling these products to the final consumer.

16.2 Dimensions of Sensitivity to Climate Change

16.2.1 *Climatic Factors*

The most relevant climatic factors directly affecting the sector "Manufacturing and Trade" are temperature and relative humidity. Those two factors strongly determine both the perceived and real environmental conditions arising in production processes. In addition, the demand for cooling (in summer) and heating (in winter) is also determined by such factors.

As the sector "Manufacturing and Trade" is highly dependent on raw materials and intermediary products from other sectors, all climatic factors that have an influence on the availability and/or the price of raw materials and intermediary products are also relevant for "Manufacturing and Trade". For example all climatic factors that have an influence on agricultural production, forestry or energy cause

feedback effects on “Manufacturing and Trade”. Furthermore, climatic indicators that change the consumption behaviour and the preferences of final consumers, such as the number of hot or rainy days, can also have significant impacts on the sector “Manufacturing and Trade”. Finally, all climatic factors that impact supply infrastructure in any way are highly relevant. These might include the growing presence of temperature extremes, heavy precipitation or storms. In addition to the purely national interlinkages, the international trade relations are also important for companies. According to the WTO and UNEP (2009, p. 64), extreme weather events outside Austria could also damage relevant infrastructure (e.g. through the closure of ports). Similarly, a severe drought could affect inland waterways, such as the Rhine, and thus raise the costs of Austrian international trade.
The subsector manufacturing is heavily dependent on raw materials, resources and intermediary products from other sectors. Specifically, the dependence on raw materials such as water, timber and energy, or on intermediate products from agriculture, is particularly high in several branches. In addition, the subsector manufacturing provides huge quantities of products to other sectors and to various areas of final demand. This is why it also depends on other downstream sectors and final consumers. As the subsector trade is one of them it is also highly affected by changes in manufacturing. The most important non-climatic factors determining the sector “Manufacturing and Trade” are thus the demand and supply interlinkages across the individual branches within manufacturing. Other non-climatic factors exerting a strong influence on “Manufacturing and Trade” are the development of the transport infrastructure, the availability of raw materials and intermediate inputs (which may influence their prices; e.g. price for water, oil, timber), the development of the work force (e.g. its size and age structure), the potential for innovation within the branches and their ability to adapt to various European level policies and regulations (e.g. EU ETS).

For an overview of the demand and supply interlinkages with other sectors, categorized as highly and moderately climate sensitive, see Supplementary Material Table 16.1 in the Supplementary Material.

16.3 Exposure to Climatic Stimuli and Impacts Up to Now

16.3.1 Past and Current Climatic Exposure and Physical Impacts

Hübler and Klepper (2007, p. 39) provide an overview of past investigations of the effects of climatic stimuli on working productivity. Bux (2006, p. 10ff.) provides an overview of the physical and physiological foundations of worker thermal comfort. Various studies (e.g. Parsons 2003; Seppänen et al. 2006; Hellwig 2004) show that reductions in working productivity as a reaction to thermal discomfort lie somewhere between 3 and 12 % for temperatures of 26 °C to 36 °C, although the figures are subject to a certain degree of uncertainty (Bux 2006, p. 19f.). Hübler and Klepper (2007, p. 39f.) review and analyse various studies that show significantly higher reductions in various performance parameters as a result of changing climatic conditions. E.g., office workers are most productive at 23 °C room temperature, while higher temperatures lead to reductions in working productivity. A room temperature of 30 °C can lead to a decrease in productivity (e.g. in writing speed) of 30 %. Mental achievement potential at 30 °C can also decrease by up to 40 %. Dunne et al. (2013) provide details of the currently observed labour productivity losses in an international context. They point out that humidity is already reducing people's working capacity by 10 % during peak months of heat stress around the world, and this is likely to grow to 20 % by 2050.

Up to now no specific or comprehensive studies have been undertaken to investigate the exposure of Austrian workers to climatic stimuli or to quantify the respective impacts of today's climate.

Air conditioning can reduce the effect of climatic conditions on indoor workers in the summer months. If that is achieved, workers may not be affected by heat strain and thus may not be faced with productivity losses. In contrast, companies are likely to face higher energy costs where the cooling of buildings is already available, and/or additional investment costs where it is not. In the winter months companies could be affected beneficially, as higher temperatures could reduce the energy needed to maintain optimum worker comfort or productivity. Such effects are discussed in Chap. 13 (Buildings) and will not be discussed further in this chapter.

16.3.2 Impact Chains up to Socioeconomic System

There are various impact chains that are relevant for "Manufacturing and Trade". Table 16.1 provides an overview and a summary of the various impact chains, and also lists which economic assessment has been chosen. As the productivity losses of workers (due to changing working conditions) are the only impact chain that can be quantified consistently across all the branches of "Manufacturing and Trade", all other impact chains (described extensively in the Supplementary Material) are simply analysed qualitatively. As a result, the numbers presented in this chapter definitely underestimate the climate change impacts for "Manufacturing and Trade". The qualitative results of changes in production processes, the dependence on supply chains and infrastructure, and the changes in private consumption behaviour are summarized in Supplementary Material Table16.16.

16.3.2.1 Productivity Changes (Due to Changing Working Conditions)

Monetizing the effects of the health-related problems of workers involves two kinds of costs. While direct costs (such as the cost of medical care or pharmaceutical products) involve a monetary transaction, indirect costs (or costs of lost

Table 16.1 Impact chains for "Manufacturing and Trade"

Climate change parameter	Impact chain	Quantified in this chapter
Temperature: Increase of daily temperature *Relative humidity*	*Productivity losses of workers* →hotter working environments → higher core body temperature → higher need for rests → lower productivity	Yes
	Prolongation of the working season → increase in temperature → possibility of working longer outdoors before the winter break	No
	Production process → necessity to keep process environment constant → higher costs for cooling and/or keeping conditions constant (e.g. humidity)	No
	Cooling and Heating → reduced heating in the winter months → increased cooling in the summer months	No
See chapters: e.g 8 *(Agriculture)*, 9 *(Forestry)*, 12 *(Water)*	*Supply chain* → dependence of "Manufacturing and Trade" on raw products and intermediate inputs → decreases in quantity and/or increases in prices can limit production possibilities	No
See chapter: 13 *(Buildings)* 15 *(Transport)*	*Infrastructure* → damage to internal infrastructure, e.g. to production halls, warehouses, salesrooms → replacement → damage to external infrastructure, e.g. to the transport network → problems obtaining intermediate products and raw materials → problems of delivery to final consumers	No
Temperature *Days with rain/snow*	*Private consumption* → higher temperatures → shift in consumption behaviour → more demand for beverages and climate-friendly technologies → decrease in demand for sweets and alcoholic beverages → increased/reduced number of shopping trips	No

productivity) result from worker absence due to illness, or from the fact that some workers may not be able to fully exploit their potential. It is possible to split indirect costs into two components, namely absenteeism and presenteeism.[1] "While the former is defined as the numbers of work days missed from the workplace or non-work activities, presenteeism illustrates the extent to which an employee is functioning while working" (Patel et al. 2006, p. 65). It is not only important to consider direct costs and costs of absenteeism, but also to investigate productivity losses arising as a result of presenteeism. Kjellstrom et al. (2009a, p. 49) argue that presenteeism, i.e. "the slowing down of work as a defence mechanism during severe heat exposure is labelled 'autonomous adaptation' by climate change researchers". However, while still allowing for autonomous adaptation, the concept of presenteeism is also looked at in more detail in the present study, since the wider effects covering the impact on employers and on society are of importance here. In cases of presenteeism not only do employees face costs, an economic burden is also imposed on employers and on society as a whole. In most cases, this is neither perceived nor taken into consideration (Yang and Liern 2009, p. 339). Consequently the present chapter focuses on investigating the effects of presenteeism, for indoor as well as for outdoor work [for direct costs and absenteeism see Chap. 11 (Human Health)]. Besides the losses that arise through a lower working productivity, another aspect (not further discussed here) is the loss of utility faced by workers because of less comfortable work conditions (e.g. higher temperatures).

In addition to productivity losses in the summer months, climate warming also increases the time span of the working season for outdoor work (beginning earlier in spring, ending later in fall). For the sector "Manufacturing and Trade", however, this effect is supposed to be rather small as only 3.8–17.2 % of work (depending on the working intensity and subsector) is performed outdoors. Longer working seasons are thus not considered any further in this chapter. Nevertheless, for sectors other than "Manufacturing and Trade", the consequences are likely to be significant.

16.3.3 Economic Impacts Up to Now

The present authors are aware of no previous studies on the productivity impacts of climate change observed for the Austrian sector "Manufacturing and Trade". In the Supplementary Material indications for productivity losses within Austria are given, based on the method described in Sect. 16.4.3 below. These calculations are used in this study as reference values for the analysis of future impacts.

[1] The concept of presenteeism is used quite differently in international literature. Johns (2010, p. 20ff.) provides various definitions of this concept. In this study presenteeism refers to a situation in which workers attend work while they are ill or feel unhealthy and as a consequence cannot achieve their full potential.

16.4 Future Exposure To and Impacts of Climate Change

The most relevant climatic indicators directly affecting the sector "Manufacturing and Trade" are temperature and relative humidity. In the following these indicators are measured by the WBGT (Wet Bulb Globe Temperature) index, which is a heat strain index describing the exposure of people to heat. According to ISO-Standard 7243 (ISO 1989) this index is regularly derived by special measuring instruments used directly in companies and/or production facilities. In the following, and according to Kjellstrom et al. (2009b), the index is approximated by the 24-h average temperature and the 24-h average relative humidity measured at meteorological measurement stations across Austria. To consider the significant differences between various regions within Austria, temperature data based on NUTS-3 regions is used within this chapter. A limiting factor is that for relative humidity, only Austrian-wide average data was available.

16.4.1 Mid Range Climatic Scenario for Manufacturing and Trade

In developing a mid-range climate scenario, it is important to use data covering information for every single day of a specific year. This is necessary as huge differences may arise, especially with respect to daily temperature values. While this has been accomplished in terms of temperature data, with respect to measures of relative humidity, only monthly average data is available. Although this causes some imprecision, the effects are relatively small as the climate change signals for relative humidity are also small. Second, as there are huge differences in climate parameters across the reference period (2003–2012), a mid-range scenario is calculated on the basis of each of the years separately in order not to lose information on single days for specific years. For this purpose the signal of absolute temperature change and the signal for the relative change of relative humidity are used. In climate scenarios these values are available for the monthly average only. We therefore assume that variability is constant, and then superimpose the absolute changes in the monthly average on the time series of daily data for the past. For some days of the reference period (2003–2012), data for a few single days is missing in the data set. For such days linear interpolation is used based on the values of the preceding and the following day. In a final step the average of the calculations based on the years of the reference period is then taken to derive the average mid-range scenario employed in the present chapter. The advantage of this method is that the calculation takes all days exhibiting a very high WBGT index value into account. This reduces loss of information on single days of high heat exposure in the various years of the base period. While in many scenarios variability of climate parameters is assumed to increase, in the present study climate variability was assumed to remain constant in future periods. The extent to which this limits the validity of the analysis must remain uncertain.

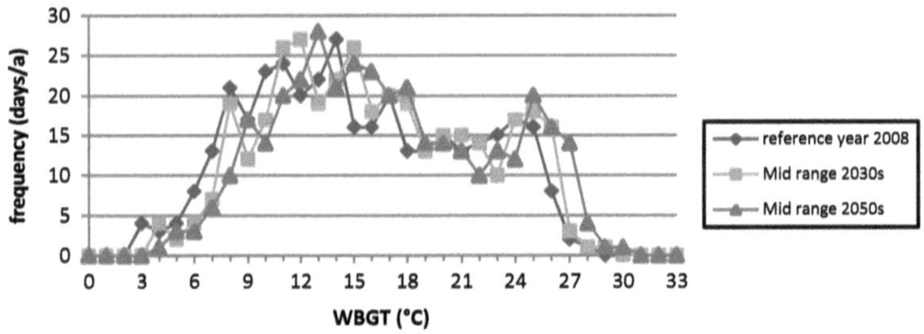

Fig. 16.1 Frequency distribution of the estimated WBGT (mid-range climate scenario) for one exemplary NUTS 3 region (Vienna) for 2008, 2030s and 2050s (*source*: own calculations based on climatic data as prepared for the COIN project)

Figure 16.1 provides an illustrative comparison of values for one specific NUTS-3 region (Vienna). The data for the WBGT index for a particular year (2008) of the base period (2003–2012) is compared to the corresponding data for the mid-range climate scenarios for the 2030s and the 2050s.[2] . It is evident from Fig. 16.1 that the number of days with high WBGT values increases by the 2030s and to an even greater extent by the 2050s.

16.4.2 High and Low Range Climatic Scenarios for Manufacturing and Trade

To identify the low-range (high-range) climate scenarios, the standard deviations of temperature are subtracted from (added to) the temperature values of the mid-range scenario. The fact that no data on the standard deviations of the relative humidity is available does, however, result in specific distortions, i.e. the predicted effects of the low-range scenario are slightly overestimating actual effects, while the high-range scenario slightly underestimates actual effects. As was the case for the mid-range climate scenario, we again assume that the standard deviations are the same for every day in each month.

Figure 16.2 shows the frequency distribution of WBGT values for the high-range climate scenario compared to the one particular year (2008) of the base period (2003–2012). Again this figure depicts data for a specific NUTS-3 region, AT130 Vienna. Figure 16.2 shows the development of the number of days with a high WBGT value for the high-range climate scenario, e.g. when we assume an adverse development of the climatic factors. Compared to 2008 and also to the mid-range climate scenario depicted in Fig. 16.1, the number of days with a high WBGT value increases in the 2030s, and even more so in the 2050s.

---

[2] In the following the term "2030s" will be used as a synonym for the time period 2016–2045 and the term "2050s" for the time period 2036–2065. This improves general readability of text and figures.

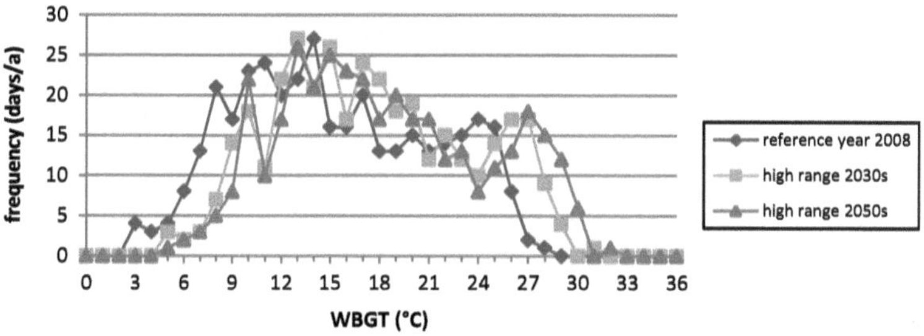

Fig. 16.2 Frequency distribution of the estimated WBGT (high-range climate scenario) for one exemplary NUTS3 region (Vienna) for 2008, 2030s and 2050s (*source*: own calculations based on climatic data as prepared for the COIN project)

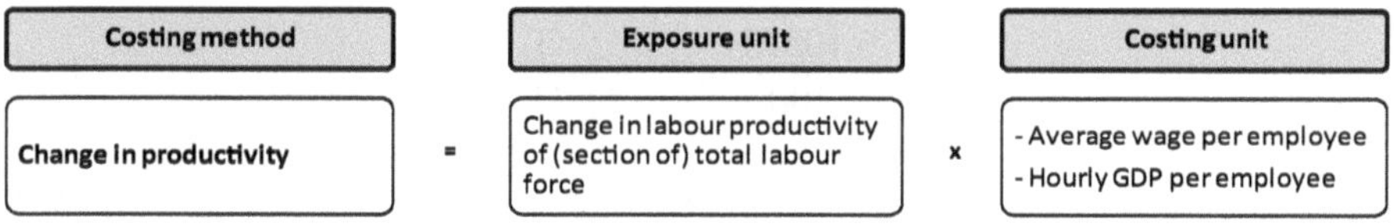

Fig. 16.3 Costing method applied and respective measurement units for "Manufacturing and Trade: labour productivity losses"

16.4.3 Specific Method(s) of Valuation and Their Implementation Steps

Figure 16.3 gives an overview of the costing method applied for evaluating the productivity losses of workers in "Manufacturing and Trade". The change in labour productivity is evaluated using two alternative methods, the average wage per employee on the one hand, and the hourly GDP per employee on the other.

Due to large differences in the climatic data across Austrian NUTS-3 regions, substantial differences in the calculated WBGT values occur. This indicates that substantial differences in productivity losses could also occur across Austrian NUTS3-regions. Thus, all calculations for each of the 35 Austrian NUTS-3 regions were carried out separately. Finally, the various productivity losses of each NUTS-3 region are analysed separately for every branch in order to derive an adequate evaluation given the regional distribution of branch activities across Austria. In a first step the meteorological data on daily dry bulb air temperature and daily relative humidity is transformed into an approximate measure of WBGT (outdoor as well as indoor) following a formula provided by Kjellstrom et al. (2009b, p. 221) "assuming moderately high heat radiation levels in light wind conditions". Thus:

$$\mathrm{WBGT} = 0.567\mathrm{T}_a + 3.94 + 0.393\mathrm{E} \tag{16.1}$$

with E given as

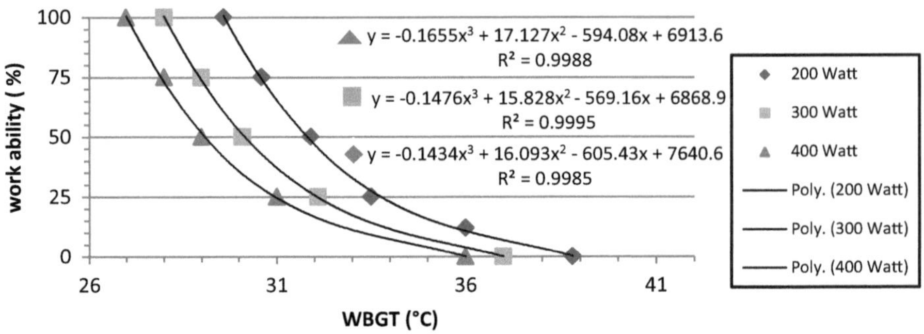

Fig. 16.4 Relationship between ability to work and WBGT [*source*: own calculations based on Kjellstrom and Dirks (2001) and WHO (2009, p. 12)]

$$E = \frac{6.105RH}{100} e^{17.27 \frac{T_a}{23.77 - T_a}}, \tag{16.2}$$

where T_a describes the 24-h average dry bulb temperature and RH the 24-h average relative humidity in percent, assuming that the "WBGT values calculated from 24-h values would represent the daytime mean WBGT outdoors" (Kjellstrom et al. 2009b, p. 221).

According to Kjellstrom et al. (2009b, p. 221) outdoor WBGT values can be adjusted to create indoor WBGT values by subtracting the impact of exposure to the sun. This means that indoor WBGT values are equal to outdoor WBGT values minus 4. From the indoor and outdoor WBGT values the productivity losses for different working intensities (200 W, 300 W and 400 W) are calculated for each day according to the functional relationships shown in Fig. 16.4.

As changes in work productivity depend on labour intensity, we use occupational statistics from "Manufacturing and Trade" and labour intensities to determine the labour intensity shares. This data is gathered at the disaggregation level of NUTS-3. Employing the climate change scenarios we get—by NUTS3 region—losses in effective hours worked per day. These are evaluated economically (see below) in one of two ways, and aggregated across the country so as to arrive at a monetary evaluation of labour productivity loss.

In order to convert productivity losses into monetary units the Human Capital Approach (HCA, also called Lost Wages Method, e.g. Berger et al. 2001) is used. The costs of 1 hour of lost work-time should equal the worker's hourly wage rate. The loss is thus calculated as the product of missed work days and daily salaries (or lost work hours and hourly wages for this specific case). The HCA was initially developed to monetise productivity losses due to absenteeism—this is a situation in which loss of productivity occurs when people are absent due to illness. Recently the method has also been extended to evaluate cases of 'presenteeism'—this is a situation in which workers are relatively less productive during the working day, for example, as a result of increased workplace temperatures. The HCA method is often preferred owing to its computational ease, its intuitive nature, and its consistency with microeconomic theory. The HCA builds on the assumptions of perfectly competitive labour markets (meaning that they are cleared and no involuntary unemployment exists) and on the assumption that workers are paid according to their marginal productivity, meaning that their wages display their marginal contribution to a firm's output (Mattke et al. 2007, p. 214; Patel et al. 2006, p. 68ff).

An alternative method to convert the productivity losses into monetary terms (also used here), is to use a value derived from the GDP per labour force member, representing the loss to society (Kovats et al. 2011). In our case assuming 225 effective working days per year, each worker contributes 309.26 €/day (or 38.65 €/h) to the GDP (for the base year 2008). Summing up all lost working hours of the different branches and multiplying this by the 38.65 € results in the total yearly productivity loss in monetary terms.

16.4.4 Range of Sectoral Socio-Economic Pathway Parameters that Co-determine Climate Impact

Various employment developments are investigated in order to assess socio-economic scenarios. This accords with the information presented in Chap. 6 (SSP) (particularly in Tables 6.1 and 6.2). We thus calculated employment growth-factors for the different socioeconomic scenarios, as depicted in Table 16.2. Those factors increase the aggregate hours worked per branch for the 2030s and 2050s. Concerning the relative proportions of indoor and outdoor workers a constant share is assumed here based on the proportions given in the reference scenario.

In addition, the results strongly depend on assumption concerning how strongly the productivity losses of workers influence the productivity of the other production factors (e.g. capital; less productive workers may mean less productive machine operation and thus lower capital productivity). In the socioeconomic scenario depicting diminishing climate impact we assume that labour productivity changes have no effect on the productivity of the other production factors (this is equivalent to the HCA). In contrast, in the so called impact enhancing socioeconomic scenario we assume that worker productivity losses induce a proportional decrease in the productivity of all other production factors (i.e. reduce total factor productivity). For the reference socioeconomic development we use an average of these two approaches.

Another development that might influence the impact of productivity changes is the level of technological development assumed for 2050. A realistic assumption might be that by 2050, at least for some sectors, the availability of more capital

Table 16.2 Employment growth factors (2008–2030s; 2008–2050s) for different socioeconomic scenarios

Diminishing 2030s	1.027761	Reference 2030s	1.029850	Enhancing 2030s	1.061791
Diminishing 2050s	1.025373	Reference 2050s	1.044776	Enhancing 2050s	1.119701

Source: own calculations based on Tables 6.1 and 6.2 in Chap. 6 (SSP)

equipment would relieve manual workers of the need to work as hard as before. This would shift downward the percentage values for working intensities used in this study.

16.4.5 Monetary Evaluation of Impacts

16.4.5.1 Data

In order to achieve more precise results, i.e. beyond those resulting from the use of climatic data and socioeconomic assumptions, further economic data for Austria is necessary. The Supplementary Material provides an extensive description of the data used in the present chapter. This covers specific data on the following areas: employment (occupational groups vs. subsectors), relative employment (occupational groups vs. subsectors), the classification of occupations according to work intensities, the share of outdoor workplaces, the hours worked per branch, wages (occupational groups vs. subsectors) and employment by NUTS-3 region and by economic activity.

16.4.5.2 Direct Sector Impacts (Costs and Benefits) Without Feedback Effects from Other Sectors

Figure 16.5 shows the productivity losses of workers for the different climate and socioeconomic scenarios for the 2030s and 2050s relative to the baseline. The horizontal axis depicts the various climate scenarios, while the impact of the

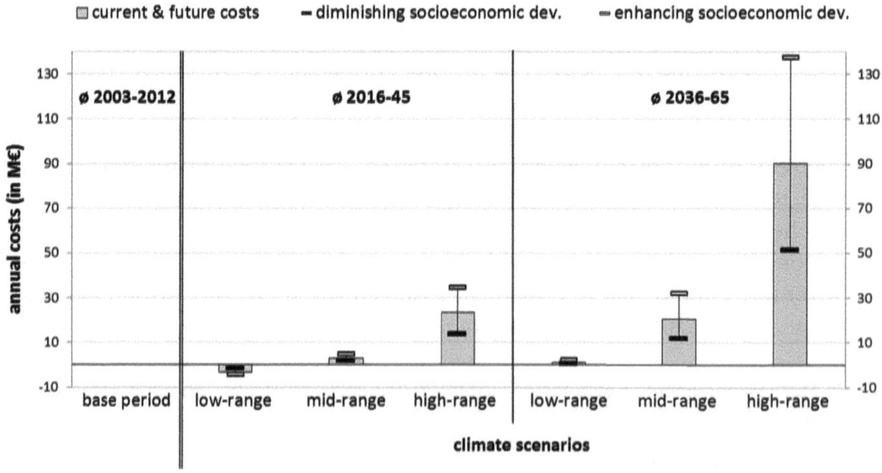

Fig. 16.5 Productivity losses of workers for different climate and socioeconomic scenarios for 2016–2045 and 2036–2065

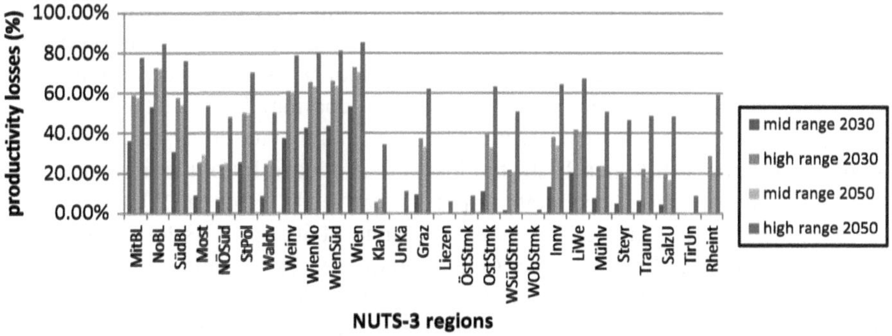

Fig. 16.6 Maximum productivity losses for specific days (here: 400 W outdoor work) [*source*: own calculations based on climatic data as prepared for the COIN project, socioeconomic and economic data (as given in detail in the Supplementary Material)]

different socioeconomic scenarios is depicted using vertical lines ending in black and white bars. According to Fig. 16.5, overall, the choice of the climate change scenario has a greater cost impact than the socioeconomic factors. However, particularly under the high-range climate scenario, socioeconomic factors can still substantially influence the costs accruing to the Austrian economy.

These results are summarized together with all different combinations of climate change scenarios and socioeconomic scenarios in Supplementary Material Table 16.11 in the Supplementary Material. The costs of productivity losses on a sector and branch-level are depicted in Supplementary Material Figures 16.6, 16.7 and 16.8.

Figure 16.6 depicts the maximum productivity losses occurring on certain days within 1 year for the different scenarios investigated within this chapter. The values refer to outdoor activities with a working intensity of 400 W. Indoor work (assuming no additional cooling) and work activities with a lower working intensity clearly show lower productivity losses. The high-range climate scenario for the 2030s and the mid-range and the high-range climate scenarios for the 2050s show substantial productivity losses. The number of regions affected increases over time as do the maximum losses occurring within 1 year. In some regions the values for productivity losses exceed 80 % on certain days.

For the analysis the productivity losses were weighted across a full year, and the respective shares of indoor and outdoor work and the shares of work intensities were included. The productivity losses increase both for the 2050s (compared to the 2030s) and in the high-range climate scenario (compared to the low-range and mid-range climate scenario). Although the values (as depicted in Supplementary Material Figure 16.5) seem to be very small in nearly all NUTS regions and for all scenarios, the subsequent analysis shows that such productivity losses can nevertheless lead to clear costs. Thus, the highest productivity losses are to be expected in the following regions (in descending order of magnitude): Wien, Nordburgenland, Wiener Umland/Südteil, Wiener Umland/Nordteil and Mittelburgenland. The losses in Sankt Pölten, Linz-Wels, Oststeiermark and Graz are clearly lower, but still significant in terms of yearly weighted average productivity losses.

To validate the results of the present study two mutually independent sensitivity analyses were undertaken. The productivity losses calculated here cover outdoor as well as indoor work. Productivity losses for indoor work, however, might not be very likely as these are often offset by additional cooling. The sensitivity analyses show that the monetary losses presented here are mainly caused by various impacts on outdoor work. While nearly all the losses are induced by outdoor work for the scenarios with low and mid climate impacts (nearly 100 % for the 2030s, 99 % for low-range 2050s and about 87 % for mid-range 2050s), this is less true for the scenarios with high climate impact (where outdoor work accounts for 90 % of the losses for the 2030s, but only for 70 % for the 2050s). The second sensitivity analysis looked at the impact of technological developments on capital equipment per workplace, and thus on reductions in work intensity. We assumed shifts of 5 to 20 % for the different scenarios (5 %, 10 % and 15 %, respectively for the enhancing, the reference and the diminishing socioeconomic scenario of the 2030s and 10 %, 15 % and 20 % respectively, for the 2050s). The results of this sensitivity analysis show that the estimated losses are quite insensitive. Reducing work intensity by 10 %, leads to productivity losses of about 95 % compared to those presented in this chapter. A reduction of 20 % leads to productivity losses of 88 %, and a reduction of 5 % has almost no effects with over 97 % of the previous losses still remaining.

16.4.5.3 Macroeconomic Effects of Labour Productivity Effects in Manufacturing and Trade

In the macroeconomic assessment below, we analyse the impacts of changes in labour productivity due to hotter and more humid climatic conditions, and thus a higher WBGT index in the summer. To depict this effect on presenteeism, we assume that only existing working contracts are fulfilled—but now with less productivity (due to productivity losses during some time periods of the year). In our analysis we thus fix the employment level in the climate change (CC) scenarios to the level in the baseline (BASE) scenario. This model assumption reflects the fact that employers are not expected to hire additional workers as a reaction to changing climate conditions during the summer months. Note that we do not quantify increased productivity during the winter season (due to milder winters and hence longer periods in which outdoor work can be undertaken). Moreover, effects on productivity are only modelled for the manufacturing and trade sectors, and not for other sectors which may also experience reduced productivity (e.g. agriculture, service sectors). As productivity decreases only in hot summer periods, the annual productivity losses are in general small compared to the base period, ranging from +0.15 to +0.005 %. Effects are in general stronger for the period 2036–2065 than for the period 2016–2045, and they are also stronger for the high-range climate change (CC) scenario with enhancing socioeconomic development than for the mid-range CC scenario with reference socioeconomic development. In the low-range CC scenario with diminishing socioeconomic development, there are even productivity gains in period 2016–2045 compared to the base period as a result of the less hot and less humid summer periods. The specific values employed for productivity changes in the macroeconomic model are given in Supplementary Material Table 16.12. The differences between the two subsectors stem from differences in the respective shares of outdoor workers and from the unequal shares of work intensities across sectors. While for the low-range CC scenario with diminishing socioeconomic development, only the productivity losses of workers are investigated (compare valuation with the HCA), for the high-range CC scenario with enhancing socioeconomic development we assume that the productivity losses of workers induce additional proportional productivity losses in other production factors (compare valuation with the GDP per employee approach). For the mid-range CC scenario with reference socioeconomic development we take an average of these two approaches.

Table 16.3 gives an overview of the sectoral effects for the mid-range CC scenario with reference socioeconomic development relative to the baseline scenario (i.e. without climate change, but with reference socioeconomic development). All effects are given as average changes of annual values in million euros (M€) relative to the baseline scenario in the respective periods 2016–2045 and 2036–2045. The productivity losses in the manufacturing and trade sectors lead to lower output. The impact on the whole economy, especially in the second period, is particularly strong. Due to reductions in output, manufacturing and trade demand

Table 16.3 Sectoral and total effects of quantified climate change impacts with mid-range CC and reference socioeconomic development in sector "Manufacturing and Trade", average annual effects relative to baseline (for periods 2016–2045 and 2036–2065)

	Ø 2016–2045			Ø 2036–2065		
Changes in M€ p.a. relative to baseline	Gross output value	Intermediate demand	Gross value added	Gross output value	Intermediate demand	Gross value added
Gaining sectors	**+0.0**	**+0.0**	**+0.0**	**+0.0**	**+0.0**	**+0.0**
all other gaining sectors	+0.0	+0.0	+0.0	+0.0	+0.0	+0.0
Losing sectors	**−17.5**	**−9.2**	**−8.2**	**−164.3**	**−87.0**	**−77.3**
Real estate activities	−2.2	−0.7	−1.6	−21.0	−6.2	−14.8
Construction	−1.3	−0.8	−0.5	−12.5	−7.8	−4.7
Wholesale and retail trade	−1.3	−0.6	−0.7	−12.2	−5.4	−6.9
Accommodation	−1.1	−0.4	−0.7	−10.5	−3.7	−6.8
all other losing sectors	−11.5	−6.8	−4.7	−108.0	−63.9	−44.1
Total effect (all sectors)	**−17.5**	**−9.2**	**−8.2**	**−164.3**	**−87.0**	**−77.3**

Note: baseline scenario = reference socioeconomic development without climate change
Climate change scenario = reference socioeconomic development and mid-range climate change
Quantified impact chains: [productivity of workers – mid-range]

fewer inputs from other sectors and hence the feedback effects on other sectors are negative as well. The strongest negative effects, both in terms of output and value added, emerge in sectors with a high share of outdoor labour such as real estate activities, construction, the wholesale and retail trade, and the accommodation sector. While effects on value added are small for the mid-range climate change scenario with reference socioeconomic development (decreasing by 8 million euros for 2016–2045, and by 77 million euros for 2036–2065), the corresponding effects are much stronger for the high-range climate change scenario with enhancing socioeconomic development (decreasing by 83 million euros for 2016–2045, and by 423 million euros for 2036–2065—see Supplementary Material Table 16.13). Moreover, one half of the effect on value added and GDP comes from reduced output quantities and the other half from reduced real prices.

For the low-range scenario for 2016–2045, the productivity of workers increases slightly for manufacturing and trade sectors. That is why GDP increases by 7 million euros for 2016–2045 and welfare by 4 million euros, relative to the baseline scenario without climate change. As we assumed that employers do not hire additional workers, there are no effects of climate change impacts in "Manufacturing and Trade" on the unemployment rate. In the mid-range scenario, GDP and welfare decrease in both periods, but considerably stronger in the second period than in the first. While in the mid-range scenario GDP decreases by 89 million euros and welfare by 54 million euros for 2036–2065, the corresponding effects in the high-range scenario are significantly larger, i.e. a decrease in GDP of 485 million euros, and of welfare by 296 million euros. These results are summarized in Table 16.4.

Mid-range climate change impacts have a relatively small negative effect on the government budget for 2016–2045 (a loss of 2.8 million euros), but this loss increases almost by a factor of 10 for 2036–2065 (26.8 million euros). For 2036–2065 labour tax intake decreases by 4.8 million euros, and production taxes, capital taxes and value added taxes decrease by 2.7 million euros, 7.8 million euros and 10.7 million euros. This causes a total decline of government revenues of 26.8 million euros. The reasons for this are the relatively lower production levels of the economic sectors and the fall in demand by private households. As we assume that government consumption has to be held constant at the baseline level of no climate change, the government thus has to reduce transfers to households by 26.8 million euros. Again, effects for the high-range scenario are much stronger. Here there is a total decrease in revenues of 173.4 million euros for 2036–2065, which again needs to be offset by a cut in transfers to households. These effects are depicted in detail in Supplementary Material Tables 16.14 and 16.15.

16.4.6 Qualitative Impacts (Non-monetarised)

The qualitative climate change impacts by branch in "Manufacturing and Trade" are given in the Supplementary Material Table 16.16. This table also provides

Table 16.4 Quantified climate change impacts in "Manufacturing and Trade" on GDP and welfare across different climate and socioeconomic scenarios, average annual effects relative to baseline (for periods 2016–2045 and 2036–2065)

		2016–2045			2036–2065		
	Socioeconomic scenarios						
Changes in M€ p.a. relative to baseline	Climate scenarios	Diminishing	Reference	Enhancing	Diminishing	Reference	Enhancing
	Climate scenarios						
GDP (changes in M€)	High			*–95*			*–485*
	Mid		**–9**			**–89**	
	Low	*+7*			*–3*		
Welfare (changes in M€)	High			*–58*			*–296*
	Mid		**–6**			**–54**	
	Low	*+4*			*–2*		

Note: Empty cells were not quantified

information on further literature that deals with the topic of climate change impacts for “Manufacturing and Trade”.

16.4.7 Sector-Specific Uncertainties

Although the present analysis of climate change impacts on the sector “Manufacturing and Trade” has attempted to achieve robust findings, several limitations and uncertainties remain. In the following, the most important uncertainties are listed and briefly discussed:

16.4.7.1 Climate data

- The most crucial assumption regarding the climate data is that of constant variability in climate parameters. This may be a limiting factor in the present study. In practice, the variability of climate parameters often increases. An increase in variability would lead to a wider range of results than those depicted in this chapter.

16.4.7.2 Economic Data

- In some areas the use of more detailed data would have led to more precise results. Regarding data on employment and wages the only distinction made is that between subsectors. It would, however, be more reasonable to differentiate between branches. Assuming that the different branches show the same proportion of different occupational groups, and that the wages of these occupational groups are the same in all different branches, is not very realistic. Furthermore, it was assumed that the relative weights for calculating the proportions of hours worked within the single NUTS-3-regions merely differ by sector, although to be more realistic they should be branch-specific.
- Several assumptions also had to be made when calculating the productivity losses of workers. The classifications of the different occupational groups according to different working intensities and skill-levels are based on a comprehensive literature review, but are nevertheless to some extent subjective.
- The employment growth factor used for the different socio-economic scenarios is assumed to be the same within for the whole of Austria and across the different branches, even though location and/or branch-specific differences are quite likely.
- Within this chapter only the effects of productivity changes within the sector “Manufacturing and Trade” are analysed. Although this indicates the possible climate damage occurring due to worker productivity losses, to gain a more complete picture for Austria a similar investigation encompassing (all) other sectors and related macroeconomic effects is required.

16.4.7.3 Other

In analyzing worker productivity losses several effects were ignored even though their potential impact on model results could not be ruled out. These include:

- The impact of the age structure in Austria on worker acclimatization: If the average population ages and there is thus a need for a longer working life then the effects of higher temperatures will be larger than those estimated above.
- The intensity of holidaying in summer: If there is an increase in the number of workers going on holiday in the summer, the estimated average productivity losses per year may be lower than those stated above.
- The impact of longer working seasons for outdoor jobs: Although this effect is not significant for "Manufacturing and Trade" (as the percentage of outdoor work is very low), for other sectors this may be a crucial issue.
- The impact of a change in the intensity of air-conditioning: The model estimates of productivity losses for indoor work assume no additional air-conditioning is used in the future. While this may be realistic for some workplaces, it is definitely not likely to be true for all. In fact, in those workplaces where air-conditioning systems are already in use, energy consumption is much more likely to increase.

16.4.8 Relevance for Other Sectors

The effects within "Manufacturing and Trade" are mainly relevant for all upstream sectors and those sectors that provide intermediary products and raw material for manufacturing and trade companies. Where companies face significant changes in the price of necessary inputs, they may change their production techniques and thus need other and/or additional inputs. This can cause significant changes in demand in other economic sectors, such as agriculture, forestry or the energy sector.

Furthermore all downstream sectors are highly influenced by developments in "Manufacturing and Trade". Changes in the production quantities and/or price changes of certain products can have significant impacts on sectors demanding products from "Manufacturing and Trade".

16.5 Summary of Climate Costs for Manufacturing and Trade and Conclusions

The sector "Manufacturing and Trade" is exposed to climate change in several ways. The two most direct forms of impact are likely to be a reduction in worker productivity due to higher temperatures, together with an increase in relative humidity. These result in higher heat exposure and thus lead to associated adjustments in production processes. As the sector "Manufacturing and Trade" is highly dependent on sectors facing potential damage from climate change (agriculture, forestry, energy) any necessary external adjustments are also likely to have a significant impact on "Manufacturing and Trade", e.g. related infrastructure damage. Furthermore, it is very important for manufacturing facilities and wholesale and retail traders to adapt to changes in consumer preferences. The productivity losses of workers are the only impact chain quantified within this chapter. All other impact chains are discussed qualitatively and the potential directions of costs and benefits are indicated.

Evaluation of worker productivity losses shows that these cover a considerable range for the mid-range climate scenario and are subject to possible deviation depending on whether higher or lower climatic stimuli are considered. Relative to the baseline, the mid-range climate scenario with reference socioeconomic development shows costs of about 3 million euros for the 2030s and about 21 million euros for the 2050s. Assuming weaker climatic stimuli and impact diminishing socioeconomic development for the 2030s, positive effects of about 2 million euros are possible, but even in this scenario they would turn negative (reaching about 1 million euros) for the 2050s. In contrast, stronger climatic stimuli (high-range climate scenario) with enhancing socioeconomic development leads to considerably higher costs of 35 million euros for the 2030s and about 138 million euros for the 2050s. As a result of the feedback effects on other sectors, the macroeconomic effects are considerably higher than the direct sector effects. Productivity losses in the manufacturing and trade sector cause a lower output in these subsectors. As a consequence, other dependent sectors face shortfalls in intermediate inputs. For the 2050s in the mid-range climate scenario, GDP and welfare decrease by 89 million euros and 54 million euros respectively. These corresponding figures for the 2050s in the high-range climate scenarios show a much higher impact, with GDP and welfare decreasing by 485 million euros and 296 million euros respectively. Potential impacts on government budgets were also quantified. It was found that the government budget decreases by 27 million euros in the 2050s in the mid-range climate scenario, and by 173 million euros in the 2050s in the high-range climate scenario.

The impacts on the whole economy for the mid-range climate scenario are relatively low. This simply reflects the fact that this chapter only investigates the effects of productivity losses within the sector "Manufacturing and Trade". If all other sectors affected were considered as well, the costs arising due to worker productivity losses would be larger. In addition to the uncertainties regarding the climatic, socioeconomic and economic data used in this study, it needs to be mentioned that several other factors, which were ignored here, are also likely to influence results. For example, consideration of summer vacation intensity is likely to have an impact on productivity losses. Furthermore, it is imaginable that rising temperatures would lead to an extension of the outdoor working season. While this effect is relatively low for the sector "Manufacturing and Trade" (as the percentage of outdoor work is relatively low), its impact is likely to be much more significant for other sectors. A rather unrealistic—but possible—scenario would also entail some kind of "cultural adaptation" such as a transition to siestas during the hottest midday hours.

The previous paragraph, as well as Sect. 16.4.6, indicates the main uncertainties with respect to the approach used here to evaluate the productivity losses due to climate change. Although further research to overcome some of these limitations is obviously necessary, the present chapter clearly shows that it is important to consider productivity effects. Furthermore, it is necessary to develop methods to quantify those effects which here have only been dealt with qualitatively. These other impact chains may also have significant monetary effects on the Austrian economy.

Acknowledgements The authors thank five anonymous reviewers for their comments on earlier versions of this chapter, as well as Timo Goeschl, Udo Ebert, Michael Rauscher, Christoph Böhringer and the participants of the annual meeting of the environmental section of the Verein fuer Socialpolitik 2014, and the Scientific Advisory Board of the COIN project for all of their comments. All remaining errors are ours.

References

Berger ML, MurrayJ F, Xu J, Pauly M (2001) Alternative valuations of work loss and productivity. J Occup Environ Med 43:18–24

Bux K (2006) Klima am Arbeitsplatz—Stand arbeitswissenschaftlicher Erkenntnisse—Bedarfsanalyse für weitere Forschungen. Forschung Projekt F 1987, Bundesamt für Arbeitsschutz und Arbeitsmedizin, Dortmund

Dunne JP, Stouffer RJ, John JG (2013) Reductions in labour capacity from heat stress under climate warming. Nat Clim Chang 3:563–566

Hellwig RT (2004) Natürlich behaglich—Natürliche Lüftung und Behaglichkeit—Gegensätze? Fachjournal für Gebäudetechnik und Innenraumklima (formerly: Gesundheits-Ingenieur) 125(5):237–244

Hübler M, Klepper G (2007) Kosten des Klimawandels Die Wirkung steigender Temperaturen auf Gesundheit und Leistungsfähigkeit. Studie im Auftrag des WWF Deutschland. Frankfurt am Main, Eine

ISO, International Standards Organization (1989) Hot environments—estimation of the heat stress on working man, based on the WBGT-index (wet bulb globe temperature), ISO Standard 7243, Geneva

Johns G (2010) Presenteeism in the workplace: a review and research agenda. J Organ Behav 31: 519–542

Kjellstrom T, Dirks K N (2001) Heat stress and ability to work in the context of climate change. Proceedings of the International Thermal Physiology Symposium, University of Wollongong, Wollongong, September 2–6, 2001

Kjellstrom T, Holmer I, Lemke B (2009a) Workplace heat stress, health and productivity—an increasing challenge for low and middle-income countries during climate change. Glob Health Action, Special Volume 2009. http://www.globalhealthaction.net/index.php/gha/article/viewFile/2047/2539. Accessed 05 Dec 2014

Kjellstrom T, Kovats RS, Lloyd SJ, Holt T, Tol RSJ (2009b) The direct impact of climate change on regional labor productivity. Arch Environ Occup Health 64(4):217–227

Kovats S, Lloyd S, Hunt A, Watkiss P (2011) The impacts and economic costs of climate change on health in europe. Technical Policy Briefing note 5: Health. http://www.climatecost.cc/images/Policy_Brief_5_Climatecost_Health_Summary_Results_vs_5_draft_final_web.pdf http://www.climatecost.cc/images/Policy_Brief_5_Climatecost_Health_Summary_Results_vs_5_draft_final_web.pdf. Accessed 10 Nov 2013

Mattke S, Balakrishnan A, Bergamo G, Newberry SJ (2007) A review of methods to measure health-related productivity loss. Am J Manag Care 13(4):211–217

Parsons KC (2003) Human thermal environments. The effects of hot, moderate and cold Environments on human health, comfort and performance—the principles and the practice. Taylor and Francis, London

Patel N, Lofland J H, Pizzi L T (2006) Chapter 4: health related productivity. In: Pizzi T L, Lofland J H (eds) Economic evaluation in U.S. health care: principles and applications, Jones and Bartlett Publishers, Sudbury

Seppänen O, Fisk WJ, Lei QH (2006) Effect of temperature on task performance in Office environment. Lawrence Berkeley National Laboratory Environmental Energy Technologies Division Indoor Environment Department. http://indoor.lbl.gov/sites/all/files/lbnl-60946.pdf. Accessed 5 Dec 2014

Statistics Austria (n.y.a) Klassifikationsdatenbank, Erläuterungen zur ÖNACE 2008. http://www.statistik.at/kdb/downloads/pdf/Erlaeuterungen_zur_OENACE_2008.pdf. Accessed 8 Apr 2013

WHO, World Health Organization (2009) Climate change exposures, chronic diseases and mental health in urban populations: a threat to health security, particularly for the poor and disadvantaged. Technical Report, Japan

WTO, UNEP, World Trade Organization, United Nations Environment Programme (2009) Trade and climate change. WTO Publications, Geneva

Yang J, Liern B (2009) The impact of presenteeism in workplace: an empirical study. Int J Manag Enterprise Dev 7(4):339–357

Chapter 17
Cities and Urban Green

Wolfgang Loibl, Tanja Tötzer, Mario Köstl, Stefan Nabernegg, and Karl W. Steininger

Abstract Cities are known to induce so called "urban heat island" effects. Therefore climate change will have a significant impact in urban environments upon thermal comfort. As urban green can mitigate local temperature peaks, green space is an essential feature in cities and one option to prevent decline of thermal comfort and related effects. Direct climate induced damages on urban green overlap with urban environmental stressors which are judged currently to be more critical than climate damages. Indirect climate induced damages of urban green will enforce subsequent negative effects of local temperature increase in cities: e.g. on health, tourism and urban economy which are difficult to delimit and quantify. The one robust option to quantify climate change damages used in this chapter is the preventative cost approach, i.e. damages are monetized by the level of costs that measures would imply to *prevent* increasing urban heat islands (here focusing on construction and maintenance of additional urban green).

Expansion of urban green is triggered by settlement growth that preserves appropriate urban green shares and—potentially—by explicit policy to counter local temperature increase in urban environments in the future. Both issues are considered here. Green space expansion because of urban growth in Austria's six larger cities is assumed to reach 144 ha (4.7 %) from 2011 till 2030 and 62 ha (2 %) from 2031 till 2050. Adapting additionally to climate change would result in more expansion: 195 ha (6.4 %) between 2011 and 2030 and 143 ha (4.7 %) between 2031 and 2050 reaching a total of 11 % urban green growth by 2051. Annual investment costs for new parks are estimated at 119 million euros for the period 2011–2030, and 93 million euros for 2031–2050 respectively. Annual costs for maintaining these additional parks are estimated at 7.6 million euros till 2030 and

W. Loibl (✉) • T. Tötzer • M. Köstl
Energy Department, Austrian Institute of Technology, Vienna, Austria
e-mail: wolfgang.loibl@ait.ac.at; tanja.toetzer@ait.ac.at; mario.koestl@ait.ac.at

S. Nabernegg
Wegener Center for Climate and Global Change, University of Graz, Graz, Austria
e-mail: stefan.nabernegg@uni-graz.at

K.W. Steininger
Wegener Center for Climate and Global Change, University of Graz, Graz, Austria

Institute of Economics, University of Graz, Graz, Austria
e-mail: karl.steininger@uni-graz.at

K.W. Steininger et al. (eds.), *Economic Evaluation of Climate Change Impacts*, Springer Climate, DOI 10.1007/978-3-319-12457-5_17

13.4 million euros till 2050. Such preventative costs are an approximation that can be considered as lower bound of thermal discomfort due to climate change in the six larger Austrian cities.

17.1 Introduction

The cities' "urban heat island" (UHI) effect is characterised by higher temperatures in urban areas compared to rural surroundings (Garstang et al. 1975; Landsberg 1981; Oke 1982). Growth of population and thus heat load caused by cumulative anthropogenic heat emission and growing heat storage due to increasing building surface and volume will contribute to further intensification of heat islands. To mitigate those effects green spaces can be of high importance. They avoid higher irradiance, mitigate local day temperature and thus improve thermal comfort at the local scale outdoor and indoor (e.g. Loibl et al. 2011b, 2014; Orehounig et al. 2014).

Direct costs of climate induced damages on urban green overlap with further urban location-related stress factors (air pollution, water shortage due to low infiltration and poor soil volume, soil pollution etc.) which are judged more critical than climate effects (personal communication: park management departments in Vienna and Innsbruck). Indirect cost of damages released through urban heat islands (heat stress, consequences thereof, cooling expenses, health impacts), are again difficult to quantify. The one robust option to quantify climate change damages used in this chapter is the preventative cost approach, i.e. damages are monetized by the level of costs that measures to prevent increasing urban heat islands would imply: costs for measures to enhance thermal comfort in urban green spaces ranging from expanding old and constructing new parks as cooling facility to conversion of streets to alleys providing more shade in sunny areas. Creating additional parks could also lead to unsealing of surfaces which increases drainage area and mitigates flood risk.

Expansion of urban green is not the only adaptation measure to mitigate urban heat island effects. There exist others, such as greening of roofs and walls of buildings, but these have to be explored in detail and are dependent on building owners and are too complex to be quantified for an entire city, and thus do not easily offer themselves to serve as a basis for the preventative costing approach.

17.2 Dimensions of Sensitivity to Climate Change

Urban climate conditions differ distinctively from general climate conditions. In urban environments climate change will have a significant impact upon levels of thermal comfort. As duration and intensity of heat episodes are expected to grow and urban population and thus the cities will further grow in the future (ESPON 2010) more additional anthropogenic heat will be emitted. Both trends are expected to enforce amplification of heat islands.

17.2.1 Climatic Factors

Main factors for urban heat islands are high irradiance trapping, accompanied by nocturnal cloud cover and low wind speed, which both reduces nocturnal cooling through flux of urban heat load to upper atmospheric layers (c.f. Oke 1982).

Oke (1982) describes the impact of different surface conditions on urban heat island effects by comparing urban and rural energy balances: green open spaces, occupied by vegetation, build one end of the urban surface spectrum: gardens, parks, and cemeteries are likely to have water storage capacities equal to those of rural areas. Following Oke (1982, p. 5), "rural heat energetics are driven by the surface net radiant flux density dominated by short-wave radiation exchange during sunshine hours and long-wave radiation during night hours releasing nocturnal cooling. The other end of the urban surface spectrum builds paved roads, places and buildings, partitioning radiant energy into sensible heat".

Figure 17.1 depicts the thermal structure of the urban atmosphere layer during clear summer weather by day and by night. Schematic profiles of potential temperature (Θ), the depths of the urban and rural internal boundary layers (- -) and the daytime mixed layer (-.-) are included.

17.2.2 Non-Climatic Factors

When dealing with cities, urban green spaces and heat exposure, we have to consider the influence of the built environment and the influence of the open spaces between. Building- and open space-properties have certain effects on urban microclimate.

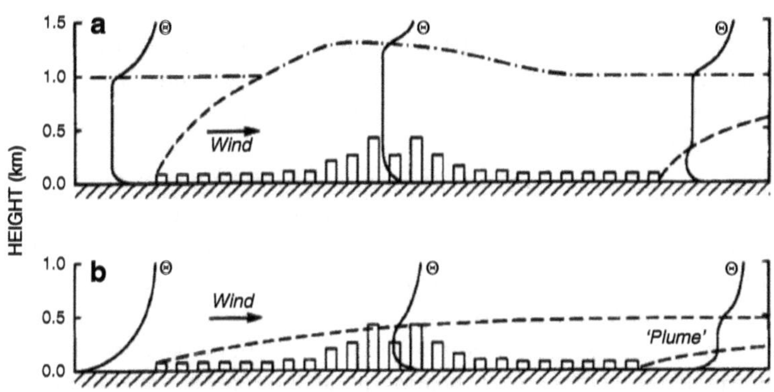

Fig. 17.1 Thermal structure of the urban atmosphere in a large mid-latitude city during summer by day (*top*) and by night (*bottom*). *Source*: Oke (1982)

These are dependent on properties like building extent, height and orientation to the sun, further open space extent and surface characteristics and finally their distribution in the city. They influence irradiance, reflection and absorption, trapping heat load as well as shading, ventilation and evaporation, which support cooling (Oke 1987).

Kleerekoper et al. (2012) summarize the non-climatic factors of urban heat island effects as follows:

- Absorption of short-wave radiation from the sun in low albedo (reflection) materials and trapping by multiple reflections between buildings and street surface.
- Obstruction of the sky by buildings, resulting in a decreased long-wave radiative heat loss from street canyons. The heat is intercepted by the obstructing surfaces, and absorbed or radiated back to the urban tissue.
- Decreased turbulent heat transport from narrow street canyons by reduction of wind speed.
- Increased heat storage by building materials with large thermal admittance and a larger surface area compared to rural areas.
- Decreased evaporation from urban areas due to less permeable surfaces and vegetation. So more energy is put into sensible heat and less into latent heat.
- Air pollution in the urban atmosphere absorbs and re-emits long-wave radiation.
- Anthropogenic heat is actively released by combustion processes such as traffic, heating and production.

17.2.3 Identification of Potential Damage Combinations

The factor combinations are layout of built area and related open spaces blended with temperature increase, accompanied by heat periods with higher temperature peaks during the day and lack of cooling during the night. Microclimate measurements and simulations prove the influence of location on the temperature regime driving urban heat islands with typically 2–3 °C higher temperatures than rural surroundings (c.f. Knight et al. 2010). Studies of urban green of a few hectares demonstrate frequently cooling effects of green spaces and—vice versa heating effects of built up and paved urban area (e.g. Gill et al. 2007; Oliveira et al. 2011; Loibl et al. 2014).

Cooling of small green spaces can even act over longer distances beyond the green if it is accompanied by appropriate urban fabric design. Figure 17.2 depicts microclimate simulation results for a 500 × 500 m area with a neighbourhood park in a densely built up area in Vienna (left). The right image shows the local temperature distribution. The 2000 m^2 park area releases a distinct temperature "sink" of 3–6 °C against the paved road area in the south. The green area prevents neighbouring street canyons from heating up as the road in the south does.

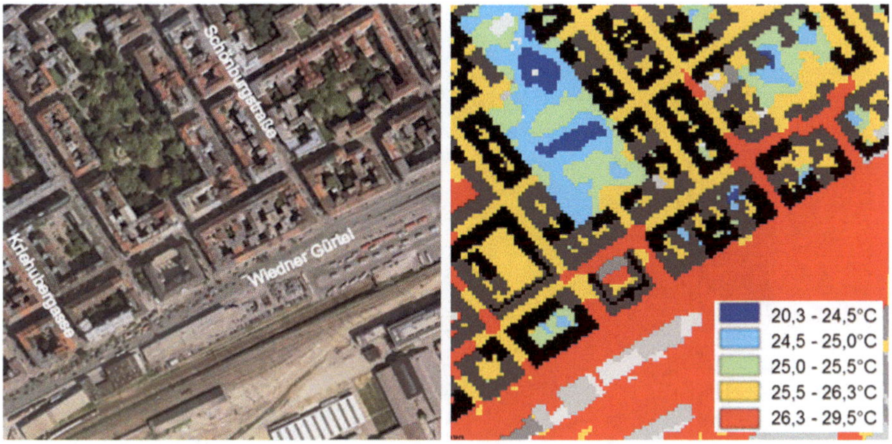

Fig. 17.2 *Left*: aerial photo of the block layout with park areas. *Right*: simulated average 2 m-temperature distribution during a hot summer day. *Source*: Loibl et al. (2014)

Thus planting of trees providing shade and appropriate surface design is important to reduce local temperature increase. Nevertheless it has to be considered, that trees can increase surface roughness reducing large scale ventilation, sky view and thus heat flux to the urban atmosphere boundary layer, slowing down nocturnal cooling.

17.3 Exposure to Climatic Stimuli

17.3.1 Past and Current Climatic Exposure

When dealing with urban issues regional climate simulation results can serve as framework to describe the general exposure pattern but an improved simulation resolution is required to explore local impact in detail. Figure 17.3 presents the exposure of Vienna to high temperatures, showing the number of summer days (>25 °C) for 100 × 100 m raster cells as result of urban climate simulations for 1981–2010 (Zuvela-Aloise and Matulla 2011). One can observe a concentration of summer day occurrences in Vienna's inner districts and less summer day numbers in areas with larger urban green shares.

Urban heat island effects are triggered by city size: Large cities with dense centres show higher temperatures than small ones. So we concentrate on the larger Austrian cities/urban regions with more than 100,000 inhabitants, where the built-up area extent makes them prone to become distinct heat islands.

Figure 17.4 shows the location of those larger cities in Austria, exposed to frequent summer day occurrences ($T_{max} > 25$ °C) during 1971–2000, as results from reclip: century Hindcast simulations (Loibl et al. 2011a). The cities have been heavily affected in the East (50–60 days), along the Danube Valley (30–40 days) and in the Southeast and South (30–35 days).

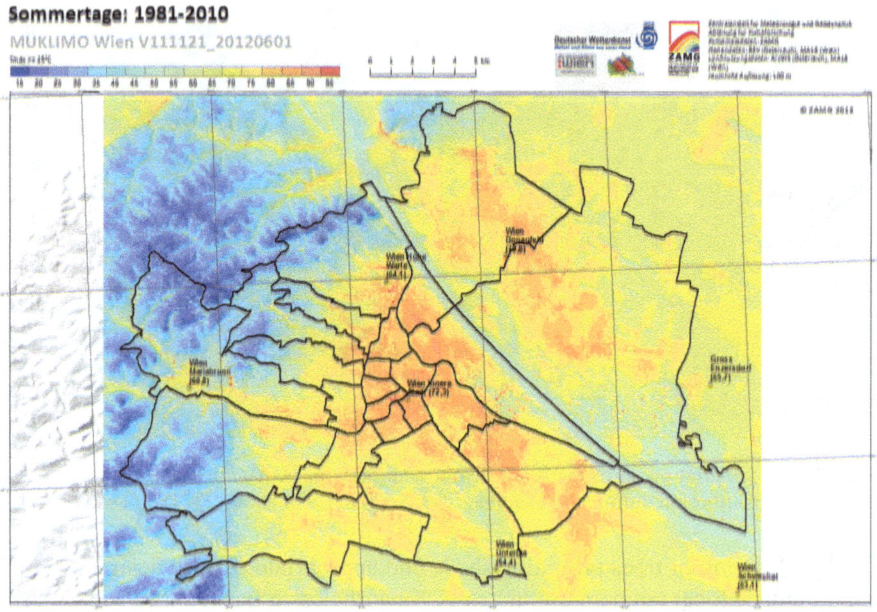

Fig. 17.3 Average number of summer days in Vienna for the period 1981–2010 simulated with the MUKLIMO_3 urban climate model. *Source*: Zuvela-Aloise and Matulla (2011)

Thus we concentrate on those cities where high heat exposure is expected also in the future: Vienna, Graz, Klagenfurt, Linz and Salzburg. Innsbruck (with currently around ten summer days) has been added to the sample as most Western larger city.

17.3.2 Impact Chains

Several impact chains addressing urban green have been identified, but only selected ones have been considered for quantification, as many direct and indirect effects are not quantifiable which is described earlier. Increase of heat and drought episodes raise the risk of urban green damage (e.g. Cregg and Dix 2001). But direct climate change effects have not been confirmed by park authorities as they judge climate stress (at least until now) not as distinct as it is covered by more severe stressors in the urban environment (personal communication: City of Vienna, MA 42—Gardening Department).

Table 17.1 gives an overview of identified impact chains affecting urban green and the prevention activities addressed as cost factor. The last column describes which of the impact chains has been quantified.

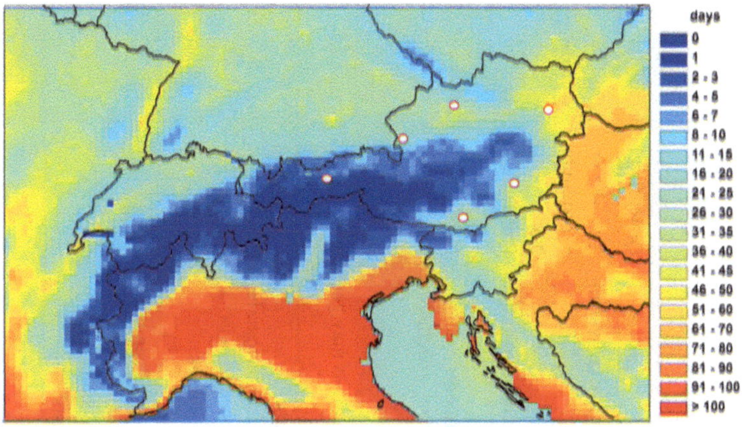

Fig. 17.4 Spatial distribution of the number of hot summer days during prior decades (1971–2000) and the location of the large province capitals. *Source*: Reclip: century-simulations, Loibl et al. (2011a)

17.3.3 Economic Impacts

As economic damages of Urban Heat Islands cannot be quantified directly, we use the indirect approach of a lower bound of these by the preventative costing approach. Thus damage cost quantification refers to preventative activities: Increase of green space to protect urban environments. The creation of additional green space and planting additional trees, adapting better to higher temperatures and drought stress (e.g. Pauleit 2003), can serve as a valuable instrument for improving climate comfort especially for the elderly or less wealthy citizens.

Additional investments can enlarge urban green in the cities:

- to secure the current local thermal comfort—
  - to keep health conditions of the inhabitants stable and
  - to secure attractiveness of the city for visitors during hot summer months,
- to mitigate urban heat islands under increasing temperatures in the future,
- to provide more drainage area, better absorbing heavy rainfall and runoff (which was not addressed as a major issue and seems less important during normal weather conditions).

Investments on urban green adaptation and maintenance to prevent climate change induced costs refer to:

- establishing new neighbourhood parks as "cooling facility",
- converting streets to alleys by planting trees,
- unsealing paved surfaces as drainage area.

Higher maintenance expenditures for urban green would allow:

- to manage green areas more intensively—e.g. by increasing irrigation,

Table 17.1 Cities and urban green related impact chains

Climate change parameter	Impact chain	Quantified in the model
Temperature increase, temperature extreme events increase Change in frequency and/or intensity of heat days and nights	**Loss of climate comfort in urban environments (Urban Heat Islands effect above overall temperature increase)**	Yes—by preventive expenditures (see below)
	Health impacts of overall temperature increase → Reduction of outdoor activities → less activity causes health effects → more health costs due to heat related heart-diseases, respiration impacts, and dehydration effects	No (but see Chap. 11)
	Loss of climate comfort for visitors and tourists comfort in urban environments → Decline in city visits during hot periods → decline in summer tourism in cities → decline in city tourism related income	No
	Heat related damage on pavements, tram rails → Heat expands asphalt and concrete pavements, rail network → pavement and rails may get damaged → additional repair costs	No
Increase of extreme temperatures and heat/drought waves: Change in frequency and/or intensity	**Improved prevention against loss of climate comfort in urban environments** → Increase of temperatures leads to less open space comfort → open green spaces have to be expanded to improve local cooling effects and may compensate heat load, trapped in built environment → green space expansion requires investments for buying and de-sealing lots → green space expansion requires higher maintenance costs	Yes
	→ Increase of temperatures leads to less open space comfort → lacking green space can be partly compensated through planting of street trees providing additional shade and cooling through evapotranspiration → tree planting requires investments for adapting technical infrastructure below ground → additional trees requires higher maintenance costs	Yes

- to better compensate vegetation damage, released through high temperatures and drought,
- to replace vegetation by varieties which better adapt to high temperatures and drought.

The expected costs for investments and additional maintenance are triggered by urban growth and by climate adaptation efforts: urban green expansion because of the expansion of the particular city (addressed as baseline scenario—see next section) and—potentially—because of prevention for heat island increase by installing additional green space achieving a higher urban green share in densely built areas (the mid-range scenario—see next section).

If such preventive measures are not made, thermal comfort may decrease and release further costs for welfare and may cause possible losses in earnings from tourism. Losses in income from city tourism due to climate effects are not observed to date in Austria at any significant level. City tourism depends on a variety of visit purposes triggered by the amount of attractions in the cities. The Austrian cities with summer festivals, all year conferences, attractive historic buildings and museums which are during hot noon hours cooler than outdoor areas, observe no or little decline of visitor numbers and thus no negative income effects. Health and welfare issues are discussed in detail in Chap. 11, tourism issues are addressed in Chap. 19, and natural hazard issues including flood risk are discussed in Chaps. 12 and 18.

17.3.4 Future Socioeconomic Change and Exposure to Climate Change

17.3.4.1 Baseline Scenario for Cities and Urban Green

Expansion of urban green is related to the expansion of built up area. The socio-economic pathways (see Chap. 6) specify further urban growth. To estimate urban green expansion it is necessary to know the urban growth expectations and the current extension of green areas. We concentrate here on green spaces embedded in the built up area and do not consider natural or agricultural areas in the cities' outskirts.

The extension of urban green areas has been extracted from official numbers from the cities' web pages. The numbers are not all comparable as some areas are indeed located outside built up areas, which cannot be extracted from official statistics. Built up, densely built up and industrial areas have been extracted by exploring CORINE land cover (EEA 2013a) and Urban Atlas data (EEA 2013b). Urban Atlas data are available for 300 urban regions with more than 100,000 habitants. Thus data is only available for five large Austrian cities: Vienna, Graz, Linz, Salzburg and Innsbruck. The land use layers of these cities have been intersected by their municipality borders to extract the cities' built up area extension. We have also added Klagenfurt to the sample because of its relevant size and climate exposure during summer. As only coarse CORINE data are available for

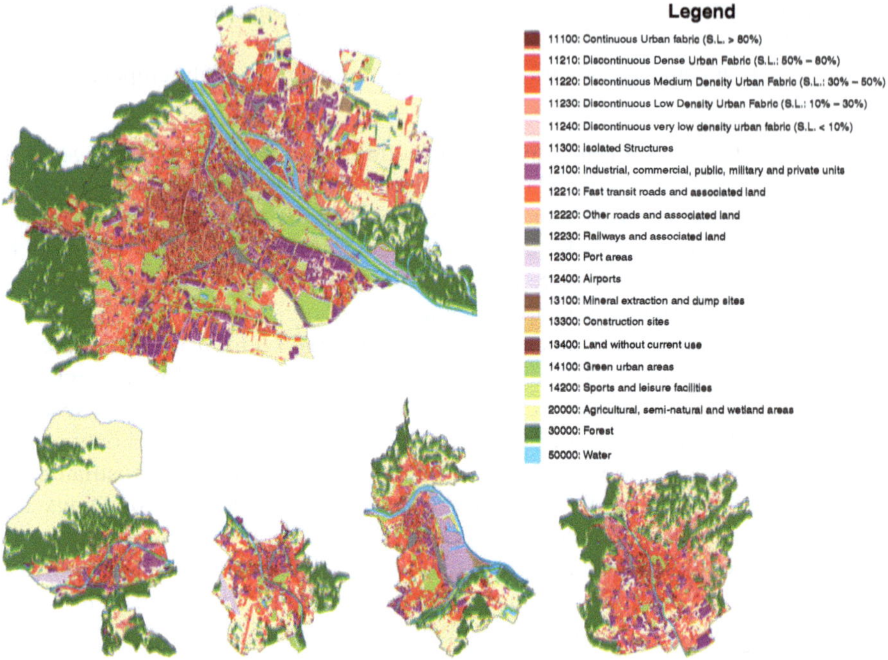

Fig. 17.5 Land use class distribution of Austria's larger cities: Vienna (*top*)—Innsbruck—Salzburg—Linz—Graz (*bottom*, *left* to *right*). *Source*: Urban Atlas data, processing—AIT

Klagenfurt, the relevant built up area density fractions have been estimated based on CORINE/Urban Atlas land use class ratios averaged from comparable cities (Salzburg, Innsbruck).

Figure 17.5 presents a compilation of the Urban Atlas maps. The maps show land use patterns typical for cities: compact centres with high built-up area density intensifying urban heat island effects.

The built up area numbers serve as start for estimating urban growth and defining green space expansion requirements. Table 17.2 compares built up and green area in the six cities.

17.3.4.2 Mid-Range Climatic Scenario for Cities and Urban Green

The changing climate triggers a steady temperature increase resulting in higher averages and more frequent heat episodes (c.f. Chap. 5). Figure 17.6 presents the changes in average maximum temperatures: From the period 1981–2010 to the 2030s the average temperature maximum during summer will be expected to increase between 1 and 1.4 °C, with highest changes in September (1.5 °C). For the 2050s the summer temperature maximum increase is expected between 2 and

Table 17.2 Area of Austria's larger cities by land use categories

City	Total area (ha)	Densely built up 2010 (ha)	Loosely built up 2010 (ha)	Total built up (incl. industry and traffic) 2010 (ha)	urban green 2010 (ha)	Urban-green in % of built up area
Graz	12,762	1,266	4,303	5,569	235	4.2 %
Innsbruck	10,492	535	1,123	1,658	120	7.2 %
Klagenfurt	11,957	1,045	1,500	2,600	215	8.3 %
Linz	9,605	1,355	2,094	3,450	400	11.6 %
Salzburg	6,584	944	1,863	2,807	200	7.1 %
Wien	41,467	8,977	11,156	20,134	1,900	9.4 %
	92,867	14,122	22,039	36,216	3,070	

Source: Urban Atlas data distinguishing between built up area density classes and cities' web-pages reporting urban green area (parks, graveyards, children playgrounds). Compilation and processing: AIT

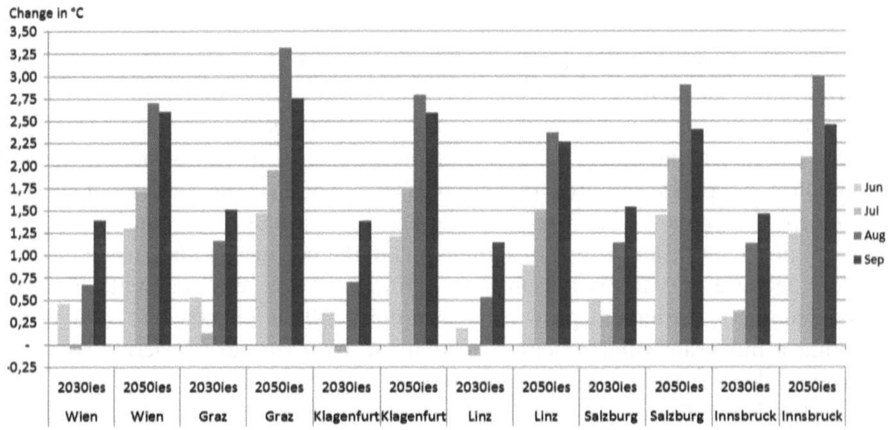

Fig. 17.6 Change of monthly maximum temperatures between 1981/2010 and the 2030s/2050s (30 year averages) in Austria's larger cities (NUTS-3 regions). *Source*: COIN. *Note*: orders of the *bars* from *left* to *right*: Jun-Jul-Aug-Sep

2.4 °C with highest increase in August (2.4–3.3 °C). The simulation results show 30 year climate averages, single years indicate higher peaks.

17.3.4.3 High and Low Range Scenario for Cities and Urban Green

We do not explore impacts of high- and low-range climate scenarios as the assumed preventive costs for the mid-range scenario are still uncertain. To illustrate impact ranges of weaker and stronger climate signals, changes are roughly related to preventative measures defining a range of urban green expansion and related maintenance. Coping with effects of weaker climatic stimuli requires less additional parks, trees and less additional maintenance, that could be set at resulting in a 50 % cut of the expansion requirements in the mid-term scenario. Coping with effects of stronger climatic stimuli requires more additional parks and related maintenance assuming to result in twice the growth factor of the mid-term climate scenario.

17.4 Monetary Evaluation of Impacts

Costing Model
As explained earlier we deal with *preventive expenditures*. The costs are related to

- investments for establishing additional green space—park enlargement and new parks,
- street tree planting in public spaces, and
- costs for maintenance of the additional green space.

Baseline Scenario: Additional Green Space due to Urban Growth
For the future an increase in population (with even higher shares of older more vulnerable population) and settlement area is expected (see Chaps. 6 and 11). Urban growth triggers enlargement of urban green aiming in the construction of new neighbourhood parks, children playgrounds and similar. The settlement growth factors for the cities' NUTS3 regions (see Chap. 6) are average numbers for the entire region with faster growth of settlements in the peri-urban than inside the city borders because of various reasons (e.g. affordable lots, environmental quality, social security). Settlement growth beyond the 2030s is expected to slow down adopting a moderate enlargement of settlement areas due to restrictive planning policies to decrease new land consumption.

The following Table 17.3 gives an overview of factors and numbers considered and derived during these calculations.

To enforce slow settlement growth within the city borders the regional growth factors have been converted into smaller city-based growth factors by relating the NUTS3 growth factors not to the entire settlement area, but to the loosely built up area of the cities. Then the additional settlement areas have been related to the cities' total built up areas to derive city-related growth factors. These new growth factors have been applied to estimate urban green expansion. All growth factors—the urban growth induced and the climate induced ones—are shown later in Table 17.4.

Table 17.3 Baseline scenario growth factors till 2030 and 2050 for NUTS3 settlement area expansion and derived growth factors for city area expansion

	NUTS3 (%)		City—area in ha						City (%)	
City	Growth factor built up 2011–2030	Growth factor built up 2011–2050	City total area (ha)	Densely built up 2010 (ha)	Loosely built up 2010 (ha)	Built up total (incl. industry and traffic) 2010 (ha)	Built up total 2030 (ha)	Built up total 2050 (ha)	City growth factor 2011–2030	City growth factor 2011–2050
Graz	17.4	24.1	12,762	1,266	4,303	5,569	6,315	6,607	13.4	18.6
Innsbruck	21.8	32.3	10,492	535	1,123	1,658	1,903	2,021	14.8	21.9
Klagenfurt	16.2	22.0	11,957	1,045	1,500	2,600	2,843	2,930	9.3	12.7
Linz	18.7	26.7	9,559	1,355	2,094	3,450	3,842	4,009	11.4	16.2
Salzburg	20.6	30.0	6,584	944	1,863	2,807	3,190	3,366	13.6	19.9
Wien	16.2	23.5	41,467	8,977	11,156	20,134	21,941	22,751	9.0	13.0
			92,821	14,122	22,039	36,216	40,034	41,682		

Source: Statistik Austria NUTS3 scenarios, Urban Atlas data, City statistics, AIT

Table 17.4 Green space embedded in densely built up area in Austria's larger cities and additional green area expansion factors as response to urban growth (left columns—baseline scenario) and as response to climate change (right columns—mid-range scenario)

	Green space (ha) 2010	Urban growth driven green space growth in % 2011–2030	Urban growth driven green space growth in % 2031–2050	Urban growth driven green space growth in % 2011–2050	Climate induced green space growth in % 2011–2030	Climate induced green space growth in % 2031–2050	Climate induced green space growth in % 2011–2050	Total green space growth in % 2011–2050
Graz	235	5.9	2.1	8.1	2.0	3.0	5.0	13.1
Innsbruck	120	3.0	1.3	4.3	1.0	1.0	2.0	6.3
Klagenfurt	215	5.8	2.7	8.5	1.0	2.0	3.0	11.5
Linz	400	7.9	3.1	11.0	1.0	2.0	3.0	14.0
Salzburg	200	6.0	2.8	8.8	1.0	2.0	3.0	11.8
Wien	1,900	3.7	1.7	5.3	2.0	3.0	5.0	10.3

17.4.1 Direct Sector Impacts (Costs and Benefits) Without Feedback Effects from Other Sectors

We refer here to preventative investments which are conversions of built up area to additional green space and planting of additional street trees described later. While the urban growth triggered urban green expansion will alter non-built up land into new parks, urban green expansion as preventative instrument will require converting already built up land into new neighbourhood parks. There are no feedback-effects from other sectors considered.

17.4.2 Costing of Potential Deployed Adaptation Measures

To adapt to climate change in urban environments it is necessary to discuss the required scope of these measures, which we do in the following to scope the dimension of potential adaptation to prevent climate change impacts of heat islands. A study on UHI effects by ZAMG has conducted simulation experiments to reduce heat exposure by enlarging the park area in Vienna. Simulation experiments were conducted by virtually expanding public park area in the city by 30 % (Zuvela-Aloise and Matulla 2011). The experiments give some hints regarding the effectiveness of such a measure. The 30 % increase suggested by (Zuvela-Aloise and Matulla 2011) is an unrealistic assumption due to lacking lots available to be converted to green space and due to costs for lots as further limiting factor. Gill et al. (2007) have suggested in their study that increasing green infrastructure area in Manchester by 10 % would result in a cooling of the surface temperature by 2 °C till 2050 (and 2.5 °C till 2080). But we can argue that increasing green space even by 10 % is not plausible, due to a lack of available lots to be converted from built up to green area.

We start here with a careful attempt assuming appropriate green space expansion till 2050 between 2 and 5 %: 2 % e.g. for Innsbruck as alpine city with little heat stress and 5 % e.g. for Vienna with high summer heat exposure (see Figs. 17.4 and 17.6). Summing up the urban growth induced green space expansion factors in the cities (4.3–11 %) the combined increase would range in the six cities between 6.3 and 14 %—numbers which cannot be easily achieved. Table 17.4 lists the additional urban green, derived for the baseline scenario to supply new built up area with neighbourhood parks, and for the mid-range scenario, converting built up area into green spaces to enhance local cooling. The expansion factors to adapt to climate change have been selected with respect to the heat exposure expectations presented in Fig. 17.6.

Figure 17.7 gives an overview of the areal expansion of green space in the addressed cities. During the years 2011–2030 the urban growth-triggered expansion is greater, during the years 2031–2050 the climate-induced expansion is larger.

Green space growth effects can be supported by converting streets into alleys through planting additional street trees. Trees are not that effective as larger parks

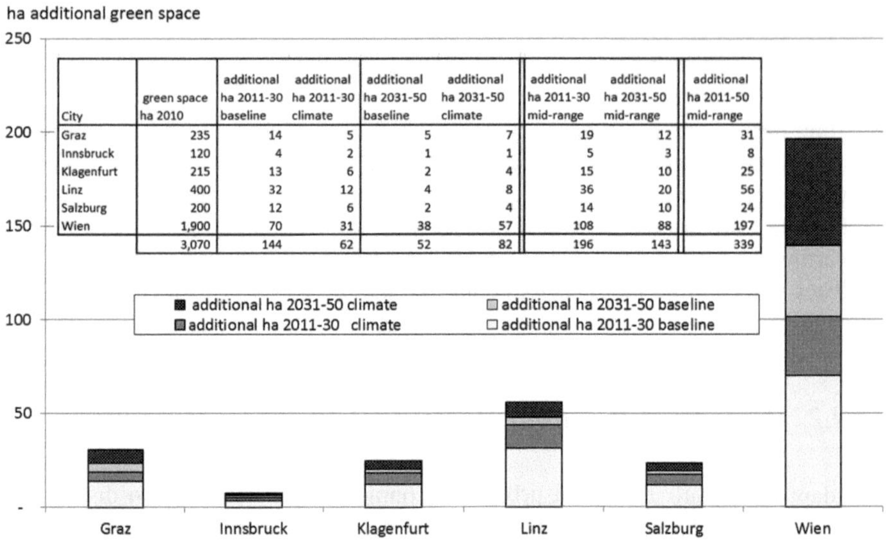

City	green space ha 2010	additional ha 2011-30 baseline	additional ha 2011-30 climate	additional ha 2031-50 baseline	additional ha 2031-50 climate	additional ha 2011-30 mid-range	additional ha 2031-50 mid-range	additional ha 2011-50 mid-range
Graz	235	14	5	5	7	19	12	31
Innsbruck	120	4	2	1	1	5	3	8
Klagenfurt	215	13	6	2	4	15	10	25
Linz	400	32	12	4	8	36	20	56
Salzburg	200	12	6	2	4	14	10	24
Wien	1,900	70	31	38	57	108	88	197
	3,070	144	62	52	82	196	143	339

Fig. 17.7 Expected expansion of urban green spaces 2011–2030 and 2031–2050 as response to urban growth (baseline scenario) and as response to climate change (mid-range scenario)

but can contribute reducing the local radiant temperature. This measure would also avoid costs for buying land which may not be available.

The estimation of required additional tree numbers is based on the current tree numbers in public places (outside parks), the built up area to be supplied and the expected temperature increase. Table 17.5 shows the current street tree numbers in the six cities (ranging between 3.6 and 7.5 trees/ha) and the increase suggestions. For additional trees we suggest a total increase by 1–3 % of the current tree number for both periods 2011–2030 and 2031–2050 depending on the cities' temperature exposure.

17.4.3 Investment Costs

All considered costs refer to preventive expenditures for constructing (and maintaining) new parks and convert streets to alleys.

Constructing new green spaces requires investments for additional lot purchase and park construction. Lot prices may differ, average numbers are (in €/m^2): Klagenfurt: 175 €, Graz: 250 €, Linz: 300 €, Salzburg and Innsbruck: 500 € each, Vienna: 1,500 € (prices from real estate market statistics). Costs for furnishing parks with plants, technical and recreation equipment are assumed with 250 €/m^2 (communication: City of Vienna, MA 42—Gardening Department). Costs for demolition and preparation of the ground are not included, neither economic losses of not building the area.

Table 17.5 Number of trees in public spaces in Austria's larger cities—current numbers and suggested increase with related increase factors

City	Street trees 2013	Street trees 2013 (ha)	Additional trees 2013–2030	Additional trees 2030–2050	Additional trees 2011–2050	Increase factor (%) 2031–2030	Increase factor (%) 2013–2050	Increase factor (%) 2011–2050
Graz	20,000	3.6	400	600	1,000	2.0	3.0	5.0
Innsbruck	11,000	6.6	110	220	330	1.0	2.0	3.0
Klagenfurt	18,000	6.9	360	180	540	2.0	1.0	3.0
Linz	18,500	5.4	555	370	925	3.0	2.0	5.0
Salzburg	21,000	7.5	210	420	630	1.0	2.0	3.0
Wien	90,000	4.5	2,700	2,700	5,400	3.0	3.0	6.0
	178,500		4,335	4,490	8,825			

Costs for planting trees are reported in Vienna to reach up to 10,000 € considering modifications in technical sub-surface infrastructure (grid, gas network, telecom wires, district heating, water and sewage pipelines) as well as destruction and repair of the pavement and 3 years of more intensive tree treatment (irrigation). Planting costs are only 1,500–2,000 €/tree, depending on plant material (personal communication: City of Vienna, MA 42—Gardening Department). Assuming that 50 % of the trees in Vienna will be planted in existing roads, requiring to adapt sub-surface infrastructure and 50 % will be planted in areas without required change of sub-surface infrastructure we assume here 5,000 € as average costs per tree in Vienna and 3,000 € for trees in other Austrian cities.

17.4.4 Maintenance Costs

As a first step the current maintenance costs per ha urban green have been extracted from the annual municipality budgets for the selected cities by summing up the relevant numbers from the particular budget sector which contains all costs (staff, material and investments) for "parks, gardens and playgrounds" ("Ansatz 8150" of the municipality budgeting system). These data—staff costs (including partly expenditures for pensions of retired employees) and costs for goods and services—have been related to the maintained park area to extract unit costs per ha.

The costs differ to some extent between the cities. The numbers range between 22,000–27,000 € and 39,000–46,000 € maintenance costs per ha in the six cities. With these ha-cost numbers the maintenance costs have been estimated for the additional urban green area to be constructed during 2011 and 2050.

17.4.5 Total Costs: Direct Sector Impacts

Tables 17.6, 17.7 and 17.8 provide summaries of the cost calculations. Urban green expansion because of urban growth is assumed to develop 144 ha new park area (4.7 %) from 2011 till 2030 and 62 ha new park area (2 %) from 2031 till 2050. Adapting to climate change requires more green space expansion: 196 ha new park area (6.4 %) between 2011 and 2030 and 143 ha new park area (4.7 %) between 2031 and 2050 reaching 11 % green space growth by 2051. Annual investment costs for new parks are estimated at 119 million euros for the period 2011–2030 and 92 million euros for the period 2031–2050, respectively. Annual costs for maintaining these new park areas are estimated at 7.8 million euros till 2030 and 13.4 million euros till 2050. The numbers refer to the mid-range climate scenario. Costs for weaker and stronger climatic stimuli have been derived by defining an urban green area expansion range and related maintenance. The costs for coping with weaker climatic stimuli (requiring less additional parks and less additional maintenance) refer to a reduced urban green area growth: 50 % of the

Table 17.6 Average annual investment costs due to additional parks (in M€)

Investments in additional parks[a]		Baseline[b]	Climate change relative to the baseline		
			Low-range	Mid-range	High-range
Ø 2011–2030	Physical units	144 ha	170 ha	196 ha	248 ha
	Costs	82	101	119	154
	Benefits	0	0	0	0
	Net effect	−82	−101	−119	−154
Ø 2031–2050	Physical units	62 ha	103 ha	143 ha	225 ha
	Costs	36	64	93	150
	Benefits	0	0	0	0
	Net effect	−36	−64	−93	−150

[a]Lot prices and park construction costs
[b]Baseline = climate-triggered economic impacts arising from a socioeconomic development according to the reference scenario, but no climate change

Table 17.7 Average annual investment costs due to additional street trees (in M€)

Investments in additional street trees[a]		Baseline[b]	Climate change relative to the baseline		
			Low-range	Mid-range	High-range
Ø 2011–2030	Physical units	0	2.167 trees	4.335 trees	8.670 trees
	Costs	0	0.5	0.9	1.8
	Benefits	0	0	0	0
	Net effect	0	−0.5	−0.9	−1.8
Ø 2031–2050	Physical units	0	2.245 trees	4.490 trees	8.980 trees
	Costs	0	0.5	0.9	1.9
	Benefits	0	0	0	0
	Net effect	0	−0.5	−0.9	−1.9

[a]Planting costs, underground preparation
[b]Baseline = climate-triggered economic impacts arising from a socioeconomic development according to the reference scenario, but no climate change

climate induced share. The costs for coping with stronger climatic stimuli (requiring more additional parks) refer to an increased urban green area growth: twice of the growth factor derived for the mid-range climate scenario.

17.4.6 Macroeconomic Costs of Financing Compensation for Climate (Change) Damages

For the economic evaluation of climate change impacts on cities and urban green two core difficulties arise. First, and as mentioned above, direct damage quantification is a task too complex for the many different types of consequences of a rise in the urban heat island effect. We therefore have chosen the preventative cost approach to quantify an order of magnitude (often considered the lower bound) of

Table 17.8 Average annual maintenance costs due to additional parks (in M€)

Maintenance additional parks[a]		Baseline[b]	Climate change relative to the baseline		
			Low-range	Mid-range	High-range
Ø 2011–2030	Physical units	144 ha	170 ha	196 ha	248 ha
	Costs	5.4	6.5	7.8	9.9
	Benefits	0	0	0	0
	Net effect	−5.4	−6.5	−7.6	−9.9
Ø 2031–2050	Physical units	206 ha	273 ha	339 ha	473 ha
	Costs	7.8	10.6	13.4	19.0
	Benefits	0	0	0	0
	Net effect	−7.8	−10.6	−13.4	−19.0

[a]Cumulative costs
[b]Baseline = climate-triggered economic impacts arising from a socioeconomic development according to the reference scenario, but no climate change

damages. Second, we neither know whether this specific road of adaptation is taken at all or even whether it is the best one. Thus, we do not have a specific activity for quantifying indirect effects across other sectors (i.e. we cannot follow the path of evaluation applied in all other chapters of this book). However, this costing method supplies one quantification of climate damages, that society (and in particular its agents) will be exposed to. Someone will have to bear these costs. If we assume, as one option, that the public sector compensates for these damages, it is of interest to quantify, what macroeconomic costs the financing of this compensation by the government will have. In the following we carry out exactly this quantification. We thus assume that the damages of climate change for cities and urban green are compensated for exclusively by the public sector and do reduce the public budget remaining for other purposes accordingly.

The resulting macroeconomic effects on the indicators of welfare, GDP and unemployment are shown in Table 17.9. The numbers indicate the earlier described impacts of the different climate scenarios in comparison to the baseline scenarios, including feedback effects between economic sectors. The effects arise from a public budget that—in the climate change scenario—is partly used for paying compensation and thus is no longer available for some tasks earlier covered. We find that a decrease for welfare and GDP occur both in the period 2016–2045 and to the period 2036–2065 for mid-range climate change as well as for the high and low climate change scenario. The changes in unemployment occur in correspondence with GDP changes (declining GDP—raising unemployment); they are quite small in most cases.

In Table 17.10 changes in government revenues and expenditures are given, which occur due to public compensation for climate damages. Comparing the mid-range climate change scenario with the baseline, we find that it is not only a shift among public expenditures that climate change damage compensation triggers, but also an additional net decline of public revenues.

Table 17.9 Effects of quantified climate change impacts in sector Cities and Urban Green on welfare, GDP and unemployment across different climate and socioeconomic scenarios, average annual effects relative to baseline (for periods 2016–2045 and 2036–2065)

		ø 2016–2045			ø 2036–2065		
Changes in M€ p.a. relative to baseline	Socioeconomic scenario	Diminishing	Reference	Enhancing	Diminishing	Reference	Enhancing
	Climate scenario						
Welfare (changes in M€)	High		−122			−198	
	Mid		**−62**			**−99**	
	Low		−32			−49	
GDP (changes in M€)	High		−47			−77	
	Mid		**−24**			**−38**	
	Low		−12			−19	
Unemployment rate (in % points)	High		+0.01			+0.01	
	Mid		**+0.01**			**+0.01**	
	Low		+0.00			+0.00	

Note: Empty cells not quantified

Table 17.10 Effects on government budget of assumed climate change damage compensation for damages via Cities and Urban Green, average annual effects relative to baseline (for periods 2016–2045 and 2036–2065)

Changes in M€ p.a. relative to baseline	ø 2016–2045	ø 2036–2065
Revenues	−11	−18
Production tax	−2	−3
Labour tax	−8	−12
Capital tax	−1	−1
Value added tax	−3	−5
Other taxes	+2	+3
Expenditures	−11	−18
Expenditure for compensating climate change damage	+40	+64
Unemployment benefits	+6	+10
Transfers to households net of other taxes	−18	−28
Government consumption	−40	−64
Government budget in baseline (p.a.)	148,839	206,157
Climate change impact on government budget	−0.01 %	−0.01 %

Note: baseline scenario = reference socioeconomic development without climate change; climate change scenario = reference socioeconomic development and mid-range climate change; Quantified impact chains: no impact chain damages quantified, but preventive cost approach applied

This total decline in public revenues by an annual 11 million euros in 2016–2035 and 18 million euros in 2036–2065 is mainly caused by lower labour tax revenues. Except for some minor tax revenues aggregated in 'Other taxes', also production, capital as well as value added tax revenues slightly decrease. On the expenditure side government has to cut consumption by the same amount as the additional cost for climate change damage compensation. Because of reduced tax revenues also net transfers to households have to decline, while unemployment payments slightly increase. Effects for the first period occur in same direction but again at a smaller scale than for the second period.

17.4.7 Qualitative Impacts (Non-monetarised)

For certain impacts there is no secure way to estimate costs (e.g. for health costs or technical infrastructure costs). Health impacts (non-monetized) are addressed in Chap. 11 for the entire population—not only the urban. Costs due to flooding are addressed in Chap. 18 but refer to repair costs and not to adaptation costs. Estimating adaptation costs would require detailed knowledge about the current capacity of the sewage systems, runoff volumes to be expected and the drainage capacity of the surface of the selected cities, which are not available.

City tourism may be also affected negatively as hot urban environments are less attractive for tourists. But there is no evidence that this effect is a general disadvantage. We assume that certain attractiveness criteria—either culture—arts—

history—or architecture and urban design can surpass climate-related disadvantages to a certain extent, depending on the city addressed.

Thus the one (suboptimal) method on climate damage cost estimation has been applied—estimating preventative costs, which is still a very uncertain topic as discussed below.

17.4.8 Uncertainties

All topics addressed in the qualitative impact section are highly uncertain. Impact and costs of urban heat island effects are uncertain too, as they cannot be directly estimated due to different accompanying-effects which are e.g. built environment characteristics, different climatic framework conditions and different indirect effects referring to health, tourism and further.

Thus we refer to preventative costs addressing an assumed expansion of urban green space to better cope with urban temperature increase. The increase of open green space within the cities as response on climate change can be judged as very uncertain. There is no possibility to relate different temperature increase estimations to growth rates for open green space expansion. As the enlargement is based on assumptions so do the cost estimations.

The expansion of open green space within the cities due to urban growth can be judged as less uncertain as it refers to expectations for settlement growth due to population growth which is quite certain as the population development refers to quite stable fertility and mortality rates. Population increase due to immigration or intra-national movement is based on expert judgment from Statistik Austria which could be verified to some extent during the last years. But for the 2030s and even more for the 2050s the expectations are vague too.

17.4.9 Relevance for Other Sectors

The topic is relevant for health and vulnerability of the population due to heat exposure. It is further relevant for city tourism as experienced in southern cities with hotter climates where in some cities a visitor number decrease is observed during the hottest months. This is not the case in cities with certain attractions like festivals, and this may be of less importance for larger cities as for a particular share of business visitors the reasons to select the respective city as travel target deviate from those of classic tourists.

17.5 Summary of Climate Costs for Cities and Urban Green

Here we deal with preventative costs, as direct costs of damages of climate change-induced rise in urban heat islands (heat stress, implications thereof, cooling demand, etc.) are too complex for quantification.

These costs refer to expansion of open urban green to mitigate urban heat island effects and improve thermal comfort through evapotranspiration and shading. The establishment of new green spaces requires investments resulting in costs to buy additional lots for neighbourhood parks and to furnish the parks with plants, technical and recreation features. An additional measure (to avoid establishing too many expensive parks) is planting trees in streets and places.

The examined cities show a built up area of 36,000 ha. Till the 2030s expansion of 3,800 ha and till 2050 a further expansion of another 1,650 ha built up land is assumed (baseline scenario). For 2030 a green space growth of 196 ha (144 ha triggered by urban growth, 52 ha by potential climate change adaptation) and till 2050 a growth of additional 144 ha (62 ha triggered by urban growth, 82 ha by potential climate change adaptation) is assumed. Taking into account the mid-range climate scenario the total investment costs for new park areas are estimated at 2,384 million euros till 2030 and at an additional 1,856 million euros till 2050, which is annual costs of 119 million euros till 2030 and 93 million euros till 2050, respectively. The costs for planting additional street trees are estimated for the mid-range climate scenario at 18.4 million euros till 2030 and at 18.9 million euros till 2050, which translates to annual costs of 0.9 million euros till 2030 and 0.9 million euros till 2050. The annual costs for maintaining 196–339 ha additional park area are estimated for the mid-range climate scenario at 7 million euros till 2030 and at 13.4 million euros till 2050.

The implementation of impacts of financing the compensation for these climate damages in the macroeconomic model (including feedback effects) shows a decrease of the welfare indicator by 62 million euros for the period 2016–2045 and 99 million euros for the period 2036–2065 (on average per year). The reduction in annual GDP is weaker (−24 million euros for the first period and −38 million euros for the second). This slowing down in the second period is triggered by the higher share of the socioeconomic impact compared to the combined socio-economic and climate change impact, as urban expansion is expected to slow down in the second period.

The costs for coping with weaker climatic stimuli (requiring less additional parks, trees and less additional maintenance) may reach the half of the climate induced mid-range numbers, the costs for coping with stronger climatic stimuli requiring more additional parks may reach twice the climate induced mid-range numbers.

References

Cregg BM, Dix ME (2001) Tree moisture stress and insect damage in urban areas in relation to heat island effects. J Arboric 27(1):8–17
EEA (2013a) CORINE land cover. http://www.eea.europa.eu/publications/COR0-landcover
EEA (2013b) Urban atlas. http://www.eea.europa.eu/data-and-maps/data/urban-atlas
ESPON (2010) FOCI (Future Orientations for Cities), various project reports. http://www.espon.eu/main/Menu_Projects/Menu_AppliedResearch/foci.html
Garstang M, Tyson PD, Emmitt GD (1975) The structure of heat islands. Rev Geophys Space Phys 13:139–165
Gill S, Handley J, Ennos R, Pauleit S (2007) Adapting cities for climate change: the role of the green infrastructure. Built Environ 30(1):97–115
Landsberg HE (1981) The urban climate. International geophysics series, vol 28. Academic Press, New York
Kleerekoper L, van Esch M, Baldiri Salcedo T (2012) How to make a city climate-proof, addressing the urban heat island effect. Resour Conserv Recycl 64(2012):30–38
Knight S, Smith C, Roberts M (2010) Mapping Manchester's urban heat island. Weather 65(7): 188–193
Loibl W, Köstl M, Züger J (2011a) Reclip: century – research for climate protection: century climate simulations, Part C – Climate scenarios: comparative analysis. ACRP Project A760437. http://reclip.ait.ac.at/reclip_century/fileadmin/user_upload/reclip_century_files/reclip_century_part_C_Aug_2011.pdf
Loibl W, Tötzer T, Köstl M, Züger J, Knoflacher M (2011b) Modelling micro-climate characteristics for urban planning and building design. IFIP Adv Inf Commun Technol 359:605–618
Loibl W, Stiles R, Pauleit S, Hagen K, Gasienica B, Tötzer T, Trimmel H, Köstl M (2014) Improving open space design to cope better with urban heat islands effects. GAIA 23(1):64–66
Oke TR (1987) Boundary layer climates, 2nd edn. Routledge, London
Oke TR (1982) The energetic basis of the urban heat island. Q J R Meteorol Soc 108:1–24
Oliveira S, Andrade H, Vaz T (2011) The cooling effect of green spaces as a contribution to the mitigation of urban heat: a case study in Lisbon. Build Environ 46:2186–2194
Orehounig K, Mahdavi A, Doppelbauer E, Loibl W, Tötzer T (2014) Projections of design implications on energy performance of future cities: a case study from Vienna. Sustain Cities Soc 12:92–101
Pauleit S (2003) Urban street tree plantings: identifying the key requirements. Proc Inst Civ Eng Munic Eng 156(1):43–50
Zuvela-Aloise M, Matulla C (2011) FOCUS-I – future of climatic urban heat stress impacts. ZAMG, Wien. http://www.zamg.ac.at/cms/de/forschung/klima/stadtklima/focus-i/

Chapter 18
Catastrophe Management: Riverine Flooding

Franz Prettenthaler, Dominik Kortschak, Stefan Hochrainer-Stigler, Reinhard Mechler, Herwig Urban, and Karl W. Steininger

Abstract Losses from natural disasters are on the rise and risk management options for lessening direct as well as indirect consequences are gaining in importance. Riverine flooding is one key concern and climate change is globally projected to increase intensity and frequency of flooding burden - albeit, due to numerous uncertainties there is only low confidence in projected changes. On the other hand, there is high confidence that today's and future losses are rising as more assets and people are moving in harm's way. The quantitative assessment of flood risk is complex, as such extreme event risk is characterized by few observations (low probability) associated with massive consequences (high impact), which by definition means substantial uncertainty around any estimates, particularly if future drivers, such as from climate change, need to be addressed as well. The methodology of choice is probabilistic catastrophe modelling, which combines hazard (flood intensity and frequency) with exposure (people and assets) and their vulnerability (susceptibility of exposed people and assets to incur losses for a given hazard). In order to properly account for uncertainty, we present three different catastrophe risk modelling approaches that outline the scope for possible changes in flood risk in Austria over the next 90 years. The analysis and findings are particularly relevant for Austria's Natural Disaster Fund, which is the primary disaster loss financing vehicle in Austria. Large uncertainties between the different approaches and various limitations restrict our general conclusion as well as a full comparison between

F. Prettenthaler (✉) • D. Kortschak
Institute for Economic and Innovation Research, Joanneum Research, Graz, Austria
e-mail: franz.prettenthaler@joanneum.at; dominik.kortschak@joanneum.at

S. Hochrainer-Stigler • R. Mechler
Risk Policy and Vulnerability, International Institute for Applied Systems Analysis, Laxenburg, Austria
e-mail: hochrain@iiasa.ac.at; mechler@iiasa.ac.at

H. Urban
Wegener Center for Climate and Global Change, University of Graz, Graz, Austria
e-mail: herwig.urban@uni-graz.at

K.W. Steininger
Wegener Center for Climate and Global Change, University of Graz, Graz, Austria

Institute of Economics, University of Graz, Graz, Austria
e-mail: karl.steininger@uni-graz.at

K.W. Steininger et al. (eds.), *Economic Evaluation of Climate Change Impacts*, Springer Climate, DOI 10.1007/978-3-319-12457-5_18

the approaches. However, we discuss possibilities to overcome these barriers in the future including suggestions how to arrive at more robust solutions in the face of such large uncertainties.

18.1 Introduction

Natural disaster impacts are worldwide on the rise, both in terms of human life and economic losses (Munich Re 2009; Swiss Re 2009; CRED 2012; IPCC 2012). In Europe, especially flooding but also wind storms and extreme temperature events seem to be increasing substantially (Hochrainer and Mechler 2013). Importantly, most severe hazards in Europe can be altered in their frequency and intensity through climate change. Hence, climate change is an important topic when it comes to determine future developments of natural extreme phenomena as well as to ways to reduce them and to spread their risk via suitable instruments.

Due to its geographical position, Austria is a country that can be hit by a multitude of different natural hazards. Both climate related hazards, such as flooding, hail and wind storms, as well as geotectonic hazards, such as earthquakes, have happened repeatedly in the past (Munich Re 2007; Formayer et al. 2001). Huge economic losses may arise from these events, especially from those that are not spatially limited, most importantly flooding (Pelling et al. 2002).

The problem of flood events may be worsening in the future due to climate change (IPCC 2012). In this context, evidence of climate change and extreme events for Austria is found and discussed in the literature (Steininger et al. 2005; Matulla et al. 2004) and the results have shown a complex picture. Concerning weather extremes, it is projected that flood risk is likely to increase in the winter season and spring, however, no general statements of changing overall risks can be made (Schimon et al. 2011; Dobler et al. 2013 and Blöschl et al. 2011). It is important to note that statistical methods for the detection of trends in extremes are currently a central topic in the extreme value statistics community (e.g. Dierckx and Teugels 2012; Chavez-Demoulin and Davison 2005). In this chapter, however, the main focus is to describe possible methods/approaches for the quantification of changes in risk for extremes as well as potential consequences (costs) of major riverine flood events (i.e. floods with a "large" spatial extent) since these are the most prominent natural disasters in Austria.

18.2 Dimensions of Sensitivity to Climate Change

Natural disaster risk is commonly defined as the probability of potential impacts affecting people; assets and the environment (see Mechler 2004). If risk becomes manifest, it may cause a variety of consequences that are commonly classified into social, economic, environmental and political categories (see Przyluski and Hallegatte 2011). They may additionally be classified as having been triggered directly by the event or as having occured over time as indirect or macroeconomic effects (Hochrainer 2006; UNISDR 2009). The standard approach to estimating natural disaster risk and potential impacts is to understand risk as a function of hazard, exposure and vulnerability (UNISDR 2005; Grossi et al. 2005; Feyen et al. 2008; Global Assessment Report 2013). Changes in each/either one of these dimensions will also result in changes in the risk (Lugeri et al. 2010; Broekx et al. 2011; Lindete et al. 2011; IPCC 2012). For the case of flood risk we provide an overview of climatic and non-climatic factors that influence losses associated with this hazard.

18.2.1 Climatic Factors

While there are different types of floods, such as urban and arroyos floods, in this chapter the focus is on riverine flooding, which is the most important type of flooding for Austria in economic terms (see CRED 2012). The major meteorological factor for major riverine flooding is precipitation, but depending on the region also temperature (snow and icemelt, snowline, etc.) has an influence on flood levels.

18.2.2 Non-climatic Factors

There are basically two types of non-climatic factors that influence the risk related to flood events. The first type can be summarized under the term river management. Under this term, one can consider all factors that have an influence on the run off behaviour of a river, like absorbing capacity of soil in the catchment, channelization and restoration of rivers, as well as flood protection measures like detention basins and levees. All these factors have an influence on the size of the flooded area in a flood event with a given precipitation structure. The second type of factors is the exposure in areas prone to floods. This exposure can usually be defined via risk maps. Risk maps depict the risk of a given area with respect to natural hazards (e.g. EU Directive 2007/60/EC demands the implementation of risk maps for flood risk). Usually risk maps depict the area that is flooded by floods with a given return period. In Austria, the 2002 floods initiated the public-private partnership zoning system HORA that provides risk maps on a countrywide scale.

18.2.3 Identification of Potential Large-Damage Combinations

Damages occur when buildings and infrastructure are flooded. There are two factors one may consider. On the one hand, there is a change in the number and intensity of floods. This factor we will relate to the change in extreme precipitation events. The other factor that has an influence on flood risk is the exposure (building and infrastructure) in areas prone to flood risk. With an increase in population we assume an increase in buildings and infrastructure. However, this does not necessarily lead to an increase in exposure since future buildings need not be built in zones prone to flood risk. This indicates that spatial planning remains extremely important to reduce future risk.

18.3 Exposure to Climatic Stimuli and Impacts up to Now

18.3.1 Past and Current Climatic Exposure and Physical Impacts

Austria has a history of flood events related to extreme precipitation events, e.g. severe flooding in 1965 and 1966 led to the creation of a national disaster fund. Also recently Austria was affected by floods: in 1997, 2002, 2005 and most recently in 2013.

18.3.2 Impact Chains up to Socioeconomic System

The impact chains start with a changed frequency and/or intensity of the climate variable triggering the disastrous events. At first there are direct losses of damaged buildings, damaged infrastructure, agricultural/forest yield, but also the workload of disaster relief forces and corresponding needed machinery is affected (see e.g. Pfurtscheller 2014). Secondly, there are also indirect effects like changes in the workload of volunteers (e.g. volunteer firefighters), which leads to changes in productive work at the volunteers' actual work places and changes in production costs for enterprises that employ volunteers (see the comprehensive Table 18.1).

It should be noted that there are also intangible (nonmarket) aspects that would need to be considered, including loss of life, affected people, loss of biodiversity and damage to ecosystems. However, these effects are difficult to be quantified in monetary terms and will not be considered in this chapter.

Table 18.1 Catastrophe related impact chains

Climate change parameter	Impact chain	Quantified in the model
Extreme precipitation events: Change in frequency and/or intensity (approximated by changing precipitation patterns)	*Flood events: Financial vulnerability of Natural disaster Fund (buildings)*	
	→ Change in damaged buildings → change in natural disaster funding for loss financing [change in cost structure of the public consumption good due to changed transfer payments]	Yes
	Flood events: Financial vulnerability of Natural disaster Fund (infrastructure, agriculture, forestry)	
	→ Change in damaged infrastructure, agricultural/forestal yield → change in natural disaster funding for loss financing [change in cost structure of the public consumption good due to changed transfer payments]	No
	Flood events: Disaster relief forces	
	→ Change in workload of disaster relief forces [change in final (public) demand] → change in machinery needed by disaster relief forces [change in final (public) demand]	No
	Flood events: Volunteers	
	→ Change in workload of volunteers (e.g. volunteer firefighters) → change in productive work at the volunteers' actual work places → change in production costs for enterprises that employ volunteers	No

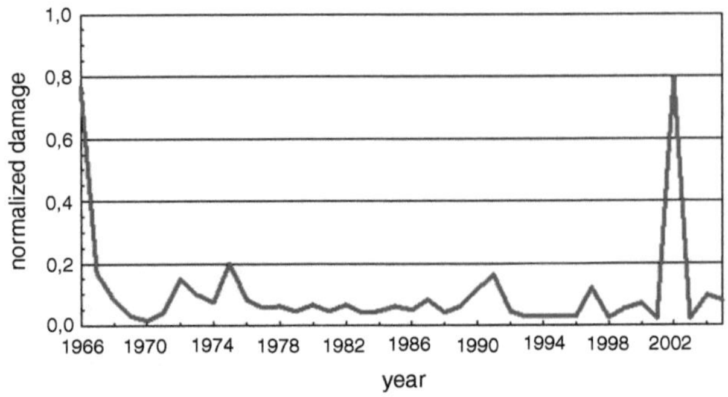

Fig. 18.1 Payments of the national disaster fund in per mill of the new construction value of the Austrian building stock, Prettenthaler and Albrecher (2009)

18.3.3 Economic Impacts up to Now

It must be stressed that economic impacts of extreme events are generally hard to estimate. The main problem lies in assigning a meaningful probability for the extreme event. Furthermore direct comparison between two events is cumbersome, since the time interval between two extreme events is often large and the economy may change considerably and one has to find meaningful normalization procedures before a comparison can be done.

We provide some examples of economic impacts of major flood events: The Oder flood event in 1997 and the European floods in 2002 inundated multiple countries and caused total losses of about 5 billion euros and 14.4 billion euros, respectively. Losses for Austria in 2002 were estimated to be around 3 billion euros (see Habersack et al. 2004). The losses due to flooding in 2005 were around 600 million euros (Bundesministerium fuer Inneres 2005). The recent 2013 flood event is estimated to have caused direct losses of about 870 million euros (Umweltbundesamt 2014).

For the management of floods also the appropriation of funds for rebuilding efforts after a national disaster is important. In Austria, a National Disaster Fund exists for this task. The fund is the main loss financing instrument for large scale losses. It was created in 1966 due to severe flooding in 1965 and 1966. Its original and main purpose was to finance prevention measures against future floods (50 % of the fund) but it also includes loss financing for physical and legal persons (see Supplementary Material for more information on the fund).

The fund provides a time series of nearly 50 years of data for actual losses occurred by floods in Austria and can therefore be seen as a good source to study the exposure of Austria to flood risk. Following Prettenthaler and Albrecher (2009), we consider the payments of the catastrophe fund to compensate the losses in private property until 2006. To account for the socio-economic changes (i.e. increasing building values) the data has been normalized by the new construction value of the Austrian building stock of every year of the time series. Further, the payments of the years 2002 and 2003 (since both can be associated with the flood of 2002) were considered as one payment. Figure 18.1 shows the results. It is interesting to observe that the floods of 2002 and 1966 have similar severity. Nevertheless, we should mention that changes in flood risk protection management (e.g. technical measures, temporary measures, prevention measures, measures of civil protection within the individual provinces, etc.) are not considered in this comparison.

18.4 Future Exposure to and Impacts of Climate Change

In this chapter we provide possible methods for the evaluation of damages caused by future riverine floods and we focus on floods with larger spatial extension where a detailed hydrological modelling is not yet feasible. Furthermore, it must be noted that we could not use all the state of the art methodologies for every detail due to limited resources available. To show the effect of model uncertainty, however, we will provide the results of three different modelling approaches. The approaches will be referred to as HORA-based, AdamCost and ClimateCost.

18.4.1 Mid Range Climatic Scenario for Catastrophe Management

The three considered methods use different Climatic scenarios. The ClimateCost method uses LISFLOOD simulations driven by 12 different GCM-RCM combinations as climate scenario inputs. The AdamCost method uses changes in flood hazard frequency over the period Ø 2010–2100 for a 100-year event for the A1B storyline based on Hirabayashi et al. (2008), which are calculated from the output of a high resolution GCM. The HORA-based method uses the daily data of precipitation from the COIN-Climate Change Data (CCD) to estimate future changes in distribution of the extreme precipitation amounts. Here we should note that the COIN-CCD is a mid-range climate scenario for the end of the century, but it is a rather dry scenario for the periods Ø 2016–2045 and Ø 2036–2065 (compare Chap. 5).

18.4.2 High and Low Range Climatic Scenarios for Flood Risk

Since the changes in precipitation pattern is not consistent in the outcome of different GCM and RCM combinations (see Feyen and Watkiss 2011), it is favorable to use more than one model and to compare the results. Since for the periods Ø 2016–2045 and Ø 2036–2065 we had no data from additional models, we cannot provide results on variability for these periods. Nevertheless, in an extension of the project to evaluate flood risk for the period Ø 2071–2100, we had results of three more GCM-RCM combinations from the ENSEMBLE project at our disposal.

18.4.3 Range of Sectoral Socio-Economic Pathway Parameters that Co-determine Climate Impact

For the evaluation of future flood risk, the main socio-economic driving factor is the value of building infrastructure in flood prone areas. For the HORA-based method, we base our projection of future building values in flood prone areas on a prediction of the future population (for details see Supplementary Material). Further, to account for an increased value of the building stock, we multiply the building stock by the relative increase of per capita income taken from the reference scenario developed in Chap. 6. The ClimateCost approach also uses population and per capita income to account for socio-economic change while AdamCost does not account for socio-economic change.

18.4.4 Specific Method(s) of Valuation and Their Implementation Steps

Generally speaking, there is currently only low confidence in projected changes in flood risk under climate change (IPCC 2012). Consequently, to assess the possible impact of climate change (including those for the national disaster fund), it is useful to compare different results from different catastrophe modelling approaches to reflect the underlying uncertainty. Each of the three approaches discussed here have their cons and pros. The HORA-based approach is the most detailed one for Austria and uses the COIN CDD. However, some relevant inputs for incorporating extreme flood events in the model (e.g. short term heavy precipitation patterns) are not fully available in the COIN CDD yet. Therefore, two other approaches are considered, which explicitly took extremes into account and therefore are used here for comparison reasons. Detailed information for each of these approaches can be found in Supplementary Material; here, we want to notice that the ClimateCost study while using a hydrological model and ensemble runs of different GCM-RCMs the results are expressed in expected damages only and the full risk information is not available. The AdamCost study uses less advanced hazard modelling approaches but provides a loss distribution; however, future changes are only included via changes in the 100-year flood return period and, therefore, large uncertainties can also be expected. A short summary of all methods is provided next (as indicated, a full discussion can be found in Supplementary Material).

The HORA-based approach builds on a method described in Prettenthaler and Albrecher (2009). In the aforementioned book two methods are used. The first method uses the data of the national disaster fund and extreme value analysis, whereas the second is based on the HORA zones. We will only use the second method because it allows us to include different climatic and socio-economic scenarios. Nevertheless, it is noteworthy that both methods provide comparable results for assessing the current risk.

The principle idea of the method is to use risk maps to evaluate the number of buildings that are affected by a flood with a given return period. The number of affected buildings together with a damage function dependent on the return period of the flood is then used to estimate the total damage. Since the marginal distribution of the return period of a flood in a small geographic entity can be assumed to be Pareto distributed, we only have to specify a dependence structure between the different geographic entities to get a simulation of a flood event for the whole of Austria. We should note that for the calculation of average annual losses, the dependence structure is unimportant. Finally, risk maps under a changed climate were created using existing maps and the change of the distribution of extreme precipitation. In this method only results on residential buildings were used.

The second data set used here is taken from the ClimateCost Project and uses results reported in Feyen and Watkiss (2011) and Rojas et al. (2013), who analysed the costs and benefits of adaptation for river flood damages in Europe using the LISFLOOD model. They focus on the mean ensemble results within the SRES A1B scenario, i.e. a medium-high emission scenario. Only direct losses on residential, agriculture, transport, commerce and industry sectors due to river flooding are considered, i.e. intra-urban flooding as well as coastal flooding are excluded. Damages are expressed in expected annual damages, but no specific loss estimates are available for specific year events on the country level.

To fill part of this gap, results from the ADAM project, specifically Lugeri et al. (2010) and Kundzewicz et al. (2010), are additionally used here (called AdamCost). They use static flood hazard maps (see Barredo et al. 2005) and attach probabilities to different flood depths, which afterwards are coupled with corresponding losses for residential, agriculture, transport, commerce and the industry sectors. The resulting GRID based loss distributions are then upscaled to the country level using a hybrid convolution approach (Hochrainer et al. 2013). In this way it was possible to derive a loss distribution on the country level. The flood curves are then changed using changes in flood hazard frequency over the period 2010–2100 for a 100-year event for the A1B storyline based on Hirabayashi et al. (2008). The additional time periods needed here (e.g. 2030) are derived by using the relative changes in expected losses within the ClimateCost study as an approximation of changes in losses over previous periods not reported in AdamCost.

18.4.5 Monetary Evaluation of Impacts

18.4.5.1 Direct Sector Impacts (Costs and Benefits) Without Feedback Effects from Other Sectors

To manage the economic impacts of future flood events appropriate financial funds have to be provided. Two important quantities are relevant for such a fund. These two quantities are the average damages caused by flood events as well as the amount of money that the fund should capitalize to be able to compensate the damage of a future flood event.

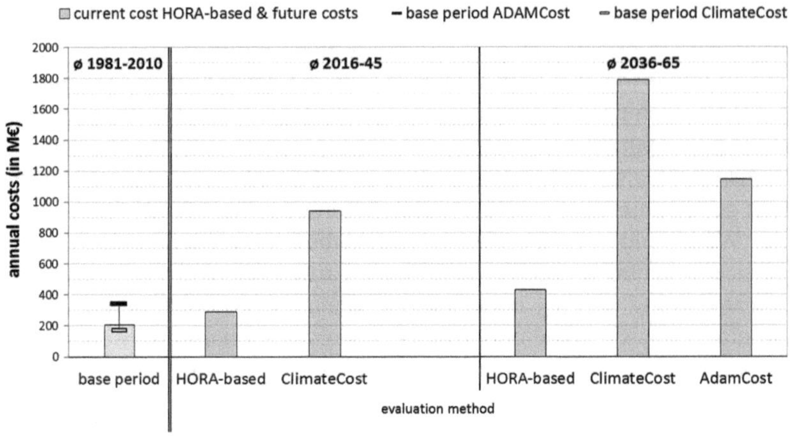

Fig. 18.2 Average annual losses for riverine flooding for different periods (in M€)

Table 18.2 Average annual climate-triggered economic impacts (annual losses) for the sector catastrophe management up to now (in M€)

Economic impact in the base period		HORA-based	ClimateCost	AdamCost
Ø 1981–2010	Costs	207	313	[149–195]
	Benefits	–	–	–
	Net effect	−207	−313	−[149–195]

For the AdamCost approach, some uncertainties in the method were considered and hence an interval is provided

The answer to the first question relates to the expected damage of a flood event. This corresponds to the fair premium in actuarial science and is needed to keep the fund positive in the long run. In Table 18.3 damages are provided for the aforementioned different methods expressed as difference to the baseline. Since for different methods we get different baselines we next provide the average annual damage for the different baselines. For the HORA-based method the baseline is 288 million euros for Ø 2016–2045 and 405 million euros for Ø 2036–2065 and for the ClimateCost method the baseline is 479 million euros for Ø 2016–2045 and 820 million euros for Ø 2036–2065. Since AdamCost does not consider changes in the socioeconomic variables, the baseline is the same as the base period. In Fig. 18.2 we provide the total cost for the different periods and methods [the numbers sum damages up to now (Table 18.2), baseline damages (Supplementary Material Table 18.9), and climate triggered damages (Table 18.3)]. The average damage for the period nearly triples in comparison to the period Ø 2016–2045 in the ClimateCost approach and then nearly doubles for the period Ø 2036–2065. The AdamCost method has an even bigger increase for the period Ø 2036–2065. In the HORA-based approach the loss only slightly increases in the period Ø 2016–2045 (only because of socioeconomic changes), and then increases again for the period Ø 2036–2065. However, the increase is by far not as strong as in the case of the other

Table 18.3 Average annual climate-triggered economic impacts (annual losses) for riverine flooding

Future economic impact		Climate change triggered additional damage		
		HORA-based	ClimateCost	AdamCost
Ø 2016–2045	Costs		461	–
	Benefits	8	–	–
	Net effect	8	−461	–
Ø 2036–2065	Costs	25	966	[848–1100]
	Benefits	–	–	–
	Net effect	−25	−966	−[848–1100]

Difference to baseline scenario in the future (in M€)
For the AdamCost approach, some uncertainties in the method were considered and hence an interval is provided

two methods. We should note that a large part of the differences can be explained by different statistical methods of implementing climate change, different socio-economic assumptions, different spatial resolutions considered for the evaluation of the extent of floods and the consideration of different damages (damage to building stock in HORA-method vs. direct costs losses associated with river flood damages on residential, agriculture, transport, commerce and industry sectors in the two other methods, although damages to building stock is a major part of the overall damages). It should also be noted that the results are showing high uncertainties and therefore these results should be treated with caution. More analysis is required to give robust estimates on costs due to extremes in the future (see the discussion in the uncertainty section).

Averages do not serve well to represent catastrophes and hence probabilistic low frequency events need to be analysed too. This is a tremendous task that incorporates a lot of difficulties and problems (see IPCC 2012). Nevertheless, with our given models we will attempt to tackle the second question stated above, i.e. what is the maximal return period of a flood such that a fund still can cope with the consequences? In other words, what is the expected damage of a flood with a given recurrence level, usually denoted as Value at Risk. An alternative approach that will not be further investigated in this paper would be to replace the Value at Risk with the expected shortfall (the expected damage of a flood with a return period of at least a given number of years). As recurrence time we have chosen to use a flood with return period 100 years in the considered period or equivalently the 99 % quantile of the underlying flood damage distribution. These results are only available for the AdamCost und HORA-based approach. We see that also the expected values differ significantly; however, the effect is less on the quantiles, especially if we correct for the different socio-economic changes (see Table 18.4).

18.4.5.2 Macroeconomic Effects

The impact chain "additional flooding damages" is implemented in the macroeconomic model via three channels, which are listed here and depicted in more

Table 18.4 The 99 % quantile of the estimated flood damage distribution (in M€)

Period/Method	HORA-based	ClimateCost	AdamCost
Ø 1977–2006	3,220	–	2,350
Ø 2016–2045	4,356	–	3,718
Ø 2036–2065	6,894	–	4,948

detail in Supplementary Material Table 18.2 for the Climate Cost method and in Supplementary Material Table 18.3 for the HORA-based approach.

First, using the Climate cost data (HORA-based approach data) consumer demand for products that have been destroyed by additional flooding is on annual average increasing by 365.70 million euros (43 million euros) for the 2016–2045 period and 860 million euros (130 million euros) for 2036–2065 period. Second, the real estate sector needs to reconstruct residential buildings that were affected by flooding. This leads to additional average annual investment in construction, 261 million euros (30 million euros) for 2016–2045 and 613 million euros (93 million euros) for 2036–2065, which is assumed to increase the overall investment level of the economy. To ensure balancing on the capital market, savings by consumers need to increase to meet higher investments. Third, regarding financial coverage of damages to private households, half of the additional damage costs are financed by the government (covered by the natural disaster fund) and the remaining half is covered out of private household budget.

Table 18.5 depicts the sectors gaining and losing after macroeconomic feedback consideration for the ClimateCost data. For 2016–2045 the sectors gaining show an average annual increase of gross output value of 445 million euros (caused by an increase of 268 million euros in intermediate demand and 177 million euros in value added). The losing sectors have to face a decrease in gross output value of 379 million euros (−129 million euros intermediate demand and −250 million euros value added). Summed up, this leads to a decrease of average annual value added of 73 million euros for 2016–2045. For 2036–2065 the value added decreases by an average annual 138 million euros. In both periods, the sectors construction, trade and repair of motor vehicles and rest of manufacturing are the sectors with the largest gains. This is caused by a higher (direct or indirect) demand for their supply to rebuild destroyed goods. The losses in some of the other sectors originate from reduced demand by consumers and reduced government consumption to finance the additional flooding damages. Reduced government consumption especially affects the contraction of sectors public, rest of services and health, as core services are demanded by government. For the HORA-based approach data, total gross value added is increasing by 1 million euros for 2016–2045 and decreasing by 4 million euros for 2036–2065. The detailed composition is depicted in Supplementary Material Table 18.4.

With respect to welfare and GDP impacts, the effects on the latter are considerably smaller than those on the former, as additional flooding damages partly cause a shift and not a reduction in consumption (and therefore in production). Welfare is corrected for additional consumption that can be classified as reparation payments

Table 18.5 Sectoral and total effects of quantified climate change impacts in sector "Catastrophe Management" (ClimateCost data), average annual effects relative to baseline (for periods 2016–2045 and 2036–2065)

	2016–2045			2036–2065		
Changes in M€ p.a. relative to baseline	Gross output value	Intermediate demand	Gross value added	Gross output value	Intermediate demand	Gross value added
Gaining sectors	**+445.3**	**+268.1**	**+177.2**	**+788.9**	**+472.8**	**+316.1**
Construction	+226.4	+134.1	+92.3	+378.1	+224.0	+154.1
Trade and repair of motor vehicles	+63.7	+30.2	+33.5	+135.7	+64.3	+71.4
Rest of manufacturing	+48.0	+30.9	+17.1	+89.5	+57.8	+31.7
All other gaining sectors	+107.1	+72.9	+34.2	+185.6	+126.7	+58.9
Losing sectors	**−379.1**	**−129.4**	**−249.7**	**−680.0**	**−225.9**	**−454.1**
Public	−90.3	−28.3	−62.0	−190.0	−59.5	−130.5
Rest of services	−77.1	−18.9	−58.2	−152.2	−37.1	−115.0
Health	−82.6	−27.1	−55.4	−161.3	−53.0	−108.4
All other losing sectors	−129.1	−55.0	−74.0	−176.5	−76.3	−100.2
Total effect (all sectors)	**+66.2**	**+138.7**	**−72.5**	**+108.9**	**+246.9**	**−138.0**
Welfare			**−613**			**−1001**

Note: baseline scenario = reference socioeconomic development without climate change; climate change scenario = reference socioeconomic development and mid-range climate change. Quantified impact chains: [flooding damages]

not increasing welfare but only re-establishing the previous level of stocks. For the ClimateCost data (HORA-based approach data) welfare on average is decreasing (increasing) by 613 million euros (11 million euros) p.a. for 2016–2045 and decreasing by 1001 million euros (26 million euros) p.a. for 2036–2065. GDP losses (gains) are 44 million euros (1 million euros) p.a. for 2016–2045 and losses of 58 million euros (2 million euros) p.a. for 2036–2065. The effects on unemployment are rather small for the ClimateCost data (increase of 0.01 % points for 2016–2045 and 0.02 % points for 2036–2065) and negligible for the HORA-based approach data. Not indicated by the unemployment rate are sectoral shifts of employment caused by changes in the consumption shares of different goods (for the detailed effects on GDP, welfare and unemployment see Supplementary Material Tables 18.5 and 18.6).

For the government budget the impact of climate change triggered increasing flooding are for the ClimateCost data (HORA-based approach data) as follows: A decrease (increase) by 250 million euros (4 million euros) p.a. occurs for 2016–2045 and a decrease by 541 million euros (14 million euros) for 2036–2065. The main reasons are both a decrease in tax revenues due to economic contraction and the public share in financing of flooding damages to households. The detailed decomposition of the government budget is depicted in Supplementary Material Tables 18.7 and 18.8.

18.4.5.3 Sector-Specific Uncertainties

One large uncertainty in the analysis of future flood events lies in the future pattern of extreme precipitation events. Unfortunately, the prediction of future precipitation is not stable in the currently used models. As noted by Feyen and Watkiss (2011), variations across different models could change the magnitude of the flood risk, even altering the sign of the change on the local level. A further large uncertainty in the currently used models is the number of affected buildings for a flood with a given return period. For example while AdamCost and HORA-based modelling both use risk maps to identify the exposure, the methods are still different. The AdamCost uses a relative wide grid of 1×1 km and CORINE Land Cover data. The HORA-based method uses the risk maps from the HORA project, which are evaluated with a finer spatial resolution and gridded data on the building stock. For both methods an additional uncertainty is the distribution of buildings inside the different risk zones (compare supplementary material). We have to mention that both methods can be considered as rather rough concerning the spatial resolution. Here we must mention on the (rather mild effect of) spatial resolution on discharge (Booij 2005) and on the (more significant) effect of spatial resolution on flooded areas (Adams et al. 2007).

Given all these uncertainties, also in the projected evolution of socio-economic development, we strongly advocate that relative changes should be considered rather than absolute values. Also, changes in the distribution of exposed buildings are less predictable because they involve the future practices of spatial planning.

18.4.5.4 Relevance for Other Sectors

This chapter considered the monetary effects of extreme events like floods. We highlighted that in such a situation it is not only important to consider mean annual losses but also the losses associated with extreme events. The ideas of the method of this sector are also relevant for other sectors where rare events have to be considered.

18.5 Exposure to Climate Change for the Period Ø 2071–2100

After the analysis of the impacts of climate change on riverine floods for the periods Ø 2016–2045 and Ø 2036–2065, there was the additional question of how flood risk evolves towards the end of the century. Hence we also analysed the impacts for the period Ø 2071–2100. For this analysis we used the HORA-based method. In addition to the COIN CCD we had the results on precipitation for a collection of three further downscaled climate model runs. As mid-range climate scenario we used the COIN-CCD, for the high- and respectively low-range climate scenario we used the one with the biggest, respectively lowest average annual damage, and finally we used as baseline a scenario with the current climate (Ø 1977–2006). To account for socio-economic change, we extrapolated the scenarios that we used for the periods Ø 2016–2045 and Ø 2036–2065 to the period Ø 2071–2100 and corrected the results on projected data for the population of the period Ø 2071–2100 which is taken from the IIASA SSP Database. The Database provides population forecasts for different social economic pathways (SSP1–SSP5) (compare Chap. 6). We used three scenarios which are i) reference: we used the SSP2 population forecast for the correction of the extrapolation; ii) (impact) enhancing: we used SSP5 population forecast for the correction of the extrapolation; iii) diminishing: we used the SSP2 population forecast for the correction of the extrapolation but allowed a growth of the building stock only outside the risk zone HQ200. For details on the used methods we refer to the Supplementary Material. Further, to account for increased values of building values, we multiplied the building values by the relative change in per capita income. For SSP2 population forecast, this resulted in a multiplication by 2.69 and for SSP5 population forecast a multiplication by 3.66.

In Table 18.6 we provide results of the simulations. Beside the average annual damage we also provide the 90 %, 95 % respectively 99 % quantiles of the damage distribution. In the mid-range climate change scenario, the average annual damage increases by 38 % relative to the baseline. Nevertheless, we can see that under the low-range climatic scenario there is no significant change in the distribution for the

Table 18.6 Impacts of climate change for the period Ø 2071–2100 (in M€)

Projected future damage			No climate change (baseline)	Climate change: Low-range	Mid-range	High-range
Avarage annual damage	Socioeconomic development	Diminishing	527	527	733	1,332
		Reference	575	570	795	1,435
		Enhancing	1,117	1,117	1,545	2,764
90 %-quantile		Diminishing	1,228	1,222	1,762	3,105
		Reference	1,351	1,338	1,932	3,379
		Enhancing	2,599	2,599	3,756	6,476
95 %-quantile		Diminishing	2,805	2,729	3,831	7,259
		Reference	3,089	2,979	4,210	7,973
		Enhancing	6,008	5,810	8,167	15,339
99 %-quantile		Diminishing	8,183	8,081	11,200	19,654
		Reference	9,107	8,870	12,223	21,158
		Enhancing	17,587	17,305	23,708	40,511

damage, whereas for the high-range scenario the annual average damage more than doubles. We should note that we only used the output of four climate change scenarios and hence real uncertainty might be even higher. Further changes in the quantiles are similar to changes in the average annual damage. If we consider the differences between the different used socio-economic scenarios, we can observe that especially for the enhancing scenario the average annual damage increases. Here we should note the importance of spatial planning, since the difference in the average damage between the different scenarios can be basically explained by a different distribution of building stocks.

To compare our results also with other results, we provide the results from the ClimateCost study for this time period. In this study, the average annual damage with climate change is estimated to be around 2.3 billion euros annually for Austria with a baseline of 1.3 billion euros annually (no climate change) and hence fall in the range of the results of enhancing socioeconomic change.

18.6 Summary of Climate Costs for Catastrophe Management and Conclusions

This chapter investigated future losses of natural disasters with the example of riverine flood risk. We focused on the impact on the government budget via the National Disaster Fund, which is of outmost importance during disaster events as a loss financing vehicle. We assessed possible increases in the cost of such instruments in the context of climate change and emphasized the need to include not only averages within such an analysis but also to explicitly incorporate extremes. While some indications of increases in risk due to climate change were found, the results have to be treated with caution. There are many uncertainties involved and not rigorously tackled within the models presented, partly because of resource restrictions. It is even possible to find changes in the sign in losses for some other storylines and/or other assumptions used. In this light, it is essential to look at robust solutions that are beneficial for all possible futures/future scenarios, also called no-regret options. Due to the problems already found to be difficult to handle with the current schemes, it is necessary to include a range of additional instruments to decrease risk in the future and to couple them with financial risk spreading instruments where risk reduction is not feasible anymore.

It can be assumed that natural disasters cause various additional negative ripple effects throughout the socio-economic system too. To decrease such threats, the inclusion of contingent extreme risks within budget planning processes could be beneficial for enhancing the development of new forms of private-public sector partnerships as well as for creating efficient incentive systems to reduce future exposure and vulnerability.

References

Adams M, Falkner M, Franke M, Jochem A, Lochner B, Link S, Vetter M, Werthmann M (2007) Hochwasser Risiko Modellierung Innsbruck – Endbericht zum Projektmodul Geoökologie und Raumforschung 2006/2007 Technical Report: Institut für Geographie, Universität Innsbruck. http://www.uibk.ac.at/geographie/personal/michael_vetter/artikel/hq_risiko_mod_ibk_adams_et_al.pdf

Barredo JI, Lavalle C, De Roo A (2005) European flood risk mapping. EC DGJRC, 2005 S.P. I.05.151.EN

Blöschl G, Viglione A, Merz R, Parajka J, Salinasand JL, Schöner W (2011) Auswirkungen des Klimawandels auf Hochwasser und Niederwasser. Österreichische Wasser- und Abfallwirtschaft 63(1–2):21–30

Booij MJ (2005) Impact of climate change on river flooding assessed with different spatial model resolutions. J Hydrol 303:176–198

Broekx S, Smets S, Liekens I, Bulckaen D (2011) Designing a long-term flood risk management plan for the Scheldt estuary using a risk based approach. Nat Hazards 57:245–266

Bundesministerium für Inneres (2005) Kurzantrag zur Mobilisierung des Solidaritätsfonds der Europäischen Union (EUSF)

Chavez-Demoulin V, Davison A (2005) Generalized additive modelling of sample extremes. Appl Stat 54(Part 1):207–222

CRED (2012) The OFDA/CRED international disaster database. Université Catholique de Louvain, Brussels, Belgium. www.em-dat.net

Dierckx G, Teugels J (2012) Trend analysis of extreme values 101–108. In: Barbour A, Chanand HP, Siegmund D (eds) Probability approximations and beyond. Springer, New York

Dobler C, Bürger G, Stötter J (2013) Simulating future precipitation extremes in a complex Alpine catchment. Nat Hazards Earth Syst Sci 13:263–277

Feyen L, Barredo JI, Dankers R (2008) Implications of global warming and urban land use change on flooding in Europe. In: Feyen L, Shannon K, Neville M (eds) Water and urban development paradigms – towards an integration of engineering, design and management approaches. CRC Press, Balkema, Leiden, pp 217–225

Feyen L, Watkiss P (2011) The impacts and economic costs of river floods in Europe and the costs and benefits of adaptation. Summary of sector results from the EC RTD ClimateCost Project

Formayer H, Eitzinger S, Nefzger H, Simic S, Kromp-Kolb H (2001) Studie: Auswirkungen einer Klimaveränderung in Österreich: Was aus bisherigen Untersuchungen ableitbar ist. Global 2000 Österreich. http://www.accc.gv.at/pdf/global2000.pdf

Global Assessment Report (2013) From shared risk to shared value. UNISDR. http://www.preventionweb.net/english/hyogo/gar/2013/en/home/index.html

Grossi P, Kunreuther H, Windeler D (2005) Catastrophe modeling, a new approach. Springer, New York

Habersack H, Bürgel J, Petraschek A (2004) Analyse der Hochwasserereignisse vom August 2002 – FloodRisk. Synthesebericht. Bundesministerium für Land- und Forstwirtschaft, Umwelt und Wasserwirtschaft, Wien

Hochrainer S (2006) Macroeconomic risk management against natural disasters. German University Press (DUV), Wiesbaden

Hochrainer S, Lugeri N, Radziejewski M (2013) Up-scaling of impact dependent loss distributions: a hybrid-convolution approach. Natural Hazards (Accepted)

Hochrainer S, Mechler R (2013) Assessing financial adaptation strategies to extreme events in Europe. In: Siebenhüner B, Arnold M, Eisenack K, Jacob HK (eds) Long-term governance for social-ecological change. Routledge, New York, pp 93–112

Hirabayashi Y, Kanae S, Emori S, Oki T, Kimoto M (2008) Global projections of changing risks of floods and droughts in a changing climate. Hydrol Sci J 53:754–773

IPCC (2012) Summary for policymakers. In: Field CB et al (eds) Managing the risks of extreme events and disasters to advance climate change adaptation. A special report of working groups I

and II of the intergovernmental panel on climate change. Cambridge University Press, Cambridge, pp 1–19
Kundzewicz ZB, Lugeri B, Dankers R, Hirabayashi Y, Doell P, Pinskwar I, Dysarz T, Hochrainerand S, Matczak P (2010) Assessing river flood risk and adaptation in Europe – review of projections for the future. Mitig Adapt Strat Glob Change 15(7):641–656
Lugeri N, Kundzewicz ZB, Genovese E, Hochrainerand S, Radziejewski M (2010) River flood risk and adaptation in Europe – assessment of the present status. Mitig Adapt Strat Glob Change 15 (7):621–639
Lindete AH, Bubeck P, Dekkers JEC, de Moel H, Aerts JCJH (2011) Future flood risk estimates along the river Rhine. Nat Hazards Earth Syst Sci 11:459–473
Matulla C, Formayer H, Haas P, Kromp-Kolb H (2004) Possible climate trends in Austria in the first Half of the 21st century. OEWAW (Österreichische Wasser- und Abfallwirtschaft) 56(1–2):1–9
Mechler R (2004) Natural disaster risk management and financing disaster losses in developing countries. Verlag Versicherungswirtschaft GmbH, Karlsruhe
Munich Re Georisks (2007) NATHAN (NATural Hazards Assessment Network). http://www.munichre.com/en/ts/geo_risks/nathan/default.aspx
Munich Re (2009) Topics Geo. Natural catastrophes 2008. Analyses, assessments, positions. Munich Reinsurance Company, Munich
Pelling M, Özerdem A, Barakat S (2002) The macro-economic impact of disasters. Prog Dev Stud 2(4):283–305
Pfurtscheller C (2014) Alpine flood event in Tyrol (Austria). Nat Hazards Earth Syst Sci 14: 359–378
Prettenthaler F, Albrecher H (2009) Hochwasser und dessen Versicherung in Österreich. Verlag der Österreichischen Akademie der Wissenschaften, Wein
Prettenthaler F, Amrusch P, Habsburg-Lothringen C (2010) Estimation of an absolute flood damage curve based on an Austrian case study under a dam breach scenario. Nat Hazards Earth Syst Sci 10:881–894
Przyluski V, Hallegatte S (2011) Indirect costs of natural hazards. ConHaz: costs of natural hazards. WP2 final report, September 2011. http://conhaz.org/
Rojas R, Feyen L, Watkiss P (2013) Climate change and river floods in the European Union: socio-economic consequences and the costs and benefits of adaptation. Glob Environ Change 23(6): 1737–1751. doi:10.1016/j.gloenvcha.2013.08.006
Schimon W, Schöner W, Böhm R, Haslinger K, Blöschl G, Merz R, Blaschke AP, Viglione A, Parajka J, Kroiß H, Kreuzingerand N, Hörhan Th (2011) Anpassungsstrategien an den Klimawandel für Österreichs Wasserwirtschaft, Bundesministerium für Land- und Forstwirtschaft, Umwelt und Wasserwirtschaft
Steininger KW, Steinreiber C, Ritz C (eds) (2005) Extreme Wetterereignisse und ihre wirtschaftlichen Folgen. Springer, New York
Swiss Re (2009) Natural catastrophes and man-made disasters in 2008. Sigma 2/2009. Swiss Reinsurance Company, Zurich
United Nations International Strategy for Disaster Reduction (UNISDR) (2005) Hyogo declaration. World conference on disaster reduction, Kobe, Hyogo
UNISDR (2009) UN international strategy for disaster reduction, terminology section. http://unisdr.org/we/inform/terminology
Umweltbundesamt (2014) Umweltbundesamt. Hochwasser. http://www.umweltbundesamt.at/umweltsituation/wasser/hochwasser_wasser/. Accessed 1 Aug 2014

Chapter 19
Tourism

Judith Köberl, Franz Prettenthaler, Stefan Nabernegg, and Thomas Schinko

Abstract Tourism ranks amongst those sectors regarded as being highly weather and climate sensitive, since lots of tourism types and activities have a strong link to the environment and to the climate itself. During snow-poor winters, such as 1989/90 and 2006/07, several Austrian regions showed noticeable drops in tourism demand—whereas extraordinary sunny, warm and dry summers, like the one in 2003, coincided with above-average tourism demand increases in lake regions. In order to assess the potential impacts of future climate change on tourism demand in Austria, we (1) use dynamic multiple regression models to quantify the sensitivity of overnight stays towards year-to-year weather for each NUTS 3 region and various seasons, (2) apply the resulting sensitivities on climate change scenarios—based on a general tourism development scenario—and (3) transform the resulting impacts on overnight stays into monetary terms using average tourist expenditures. Outcomes suggest predominantly negative impacts on winter tourism and mainly positive impacts on summer tourism, with the net impact being negative. Finally we (4) evaluate the effects of the negative tourism impacts in a macroeconomic CGE model. Resulting spillover effects to other economic sectors as well as changes in GDP and welfare are found to be even higher than the impacts on tourism. There are considerable uncertainties however, not only with respect to climate change scenarios, but also for instance regarding future tourist preferences and weather/climate sensitivities.

J. Köberl (✉) • F. Prettenthaler
Institute for Economic and Innovation Research, Joanneum Research, Graz, Austria
e-mail: judith.koberl@joanneum.at; franz.prettenthaler@joanneum.at

S. Nabernegg • T. Schinko
Wegener Center for Climate and Global Change, University of Graz, Graz, Austria
e-mail: stefan.nabernegg@uni-graz.at; thomas.schinko@uni-graz.at

K.W. Steininger et al. (eds.), *Economic Evaluation of Climate Change Impacts*, Springer Climate, DOI 10.1007/978-3-319-12457-5_19

19.1 Introduction

Tourism represents a highly important economic sector in many parts of the world, generating income and employment and representing one of the fastest growing economic sectors globally. In 2012, worldwide tourism directly contributed 2.9 % to global GDP. Taking indirect and induced effects into account as well, the sector's contribution comprised 9.3 % of global GDP (WTTC 2013).

Tourism also plays an important role in the Austrian economy. In 2012 it generated 17.94 billion euros in direct value added and hence contributed 5.8 % to Austria's GDP. Taking indirect effects into account as well, the sector's contribution amounted to 22.82 billion euros or 7.4 % of total value added (Statistics Austria and WIFO 2014).

19.2 Dimensions of Sensitivity to Climate Change

The tourism industry ranks among those sectors that are regarded as being highly weather and climate sensitive, since lots of tourism types and activities—e.g. ski tourism, beach and lake tourism, or hiking tourism—have a strong link to the environment and to the climate itself (UNWTO-UNEP-WMO 2008). On the supply side, climate co-determines a region's basic suitability for offering particular tourism types or activities. On the demand side, weather (forecasts) and climate may influence a tourist's decision-making process about destination choice and when to travel. Thus, climate represents a principal driver of seasonality in tourism demand (Cooper et al. 2008). Moreover, the actual weather experienced during holidays may affect tourists' satisfaction and enjoyment and—given sufficient flexibility—even cause them to extent, shorten or cancel their vacation. Besides supply and demand, weather and climate also affect important aspects of tourism operations, including operating costs (heating, cooling, artificial snow production, irrigation, etc.), activity planning and infrastructure (Scott and Lemieux 2010). Due to the importance of weather and climate for tourism supply, demand and operations, changes in climate may directly affect tourism in various ways. Additionally, climate change may also affect tourism indirectly through impacts on environmental resources that represent important factors for tourism, such as biodiversity, landscape aesthetics or water quality and availability (UNWTO-UNEP-WMO 2008).

19.2.1 Climatic Factors

Various climatic factors are relevant for the economic performance of the tourism industry, since different tourism types require or benefit from distinct weather and climatic conditions. Snow-based tourism types, for instance, require at least sufficiently cold temperatures (for artificial snow production) or better yet, sufficiently cold temperatures together with precipitation. Hence, insufficient snow conditions may lead to noticeable demand reductions in snow-based tourism and economic losses (Hamilton et al. 2007; Dawson et al. 2009; Töglhofer et al. 2011; Steiger 2011). Cloudiness and wind speed can also affect winter tourism demand (Falk 2013; Shih et al. 2009). Moreover, inadequate amounts of natural snow and/or higher temperatures may increase the need and costs of artificial snowmaking. Beach or lake tourism, on the other hand, requires sufficiently high temperatures together with dry conditions. Additionally, it generally benefits from sunshine and the absence of strong wind. Thus, losses to this tourism type may arise from cold, cloudy and rainy weather (Castellani et al. 2010; Moreno 2010), but also from temperatures regarded as being too hot (Rutty and Scott 2010). For hiking and nature-based tourism, precipitation seems to be the dominating climatic factor (Scott et al. 2007), whereas urban tourism might be negatively affected by high temperatures (UNWTO-UNEP-WMO 2008).

19.2.2 Non-climatic Factors

Tourism is a strongly demand-driven sector. Hence, the economic consequences of climate change for particular tourism regions will not only depend on the intensity of climate change itself, but also on non-climatic factors influencing tourists' decision making processes on when and where to go, such as the ability (and willingness) of tourists to adjust their travel date (EEA 2012). Assume that a region dominated by alpine winter sports tourism faces a temporal shift in cold temperatures and snowfall away from current peak to off-peak seasons. If tourists are able and willing to adjust their travel date while deciding on the same destination, the region might not face significant economic losses. If on the other hand, tourists are not able (e.g. due to holiday regulations) or willing to adjust their travel date, but rather choose another destination, the region may suffer from considerable losses. Further non-climatic factors include tourists' preferences and sensitivities towards weather and climate. These may change over time due to changes, for example, in demography [preferences towards weather and climate vary with age; see Lise and Tol (2002)] or preferred tourism activities (different tourism activities show distinct weather and climate sensitivities).

19.2.3 *Identification of Potential Large-Damage Combinations*

Potential large-damage combinations include temporal shifts of "favorable" climatic conditions from current peak to off-peak seasons together with tourists' inflexibility or unwillingness to adjust their travel date. A change in tourists' preferences resulting in remarkably higher weather sensitivities together with a pronounced change to more "adverse" climatic conditions represents another combination potentially leading to large economic damages. Summarizing, a crucial factor in determining if climate change will cause large damages to particular tourism regions is the way tourists will (be able to) adapt to these changes.

19.3 Exposure to Climatic Stimuli and Impacts Up to Now

19.3.1 *Past and Current Climatic Exposure and Physical Impacts*

Variations in climatic factors may affect the performance of the tourism sector. What follows are some examples for past impacts on tourism demand due to variations in climatic conditions. For past trends in climatic conditions see the online Supplementary Material.

19.3.1.1 Winter Season

Particularly snow-poor winters within recent decades, including the 1989/1990 and 2006/2007 seasons, had noticeable impacts on Austrian winter tourism demand. According to Töglhofer et al. (2011), the growth rate of overnight stays in Austrian ski areas dropped by 8.1 % points in the warm and snow-poor 1989/1990 winter season, when the number of snow days[1] was 22 % below long-term average. In the 2006/2007 winter season, a reduction in the number of snow days by 29 % compared to average conditions was accompanied by a 2.7 % point decrease in the growth rate of overnight stays. Each time, decreases in the growth rates of overnight stays were more pronounced in lower-lying areas whereas no noticeable changes were observed for higher-lying areas. Similar effects were found by Steiger (2011), who investigated Tyrolean overnight stays in the record warm and snow-scarce 2006/2007 winter season. Due to a decrease in overnight stays by 3 % relative to the preceding 3 years he estimated the economic losses of this snow-scarce season to amount to 104 million euros. The highest losses were experienced by districts with mainly low-altitude or higher located but small ski areas, whereas

[1] Days with at least 1 cm snow depth.

districts with large to extra-large ski areas at mid to high altitudes showed constant or even increasing overnight stays. Gains were also observed in the provincial capital, Innsbruck, which has the most developed offerings of non-skiing products like culture and congress tourism within Tyrol.

19.3.1.2 Summer Season

In 2003, tourism in Austria was exposed to the hottest summer since the beginning of regular recordings. Comparing summer overnight stays in 2003 to average summer overnight stays in 2002 and 2004, Fleischhacker and Formayer (2007) found a nationwide increase of 1.8 %, with the rise in domestic overnight stays (2.7 %) being almost twice as high as the rise in foreign overnight stays (1.4 %). Single tourism types were seemingly able to benefit over-proportionally from this extraordinary summer, including lake tourism (+4.4 %) and tourism in nature reserves (+2.4 %). In contrast, health and wellness tourism (−0.2 %) as well as urban tourism (−0.6 %) experienced losses compared to the average figures of 2002 and 2004.

19.3.2 Impact Chains up to Socioeconomic System

Several impact chains of climate change on tourism have been identified and are listed in Table 19.1, which makes no claims of being complete. Due to limited resources and/or (too) high assessment uncertainties, some of the presented impact chains could not be quantified within the current project.

Regarding winter tourism, a change in (natural) snow conditions may change tourism demand in regions offering snow-based tourism types. Consequently, the tourism sector's demand in products and services of upstream industries (e.g. energy sector, food sector, construction sector, etc.) would change as well. A similar impact chain, albeit triggered by different climatic factors, holds for summer tourism. Changes in precipitation and/or temperature conditions may change tourism demand in regions focused, for example, on hiking, mountain biking or lake tourism. Demand in urban tourism might also be affected by changing temperatures (e.g. by an increase in hot temperatures). The impact chains mentioned may not only be directly triggered by changes in climatic factors, but also indirectly by climate caused changes in environmental resources important for tourism. Besides, changes in temperature and precipitation conditions may affect the tourism sector's water and energy demands by altering its need for irrigation (e.g. golf courses, hotel facilities, etc.), heating and cooling or artificial snowmaking, thus modifying the sector's cost structure. Moreover, changes in the frequency and intensity of extreme events, including floods and mass movements, are likely to change the frequency and intensity of destroyed tourism facilities and/or transport infrastructure leading to business interruptions.

Table 19.1 Tourism related impact chains

Climate change parameter	Impact chain	Quantified in the model
(Natural) snow: Change in ski area's (mean) snow conditions	*Change in winter tourism demand* → Change in winter overnight stays [change in final demand] → change in tourism sector's upstream demand (e.g. energy, food, construction, etc.)	Yes
Precipitation: Change in (mean) precipitation conditions	*Change in summer tourism demand* → Change in summer overnight stays [change in final demand] → change in tourism sector's upstream demand	Yes
	Change in tourism sector's water demand → Change in irrigation needs [change in cost structure] → change in tourism sector's water demand	No
Temperature: Change in (mean) temperature conditions	*Change in summer tourism demand* → Change in summer overnight stays [change in final demand] → change in tourism sector's upstream demand	Yes
	Change in environmental resources important for tourism (demand) → Change in e.g. biodiversity, landscape aesthetics (such as glaciers), water quality, water availability, etc. → change in tourism demand [change in final demand] → change in tourism sector's upstream demand	No
	Change in tourism sector's water and/or energy demand → Change in e.g. irrigation, heating & cooling, or artificial snowmaking needs [change in cost structure] → change in tourism sector's water and/or energy demand	No
Extreme events: Change in frequency/intensity of e.g. floods, mass movements, etc.	*Business interruption* → Change in frequency/intensity of destroyed tourism facilities and/or transport infrastructure → change in frequency/intensity of business interruption → change in tourism demand [change in final demand] → change in tourism sector's upstream demand	No

19.3.3 Economic Impacts Up to Now

Some examples of past physical and/or economic impacts on tourism due to year-to-year variations in weather conditions have been quoted in Sect. 19.3.1. Averaged over a longer period of time however, e.g. 30 years, gains and losses due to climate variability are likely to compensate each other to a high degree—at least in the absence of very extreme events. Hence, in the analyses that follow we focus on impacts caused by a change in average climatic conditions rather than by a change in climate variability. We therefore refrain from providing comprehensive estimates on average annual tourism gains and losses due to climate variability in the base period 1981–2010.

19.4 Future Exposure to and Impacts of Climate Change

19.4.1 Mid-range Climatic Scenario for Tourism

For our analyses on climate change impacts and costs of inaction, we draw on the COIN climate change data (COIN CCD), which projects an increase in mean annual temperatures of +1.05 °C (+2.02 °C), a change in annual precipitation sums of +1.4 % (−2.3 %) and a change in wet days[2] of +2.1 % (−3.5 %) between the base period 1981–2010 and the first (second) scenario period 2016–2045 (2036–2065). Regarding precipitation sum and wet days, COIN CCD indicates an increasing trend for the winter half-year and a decreasing trend for the summer half-year. Whereas in the first scenario period precipitation gains during the winter half-year dominate annual net effects, in the second scenario period the expected decline in summer precipitation becomes the dominating effect.

Regarding snow data, COIN CCD differentiates between four different elevations: 500, 1,000, 1,500, and 2,000 m. Depending on the elevation class considered, mean annual snow depth is projected to change by +1 to −21 % (−13 to −37 %) between base and first (second) scenario period, whereas the annual number of snow days is expected to change by −12 to −18 (−21 to −35) days. Within the following analyses, we consider snow conditions in ski areas and their impacts on winter overnight stays at NUTS 3 level. Figure 19.1 shows the change in the annual number of snow days on NUTS 3 level for the altitude class representative of the ski areas within the considered region.[3] For further details see Supplementary Material.

---

[2] Days with at least 1 mm precipitation.

[3] To decide on which altitude class (500, 1,000, 1,500 and 2,000 m) is representative for the ski areas within a NUTS 3 region, we form a transport capacity weighted (TCW) average over the mean altitudes of all ski areas within a NUTS 3 region that have more than five transport facilities or at least one cable car (Töglhofer 2011). TCW mean altitudes up to 749 m are allocated to elevation class 500, TCW mean altitudes between 750 m and 1,249 m to elevation class 1,000, etc.

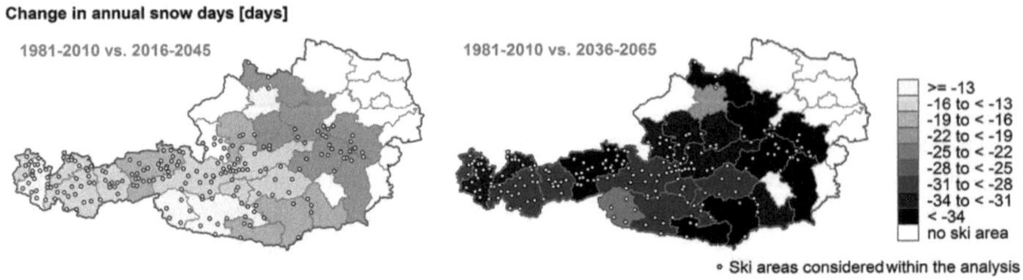

Fig. 19.1 Change in snow conditions as projected by COIN CCD at the altitude class representative of the regions' ski areas

19.4.2 High and Low Range Climatic Scenarios for Tourism

To represent (at least part of) the uncertainty range related to climatic scenarios, we additionally consider climate change data resulting from four different regional climate models of the ENSEMBLES family (http://www.ensembles-eu.org) (CNRM-RM4.5, ETHZ-CLM, ICTP-REGCM3, and SMHI-RCA), which are all based on the A1B emission scenario. Data from these four models have been edited within the ACRP-funded project "Adaptation to Climate Change in Austria" (ADAPT.AT) for the period 1951–2050 and have already been used for Austrian climate change impact assessments in Köberl et al. (2011). Due to the limited time span of edited data available, low and high-range climatic scenarios can only be derived for the first scenario period, i.e. 2016–2045. They are defined in such a way that the low-range scenario tends to cause the lowest negative (or highest positive) net impacts, whereas the high-range scenario is associated with the highest negative (or smallest positive) net impacts. Hence compared to the mid-range climatic scenario the low-range scenario represents warmer and dryer summers as well as snowier winters, whereas the high-range scenario is defined to represent colder and wetter summers as well as snow-poorer winters.

19.4.3 Specific Method(s) of Valuation and Their Implementation Steps

Various studies deal with the impacts of climate change on tourism in Austria (e.g. Breiling and Charamza 1999; Rudel et al. 2007; Steiger and Abegg 2013). Many of them focus on the supply side by examining the change in the climatic potential for particular tourism types, but do not explicitly take the relationship between weather/climatic conditions and tourism demand into account. However, since tourism is a strongly demand-driven sector, quantifying this relationship seems an essential task for assessing the (monetary) impacts of climate change and the costs of inaction. Hence, in order to assess direct impacts of climate change

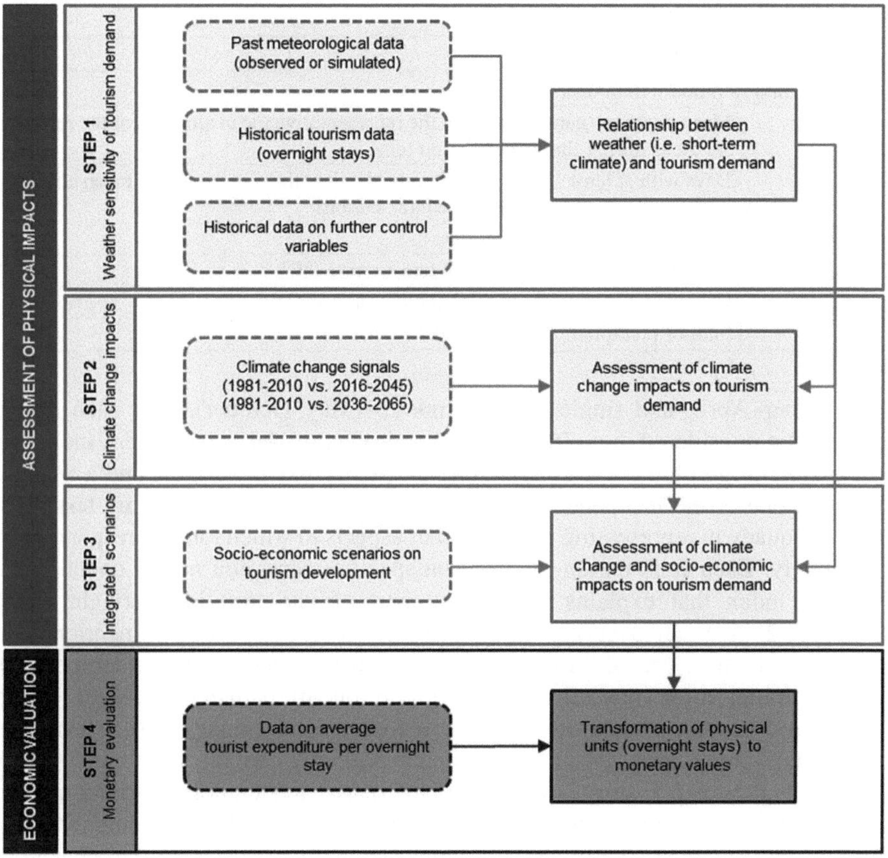

Fig. 19.2 Valuation of (direct) climate change impacts on tourism demand

on tourism demand (represented by overnight stays), we make use of a four-step-procedure illustrated in Fig. 19.2.

The first three steps comprise of physical impact assessments, where impacts are measured in overnight stays; whereas the last step includes the transformation from physical into monetary units.

STEP 1: Weather Sensitivity of Tourism Demand

In the first step, the sensitivity of tourism demand towards weather variability is quantified based on historical data for the period 1974–2006 and the method of multiple regression analysis. Different tourism types may show different sensitivities towards different weather or climatic aspects. Hence, analyses are carried out for each Austrian NUTS 3 region as well as being separated into winter season

Table 19.2 Tested weather/climatic indices

Abbreviation	Explanation
Weather indices tested within winter analyses:	
S_{mean}	Mean depth of (natural) snow at the representative mean altitudes of the region's ski areas during the winter season (cm)
S_{days}	Days with at least 1 cm (natural) snow depth at the representative mean altitudes of the region's ski areas (days/winter season)
Weather indices tested within summer analyses:	
T_{mean}	Monthly average of daily mean temperature (°C)
R_{days}	Days with at least 1 mm precipitation (days/month)
R_{sum}	Sum of precipitation (mm/month)

(November–April) and single summer months (May–October).[4] For each region and season considered, a multiple linear regression model is estimated, including (the natural logarithm of) overnight stays as the dependent variable and a weather index as one of the independent variables. Various weather indices are tested for their adequacy in representing those weather aspects to which tourists respond most sensitively. Each final region- and season-specific regression model contains the weather index that explains the biggest part of variation in overnight stays. Table 19.2 gives an overview of the weather indices tested. As mentioned in Sect. 19.2.1, there are additional meteorological parameters besides temperature and precipitation (or snow) that may influence tourism demand. However, due to the limited number of data observations available for the analyses ($n = 33$), each final region- and season-specific regression model only contains the weather index with the highest explanatory power. Given the spatial and temporal resolution of the analyses together with the tourism demand indicator applied, we assume temperature or precipitation conditions (including snow) to exhibit higher explanatory powers than, for example, humidity, wind speed or sunshine hours.

Data on meteorological parameters stem from the EWCR-Weather-Data-Set (Themeßl et al. 2009), which in turn is based on data from the Austrian Central Institute for Meteorology and Geodynamics (ZAMG). It includes several temperature and precipitation indices on a monthly basis for each Austrian municipality. We aggregate them from municipal to NUTS 3 level by forming the median over all

[4] Since some tourism types are restricted to particular times of the year (e.g. lake or skiing tourism) and Austrian NUTS 3 regions show different priorities with respect to tourism types, differentiating between NUTS 3 regions and months/seasons represents one way of accounting for potential sensitivity differences in tourism types. We tested two different temporal resolutions by conducting analyses (1) for each single month and (2) for winter and summer season. Regarding the winter half-year, analyses carried out on a seasonal basis revealed significant snow dependencies for a higher number of regions and more intuitive results than analyses carried out on a monthly basis. Concerning the summer half-year, analyses conducted on a seasonal basis indicated hardly any significant weather dependencies, contrary to monthly analyses. Hence, we finally used a seasonal resolution for winter and a monthly resolution for summer tourism analyses. Methods and results are only described for these final settings.

data points within a NUTS 3 region. Additionally, the data set includes monthly snow data for the representative mean altitudes[5] of 202 Austrian ski areas (Töglhofer 2011) from a simple snow cover model (Beck et al. 2009). We aggregate snow data from ski area to NUTS 3 (and national) level by forming weighted averages, with the ski areas' transport capacities serving as weighting factors. A higher weighting is therefore given to snow conditions in bigger ski areas. In the case of NUTS 3 regions that do not include any considered ski region, we use snow data aggregated from ski area to national level in order to account for the possibility of regions with predominantly non-snow-based tourism types (e.g. wellness & thermal spa) benefitting from poor overall snow conditions.

Before continuing with the methodology description, we want to shortly discuss relevance, adequacy and limitations of the weather indices tested. Firstly, only considering natural snow depths for quantifying the snow sensitivity of tourism demand is somehow suboptimal in light of current snowmaking coverage.[6] However, actual past total snow depths are hard to reconstruct, since this would require information on how long, to what extent and with which technology snowmaking has been utilized in Austria's single ski areas. Secondly, using a threshold of 1 cm snow depth for constructing the index S_{days} instead of the frequently applied 30 cm (e.g. Steiger and Abegg 2013) is due to limitations of the snow cover model deployed in the generation process of the EWCR-Weather-Data-Set. As pointed out in Töglhofer (2011, p. 64) "[...] *the model performs better with lower threshold definitions and higher ones may be more vulnerable to biased model outputs*". Hence, in light of the snow data's limitations and particularities we follow Töglhofer (2011) in preferring a threshold of 1 cm snow depth for quantifying the snow sensitivity of winter tourism demand. Thirdly, the relevance of S_{mean} might seem questionable from a theoretical point of view. By taking averages over a whole winter season, critical snow conditions during particular periods (e.g. Christmas) may be masked by high snow depths during other periods. Moreover, after exceeding a certain threshold, further variations in snow depths may be irrelevant for skiers' behaviours. Nevertheless, when empirically testing the suitability of several snow indices for measuring weather sensitivities/risks in the skiing industry, Töglhofer (2011) found S_{mean} to rank among those suitable. The weather indices tested within summer season analyses rank among those quite common in the literature (see e.g. Agnew and Palutikof 2006; Castellani et al. 2010; Rosselló-Nadal et al. 2011). Nonetheless, they also encounter limitations, such as the masking of potential extreme events due to the use of averages (T_{mean}) and sums (R_{sum}).

To control for other influencing factors besides weather conditions, the inclusion of further explanatory variables is tested. Due to the limited number of observations

---

[5] Mean altitudes of all the ski area's transport facilities (except drag lifts), weighted by transport capacities.

[6] Almost 60 % of Austrian ski slopes are equipped with snowmaking facilities (Professional Association of the Austrian Cable Cars 2013).

(n = 33), we restrict the amount of explanatory variables simultaneously entering the regression model to four. The additionally tested variables include:

- **(Natural logarithm of) lagged overnight stays**: Overnight stays not only enter the regression model as the dependent but also as an explanatory variable, albeit lagged by one period. Taking such dynamic effects into account decreases the risk of spurious regressions and allows the consideration of expectations and habit persistence of tourists (Song and Witt 2000). Data stem from Statistics Austria.
- **(Natural logarithm of) GDP per capita**: The gross domestic product per capita of the most important tourist-sending countries, weighted by the countries' shares in overnight stays, is used to approximate income levels. Data originate from the OECD. Since the original index turned out to be integrated of order 1, we use its first differences for regression analyses.
- **Easter**: The dummy variable "Easter" indicates if the holy week falls mainly into March. The timing of Easter is expected to influence tourism demand for two reasons: (1) it co-determines ski season length as most ski areas usually close shortly afterwards and (2) the later Easter falls, the higher the probability of either poor/insufficient snow conditions and/or lack of motivation for skiing holidays.
- **Feast days**: The variable "feast days" indicates the number of feast days falling on a week day.
- **Year**: The variable "year" represents the year of the observation and serves the purpose of capturing unexplained trends.

For each considered region and season, various model specifications are tested, differing with respect to the kind of weather index applied as well as the kind and total number of explanatory variables included (see Supplementary Material for further details). The final model specification is selected based on both, the fulfillment of various diagnostic tests—including normally distributed residuals and the absence of functional form misspecification—and the Bayesian Information Criterion (BIC) (Schwarz 1978). If the BIC decides on a final model that does not include a weather index, or the estimated coefficient of the finally selected weather index does not fulfill the criterion of statistical significance at the 10 % level, we assume the weather sensitivity of tourism demand in the considered region and season to be negligible, i.e. zero.

STEP 2: Climate Change Impacts on Tourism Demand

After quantifying how sensitively overnight stays respond to changes in particular weather indices (STEP 1), the impacts of long-term average changes in these weather indices are assessed. For this purpose, the region- and season-specific weather sensitivities are applied to climate change signals (1981–2010 vs. 2016–2045 and 1981–2010 vs. 2036–2065). Results of STEP 2 show the pure impacts of changing "average weather" conditions without considering any socioeconomic changes—and are given as percentage change in overnight stays.

STEP 3: Integrated Scenario

STEP 3 additionally takes scenarios on future tourism development into account. Future scenarios on the region- and season-specific evolution of overnight stays are based on the extrapolation of past trends into the future, using ETS (**E**xponen**T**ial **S**moothing) and ARIMA (**A**uto**R**egressive **I**ntegrated **M**oving **A**verage) models. These scenarios indicate an increase of nationwide annual overnight stays by 17 % (39 %) between 2008 and 2030 (2050). Assuming that these tourism development scenarios do not account for climate change, overnight stays projected for 2030 and 2050 are subsequently corrected for the climate change impacts quantified in STEP 2. Comparing climate-change-corrected to uncorrected future overnight stays indicates the impacts of climate change under consideration of tourism development.

STEP 4: Monetary Evaluation

In the last step, physical impacts are translated into monetary terms using average tourist expenditure per overnight stay. According to T-MONA (Tourismus MONitor Austria), tourists spent 135 € per winter overnight stay and 108 € per summer overnight stay on average in 2009 (Töglhofer 2011). As in the entire study, all prices are measured in real terms. Regarding the future development of (real) tourist expenditures per overnight stay, we assume a growth rate of 0.8 % per annum. Since the derived scenario on the evolution of overnight stays suggests an annual nationwide growth rate of about 0.8 %, this results in a growth rate of total (real) tourist expenditures of 1.6 %, which is comparable to real GDP growth as assumed by the SSP (see Chap. 6). Figure 19.3 summarizes the costing method applied within tourism.

19.4.4 Range of Sectoral Socio-economic Pathway Parameters That Co-determine Climate Impact

The climate change independent evolution of both overnight stays and (real) tourist expenditures per overnight stay co-determine climate impacts by co-determining the overall tourism volume exposed to climate change. Both parameters are influenced by various factors, including the economic development in important tourist-sending countries, the evolution of transportation costs, the alteration of tourists' preferences, etc. With the future evolution of both variables being highly uncertain, we carry out some sensitivity analyses by assuming a reduction

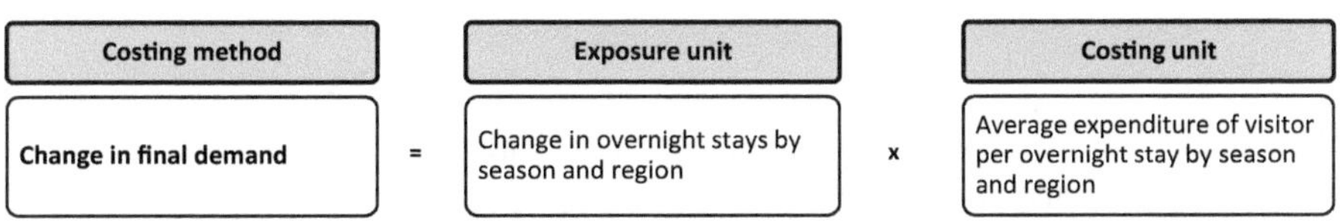

Fig. 19.3 Costing method applied and respective measurement units for tourism

Table 19.3 Average annual climate-triggered economic impacts on tourism demand arising from socioeconomic development and climate change in the future (in M€)

Future economic impact relative to Ø 1981–2010		Climate change Low-range	Mid-range	High-range
Ø 2016–2045	Costs	75	104	213
	Benefits	54	37	15
	Net effect	−21	−67	−199
Ø 2036–2065	Costs	n.a.	316	n.a.
	Benefits	n.a.	106	n.a.
	Net effect	n.a.	-210	n.a.

Not adjusted for rounding differences

(increase) in the growth of both parameters by 25 % compared to our reference assumptions.

Changes in socioeconomic factors may not only affect the tourism sector's exposure, but also its sensitivity towards climate change. Altered tourist preferences or changes in holiday regulations could for instance manifest themselves in altered weather sensitivities. Hence, we carry out some sensitivity analyses by assuming a reduction (increase) in future region- and season-specific weather sensitivities of tourism demand by 25 % compared to those historically observed.

19.4.5 Monetary Evaluation of Impacts

19.4.5.1 Direct Sector Impacts (Costs and Benefits) Without Feedback Effects from Other Sectors

Table 19.3 illustrates the final outcome of the applied four-step-procedure, aggregated from NUTS 3 to national and from monthly/seasonal to annual level.[7] It shows the average annual economic impacts on future tourism demand due to changes in average climatic conditions, differentiating between up to three different climate change scenarios (see Sects. 19.4.1 and 19.4.2). Note that potential impacts due to changes in climate variability are not taken into account.

Assuming socioeconomic pathway parameters as in the reference scenario[8] and a change in the climate as indicated by the mid-range scenario, average annual climate-triggered future economic losses in the tourism field are estimated at 104 million euros (316 million euros) in the first (second) scenario period, of which 101 million euros (291 million euros) are attributable to the winter season.

[7] For interim results and further details see Supplementary Material.

[8] i.e. an average annual nationwide growth rate of overnight stays of about 0.8 %, an annual growth rate of real tourist expenditures per overnight stay of 0.8 %, and weather sensitivities of tourism demand as observed in the past.

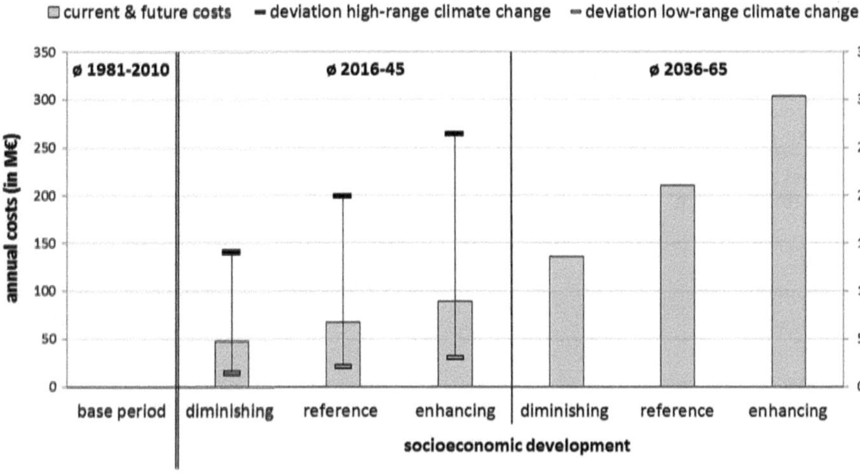

Fig. 19.4 Average annual climate-triggered costs in tourism arising from socioeconomic development and climate change (in M€)

Average annual climate-triggered future economic benefits, on the other hand, are estimated at 37 million euros (106 million euros), of which 32 million euros (90 million euros) are attributable to the summer season. Hence, an average annual future net loss of almost 70 million euros or 0.3 % (210 million euros or 0.7 %) is expected compared to a situation without climate change.

Whereas net impacts aggregated to national and annual level seem rather small due to counteracting effects, impacts on particular regions during specific seasons may be more pronounced. In the case of Carinthia, results for the winter season suggest average annual climate-triggered future economic losses of almost 3 % (over 6 %) in the first (second) scenario period. Assuming climate change according to the high-range scenario, these losses rise to almost 7 % (about 10 %).

A graphical illustration of the average annual climate-triggered economic net impacts on tourism demand is provided in Fig. 19.4, which additionally illustrates the effects of altered socioeconomic pathway assumptions (see Sect. 19.4.4). Since the analysis only considers impacts due to changes in average climatic conditions, annual net impacts in the base period equal zero.

19.4.5.2 Macroeconomic Effects

For the macroeconomic model, we first had to identify tourism relevant sectors in the Austrian Input-Output (IO) table, since this database does not contain a specific tourism sector. Based on Statistics Austria (2012) we classified the following five NACE-sectors as tourism relevant (with their tourism specific shares given in brackets): *accommodation, food and beverage service activities* (75.4 %); *travel agencies* (100 %); *creative arts and entertainment activities* (46.8 %); *libraries,*

Table 19.4 Implementation of baseline and climate change scenario for Tourism in the macro-economic model

	2008	2030		2050	
Change relative to base year (2008)	Private consumption (M€)	Baseline	Climate change	Baseline	Climate change
Change in private demand (total tourism sector)		+43.8 %	+43.3 %	+101.5 %	+100.07 %
Thereof:					
Accommodation	15,277	+43.7 %	+43.3 %	+100.8 %	+99.8 %
Travel agencies	1,413	+43.8 %	+43.3 %	+101.5 %	+100.1 %
Entertainment activities	942	+43.5 %	+43.3 %	+100.1 %	+99.4 %
Cultural activities	178	+43.5 %	+43.3 %	+100.1 %	+99.4 %
Sport activities	1,336	+43.5 %	+43.3 %	+100.1 %	+99.4 %

Note: baseline scenario = reference socioeconomic development without climate change; climate change scenario = reference socioeconomic development and mid-range climate change; quantified climate impact chains: change in summer (temperature, precipitation) and winter tourism demand (snow)

archives, museums and other cultural activities (46.8 %); *sports, amusement and recreation activities* (46.8 %).

In the first step of linking the top-down CGE model to the detailed tourism sector model, we calibrated the CGE model such that it replicated the climate change independent development of tourism demand—hereafter labeled as "baseline"—as indicated by sectoral analysis for the scenario periods 2016–2045 and 2036–2065. Since we followed a comparative static CGE modelling approach, we looked at the mean of simulated annual tourism demand for the two scenario periods, represented by the years 2030 and 2050. To replicate the sectoral model's baseline we proportionately translated these developments into changes in private final demand (cf. Fig. 19.3) for domestic tourism services, relative to the IO table base year 2008 within the CGE model. In the second step, climate change impacts—i.e. climate induced deviations from the baseline—were also modelled as demand shocks in the CGE model.[9]

Table 19.4 shows the change in private consumption in the tourism relevant sectors relative to the base year according to both the baseline scenario and the mid-range climate change scenario. While the rest of the economy is assumed to grow by 1.65 % p.a. (see Chap. 6), i.e. +43.33 % from 2008 to 2030 and +98.84 % from 2008 to 2050, growth rates in the tourism specific parts of the tourism relevant sectors reflect tourism demand development as indicated by sectoral analysis. As

[9] Note that, unlike Schinko et al. (2014), we did not consider climate change impacts on production structures of the tourism relevant NACE-sectors.

shown in Table 19.4, tourism relevant sectors show a somewhat stronger baseline growth than the rest of the economy. Climate change lowers this growth, however.

Negative climate change impacts on tourism demand, i.e. a lower private consumption of tourism-relevant services in the climate change scenario compared to the baseline (see Table 19.4), lead to a higher disposable income of private households for the consumption of other goods and services. Note that in the Austrian IO table all domestic tourism services consumed either by residents or foreign tourists are treated as if they were consumed by residents only. Thus, we had to adjust the domestic disposable income resulting from climate change impacts downwards to correct for that accounting error.[10]

Lower demand for tourism-specific services in the climate change scenario, combined with the necessary adjustment of disposable income, leads to reduced annual gross output values in tourism-relevant as well as other sectors. These annual changes are illustrated in Table 19.5 and include quantity as well as relative price effects. Total effects are negative in both scenario periods, but about three times higher in 2036–2065 than in 2016–2045. Besides the tourism relevant sectors, relatively high reductions in sectoral gross output value are also found for the sectors of food production, beverages and agriculture, since they form major intermediate inputs into the tourism relevant sectors.

Summing up across all sectors, the changes in gross value added lead to a GDP effect of −0.03 % in the first and −0.06 % in the second scenario period (without the effects of altered tax revenues and expenditures for subsidies). The vast majority of the GDP effect can be attributed to reductions in output quantity, and only a smaller share to changes in prices.

In addition to direct climate change impacts on the Austrian tourism sector we consider overall macroeconomic effects on the Austrian economy. Results for the mid-range climate scenario show that, given the model settings, impacts on the macroeconomic indicators "welfare" and "GDP"[11] are negative in both periods. Compared to the baseline, annual welfare (GDP) is 92 million euros (102 million euros) lower on average in 2016–2045 and 310 million euros (339 million euros) lower in 2036–2065. This is due to increased unemployment (0.02 % points in 2016–2045 and 0.04 % points in 2036–2065) triggered by negative climate change impacts on the Austrian tourism sector and macroeconomic feedback effects. Assuming climate change according to the low and high-range scenario (see Sect. 19.4.2), impacts on welfare (GDP) within the first scenario period range from −43 million euros to −269 million euros (−43 million euros to −271 million euros).

Overall, reduced economic output and increased unemployment rates under the mid-range climate change scenario (compared to the baseline) trigger a reduction in government budget of 38 million euros in 2016–2045 and 127 million euros in

[10] For this purpose, we applied the fraction of foreign overnight stays reported by Statistics Austria for 2008, i.e. 0.73.

[11] See Chap. 7 for characterization.

Table 19.5 Sectoral and total effects of quantified climate change impacts in sector Tourism, average annual effects relative to baseline (for periods 2016–2045 and 2036–2065)

	Ø 2016–2045			Ø 2036–2065		
Changes in M€ p.a. relative to baseline	Gross output value	Inter-mediate demand	Net value added	Gross output value	Inter-mediate demand	Net value added
Losing sectors	**−177**	**−88**	**−89**	**−590**	**−294**	**−296**
Accommodation	−58	−21	−37	−182	−66	−116
Travel agencies	−7	−5	−2	−22	−17	−5
Entertainment activities	−3	−1	−2	−8	−2	−6
Cultural activities	−0	−0	−0	−2	−1	−1
Sport activities	−4	−1	−2	−12	−5	−7
Food products	−4	−3	−1	−14	−10	−4
Beverages	−3	−2	−1	−10	−8	−3
Agriculture	−2	−1	−1	−8	−4	−3
All other sectors	−96	−53	−43	−332	−182	−150
Total effect (all sectors)	**−177**	**−88**	**−89**	**−590**	**−294**	**−296**
GDP at producer price			−0.03 %			−0.06 %
...thereof price effect			−0.00 %			−0.01 %
...thereof quantity effect			−0.02 %			−0.06 %

Note: baseline scenario = reference socioeconomic development without climate change; climate change scenario = reference socioeconomic development and mid-range climate change; quantified climate impact chains: change in summer (temperature, precipitation) and winter tourism demand (snow)

2036–2065 (see Table 19.6). Decreasing labour tax revenues and expenditures caused by higher unemployment especially contribute to the reduction in government budget. In addition, reduced GDP causes a relatively strong reduction in revenues from value added tax.

19.4.6 Qualitative Impacts (Non-monetarised)

Some climate change impacts on tourism are rather hard to quantify. This particularly pertains to impacts due to climate induced changes in tourism relevant environmental resources, including alterations in the landscape (shrinking glaciers, dried-up lakes, etc.), loss in biodiversity, or increased safety risks in alpine terrain

Table 19.6 Effects of quantified climate change impacts in sector Tourism on government budget, average annual effects relative to baseline (for periods 2016–2045 and 2036–2065)

Changes in M€ p.a. relative to baseline	Ø 2016–2045	Ø 2036–2065
Revenues	**−38**	**−127**
Production tax	−2	−6
Labour tax	−18	−59
Capital tax	−6	−18
Value added tax	−13	−43
Other taxes	−0	−1
Expenditures	**−38**	**−127**
Unemployment benefits	+20	+68
Transfers to households net of other taxes	−58	−195
Government budget in baseline (p.a.)	**149,066**	**206,459**
Climate change impact on government budget	−0.03 %	−0.06 %

Note: baseline scenario = reference socioeconomic development without climate change; climate change scenario = reference socioeconomic development and mid-range climate change; quantified climate impact chains: change in summer (temperature, precipitation) and winter tourism demand (snow)

due to melting permafrost. In addition, some of the potential climate change impacts listed in Table 19.1 could not be quantified in the present project due to resource limitations. These include climate change induced alterations in water and/or energy demand as well as losses due to business interruptions following natural disasters.

19.4.7 Sector-Specific Uncertainties

The four-step-procedure described in Sect. 19.4.3 exhibits various critical assumptions, limitations and uncertainties that have to be considered when interpreting the model results:

- **Extreme events**: The method applied focuses on changes in mean weather conditions rather than on changes in weather extremes. This may lead to an underestimation of climate change impacts.
- **"Weather memory" of tourists:** Steiger (2011) found an enduring effect of the extraordinary snow-poor winter season 2006/07 in some Tyrolean districts. Especially in the case of several consecutive periods of adverse weather conditions, this kind of "weather memory" may intensify climate change impacts considerably. However, the procedure applied does not consider such "weather memory" effects.
- **Weather sensitivities**: Climate change impacts are assessed on the basis of past weather sensitivities observed for the period 1974–2006. However, especially

with respect to natural snow conditions, sensitivities might have changed systematically over time due to the introduction and expansion of artificial snowmaking during the last decades. Using two different panel data approaches, Töglhofer et al. (2011) found some evidence that the sensitivity of overnight stays in 185 Austrian ski areas towards natural snow conditions may have decreased over time. Hence, the snow sensitivities applied for impact assessment may be somewhat overestimated. Moreover, weather sensitivities might be subject to future change, for instance due to tourists' changing preferences. Hence, applying historically observed sensitivities for assessing climate change impacts bears uncertainties.

- **Evolution of socioeconomic parameters:** The region- and season-specific future development of overnight stays and the evolution of tourist expenditures per overnight stay are affected by a whole range of factors (including costs of travel, terrorism and war, tourist preferences, etc.), and are therefore highly uncertain.
- **Tourist preferences**: Principally, we assume tourist preferences about holiday destinations, tourism types, weather/climatic conditions, etc., to remain constant over time. However, tourist preferences are actually subject to constant change. We partially account for this fact within sensitivity analyses, but overall, the future evolution of tourist preferences remains a highly uncertain factor.
- **Climate change in tourist-sending countries / competing destinations**: Climate change in tourist-sending countries or competing destinations and its impacts on tourism in Austria are not taken into account. Comparably cooler alpine destinations may benefit from increasing heat waves in nearby cities or the Mediterranean (Serquet and Rebetez 2011; Amelung and Viner 2006). Moreover, changes in the snow reliability of competing destinations may affect tourism demand in Austrian ski areas.
- **Day visitors**: Due to data availability, present analyses focus on overnight guests. However for some regions, day visitors are also of high importance and climate change may affect them too. Müller and Weber (2008), who estimate the economic effects of climate change on tourism in the Bernese Oberland (Swiss), expect climate induced impacts on revenues related to daily visitors during winter (summer) to be about one third smaller (higher) than those related to overnight stays.

19.5 Summary of Climate Costs for Tourism and Conclusions

Our analysis of potential climate change impacts on tourism demand in Austria indicates predominantly negative effects on winter tourism and mainly positive effects on summer tourism, with net impacts being negative. Although results suggest nationwide effects to be rather small, some regions may suffer from considerable impacts within particular seasons of the year. The macroeconomic evaluation shows negative effects on GDP and welfare, about 50 % higher than direct tourism impacts. The strongest negative spillover effects emerge for the food and beverage production sectors as well as the agriculture sector, because of the high relevancy of their inputs into the tourism sector. Overall, results have to be interpreted with caution, since they are subject to a range of uncertainties.

Acknowledgement The authors would like to thank the Scientific Advisory Board and the anonymous reviewers for their valuable comments and suggestions to improve the quality of the chapter.

References

Agnew MD, Palutikof J (2006) Impacts of short-term climate variability in the UK on demand for domestic and international tourism. Clim Res 31:109–120

Amelung B, Viner D (2006) Mediterranean tourism: exploring the future with the tourism climatic index. J Sustain Tour 14(4):349–366

Beck A, Hiebl J, Koch E, Potzmann R, Schöner W (2009) Institutionelle und regulatorische Fragestellungen der Bereitstellung von Wetterdaten. EWCR Working Paper No 10/2009. www.klimarisiko.at. Accessed 18 June 2013

Breiling M, Charamza P (1999) The impact of global warming on winter tourism and skiing: a regionalized model for Austrian snow conditions. Reg Environ Change 1:4–14

Castellani M, Mussoni M, Pattitoni P (2010) Air passengers and tourism flows: evidence from Sicily and Sardinia. Almatourism 1:16–28

Cooper C, Fletcher J, Fyall A, Gilbert D, Wanhill S (2008) Tourism: principles and practice, 4th edn. Pearson, Harlow

Dawson J, Scott D, McBoyle G (2009) Climate change analogue analysis of ski tourism in the northeastern USA. Clim Res 39:1–9

EEA (2012) Climate change impacts and vulnerability in Europe 2012 – an indicator-based report. EEA Report no 12/2012, Copenhagen

Falk M (2013) Impact of long-term weather on domestic and foreign winter tourism demand. Int J Tour Res 15:1–17

Fleischhacker V, Formayer H (2007) Die Sensitivität des Sommertourismus in Österreich auf den Klimawandel. StartClim2006D1, Vienna

Hamilton LC, Brown C, Keim BD (2007) Ski areas weather and climate: time series models for New England case studies. Int J Climatol 27:2113–2124

Köberl J, Gobiet A, Heinrich G, Leuprecht A, Prettenthaler F, Töglhofer C (2011) Impacts of weather variability and climate change on tourism in Austria. In: Bordsdorf A, Stötter J, Veulliet E (eds) Managing Alpine Future II – inspire and drive sustainable mountain regions, Proceedings of the Innsbruck conference, Verlag der Österreichischen Akademie der Wissenschaften, Vienna, 21–23 Nov 2011

Lise W, Tol RSJ (2002) Impact of climate on tourist demand. Clim Change 55:429–449

Moreno A (2010) Climate change and tourism – impacts and vulnerability in coastal Europe. Doctoral Thesis, Universitaire Pers Maastricht. http://arno.unimaas.nl/show.cgi?fid=20143. Accessed 18 June 2013

Müller HR, Weber F (2008) Climate change and tourism – scenario analysis for the Bernese Oberland in 2030. Tour Rev 56(3):57–71

Professional Association of the Austrian Cable Cars (2013) Factsheet – Die Österreichischen Seilbahnen in Zahlen, Winter 2012/13. www.seilbahnen.at. Accessed 15 April 2014

Rosselló-Nadal J, Riera-Font A, Cárdenas V (2011) The impact of weather variability on British outbound flows. Clim Change 105:281–292
Rudel E, Matzarakis A, Koch E (2007) Summer tourism in Austria and climate change. In: Oxley L, Kulasiri D (eds) MODSIM 2007 international congress on modelling and simulation. Modelling and simulation society of Australia and New Zealand, Canberra, pp 1934–1939
Rutty MK, Scott D (2010) Will the Mediterranean become "too hot" for tourism? A reassessment. Tour Plann Dev 7(3):267–281
Schinko T, Köberl J, Bednar-Friedl B, Prettenthaler F, Töglhofer C, Heinrich G, Gobiet A (2014) An integrated economic analysis of climate change impacts on tourism in Austria across tourism types and seasons, Graz Economics Papers 2014–10. University of Graz, Department of Economics
Schwarz G (1978) Estimating the dimension of a model. Ann Stat 6:461–464
Scott D, Lemieux C (2010) Weather and climate information for tourism. Procedia Environ Sci 1:146–183
Scott D, Gössling S, de Freitas CR (2007) Climate preferences for tourism: an exploratory tri-nation comparison. In: Matzarakis A, de Freitas CR, Scott D (eds) Developments in tourism climatology. Commission on Climate, Tourism and Recreation, Freiburg, pp 18–23
Serquet G, Rebetez M (2011) Relationship between tourism demand in the Swiss Alps and hot summer air temperatures associated with climate change. Clim Change 108:291–300
Shih C, Nicholls S, Holecek DF (2009) Impact of weather on downhill ski lift ticket sales. J Travel Res 47:359–372
Song H, Witt SF (2000) Tourism demand modelling and forecasting: modern econometric approaches. Pergamon, Oxford
Statistics Austria (2012) Tourismus-Satellitenkonto für Österreich. Statistics Austria, Vienna
Statistics Austria, WIFO (2014) Tourism satellite accounts for Austria. Statistics Austria, Vienna
Steiger R (2011) The impact of snow scarcity on ski tourism: an analysis of the record warm season 2006/07 in Tyrol (Austria). Tour Rev 66:4–13
Steiger R, Abegg B (2013) The sensitivity of Austrian ski areas to climate change. Tour Plann Dev 10:480–493
Themeßl M, Gobiet A, Töglhofer C (2009) Meteorologische Daten für lokale Impact-Studien. The economics of weather and climate risks working paper series working paper Nr 7/2009, Graz
Töglhofer C (2011) Measuring weather risk: the case of winter tourism in Austria. Studien zum Klimawandel Band VII. Verlag der Österreichischen Akademie der Wissenschaften, Vienna
Töglhofer C, Eigner F, Prettenthaler F (2011) Impacts of snow conditions on tourism demand in Austrian ski areas. Clim Res 46:1–14
UNWTO-UNEP-WMO (2008) Climate change and tourism – responding to global challenges. UNWTO, Madrid
WTTC (2013) Travel & tourism economic impact 2013. http://www.wttc.org/site_media/uploads/downloads/world2013_1.pdf. Accessed 19 Aug 2013

Part IV
Aggregate Evaluation

Chapter 20
Assessment of the Costs up to 2100 and Barriers to Adaptation

Claudia Kettner, Angela Köppl, and Katharina Köberl

A quantification of climate damages or the costs of inaction faces the inherent uncertainty of future climate scenarios and socio-economic developments. For the appraisal of the long-run cost of inaction in COIN we therefore apply the Delphi technique until 2100 that offers a qualitative assessment by recognised experts rather than quantitative results. The Delphi results suggest pronounced increases in the damage costs in the second half of the twenty-first century. For half of the sectors addressed, there is unanimous consensus among experts that climate damage costs in 2070 will be higher than in 2050. A further increase in costs after 2070 is expected for the majority of sectors. Economic and social developments are considered the most important cost drivers in the long run. Despite this judgement, however, uncertainty of future social, economic and thus cost development is rated considerably high. Extreme events might be key determinants of the long-term cost of inaction.

20.1 Introduction

Climate change is one of the most exigent problems our society faces. According to the most pessimistic scenario of the latest IPCC Assessment Report (AR5) (RCP8.5[1]; IPCC 2013), global mean surface temperature very likely increases by 2.6–4.8 °C compared to 1900 by the end of the twenty-first century. Furthermore, fundamental changes of extreme weather events and weather periods are considered very likely for the second half of the century: E.g., it is considered virtually certain that the intensity and the frequency of cold days and nights will decrease over most land areas, while the intensity and the frequency of hot days and nights will rise

With contributions of Martin König, Wolfgang Loibl, Willi Haas, Stefan Hochrainer, Lukas Kranzl, Hermine Mitter, Roman Neunteufel, Karl Steininger

[1] In the fifth IPCC Assessment Report (AR5) four different greenhouse gas concentration trajectories are used. In the RCP8.5 scenario for 2100 a greenhouse gas concentration of approximately 1370 CO_2-equiv. is assumed and radiative forcing values are assumed to increase to 8.5 W/m^2.

C. Kettner (✉) • A. Köppl • K. Köberl
WIFO—Austrian Institute of Economic Research, Vienna, Austria
e-mail: claudia.kettner@wifo.ac.at; angela.koeppl@wifo.ac.at; katharina.koeberl@wifo.ac.at

K.W. Steininger et al. (eds.), *Economic Evaluation of Climate Change Impacts*, Springer Climate, DOI 10.1007/978-3-319-12457-5_20

(cf. Chap. 5). Moreover, according to the AR5 it is very likely that heat waves and heavy precipitation events will increase just as the sea level will rise further; an increase in the intensity/duration of droughts is considered likely (IPCC 2013).

Over the past 40 years, numerous models have been developed to estimate the cost of climate change. In the beginning, with a stronger focus on identifying optimal abatement decisions, whereas in the recent past model simulations are increasingly used to assess robust adaptation strategies [see e.g. Ciscar (2009) and Watkiss (2011b) for the EU]. The quantified damage costs, however, have to be interpreted with caution because of limitations of the models employed (e.g. weak representation of extreme weather events, assumption of perfect rationality), the inherent uncertainty of future socio-economic developments and the complexity of climate change.

For the appraisal of the long-run cost of inaction in COIN we apply the Delphi technique that offers a qualitative assessment from recognised experts, rather than presenting quantitative results for that time horizon. The qualitative assessment of the long-term development is motivated by the intrinsic uncertainties with respect to long-run developments in terms of dimensions such as socio-economic structural change, prices and technology. The Delphi technique is well-suited for assessing future developments under uncertainty (e.g. Okoli and Pawlowski 2004; Yang et al. 2012). The main part of the Delphi survey in the COIN project deals with a qualitative appraisal of possible long-term (2070/2100) sectoral economic impacts of climate change. The second part addresses barriers to adaptation on the sectoral level. In order to avoid substantial climate damage costs in the long run it is essential to overcome possible barriers for adaptation.

The chapter is structured as follows: We start out with a survey of estimates of the long-term cost of inaction at global and EU level. In Sect. 20.3, we describe the Delphi approach that was employed for the qualitative appraisal of the long-term climate damage costs in Austria. This is followed by the presentation and discussion of the results of the Delphi survey with respect to the development of the cost of inaction, the relevant cost drivers and the uncertainty underlying these developments as well as the sectoral barriers to adaptation. The last section concludes.

20.2 Overview of Estimates of the Long-Term Cost of Inaction at Global and EU Level

Since the beginning of the 1990s, much effort has been put into economic modelling and into the quantification of the (direct and indirect) effects of climate change on socio-economic systems. The first economic analyses of climate change impacts focused on the US [see IPCC (1996), for an overview]. Only a few years later, the first global estimates of the costs of climate change were published. Based on Tol (2011), Table 20.1 provides an overview of selected studies analysing global climate damage costs including the assumed changes in temperature which can

Table 20.1 Global estimates of the cost of inaction

Study	Warming (°C)	Impact (% GDP)	Regional disaggregation	Sectoral/thematic disaggregation
Nordhaus (1994a)	3.0	−4.8		
Nordhaus (1994b)	2.5–3.0	−1.3		
Fankhauser (1995)	2.5	−1.4	✔	✔
Tol (1995)	2.5	−1.9		
Nordhaus and Yang (1996)	2.5	−1.7	✔	
Nordhaus and Boyer (2000)	2.5	−1.5	✔	✔
Tol (2002)	1.0	2.3	✔	✔
Maddison (2003)	2.5	−0.1	✔	
Rehdanz and Maddison (2005)	1.0	−0.4	✔	
Nordhaus (2006)	2.5	−0.9	✔	
Stern (2006)	2–3	0– −3		
Tol (2013)	3.5	−1.2	✔	✔

Source: Adapted from Tol (2011)

be interpreted as a measure of the intensity of mitigation efforts and the economic impact in terms of changes in global GDP.[2] Although not stated explicitly in the early work, it can be supposed that all studies refer to long-term costs of climate change reflected in the assumed temperature increases (Tol 2011).

The studies follow different approaches in the estimation of the costs of climate change. The first study, Nordhaus (1994a), estimated climate damage costs based on an expert survey in which 20 inter-disciplinary experts were asked to approximate the total long-term economic impact of climate change. Other studies like Fankhauser (1995), Nordhaus (1994b), Nordhaus and Yang (1996), Nordhaus and Boyer (2000) and Tol (1995, 2002, 2013) use enumerative methods to assess climate damage costs, i.e. they base the analysis of costs on estimates of the physical impacts of climate change in certain sectors obtained from natural sciences. Another approach is followed by Maddison (2003) or Nordhaus (2006): They use a 'statistical approach' (Tol 2011) which relies exclusively on observed variations in household consumption or output due to climatic differences between regions to estimate the future effects of climate change. The RICE/DICE models as used in Nordhaus (1994b), Nordhaus and Yang (1996) and Nordhaus and Boyer (2000) as well as the FUND model as used in Tol (1995, 2002, 2013) follow a welfare optimization or optimal growth approach, while the PAGE model as used in the Stern Review can be classified as a simulation model for climate damage where

[2] The Stern Review (Stern 2006) analyses the costs of climate change in the absence of mitigation. Assuming a temperature increase of 5–6 °C, climate damage costs in the range of 5–10 % of global GDP are estimated.

central economic parameters are exogenous (Stanton et al. 2009; van Vuuren et al. 2009).

Nordhaus (1994a) and Stern (2006) present aggregate global estimates of climate damages whereas Nordhaus and Yang (1996), Maddison (2003), Rehdanz and Maddison (2005) and Nordhaus (2006) also provide regionally disaggregated results. Nordhaus and Boyer (2000) and Tol (2002, 2013) additionally present results for selected sectors. These detailed estimates are, however, difficult to compare as regional and sectoral/thematic disaggregation levels differ significantly between the studies. Even on an aggregate level, one can already observe pronounced differences in the model results on global climate damage costs. For an assumed temperature increase of 2.5 °C compared to pre-industrial levels, estimates of global damage costs e.g. range between 0.1 and 1.7 % of global GDP. There is, however, another caveat when comparing the aggregate cost estimates in the different studies: In some studies results refer to discounted[3] future damage costs, while in other analyses they represent damage costs in a certain future year. In Maddison (2003) or Rehdanz and Maddison (2005) e.g. no discounting is applied just as in the enumerative, sectoral approaches by Frankhauser (1995). The cost of climate change in Nordhaus and Boyer (2000), Nordhaus and Yang (1996), Nordhaus (1994b), Stern (2006) and Tol (1995, 2013), in contrast, represent discounted future damage costs. As pointed out e.g. by Tol (2010) or Pindyck (2013) the discount rate applied to future costs and benefits represents one of the most important factors influencing the estimates of climate damage costs. Other factors that determine differences in the cost estimates of the models include the impact functions used, the use of equity weights and risk aversion [see Watkiss (2011a) for a detailed discussion].

For the European Union, the long-term cost of inaction at sector level were recently analysed in two major research projects: PESETA (Ciscar 2009) and ClimateCost (Watkiss 2011b). Both projects follow a sectoral bottom-up approach to estimate the cost of inaction until 2100, i.e. the analysis is based on sectoral physical impact models.[4] ClimateCost follows a scenario-based approach while PESETA assesses "the effects of future climate on today's economy" (Ciscar et al. 2012).

The estimated costs vary significantly between sectors and different climate and socio-economic scenarios (see Table 20.2): In the PESETA project, the model results for the agriculture sector e.g. show a range of effects of inaction in the EU in 2100[5] between a yield increase of 3 % (based on a climate scenario with a

---

[3] The motivation for discounting is based on the assumption that future costs and benefits are valued less than current costs and benefits. Whether this assumption is valid on societal level is, however, disputed. Furthermore, in the case of discounting the choice of a particular discount rate has to be considered arbitrary as it lacks theoretical underpinning (see e.g. van den Bergh 2004).

[4] Similar to the approach chosen in the COIN project for estimating the cost of inaction until 2050, PESETA furthermore applied a CGE model to assess the direct and indirect impacts of climate change throughout the economy.

[5] I.e. for an average year in the period 2070–2100.

Table 20.2 EU estimates of the cost of inaction

Study	Warming (°C)	Sectoral impact	
PESETA	2.5	Agriculture	Increase in crop yields by 3 %
		River floods	Additional economic damage of 8 billion euros p.a.
		Coastal systems	Total residual damage costs of 10 billion euros p.a.
		Tourism	Increase in expenditure receipts by 2 billion euros p.a.
		Health	Increase in heat-related death (VOLY: 27 billion euros, VSL: 65 billion euros)
			Decrease in cold-related death (VOLY: 48 billion euros, VSL: 112 billion euros)
	3.9	Agriculture	Reduction of crop yields by −2 %
		River floods	Additional economic damage of 12 billion euros p.a.
		Coastal systems	Total residual damage costs of 45 billion euros p.a.
		Tourism	Increase in expenditure receipts by 3 billion euros p.a.
		Health	Increase in heat-related death (VOLY: 9–51 billion euros, VSL: 21–119 billion euros)
			Decrease in cold-related death (VOLY: 87 billion euros, VSL: 204 billion euros)
ClimateCost	3.4	Sea level rise	Economic damage costs of 25 billion euros (19.3 billion euros -37.2 billion euros); 156 billion euros with high sea level rise
		River floods	Economic damage costs of 359 billion euros
		Energy	Increase of electricity use for space cooling by 3 %
			Decrease of heat energy demand by 22 %
		Health	Increase in heat-related death (VSL: 146 billion euros)

Source: Ciscar (2009) and Watkiss (2011b); own illustration

temperature increase of 2.5 °C) and a yield decrease of −2 % (based on a climate scenario with a temperature increase of 3.9 °C). On the regional level, one can find even higher differences. Similar bandwidths of estimates were derived for the other sectors. The modelling results of the ClimateCost project also show a high range of variation, depending on the assumed climate and socio-economic development. With respect to sea level rise, the estimated cost of inaction in Europe e.g. range between 19.3 billion euros and 37.2 billion euros for the baseline climate change scenario and 156 billion euros in a scenario with high sea level rise.

The approaches chosen in PESETA and ClimateCost differ fundamentally and the sectoral disaggregation used in PESETA and ClimateCost does only partially overlap; therefore the two studies are difficult to compare. With respect to the evaluation of flood damage, one can, however, observe pronounced differences in the findings that may result from different underlying climate scenarios but also from different socio-economic pathways and system boundaries and—most importantly—from the different models used in the projects.

20.3 The COIN Delphi Survey for the Qualitative Appraisal of the Long-Term Cost of Inaction in Austria: Methodological Approach

The overview of cost estimates in Tables 20.1 and 20.2 shows a considerable spread stemming predominantly from diverse modelling approaches, system boundaries and the assumptions on climate scenarios and socio-economic development. One way to avoid detailed assumptions as required by Integrated Assessment Models is to apply a qualitative approach for the assessment of long-term climate damages.

The Delphi technique as applied here was developed as a forecasting tool under uncertainty by the RAND Corporation in the early 1960s (Häder 2009). It is a systematic, multi-stage survey method with feedback and is often used as a tool to assess future events, trends or technological developments. The Delphi is based on an expert panel and aims at integrating the judgments of experts from different disciplines or with different viewpoints. Here we apply the Delphi survey for judging the long-term cost of inaction and assessing relevant sectoral barriers to adaptation in Austria.

A wide range of different Delphi approaches exists [see e.g. Häder (2009) for an overview of typologies of Delphi surveys]. The aim of the Delphi method in general is to elaborate on which questions there is consensus within the expert group and on which questions there is dissent. Anonymity of the participants and the iterative process with systematic feedback ensures that the relevant knowledge of each respondent is taken into account, irrespective of his reputation.

The COIN Delphi survey was conducted online with the (sectoral) experts of the project consortium. The motivation for this approach is to ensure consistency between the qualitative long-term appraisal and the quantitative estimates for 2050 on the one hand and the tight timeline of the project on the other hand. For each sector, between three and seven project experts answered the questionnaire (Table 20.3). The small sample size for the individual sectors limits the possibility for the use of advanced statistical methods in the analysis of the Delphi answers. This disadvantage could be alleviated by the fact that the experts involved have profound knowledge of the methods and results of the quantitative appraisal of cost of inaction for 2050 which contributes to the coherence of the judgments.

The questionnaire covers two thematic areas: The main focus of the COIN Delphi survey is the qualitative appraisal of possible long-term (2070/2100) sectoral economic impacts of climate change as a complement to the quantified results for 2050. The second part gathers information on barriers to adaptation on the sectoral level. Due to the tight timeline of the overall project, it was decided that the

Table 20.3 Number of respondents in the Delphi survey by sector

Sector	Respondents
Agriculture	7
Forestry	7
Ecosystem Services: Pest Control and Pollination	4
Human Health	4
Water Supply and Sanitation	4
Electricity	6
Buildings: Heating and Cooling	3
Transport	4
Manufacturing and Trade: Labour Productivity Losses	3
Cities and Urban Green	6
Catastrophe Management: Riverine Flooding	7
Tourism	5

COIN Delphi survey should comprise only two rounds. The results of the survey (Sect. 20.3) show, however, already stable valuations for the two rounds and high consensus among sector experts; a further Delphi round would not have delivered additional new insights. Compared to the classic Delphi approach, the COIN Delphi survey deviates in the sense that it also used elements of a Group Delphi [see Schulz and Renn (2009)] in order to improve the validity of results: During a project workshop, the experts involved in the Delphi spent some hours discussing the issues covered. This helped ensure that the experts had a common understanding of the survey for the second Delphi round reconsidering the identified effects and their uncertainty.

The quantitative sectoral estimates for 2050, which are summarised in Table 20.4, constitute the starting point for the qualitative appraisal of the long-term economic cost of inaction. On this basis for 2050, the COIN experts were asked in a first step to give their qualitative judgement on how the economic impacts in 2070 and 2100 differ from 2050. In a second step, the sector experts were asked to evaluate the importance of different categories of cost drivers (climatic, economic, social, technological and policy drivers)[6] for the development of climate damage costs, complemented by their assessment on the certainty of the development of the cost drivers. For each aspect, the questionnaire includes a closed rating question and an open question where the experts were asked to argue their evaluations. For the sectors health and ecosystem services, no quantification of the economic cost of inaction in 2050 has been carried out. Therefore, for these two sectors the long-term qualitative appraisal was based exclusively on open questions.

[6] In the first Delphi round, only the categories 'climate drivers', 'economic drivers' and 'social drivers' were differentiated. Based on feedback by the experts, for the second round of the Delphi the categories 'technological drivers' and 'policy drives' have been added as these were considered key drivers in selected sectors.

Table 20.4 The quantified cost of inaction for 2050 for selected quantified impact chains

Sector	Impact quantified	Cost of inaction in million euros	Impact chain(s)
Agriculture	Change in output	−267	• Prolonged growing season → increase in biomass production • Increase in evapotranspiration and higher evaporation rate → increase in water demand → decrease in biomass production in water-limited regions • Frost damage → decrease in biomass production or yield loss • Change in precipitation → change in biomass production and biomass productivity • Shift in the seasonal precipitation distribution → Limited water availability in growing season • Summer wet periods → crop damages and damages of soil structure
Forestry	Change in investment Change in output	540	• Bark beetle damages • Storm damages
Ecosystem Services: Pest Control and Pollination	(Qualitative appraisal)		• Adverse effects on pest control • Adverse effects on pollination
Human Health	(Non-monetary appraisal)		• Increase in high temperature-related mortality [deaths and years of life lost (YLL)] • Increase in food-borne diseases: Salmonellosis • Climate-sensitive non-communicable diseases: Additional allergies caused by increased spread of ragweed
Water Supply and Sanitation	Change in investment Change in repair costs	2.2	• Lower yield of springs groundwater/surface water recharge • Increase of days with higher turbidity of spring water resources • Infrastructure damage through landslide/mudflow events or flood events • Change of withdrawal • Increase of depth of drying cracks in the soil • Increase in wastewater volume during winter

			• Increase of sewer flooding • Increase of sedimentation in the sewer system during dry weather flow
Electricity	Change in output Change in investment	−51	• Change in hydropower generation due to change in river run-off levels • Increase in cooling peak load and required peak load capacities • Change in electricity generation mix due to shifts in demand and supply
Buildings: Heating and Cooling	Change in demand	−334	• Decrease in heating demand • Increase in cooling demand
Transport	Change in investment	0.69	• Damages to transport infrastructure due to increase in floods, landslides and mudflows
Manufacturing and Trade: Labour Productivity Losses	Productivity loss	11.9	• Productivity loss of workers due to hotter working environments
Cities and Urban Green	Change in investment	42.2	• Additional investments for green spaces • Additional maintenance costs of green spaces
Catastrophe management: Riverine Flooding	Change in public expenditure	53.0	• Increase damage of buildings → higher natural disaster funding for loss financing
Tourism	Change in demand	237	• Decrease in winter tourism demand due to decrease in (natural) snow • Increase in summer tourism demand due to reduced precipitation and higher temperature

The second part of the questionnaire addresses which barriers for adaptation the respondents rate most relevant for their sectors of expertise. In the first round of the Delphi survey, experts were asked to name the three most relevant barriers for each sector. For the second round of the Delphi, we categorised the stated barriers into five groups (institutional barriers, informational barriers, financial barriers, social barriers and other barriers) for which the sector experts judged the relevance of each group. In addition, the experts evaluated the likelihood at which the respective barriers could be overcome in the future as well as the lead times for adaptation. The section on barriers contains a closed rating question for each aspect and an open question where the experts were asked to give more details on their evaluations.

20.4 Results of the Qualitative Appraisal of the Long-Term Cost of Inaction in Austria

In the following we discuss the results of the COIN Delphi survey, i.e. the qualitative assessment of the cost of inaction in 2070 and in 2100 compared to 2050. As argued above, the qualitative assessment of the long-term cost of inaction is motivated by the intrinsic uncertainties for long-run developments such as demographic change, structural economic change, prices, technology and economic growth. We summarise the Delphi results on the importance and certainty of relevant cost drivers as well as on the relevant sectoral barriers for adaptation and their persistency. For each aspect, the ranked results from the analyses of the closed questions of the Delphi survey are supplemented with summarised narrative information from the experts. Furthermore, in this chapter the sector experts provide additional information that underpins the ratings.

20.4.1 The Cost of Inaction in 2070 and 2100

COIN quantifies the cost of inaction in 2050 that arise out of ten sectors for which the qualitative assessment for 2070 and 2100 from the COIN Delphi survey are presented in Fig. 20.1. As can be seen in the figure, experts value the costs higher for each sector in the very long run compared to 2050, with higher increases by the end of the twenty-first century. These expected cost increases are illustrated in the figure by positive values for all sectors and time periods; higher values for 2100 point to even higher climate damage costs. For six out of the ten sectors (agriculture, water, buildings, transport, manufacturing and trade, cities), there is unanimous consensus among experts that climate damage costs in 2070 will be higher than in 2050.

A detailed breakdown of the cost estimates derived in the Delphi survey for 2100 is provided in Fig. 20.2. For seven sectors—agriculture, electricity, water, buildings, cities and urban green, manufacturing and trade, transport—the respondents

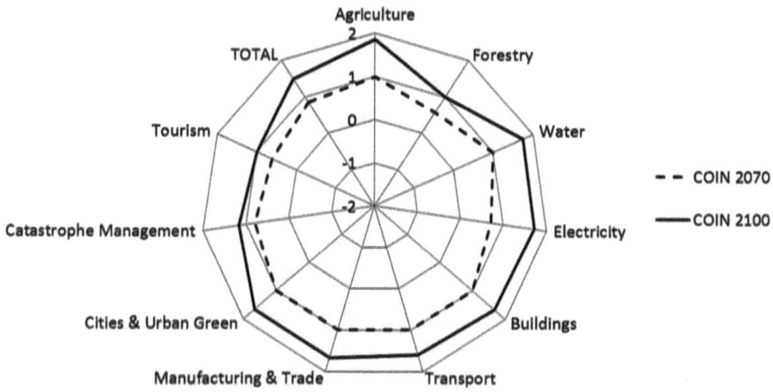

Fig. 20.1 Delphi results for cost of inaction in 2070 and in 2100 compared to 2050—sector means. *Source*: Own calculations. *Note*: For the illustration of the results a weighting procedure is applied. For the evaluation of cost development in 2070, the experts could choose between the categories 'higher than in 2050' (weighting factor: 1), 'same as in 2050' (weighting factor: 0), 'lower than in 2050' (weighting factor: −1); for the evaluation of costs in 2100 two additional categories were added: 'even higher than in 2070' (weighting factor: 2) and 'even lower than in 2070' (weighting factor: −2)

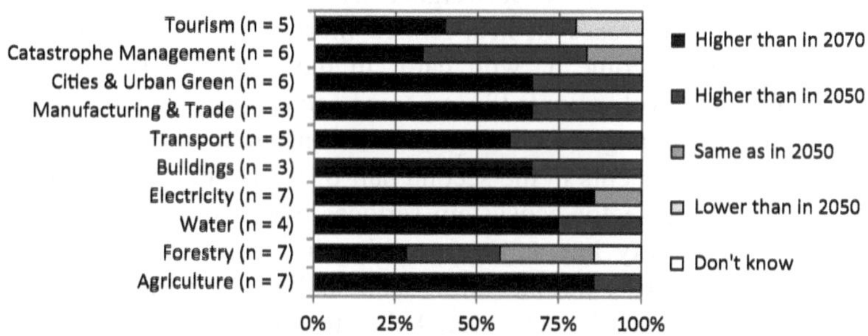

Fig. 20.2 Delphi results for cost of inaction in 2100 compared to 2050. *Source*: Own calculations

expect high dynamics in cost development, i.e. for these sectors more than 50 % of the experts consider the increase in the cost of inaction in 2100 more pronounced than in 2070. The highest cost increases compared to 2050 are expected for the agriculture sector, i.e. the net benefits estimated for the sector until 2050 are in general considered to 'turn into net costs until 2070 and—even sharper—2100' due to a stronger climate signal.

For the sector catastrophe management, half of the experts estimate similar climate damage costs for 2100 as in 2070 and one third believes that the costs will even increase towards 2100. For the forestry sector, experts' judgements are mixed: One third expects higher cost of inaction by the end of the century than in 2070, one third expects climate damage costs in 2100 and 2070 to be of the same magnitude and one third believes that the cost in 2100 will not exceed those of 2050. The spread of evaluations may be due to the fact that the impacts of climate change on the Austrian forestry sector will be largely determined by the tree species composition—especially spruce forest might be affected by the direct and indirect (pests etc.) effects of climate change—and that experts have different expectations with respect to the long-run tree species distribution. For the sector tourism, the majority of the experts estimate the cost of inaction to rise in the long run. However, the appraisal also shows differences in the narratives to the judgements. One expert argues that the costs in 2100 would be lower than in 2050. The reason is that the cost intensive transformation of the winter tourism sector will occur mainly in the next 50 years, resulting in lower costs in the second half of the century compared to 2050. The expert therefore concludes that 'in 2070 and even 2100 the positive effects of climate change (compared to 2050) will be the main factor for tourism in Austria' (cf. Chap. 19).

It was already pointed out above that the COIN project does not quantify the climate damage costs in 2050 for the sectors health and ecosystems services. From the qualitative long-term appraisal increasing costs of inaction are expected in the second half of the century. With respect to the health sector (cf. Chap. 11), experts note that 'events like mass movements and floods would potentially claim more victims than in the first half of the twenty-first century [and that] the psychological impacts of more frequent extreme events connected to private losses and victims might be substantial'. Potential benefits of climate change, i.e. less casualties due to cold waves, would probably be outweighed by more frequent and intense heat waves leading to an increase in heat-related mortality. One expert argues that climate change will become a 'wild card' in the second half of the century: e.g. socio-economic developments like an aging society will increase vulnerability and stronger climate change (including an increasing number of extreme events) will shape cost developments in the second half of the twenty-first century.

With respect to the sector ecosystem services, species range movements and decoupling of community interactions resulting in an impairment of ecosystem services are expected to continue after 2050 in line with intensifying climate change. This trend might be aggravated due to socio-economic developments such as an increased pressure on non-used arable land resulting from higher food demand (cf. Chap. 10).

20.4.2 Relevant Cost Drivers

In addition to the qualitative long-term cost appraisal the Delphi aims at assessing the relevant cost drivers. The expert appraisal for the years 2070 and 2100 is similar, therefore only the results for 2100 are reported here.

The questionnaire differentiates between five categories of cost drivers: climate, economic, social, technological and policy drivers. Figure 20.3 illustrates the importance of the drivers for the COIN sectors, with positive values pointing to a higher relevance of the driver. As depicted in Fig. 20.3, for most sectors all categories are considered to have important effects on the climate damage costs in 2100. On average over all sectors, climate, policy and technological drivers are valued of comparably lower relevance for the long-run cost of inaction, while

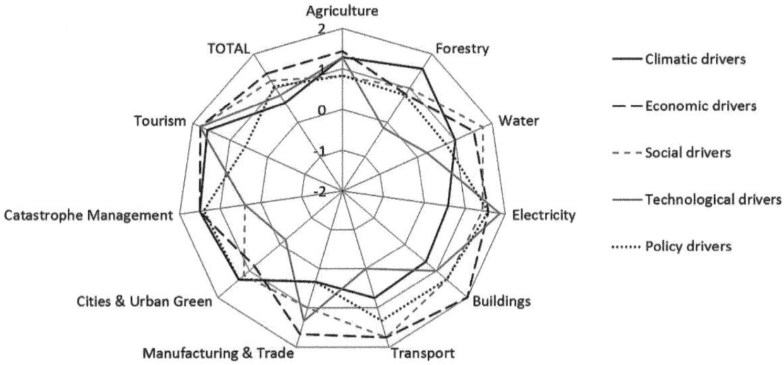

Fig. 20.3 Importance of cost drivers in 2100—sector means. *Source*: Own illustration. *Note*: For the illustration of the results a weighting procedure is applied. For the evaluation of the importance of the different categories of cost drivers, the experts could choose between the categories 'very important' (weighting factor: 2), 'important' (weighting factor: 1), 'less important' (weighting factor: −1) and 'not important' (weighting factor: −2)

economic and social drivers is given higher importance. Some sectors deviate, however, considerably from this common trend: E.g. in the sectors electricity and tourism, technological drivers are rated most important; this refers to the development of renewable electricity technologies, cooling technologies and the electricity grid and respectively to the development of new technologies for snow-making. For forestry and cities and urban green, in contrast, the category climatic drivers is given the highest importance, since damage costs are estimated to increase with the climate signal. In the water sector social drivers are identified as most important as they trigger the demand for water and water supply. In manufacturing and trade, it is economic drivers that determine both the size of this sector and technical equipment to mitigate heat stress for workers and related productivity losses. An overview of the relevant cost drivers by sector is provided in Table 20.5. Across most sectors, the occurrence of extreme weather events is mentioned as a central determinant for climate damage costs in the long run.

For the sectors health and ecosystem services the importance of the different categories of drivers was assessed in open questions. For the health sector (cf. Chap. 11), social and economic factors are considered the key drivers, while the effect of climate damage stimuli on the cost of inaction is expected to be of minor importance. Socio-economic factors will e.g. shape the market penetration of cooling systems or the development of flood-protection, determining the distribution of health impacts along the social stratification of society.

For the sector ecosystem services (cf. Chap. 10), the changing climate signal is considered a very important driver for the long-term cost of inaction. Still, socio-economic drivers like an increasing demand for agricultural products and the resulting pressure on currently non-used arable land are also judged to be important as the amount of non-used arable land in the landscape determines the integrity, ecological niches for certain species and connectivity of ecosystems. The development of the future cost of inaction in the sector ecosystem services is therefore closely interrelated with developments in the agriculture and forestry sector.

Table 20.5 Development of the sectoral cost of inaction in 2100 and key drivers

Sector	Cost development[a]	Important drivers
Agriculture	Even higher costs (4.9)	Economic drivers such as changes in economic conditions (e.g. WTO regulations, CAP reform), technological progress, loss of agricultural land and soil fertility, increasing demand for agricultural products including food, renewable energy, green bio-refinery products, and changes in world markets prices are very important drivers; Social drivers like population growth and consumer behaviour (e.g. changing food preferences) will also determine future costs
Forestry	Higher costs (4.0)	The sectoral cost of inaction (costs for forest protection, pest control and restoring infrastructure) will generally increase with climate change
Ecosystem Services: Pest Control and Pollination	Higher costs	Climate is considered an important driver as it triggers species movement; Economic drivers are important because they will put pressure on non-arable landscape compartments
Human Health	Even higher costs	Climate drivers are considered very important since heat waves, flood extremes and climate-induced mass movements will increase damage; furthermore, heat-induced mortality, infectious diseases distributed by yet non-endemic species and allergic reactions due to extension of non-endemic allergen species may increase
Water Supply and Sanitation	Even higher costs (4.8)	Social drivers are most important as they trigger the demand for water and water supply, but water demand will be driven by climate change as well (during heat waves and for irrigation during droughts). A stronger climate signal furthermore adds to a deterioration of the resource situation, infrastructure maintenance and restoration costs as well as the demand situation
Electricity	Even higher costs (4.7)	Technological, social and economic drivers are very important as they determine the electricity demand for air-conditioning as well as the development of the infrastructure and the electricity grid; Policy stimuli are important as they drive long-term investments in the electricity sector
Buildings: Heating and Cooling	Even higher costs (4.7)	Economic and technological drivers are very important as they determine the energy demand for buildings; Cooling costs will strongly increase; maintenance costs can increase due to more

		frequent extreme events that damage the building stock; Reduction in heating energy costs will decline due to the more and more efficient building stock with lower energy demand for heating
Transport	Even higher costs (4.6)	Economic drivers are very important as an extension of the road network will result in increasing maintenance costs; Policy is also an important driver as it determines how settlement and mobility infrastructure will develop
Manufacturing and Trade: Labour Productivity Losses	Even higher costs (4.7)	Economic and social drivers are most important, as being able to reduce risk and thus costs; Outdoor work is much more affected than indoor work; therefore, the share of the construction sector is relevant; Older workforce has a reduced ability to acclimatise to new conditions like a hotter work environment
Cities and Urban Green	Even higher costs (4.7)	Economic drivers are important since the costs for expanding green space in urban areas will increase due to higher lot prices and higher investment costs; Policy is also an important driver as it controls spatial planning and building standards
Catastrophe Management: Riverine Flooding	Even higher costs (4.2)	Climate is an important driver as it likely will change weather variability and extreme events; Economic development is important too since this will also determine the value of buildings and construction costs; migration patterns will trigger settlements in vulnerable regions; Policy—i.e. spatial planning—will also be highly important
Tourism	Higher costs (4.0)	Climate is important as less snow affects the winter tourism. Costs will rise because skiing slopes and mountain bike routes need large investments

[a]Numbers in brackets refer to the weighted evaluations of cost developments in 2100 as presented in Fig. 20.1

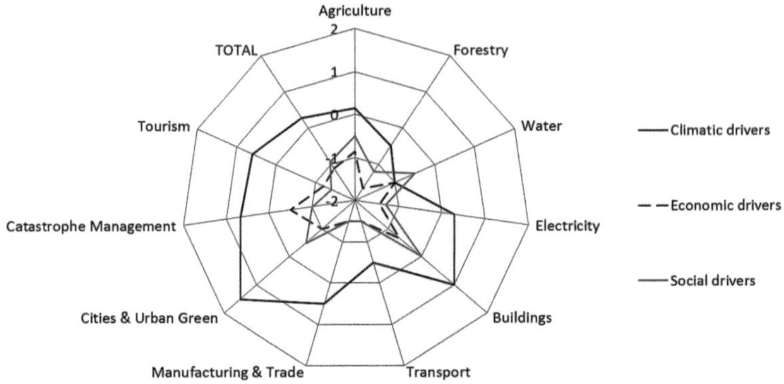

Fig. 20.4 Certainty of cost drivers 2100. *Source*: Own illustration. *Note*: For the illustration of the results a weighting procedure is applied. For the evaluation of the certainty of cost drivers in 2100, the experts could choose between the categories 'very certain' (weighting factor: 2), 'certain' (weighting factor: 1), 'uncertain' (weighting factor: −1) and 'very uncertain' (weighting factor: −2)

Figure 20.4 illustrates how confident the sector experts are with respect to the certainty of the different cost drivers in 2100; positive values in the diagram indicate a higher degree of certainty. All cost drivers are rated as rather uncertain across all sectors, with uncertainty slightly increasing towards the end of the twenty-first century. The uncertainty regarding climate drivers is, however, considered to be lower compared to the uncertainty related to social and especially economic drivers. Yet, not all aspects of climate change are considered equally certain: While temperature development is judged rather certain in the second half of the century, uncertainties prevail regarding the development of precipitation sums and distributions. Especially high uncertainty exists regarding the occurrence of extreme events caused by climate change. Economic and social drivers are closely interrelated and subject to high uncertainty even in the medium run. Until the end of the twenty-first century, socio-economic systems are likely to look completely different than today. The uncertainty related to these drivers is hence even more pronounced than those relating to climate change.

20.4.3 Barriers for Adaptation to Climate Change

The qualitative appraisal for the second half of the twenty-first century points at significant costs of inaction in most sectors. In order to avoid these costs it is essential to overcome the barriers for adaptation. This section describes the experts' valuation of different categories of barriers in the 12 COIN sectors. We distinguish between four different categories: institutional barriers, informational barriers (including cognitive barriers and uncertainty), financial barriers and social barriers. Institutional barriers refer to policies and administrative barriers such as potential conflicts between regulations, the short-term perspective of policy makers or

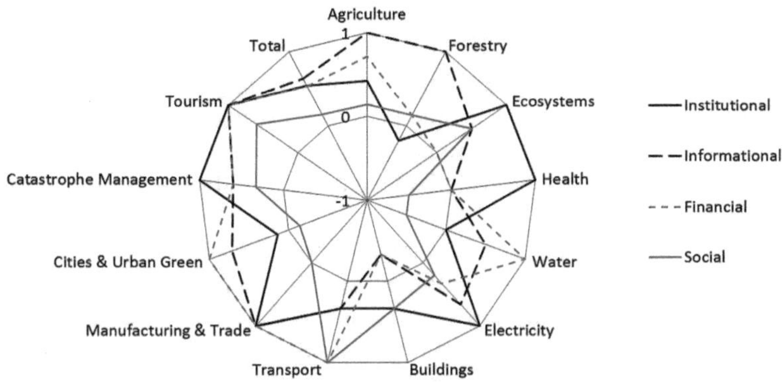

Fig. 20.5 Importance of adaptation barriers—sector means. *Source*: Own illustration. *Note*: For the illustration of the results a weighting procedure is applied. For the evaluation of the importance of the different barriers, the experts could choose between the categories 'important' (weighting factor: 1) and 'not important' (weighting factor: −1)

obstacles to changes of current administrative arrangements. Informational, cognitive and uncertainty barriers include limited knowledge of climate change and its possible impacts, the lack of relevant data as well as uncertainty about climate change that hamper action. Financial barriers comprise constraints of public budgets as well as of private ones. Social barriers refer to limited societal acceptance of necessary adaptation measures, e.g. resettlement, and other behavioural barriers like rigidity in consumption preferences.

The different categories of barriers for the COIN sectors and their importance are depicted in Fig. 20.5. In the figure, positive values express that a barrier is relevant for a certain sector; a value of +1 indicates that a barrier is valued important by 100 % of the sector experts. For the sectors ecosystem services, health, electricity, manufacturing and trade, catastrophe management and tourism, institutional barriers are considered an important factor by all respondents. Informational barriers are rated particularly important in the sectors agriculture, forestry, manufacturing and trade and tourism and are diverse: Stated examples range from a lack in relevant data that would be necessary to substantiate decisions (e.g. in the sectors ecosystem services, water and transport) over the uncertainty about climate change and extreme events to the volatility of the market in the agriculture sector. Financial barriers are regarded as an important obstacle to adaptation in the sectors water, transport, manufacturing and trade, cities and urban green and tourism where large investments have to be triggered for the adaptation to climate change. Social barriers are considered most important in the sector transport. This refers to the fact that a resettlement from certain Alpine valleys, where the cost of road maintenance will significantly increase with climate change, might not be seen as a socially acceptable adaptation measure.

Experts' evaluations of the probability to overcome the adaptation barriers are illustrated in Fig. 20.5. Here, positive values indicate that it is likely that the sectoral barriers will be overcome while negative values express that the likelihood of overcoming the barriers is low. With respect to the persistency of the barriers, on average, informational barriers are considered to be overcome most likely. In the sectors health and water the probability of overcoming informational barriers is judged particularly high. In the sectors ecosystem services and buildings where a lack of awareness is considered an important barrier to adaptation, in contrast, 75 % and 50 % of the experts respectively believe that it is unlikely that the barriers will be overcome.

With respect to social barriers, the Delphi survey delivers mixed evidence across sectors: In the sector water the likelihood for overcoming social barriers is rated comparably high. This might, however, be due to the fact that for the water sector social barriers are not perceived as an important issue (see Fig. 20.5 above). For the sectors ecosystem services, transport and tourism the opposite is true, i.e. most experts perceive social barriers like limited social acceptance of resettlement or rather fixed consumption preferences as an important issue but think that they are likely to persist.

The Delphi results for institutional barriers also present a rather pessimistic view across all sectors concerning the likelihood for overcoming them; for the agriculture sector e.g. 60 % of the respondents believe that it is (very) unlikely that these barriers can be overcome in the future. Reported institutional barriers include lacking cooperation between institutions in the health sector, the large number of decision makers involved (electricity; buildings) and policymakers' lack of interest in the subject (ecosystem services; cities and urban green).

Mixed results can be observed with respect to financial barriers: In the sectors manufacturing, cities and urban green and tourism they are considered an important category of barriers that is unlikely to be overcome. In the water and transport sector, the importance of financial barriers is also rated high but the barriers are expected to be less persistent (Fig. 20.6).

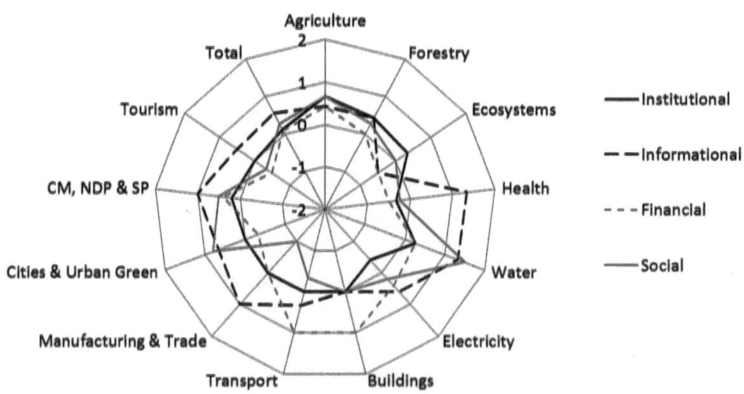

Fig. 20.6 Likelihood of the barriers to be overcome—sector means. *Source*: Own illustration. *Note*: For the evaluation of the likelihood that the sectoral barriers to adaptation can be overcome the experts could choose between 'very likely' (weighting factor: 2), 'likely' (weighting factor: 1), 'unlikely' (weighting factor: −1) and 'very unlikely' (weighting factor: −2)

Table 20.6 Categories of barriers for adaptation to climate change in the COIN sectors

Sector	Barriers
Agriculture	Regulation at European and national level as well as subsidies are decisive factors influencing adaptation; Comprehensive information systems should be introduced and advertised to overcome information barriers. Increasing efforts in the agricultural educational system may increase farm management and agronomic skills which are relevant to increase the adaptive capacity; Financial constraints may be overcome by public efforts; Social barriers [e.g. concerning genetically modified organisms (GMOs)] are not flexible
Forestry	The ownership structure is considered highly relevant: In contrast to large forest owners, small forest owners are often not endowed with sufficient capital and information
Ecosystem Services: Pest Control and Pollination	The limited funds for non-arable land represent one barrier to adaptation in the sector; Social acceptance of protected land areas is, however, high—even against economic interests
Human Health	Institutional barriers are most relevant as health service reforms focus on treatments. Climate change primarily requires using the potential of improved primary health care, prevention and health promotion to reduce vulnerability of people; Informational barriers—i.e. a lack of information on how to cope with high temperature—are also relevant in the context of health issues. Specifically, vulnerable groups are not reached by general information measures. They need tailor-made approaches via trusted health professionals
Water Supply and Sanitation	Financial barriers are most relevant for the water sector. As water issues—especially water supply service—are very important for mankind, the existing (mainly) economic barriers likely will be overcome
Electricity	A lack of awareness combined with little focus on long-term aspects are the main barriers for adaptation. Moreover, some impacts (e.g. shift of hydro power from summer to winter, lower energy demand for heating) are associated with benefits that reduce the focus on those areas where adaptation would be required
Buildings: Heating and Cooling	The main barriers are at builder or installer level as well as at the level of end-users or investors who are not aware of efficiency measures to reduce cooling loads; Institutional barriers such as structural issues are also an important barrier

(continued)

Table 20.6 (continued)

Sector	Barriers
Transport	Social barriers are very relevant for the sector as social acceptance for resettlement or the user pays principle is missing. Multi-level governance is also considered as an obstacle to adaptation in the transport sector
Manufacturing and Trade: Labour Productivity Losses	Financial barriers are very important as especially small companies lack financial resources for adaptation measures
Cities and Urban Green	Financial constraints are relevant as lot prices in cities are very high and especially re-greening of sealed surface and planting additional street trees is cost-intensive
Catastrophe management: Riverine Flooding	Property rights on land hamper adaptation, as federal states may not receive command over land that would be necessary to increase flood safety; Institutional deficiencies like the crowding out of private efforts prevail in the Austrian system of catastrophe management
Tourism	Financial barriers including increased costs for snow making and low own capital of companies are highly relevant

Not only the relevance of barriers, but also lead times for adaptation are expected to vary significantly between sectors. Experts for the sectors agriculture, manufacturing and trade and health estimate that it will take only up to 10 years until adaptation measures will become effective. With respect to agriculture, for instance, adaptation measures for annual crops can be implemented every year in the course of general cropping decisions. For the sectors buildings, electricity, water, cities and urban green, catastrophe management and forestry, in contrast, partly long lead times for adaptation are expected. For forestry, this is owed to the long rotation periods of crops; for the other sectors the lead times reflect the long life cycles of built structures which prevent rapid adaptation (Table 20.6).

20.5 Conclusions

Accelerating climate change poses a major challenge to our society. It is, however, difficult to assess how climate change will impact our socio-economic systems: Climate change is a complex phenomenon and the socio-economic developments are particularly in the long term subject to high uncertainty. Against this background a model evaluation of the long-term impacts faces substantial constraints. For this reason, in the COIN project we have chosen a qualitative approach for the appraisal of the long-term costs of inaction in Austria, based on a Delphi survey.

In general, the COIN Delphi results point at pronounced increases in the cost of inaction in the second half of the twenty-first century. For six out of the ten sectors analysed in the COIN project (agriculture, water, buildings, transport, manufacturing and trade, cities and urban green), there is unanimous consensus among experts that climate damage costs in 2070 will be higher than in 2050. A further increase in costs after 2070 is expected for the majority of sectors. The highest cost increases compared to 2050 are expected originating from the agriculture sector, i.e. the net benefits arising from agricultural plant productivity until 2050 are in general considered to 'turn into net costs until 2070 and—even sharper—2100' due to a stronger climate signal.

Economic and social developments are considered the most important cost drivers in the long run for most COIN sectors. Experts rate the influence of these cost drivers particularly high, despite uncertainty of future social, economic and thus cost development. Extreme events might be key determinants of the long-term cost of inaction.

The COIN Delphi delivers first results on the long-term sectoral climate damage costs in Austria. The survey is limited to the sector experts of the project consortium. While this approach ensures the consistency between the qualitative long-term appraisal and the quantitative estimates for 2050, the small sample size for the individual sectors limits the use of advanced statistical methods in the analysis of the Delphi results. For future research it would therefore be desirable to conduct a long-term appraisal of climate impacts with a larger number of experts. In this context, a detailed assessment of (subjective) uncertainty of future cost development would be worthwhile.

The results of the COIN Delphi confirm that the economic costs of climate change very likely will show an increasing trend if policy refrains from active climate policy. Such a policy needs to address emission reduction as well as adaptation to non-avoidable climate change. Despite the importance of adaptation to non-avoidable climate change, the implementation of low carbon structures of our economies and societies determines whether climate change is likely to be manageable in the long run.

References

Ciscar JC, Szabó L, Regemorter D, Soria A (2012) The integration of PESETA sectoral economic impacts into the GEM-E3 Europe model: methodology and results. Clim Change 112(1): 127–142

Ciscar JS (ed) (2009) Climate change impacts in Europe. Final report of the PESETA research project. JRS – IPTS, Luxembourg

Fankhauser S (1995) Valuing climate change. The economics of the Greenhouse. Earthscan, London

Häder M (2009) Delphi-Befragungen. Ein Arbeitsbuch. 2. Auflage. VS Verlag für Sozialwissenschaften, Wiesbaden

IPCC (1996) Climate change 1995: economic and social dimensions of climate change. Cambridge University Press, Cambridge

IPCC (2013) Climate change 2013: the physical science basis. IPCC, Stockholm.

Maddison D (2003) The amenity value of climate: the household production function approach. CSERGE Working paper GEC 98–08

Nordhaus W (1994a) Managing the global commons: the economics of climate change. MIT Press, Cambridge

Nordhaus W (1994b) Expert opinion on climate change. Am Sci 82:45–51

Nordhaus W (2006) Geography and macroeconomics: new data and new findings. Proc Natl Acad Sci USA 113(10):3510–3517

Nordhaus W, Boyer J (2000) Roll the DICE again. MIT Press, Cambridge

Nordhaus W, Yang Z (1996) A regional dynamic general-equilibrium model of alternative climate-change strategies. Am Econ Rev 86(4):741–765

Okoli C, Pawlowski S (2004) The Delphi method as a research tool: an example, design considerations and applications. Inf Manage 42(1):15–29

Pindyck RS (2013) Climate change policy: what do the models tell us? J Econ Lit 51(3):860–872

Rehdanz K, Maddison D (2005) Climate and happiness. Environ Resour Econ 52(1):111–125

Schulz M, Renn O (2009) Das Gruppendelphi: Konzept und Vorgehensweise. In: Schulz M, Renn O (eds) Das Gruppendelphi. Konzept und Fragebogenkonstruktion. VS Verlag für Sozialwissenschaften, Wiesbaden, pp 11–21

Stanton EA, Ackerman F, Kartha S (2009) Inside the integrated assessment models: four issues in climate economics. Clim Dev 1(2):166–184

Stern N (2006) The economics of climate change: the stern review. Cambridge University Press, Cambridge

Tol RSJ (1995) The damage costs of climate change toward more comprehensive calculations. Environ Resour Econ 5(4):353–374

Tol RSJ (2002) Estimates of the damage costs of climate change part I benchmark estimates. Environ Resour Econ 21(1):47–73

Tol RSJ (2010) The economic impact of climate change. Perspektiven der Wirtschaftspolitik 11(s1):13–37

Tol RSJ (2011) The social cost of carbon. ESRI Working paper no. 377, March 2011
Tol RSJ (2013) The economic impact of climate change in the 20th and 21st centuries. Clim Change 117(4):795–808
van den Bergh JCJM (2004) Optimal climate policy is a utopia: from quantitative to qualitative cost-benefit analysis. Ecol Econ 48(4):385–393
van Vuuren DP, Lowe J, Stehfest E, Gohar L, Hof AF, Hope C, Warren R, Meinshausen M, Plattner GK (2009) How well do integrated assessment models simulate climate change? Clim Change 104(2):255–285
Watkiss P (2011a) Aggregate economic measures of climate change damages: explaining the differences and implications. WIREs Clim Change 2(3):356–372
Watkiss P (ed) (2011b) The climate cost project, vol 1. Final report. Environment Institute Stockholm, Europe
Yang X-J, Zeng L, Zhang R (2012) Cloud Delphi method. Int J Uncert Fuzziness Knowl Based Syst 20(1):77–97

Chapter 21
Macroeconomic Evaluation of Climate Change in Austria: A Comparison Across Impact Fields and Total Effects

Gabriel Bachner, Birgit Bednar-Friedl, Stefan Nabernegg, and Karl W. Steininger

Abstract This chapter evaluates the aggregate macroeconomic effects of the quantifiable impact chains in ten impact fields for Austria: Agriculture, Forestry, Water Supply and Sanitation, Buildings (with a focus on heating and cooling), Electricity, Transport, Manufacturing and Trade, Cities and Urban Green, Catastrophe Management, and Tourism. First, the costing methodology used for each impact chain as well as the respective interface to implement them within the macroeconomic model are reviewed and compared across impact fields. The main finding here is that gaps in costing are mostly the consequence of insufficient data and for that reason, the two important impact fields Ecosystem Services and Human Health could not be assessed in monetary terms. Second, for the subset of impact chains which could be monetised, a computable general equilibrium (CGE) model is then used to assess the macroeconomic effects caused by these. By comparing macroeconomic effects across impact fields, we find that the strongest macroeconomic impacts are triggered by climate change effects arising in Agriculture, Forestry, Tourism, Electricity, and Buildings. The total macroeconomic effect of all impact chains—which could be quantified and monetised—is modest up to the 2050s: both welfare and GDP decline slightly compared to a baseline development without climate change. This is mainly due to (a) all but two impact chains refer to trends only (just riverine flooding damage to buildings and road infrastructure damages cover extreme events), (b) impacts are mostly redistribution of demand, while stock changes occurring as a consequence of extreme events are basically not covered and (c) some of the precipitation-triggered impacts point in opposite directions across sub-national regions, leading to a comparatively small net effect on the national scale.

G. Bachner (✉) • S. Nabernegg
Wegener Center for Climate and Global Change, University of Graz, Graz, Austria
e-mail: gabriel.bachner@uni-graz.at; stefan.nabernegg@uni-graz.at

B. Bednar-Friedl • K.W. Steininger
Wegener Center for Climate and Global Change, University of Graz, Graz, Austria

Department of Economics, University of Graz, Graz, Austria
e-mail: birgit.friedl@uni-graz.at; karl.steininger@uni-graz.at

K.W. Steininger et al. (eds.), *Economic Evaluation of Climate Change Impacts*, Springer Climate, DOI 10.1007/978-3-319-12457-5_21

21.1 Introduction

Having assessed the macroeconomic effects of climate change impacts by impact field in Chaps. 8–19, this chapter looks into the aggregate effect when all impact chains which can be assessed in terms of costs are considered jointly.

Regarding climate change impacts, Sect. 21.2 provides an overview of the so called "impact chains" considered by impact field. Moreover, there are differences both in terms of available data and modelling approach across impact fields. Section 21.2 makes these differences transparent by assessing the quality of methodology and data by impact field.

The costs of climate change for the period 2016–2045 are defined as the difference between the average annual effect in the climate change scenario (mid-range climate change and reference socioeconomic development) for the period 2016–2045 and a baseline scenario (reference socioeconomic development without climate change) for the same period. Likewise, the costs of climate change for the period 2036–2065 (a more distant future) are the average annual differences between the climate change scenario and the baseline scenario for the period 2036–2065. In Sect. 21.3, we briefly describe the underlying development of the baseline scenario in our comparative static approach (i.e. relative to the CGE model's base year 2008).

Regarding macroeconomic results with respect to the costs of climate change, the aim of this chapter is twofold: in Sect. 21.4, we draw comparisons across impact fields; in Sect. 21.5, we are interested in the overall effect of all quantified climate impact chains in Austria up to 2030 and 2050 as well as in the sectoral distribution of this effect. For both types of comparisons, it needs to be acknowledged that the number of impact chains considered is limited and that there is a substantial difference as to how broad or narrow the coverage is across impact chains and impact fields. Finally, Sect. 21.6 discusses our key findings by comparing them to results found in the literature.

21.2 Sectoral Costing Methods by Impact Field

Table 21.1 provides an overview of the impact chains by impact field and characterises the applied sectoral costing method. For some impact fields like Electricity, detailed sectoral models for Austria were used to assess the impacts of climate change, both in physical units (change in yield) and in economic units (change in profit margins, in costs, in investments, in demand). For the impact field Agriculture, land use and livestock scenarios from a sectoral model were applied. In other impact fields like Tourism or Transport, regression analyses were conducted based on Austrian data (e.g. based on overnight stays, road infrastructure damages) to derive an impact function which was then used to estimate future costs and benefits. Finally, for some fields (e.g. Catastrophe Management, Human Health,

Table 21.1 Summary of sectoral impact chains, their costing and implementation in the CGE model

Impact field	Impact chains: change in.../effect on...	Sectoral costing method	Performance of sectoral costing			Importance for CC impact costs
			Method	Impact data	Implementation	
Agriculture (agr)	Crop productivity of main crops (grain maize, winter wheat, winter rape, soybean, temporary grassland) due to changes in temperature and precipitation	Regression analysis to obtain effects on biophysical units, 1 km resolution	Good	Good	Good	Medium
	Grassland productivity of permanent grassland due to changes in temperature and precipitation	Regression analysis based on biophysical process model, 1 km resolution	Good	Good	Good	Medium
	Orchards, vineyards, vegetables production due to changes in temperature and precipitation	Not quantified			No	Low
	Livestock production due to changes in temperature and precipitation	Direct impacts not quantified, but additional capacity from increasing grassland yields considered for dairy cow production		Fair	No	Medium
	Infestation pressure of pests, diseases, and weeds	Not quantified			No	Medium
	(Sub-daily) heavy precipitation and hail events	Not quantified			No	High
Forestry (for)	Biomass productivity in commercial production forests due to changed precipitation and temperature	Expert guess	Fair	Good	Good	High
	Impact of bark beetle disturbances on productivity of commercial forests	Expert guess	Fair	Fair	Fair	High
	Impact of bark beetle disturbances to protection forests	Expert guess	Fair	Poor	Fair	High

(continued)

Table 21.1 (continued)

Impact field	Impact chains: change in.../effect on...	Sectoral costing method	Performance of sectoral costing			Importance for CC impact costs
			Method	Impact data	Implementation	
Ecosystem Services: Pest Control and Pollination (ess)	Decrease of soil moisture	Not monetised	Fair	Poor	No	High
	Reduced snow cover	Not monetised	Fair	Poor	No	High
	Prolonged draft periods	Not monetised	Fair	Poor	No	High
	Distribution changes	Not monetised	Fair	Poor	No	High
	Phenological changes	Not monetised	Fair	Poor	No	High
	Emergence of multi-voltinism	Not monetised	Fair	Poor	No	High
Human Health (hea)	Mortality for persons aged 65 and older due to heat waves and continuous temperature increases	Assessment of additional deaths + reduced DALYs in systems dynamic model, impacts not monetised	Fair	Fair	No	High
	Morbidity for persons aged 65 and older due to heat waves and continuous temperature increases	Not quantified		Poor	No	High
	Mortality and morbidity for persons aged 0–64	Not quantified		Poor	No	Medium
	Climate-sensitive communicable diseases: Food-borne diseases—Salmonellosis	Not quantified		Poor	No	Medium
	Health effects due to other extreme weather events like floods, hail, storms etc.	Not quantified		Poor	No	Medium
	Water-borne diseases	Not quantified		Poor	No	Low
	Vector-borne and rodent-borne diseases	Not quantified		Fair	No	Medium
	Climate sensitive non-communicable diseases: Allergies – caused by ragweed	monetised (treatment costs), but not included in macroeconomic assessment	Good	Fair	No	Medium

Water Supply and Sanitation (wat)	Water supply: Lower yield of springs and drying cracks in the soil	Expert guess based on extrapolation of estimated or existing data together with anticipated changes	Fair	Good	Fair	Medium
	Water supply: Lower groundwater/ surface water recharge	Expert guess based on extrapolation of estimated or existing data together with anticipated changes	Fair	Good	Fair	Medium
	Water supply: Less precipitation in the vegetation period	Not quantified		Poor	No	High
	Water supply: Turbidity of spring water	Expert guess based on extrapolation of estimated or existing data together with anticipated changes	Fair	Good	Fair	Medium
	Water supply: Increase of land slide, mudflow or flood events	Expert guess based on extrapolation of estimated or existing data together with anticipated changes	Poor	Fair	No	Medium
	Water supply: Change of resource quality	Not quantified		Poor	No	Low
	Water supply: Change of withdrawal	Expert guess based on extrapolation of estimated or existing data together with anticipated changes	Poor	Fair	No	Low
	Water supply: Drought duration in the summer	Not quantified		Poor	No	Medium
	Sanitation: Increase of wastewater volume	Expert guess based on extrapolation of estimated or existing data together with anticipated changes	Fair	Good	Fair	Low
	Sanitation: lower surface water recharge	Not quantified		Poor	No	Medium
	Sanitation: Increase of sewer flooding	Expert guess based on extrapolation of estimated or existing data together with anticipated changes	Poor	Fair	No	Medium

(continued)

Table 21.1 (continued)

Impact field	Impact chains: change in.../effect on...	Sectoral costing method	Performance of sectoral costing			Importance for CC impact costs
			Method	Impact data	Implementation	
	Sanitation: Increase of land slide, mudflow or flood events	Expert guess based on extrapolation of estimated or existing data together with anticipated changes	Poor	Fair	No	Medium
	Sanitation: Lower oxygen solubility	Not quantified		Poor	No	Medium
	Sanitation: Receiving waters temperature increase	Not quantified		Poor	No	High
	Sanitation: Wastewater temperature increase	Not quantified		Poor	No	Medium
	Sanitation: Sewer sedimentation during dry weather	Expert guess based on extrapolation of estimated or existing data together with anticipated changes	Fair	Good	Fair	Medium
Buildings: Heating and Cooling (h&c)	Reduced heating energy demand due to higher temperature in winter	Energy sector model, housing model	Good	Good	Good	High
	Higher cooling energy demand and stronger growth of air conditioning in buildings due to higher temperature in summer	Energy sector model, housing model	Good	Fair	Good	High
	Higher temperature levels in buildings and lower comfort of occupants due to higher temperature in summer	Not quantified		Poor	No	High
	Increased frequency of damage to roofs and tiles due to storm events	Not quantified		Poor	No	Medium
Electricity (ele)	Change in river run-off and water temperature in summer leading to lower availability of cooling water	Not quantified		Poor	No	Low

	for thermal and nuclear power plants, change in electricity generation mix, and/or reduction in reliability of the electricity system					
	Change in river run-off levels leading to change in overall annual hydropower generation, change in seasonal hydropower generation profile, and change in electricity generation mix	Electricity sector model (dispatch)	Good	Good	Good	High
	Change in wind and photovoltaic power generation thus change in electricity generation mix	Electricity sector model (dispatch)	Good	Poor	Fair	Low
	Increased cooling energy demand in summer leading to increased electricity demand and summer electricity peak load	Electricity sector model (dispatch)	Good	Fair	Good	High
	Decreased electricity demand and winter electricity peak load	Electricity sector model (dispatch)	Good	Good	Good	High
	Change in supply and demand profiles and change in reliability of electricity supply and change in probability of blackouts	Not quantified	–	poor	no	high
	Natural hazards leading to electricity infrastructure at risk (supply and transmission), change in reliability of electricity supply and change in probability of blackouts	Not quantified	–	Poor	No	High

(continued)

Table 21.1 (continued)

Impact field	Impact chains: change in.../effect on...	Sectoral costing method	Performance of sectoral costing			Importance for CC impact costs
			Method	Impact data	Implementation	
Transport (trn)	Road damages due to increase in floods, landslides and mudflows and decrease in snowfall and freezing precipitation	Regression analysis on past damage events and costs	Good	Fair	Good	Medium
	Rail line damages due to increase in floods, landslides and mudflows and decrease in snowfall and freezing precipitation	Not quantified		Fair	No	Medium
	Transport service interruption due to increase in floods, landslides and mudflows and increase of susceptibility to wildfires	Not quantified		Poor	No	Low
	Reduction of depth of inland waterways	Not quantified		Poor	No	Low
	Storm damage to rail lines and roads and transport service interruption due to increase of debris and broken trees	Not quantified		Poor	No	Medium
	Deformation of roads, railway tracks and airport infrastructure due to high temperatures	Not quantified		Poor	No	Low
	Passenger discomfort and higher energy consumption due to high temperature	Not quantified		Poor	No	Low
	Longer ice-free shipping season due to higher winter temperature	Not quantified		Poor	No	Low
	Reduced snowfall and ice conditions due to milder winters	Not quantified		Poor	No	Medium

Manufacturing and Trade: Labour Productivity Losses (m&t)	Productivity losses of workers	Use of impact function from international study	Fair	Poor	Fair	Low
	Prolongation of the working season	Not quantified		Poor	No	Medium
	Keeping the process environment stable	Not quantified		Poor	No	Low
	Cooling and heating of office buildings and plants	Not quantified, but included in Buildings		Poor	No	Low
	Delivery problems in the supply chain	Not quantified		Poor	No	Low
	Damages to infrastructure (office buildings, plants)	Not quantified		Poor	No	Medium
	Changes in private consumption	Not quantified, but partly included in Buildings and Catastrophe Management		Poor	No	Low
Cities and Urban Green (cug)	Loss of climate comfort for tourists in urban environments	Not quantified		Poor	No	Medium
	Heat related damage on pavements, tram rails	Not quantified		Poor	No	Low
	Improved prevention against loss of climate comfort in urban environments—investments in and maintenance of additional parks, additional tree planting	Top down estimation: cost factors related to green open space expansion and tree number increase to compensate increased built up area and accelerated urban heat exposure	Fair	Fair	No	Medium
	Other loss of climate comfort in urban environments	Not quantified		Poor	No	Medium
Catastrophe Management: Riverine Flooding (cam)	Average damage of riverine floods	Bottom up models for a spatial entity flood damages for floods with different return periods are	Good	Fair	Good	High

(continued)

Table 21.1 (continued)

Impact field	Impact chains: change in.../effect on...	Sectoral costing method	Performance of sectoral costing			Importance for CC impact costs
			Method	Impact data	Implementation	
		simulated and afterwards aggregated.				
	Change in workload of disaster relief forces	Not quantified		Poor	No	Medium
	Change in workload of volunteers	Not quantified		Poor	No	Medium
	Change in damaged buildings for storm events	Not quantified		Poor	No	High
Tourism (tsm)	Winter tourism demand	Regression analysis for overnight stays at NUTS3 level; costing based on regression and tourism satellite accounts	Good	Good	Good	High
	Summer tourism demand	Regression analysis for overnight stays at NUTS3 level; costing based on regression and tourism satellite accounts	Good	Good	Good	Medium
	Tourism sector's water & energy demand	Not quantified		Poor	No	Medium
	Business interruptions due to natural catastrophes	Not quantified		Poor	No	Medium

Cities and Urban Green) impact functions derived in international/European studies were applied to the Austrian data. When none of these approaches were available, expert guesses were used to estimate potential climate change costs (e.g. Forestry).

As can be seen from Table 21.1, only a subset of all identified impact chains could be estimated in terms of physical impacts (see column "sectoral costing method"). Moreover, a few of these quantified impact chains could not be monetised, either because it was unclear which types of cost/benefit would arise or because there was no secondary data available (see Ecosystem Services; Human Health). We decided, therefore, to include only well founded impacts into the macroeconomic assessment from impact chains which are well understood in terms of costs instead of biasing results with results from other impact chains where this is not the case.

The remaining columns of Table 21.1 indicate the quality of the sectoral costing with respect to the method applied, the impact cost data available, and the implementation interface within the macroeconomic assessment (CGE model, see Chap. 7). Whenever the sectoral costing model is based on and has been validated in different applications before, the method is assessed as *good* (column "Performance of sectoral costing: method"). If, instead, impact estimates are transferred from other studies/regions, the method is *fair*, while the method is classified as *poor* when it is solely based on expert judgment. Regarding quality of impact data (see column "Performance of sectoral costing: data"), the scale *good* refers to data which is available for many years with broad spatial coverage, while *fair* is used for data which is available for selected or some years and/or for some regions only. *Poor* data quality is used if data is only available for other countries or at the European level. Finally, implementation in the CGE model (column "Performance of sectoral costing: implementation") is said to be *good* when the derivation of cost estimates is model based (e.g. yield model, electricity dispatch model) and when there is a clear mapping of impacts into cost categories (e.g. production cost categories, demand, land/labour/capital productivity). A *fair* implementation is also based on a mapping of impacts into cost categories but there is some mismatch or ambiguity in this mapping, e.g. because sectoral models and accounts are not well represented in the CGE model. For that reason, impact costs were transferred to the macroeconomic model in terms of relative changes (% changes) instead of directly transferring absolute numbers. Finally, no implementation in the CGE model was undertaken when impacts could not be quantified and monetised.

The last column of Table 21.1 indicates how relevant the omission of impact chains is for the overall assessment of economic costs (i.e. the social costs including external costs). According to expert judgment of the project team which comprised 19 institutions from all relevant disciplines, most impact chains with high relevance for climate change in Austria could be assessed, except for the two impact fields Ecosystem Services and Human Health. Also a considerable amount of climate change impacts of medium importance for the economic costs are included in the assessment, while impacts with low damage potential are in general not part of the assessment. Note that extreme events are captured only poorly, firstly due to the CGE model's characteristics themselves (see Chap. 7) and secondly due to limited data availability. One important characteristic of the CGE model is that autonomous adaptation to price changes is happening, leading to lower costs compared to an assessment with no such adjustment processes. As a consequence, the macroeconomic costs of the impact chains which are assessed in the CGE model should be understood as a lower bound estimate for average annual climate change costs and benefits for Austria.

Moreover, it is important to note that damages are assessed at the national scale. While according to the climate scenarios for 2016–2045 and 2036–2065, changes in temperature point in the same direction across Austrian NUTS3 regions and across climate scenarios, changes in precipitation do not. As a consequence, many precipitation-triggered impacts are cancelled out across Austrian regions and hence the total effect for Austria is smaller than if changes would occur in the same direction across all regions.

In interpreting the results it is also important to consider that the current macroeconomic assessment quantifies average changes in the climatic periods 2016–2045 and 2036–2065 relative to the reference period 1981–2010 (monthly, seasonal and yearly averages) but not an exceptional year such as, e.g. a year in which a once in a century flooding occurs. Thus, macroeconomic effects represent the increase in annual macroeconomic costs averaged over several years. Second, extreme events are only covered for two impact chains (riverine flooding damage to buildings and extreme event triggered road infrastructure damages—and as mentioned in terms of annual average damage), but no other extreme events could be integrated on a sufficiently robust basis.

Regarding the comparison across impact fields, it is of high importance that different sectoral costing methods and models were used to assess the direct costs. It is well understood in the literature that different types of models may lead to different magnitudes of cost estimates, especially when some of them are bottom-up models (optimizing at NUTS3 level or lower) and others are top-down (working at the overall national scale only). However, not only is the class of models significant, but also the availability of suitable data, especially for rare events with high damage potential. While in some impact fields like natural catastrophes sufficient data is available from major flooding events in the past decade, this does not hold for other damages such as black-outs in the electricity sector. Therefore, any sectoral comparison drawn across impact fields in the following section has to be interpreted with extra care.

21.3 Implementation of Baseline and Climate Change Impacts in the CGE Model

21.3.1 Baseline 2030 and 2050 Without Climate Change

The main assumptions regarding the baseline scenario until 2030 and 2050 are the following (for a more detailed description of the shared socioeconomic pathways [SSP] see Chap. 6. For impact field specific baseline assumption see the respective chapter):

- GDP growth: According to the shared socioeconomic pathway (SSP), we assume an annual growth rate of 1.65 % until 2050. Hence, the economy in 2050 is about twice as large as in 2008. All sectors grow at the same rate.[1] Therefore the production cost structures per unit are the same in 2008 and in 2050.
- Production cost: In the electricity sector, a change in the generation mix is assumed towards a higher share of renewables and gas which leads to higher prices for electricity (see Chap. 14 for a more detailed description of how the baseline development was modelled). In accordance with the OECD-FAO agricultural outlook (OECD-FAO 2011), international price projections for 2020 underlie the production cost changes for agricultural products in 2030 and 2050 respectively (see Chap. 8 for details).
- Climate policy: To consider the effect of the European Emissions Trading Scheme in the single country CGE model for Austria, we introduce an exogenously given CO_2 emission permit price of €26.64/t CO_2 in 2030 and €41.04/t CO_2 in 2050 (according to the Current Policy scenario of the World Energy Outlook 2010, IEA 2010). Only those sectors which are covered by the current EU-wide emission trading scheme are affected by climate policy and are confronted with additional production costs for the emission of CO_2. These sectors are: Electricity, gas, steam and air conditioning supply (ELEC), Manufacture of coke and refined petroleum products (COKE), Manufacture of basic metals and fabricated metal products, except machinery and equipment (META), Manufacture of rubber and plastic products and other non-metallic mineral products (PLAS) as well as Manufacture of paper and paper products (PAPE).
- Subsidies and taxation: In sectors Agriculture as well as Water Supply and Sanitation we introduce cuts in subsidies (i.e. higher production taxes) by 2030 and 2050 to reflect the projected stepwise reduction in subsidies by 2020 and beyond. For details, see Chaps. 8 and 12.

[1] Future economic and technological development is subject to high uncertainties. Nevertheless for the construction of the baseline scenario, assumptions concerning economic growth and technological development were necessary. We therefore applied the strong but also cautious assumption of homogenous growth across all economic sectors.

21.3.2 Implementation of Impact Chains Across Impact Fields

Supplementary Material Tables 21.1–21.3 (online) show the parameter values which are used in the CGE model to represent the different impact chains. All values are expressed as the difference between the climate change impact scenario and the baseline scenario (reference socio-economic development, including sector specific policies). As explained in the model description and in the respective sectoral assessments in more detail (see Chaps. 7–19), the parameters represent changes in production costs, productivity, demand, investment, and changed public expenditures.

After implementing all of the quantified impact chains (see Sect. 21.2), results on sectoral output, GDP and welfare are obtained for the two future periods 2030 (representative for the yearly effects in the period 2016–2045) and 2050 (period 2036–2065). For GDP, the effect is decomposed into a real price effect and a quantity effect. The former describes the change of GDP which can be attributed to changes in real prices between the climate change and the baseline scenario, whereas the latter describes the change of GDP which is triggered by altered activity levels (output quantities). The sum of the respective price and quantity effects yields the total effect on GDP. As price changes of consumption goods are not relevant for a change in welfare, the correct measure for welfare is the quantity effect in isolation. In contrast, for GDP and sectoral output, both the values (price and quantity changes) and the contribution of prices and quantities to this total effect will be discussed.

21.4 Macroeconomic Effects of Climate Change: Comparison Across Impact Fields

The aim of this section is to compare the total macroeconomic effects triggered by climate change impacts in the different impact fields (see Table 21.1 for the list of the impact fields). Note that an impact field may subsume different economic sectors (e.g. impact field Tourism subsumes parts of the sectors Accommodation, Travel Agencies, Entertainment, Cultural Activities and Sports). An impact field is therefore not the same as an economic sector.

In this section, we compare across impact fields which direct and indirect effects are triggered by those impact chains that could be quantified and modelled in total in each impact field. Moreover, we analyze how these effects contribute to the total effect when all impact chains across impact fields are active simultaneously.

21.4.1 GDP and Welfare Effects Across Impact Fields and in Total

Figures 21.1 and 21.2 give an overview of the effects on GDP as well as on consumption and welfare which are shown for each of the impact fields' single model runs as well as for the combined model run "all" (see Chaps. 8–19 for detailed explanations of the macroeconomic effects triggered by climate change in each impact field in isolation).[2] All effects are average annual effects when comparing the climate change scenario (mid-range climate change and reference socioeconomic development) to the baseline scenario (reference socioeconomic development) in the two periods under consideration: Year 2030 represents the annual average effect for period 2016–2045 whereas 2050 represents 2036–2065.[3]

Before comparing the effects triggered in different impact fields, it is important to note that in many cases only some of the qualitatively identified climate impact

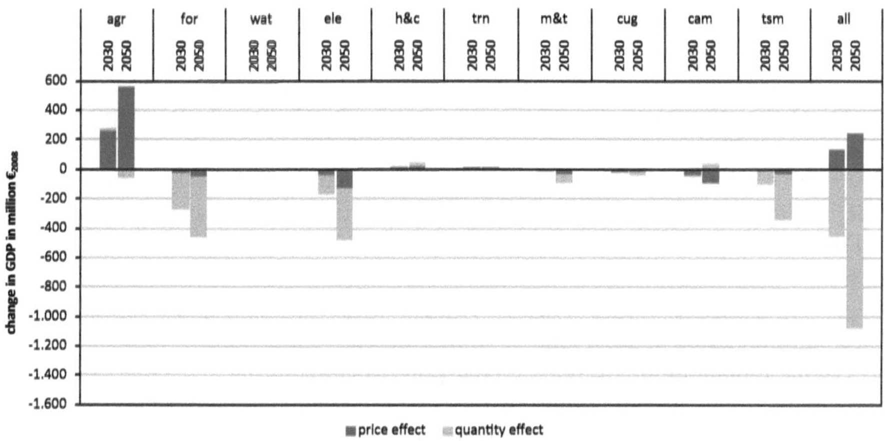

Fig. 21.1 Average annual GDP effects of mid-range climate change (relative to baseline with reference socioeconomic development) by impact field and in total (2030 = period 2016–2045; 2050 = period 2036–2065). *Impact fields*: Agriculture (agr), Forestry (for), Water (wat), Electricity (ele), Buildings: Heating and Cooling (h&c), Transport (trn), Manufacturing and Trade (m&t), Cities and Urban Green (cug), Catastrophe Management (cam), Tourism (tsm), All impact fields (all). Note: For description of quantified impact chains by impact field, see Table 21.1

[2] A sensitivity analysis was performed regarding the interaction of different impact fields. When running the model for each impact field separately and summing up the effects on GDP, we obtained a very similar result as in the combined model run. Therefore the decomposition by impact field can be carried out by taking the shares of the separate model runs.

[3] Climate impact enhancing and climate impact diminishing socioeconomic development were defined differently for each impact field and also low range and high range climatic change were used only for the key climate parameter in each impact field. As a consequence, a joint macroeconomic analysis of these various specifications across impact fields is not possible.

chains have been quantified and therefore the comparability between impact fields is limited (see Sect. 21.3 above for details).

In general the effects in 2050 are stronger than in 2030 and in most of the cases the quantity effect dominates the results. Comparing results across impact fields, there are only two impact fields for which climate change triggers positive effects on GDP. Those are Agriculture (*agr*) with relatively large positive macroeconomic effects (due to higher productivity) as well as Buildings: Heating and Cooling (*h&c*) with much smaller positive effects. Regarding the impact field Agriculture (*agr*), the positive effect of about +280 million euros in 2030 and +500 million euros in 2050 is due to increased productivity which increases value added (and thus the contribution to GDP) but the effect on GDP originating from quantity effects only is either slightly positive (in 2030) or negative (in 2050). This positive macroeconomic effect via higher prices is mainly due to productivity gains in the agricultural sector which implies that households have lower expenditures on food and thus are able to expand their consumption for other goods and services which therefore become more expensive. As a consequence, the value of overall output increases due to higher agricultural productivity. Note, however, that many impacts with eventually negative consequences for the impact field Agriculture have not been quantified (see Chap. 8 for details). The effect in *h&c* (+20 million euros in 2030 and +40 million euros in 2050) is mostly attributable to quantity effects.

The strongest negative GDP effects are caused by climate change impacts in the impact fields Electricity (*ele*), Forestry (*for*) and Tourism (*tsm*). In each of the three cases, price and quantity effects are both negative, but the price effect plays a minor role. For impacts in *ele* the effect on GDP is −170 million euros in 2030 and with −470 million euros much stronger in 2050. The effect on GDP of climate impacts in Forestry is about −270 million euros in 2030 and −460 million euros in 2050. Regarding climate change impacts in Tourism, the effect on average annual GDP is −100 million euros in 2030 and −340 million euros in 2050.

For the remaining fields Water, Transport, Manufacturing and Trade and Cities and Urban Green, it is important to stress that the comparatively small macroeconomic effects are due to the incomplete coverage of impact chains in these fields because direct costs of many relevant impacts are not available. So the low numbers reflect the uncertainty involved, and not that impacts triggered in these fields might not lead to significant macroeconomic effects as well. For more details see Sect. 21.6.

When combining all of the quantified impact chains in one model run (see the bars labelled *all* in Fig. 21.1), the quantity effect on GDP is negative, but it is compensated for partly by positive price effects, which are mainly attributable to the impacts chains of Agriculture. The effect resulting from the combination of all impact fields is a lower GDP by 330 million euros (−0.08 %) per year in 2030 whereas it is 830 million euros lower (−0.15 %) in 2050.

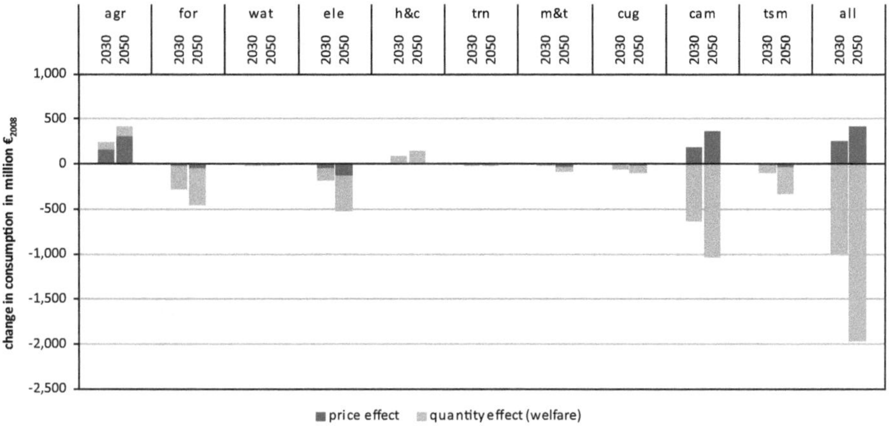

Fig. 21.2 Average annual consumption effects (price + quantity effect) and welfare effects (quantity effect) of mid-range climate change (relative to baseline with reference socioeconomic development) by impact field and in total (2030 = period 2016–2045; 2050 = period 2036–2065) *Impact fields*: Agriculture (agr), Forestry (for), Water (wat), Electricity (ele), Buildings: Heating and Cooling (h&c), Transport (trn), Manufacturing and Trade (m&t), Cities and Urban Green (cug), Catastrophe Management (cam),Tourism (tsm), All impact fields (all). *Note*: For description of quantified impact chains by impact field, see Table 21.1

Regarding the effects on consumption and welfare,[4] Fig. 21.2 gives an overview. In general the overall effect on consumption is larger than the effect on GDP. While investigating consumption effects, we differentiate again between price and quantity effect, where the latter offers one possible way to measure the actual effect on welfare. Across impact fields the direction of effects are similar to those of GDP. Concerning consumption and welfare there are two impact fields with positive effects due to climate change, namely Agriculture and Buildings: Heating and Cooling. The largest negative consumption and welfare changes emerge for the impact chains of Catastrophe Management, Electricity, Forestry, and Tourism. When combining all quantified impact chains into one model run, the effect on welfare is strongly negative: Due to climate change, average annual welfare is lower by 1 billion euros in 2030 (−0.33 %) and by 2 billion euros in 2050 (−0.48)%.

In Fig. 21.3 annual GDP and welfare effects for 2030 and 2050 are decomposed by impact field; the net effect is indicated by a black square, respectively. While the direction of effects on GDP and welfare is the same for each impact field, the different impact fields contribute differently in strength to the total GDP and the total welfare effect (i.e. when all fields are considered jointly). On the one hand, the effect triggered by impacts in Agriculture leads to smaller welfare than GDP effect,

[4] In this case we use the so-called "Hicksian equivalent variation". In this sense welfare can be interpreted as the amount of money that is needed to be added to (or subtracted from) the household's benchmark income in order to keep its utility at the same level as in the benchmark.

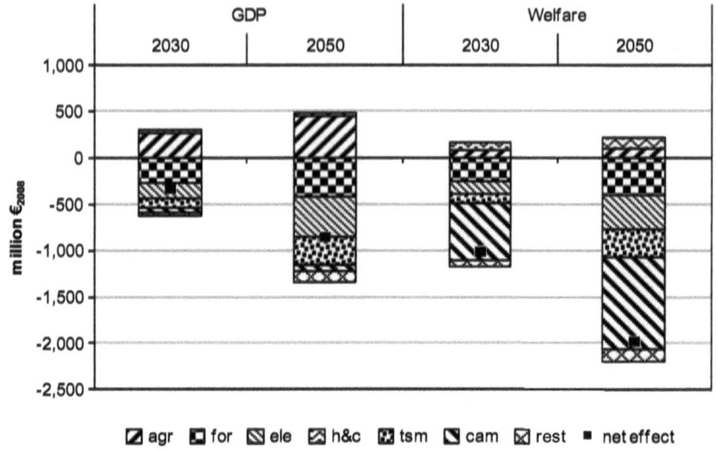

Fig. 21.3 Decomposition of annual GDP (based on quantity and price changes) and welfare effects (based on quantities) of climate change (relative to baseline with reference socioeconomic development) by impact field and in total (2030 = period 2016–2045; 2050 = period 2036–2065). *Impact fields*: Agriculture (agr), Forestry (for), Buildings: Heating and Cooling (h&c), Electricity (ele), Tourism (tsm), Catastrophe Management (cam) and rest: Water (wat), Transport (trn), Manufacturing and Trade (m&t), Cities and Urban Green (cug). *Note*: For description of quantified impact chains by impact field, see Table 21.1

as agricultural productivity increase sets consumer budget free to demand other products and thus prices rise, which shifts GDP more strongly than welfare. The effects stemming from Electricity as well as Catastrophe Management have a stronger negative effect on welfare than on GDP (e.g. rebuilding the damages after floods raises GDP while only restoring the earlier welfare level). Taking these positive and negative deviations together, the total welfare effect (i.e. when all impact fields are considered jointly) is substantially stronger than the effect on GDP (see net effect in Fig. 21.3).

21.5 Macroeconomic Effects of Climate Change: The Overall Effect of all Quantified Impact Chains

While the focus of Sect. 21.4 was the comparison across impact fields, we now focus on all quantified impact chains in total and investigate the direct and indirect effects of them across the 40 sectors of our CGE model.

21.5.1 Macroeconomic Effects of all Quantified Impact Chains in Total

Table 21.2 gives an overview of the macroeconomic feedback effects across economic sectors which emerge when all quantified impact chains of all impact fields are implemented in the model (scenario *ALL*). All effects are again given as average changes of annual values in million euros (M€) relative to the baseline scenario without climate change but with reference socioeconomic development.

Regarding sectoral effects, we look at gross value added in order to classify if a sector is "gaining" or "losing". Furthermore gross output value is given in Table 21.2 (i.e. sectoral output quantity valued at its market price). By subtracting sectoral intermediate demand from gross output value we obtain sectoral gross value added, which in turn is the contribution to GDP. The sectoral effect on value added therefore shows how the contribution to GDP changes by sector.

In terms of value added but also gross output value, there is one major sectoral winner, the construction sector. This is due to required investments for reconstructing climate change triggered damages to protective forest (impact field Forestry) and investment into additional electricity generation capacity (impact field Electricity).[5] Therefore sectoral gross value added of the construction sector rises by about +150 million euros in 2030 and by +250 million euros in 2050. In terms of gross value added (i.e. sectoral contribution to GDP), there are also positive effects for Agriculture and Food products due to higher agricultural productivity as well as for Trade and Repair of Motor Vehicles, reflecting the damages to privately owned cars originating from impact field Catastrophe Management (*cam*).

Regarding the sectoral losers we see that the public and private service sectors are negatively affected due to climate change impacts (due to higher public sector expenditures on Catastrophe Management as well as lower net income by households), as well as the energy sector (especially Electricity), and Accommodation (due to impacts on Tourism).

Summing up the effects on gross value added for all sectors gives the effect on GDP[6] which is −300 million euros in 2030 and −800 euros in 2050, leading to a lower economic growth rate by −0.08 %-points p.a. in 2030 and by −0.15 %-points in 2050. By looking at sectoral value added we see that positive and negative effects cancel each other out partly. Whereas the losing sectors lower GDP by about −500 million euros in 2030 (−1,100 million euros in 2050) the gaining sectors dampen this effect as they contribute more to GDP in the climate change scenario by about +190 million euros (+320 million euros in 2050). It is important to note that the effect on welfare is three to four times stronger than the effect on GDP (−0.33 % in

[5] Repair of roads and required additional investment in the water sector were also implemented but contribute much less to cost increases.

[6] Note that the sum of all sectoral effects on value added has to be corrected by indirect taxes and subsidies to obtain the actual effect on GDP.

Table 21.2 Sectoral and total effects of all quantified climate change impacts in M€$_{2008}$, average annual effects relative to baseline (for periods 2016–2045 and 2036–2065)

	Ø 2016–2045			Ø 2036–2065		
Changes in M€ p.a. relative to baseline	Gross output value	Intermediate demand	Gross value added	Gross output value	Intermediate demand	Gross value added
Gaining sectors	**+346**	**+152**	**+194**	**+729**	**+408**	**+321**
Construction	+369	+220	+149	+618	+368	+251
Trade and repair of motor vehicles	+57	+27	+30	+116	+55	+61
Agriculture	−6	−16	+10	−5	−15	+9
All other gaining sectors	−73	−78	+5	−0	−0	+0
Losing sectors	**−1,103**	**−606**	**−496**	**−2,518**	**−1,399**	**−1,119**
Rest of services	−103	−25	−78	−214	−52	−163
Public services	−109	−34	−75	−222	−69	−153
Health	−97	−32	−65	−189	−61	−128
Accommodation	−103	−39	−65	−258	−95	−164
Electricity	−288	−246	−43	−594	−514	−80
Wholesale and retail trade	−64	−26	−38	−168	−68	−100
Real estate activities	−29	−9	−20	−75	−22	−53
All other losing sectors	−309	−196	−113	−797	−518	−279
Total effect (all sectors)	**−756**	**−454**	**−302**	**−1,789**	**−991**	**−798**
GDP at producer price			−0.08 %			−0.15 %
Thereof price effect			+0.03 %			+0.04 %
Thereof quantity effect			−0.11 %			−0.19 %

	Change (M€)		Change (%)	Change (M€)		Change (%)
Welfare	**−995**		**−0.33 %**	**−1,955**		**−0.48 %**

Note: For description of quantified impact chains by impact field, see Table 21.1

2030 and −0.48 % in 2050) as we correct for climate change induced "forced" consumption which does not enhance welfare (but GDP).

To investigate whether a sector is growing stronger or weaker or is even shrinking due to climate change (relative to the baseline), the effects on gross output value are less helpful as price effects may cancel out quantity effects. Hence, we are now interested in the effects on sectoral output quantities in isolation (no price effects included) which corresponds to sectoral activity. Figure 21.4 gives the sectoral changes in output decomposed into quantity and price effect in M€ for

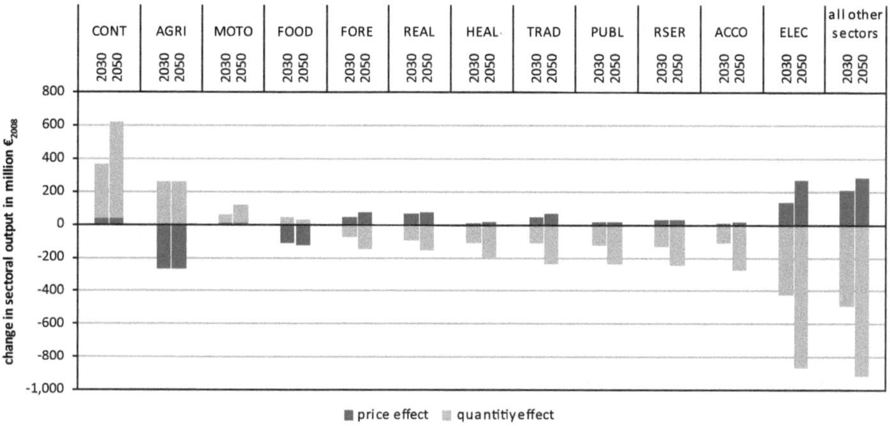

Fig. 21.4 Average annual effect of all quantified impact chains on output (quantity and price effects) by sector compared to the baseline (2030 = period 2016–2045; 2050 = period 2036–2065). *Note*: For description of quantified impact chains by impact field, see Table 21.1

selected sectors.[7] Sectors CONT (Construction), AGRI (Agriculture), MOTO (Trade and Repair of Motor Vehicles) and FOOD (Food Products) expand their output in quantities and thus grow, whereas all other economic sectors shrink. The output increase in Construction is the result of additional investment necessary due to climate change, such as for catastrophe management, electricity supply but also for water and transport infrastructure. The top "losers" in terms of output quantity are ELEC (Energy including Electricity), ACCO (Accommodation), RSER (Rest of Services), TRAD (Trade), PUBL (Public Services), HEAL (Health), REAL (Real Estate) and FORE (Forestry). In general the effects are stronger in 2050 than in 2030 with a total effect of −1.0 billion euros in 2030 and −2.3 billion euros in 2050.

21.5.2 Effects on Public Budget

The effects of the quantified climate change impact chains on public budgets is depicted in Table 21.3. Starting with revenues, we see that climate change reduces the annual average budget by 230 million euros in 2030 (period 2016–2045) and by 500 million euros in 2050 (period 2036–2065). This is mainly attributable to lower labour tax revenues as unemployment increases. Next to that, lower production taxes (originating from lower economic activity and higher subsidies to Forestry and Water to deal with climate change impacts) as well as lower value added tax (originating from less consumption) contribute strongly to the negative effect on tax

[7] All winning sectors as well as all losing sectors with losses larger than 100 million euros in 2050 are shown separately.

Table 21.3 Effects of all quantified climate change impacts on annual government budget in M€$_{2008}$

	Ø 2016–2045			Ø 2036–2065		
Changes in M€$_{2008}$ p.a.	Baseline	Climate change	Climate change impact	Baseline	Climate change	Climate change impact
Revenues	**+150,460**	**+150,230**	**−230**	**+206,548**	**+206,056**	**−492**
Production tax	+18,351	+18,258	−93	+25,676	+25,518	−158
Thereof subsidies to forestry	−59	−144	−84	−77	−199	−122
Thereof subsidies to water/waste	+140	+140	−0	+151	+151	−0
Labour tax	+87,144	+87,022	−123	+119,829	+119,573	−256
Capital tax	+19,816	+19,824	+8	+26,860	+26,852	−8
Value added tax	+28,932	+28,902	−30	+39,495	+39,441	−54
Other taxes	−3,783	−3,776	+8	−5,312	−5,329	−17
Expenditures	**+150,460**	**+150,230**	**−230**	**+206,548**	**+206,056**	**−492**
Government consumption	+75,971	+75,700	−271	+103,813	+103,266	−547
Compensation for *cug*	+87	+128	+40	+44	+107	+64
Compensation for *cam*	+83	+313	+231	+253	+737	+483
Unemployment benefits	+4,083	+4,253	+170	+5,793	+6,117	+323
Transfers to households net of other taxes	+70,236	+69,836	−400	+96,645	+95,830	−815
Climate change impact on government budget			**−0.15 %**			**−0.24 %**

Note: For description of quantified impact chains by impact field, see Table 21.1

revenues. As annual revenues are lower in the climate change scenario, annual expenditures have to be lower by the same amount (by assumption public deficit does not increase and therefore expenditures have to adjust to revenues). However, government spending has to increase to cover higher unemployment benefits but also to finance needs for the impact fields Cities and Urban Green (*cug*) and Catastrophe Management (*cam*). Hence, to balance expenditures and revenues, transfers to private households are cut by 400 million euros in 2030 and by 820 million euros in 2050.

Summing up, the average annual public budget decreases due to climate change by 0.15 % in 2030 and by 0.24 % in 2050. This amount is equivalent to 1.2 % (1.8 %) of capital tax revenues in 2030 (2050) or equivalent to 0.8 % (1.2 %) of value added tax revenue and could therefore also be compensated for by raising tax rates in that order of magnitude.

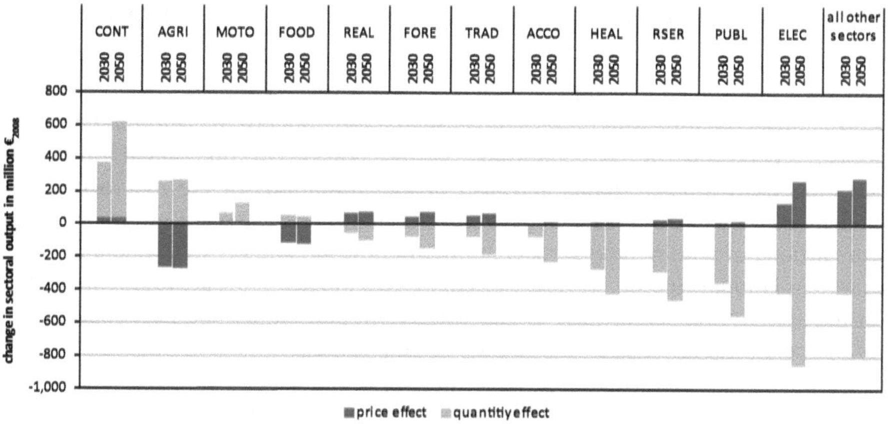

Fig. 21.5 Average annual effect of all quantified impact chains on output quantities in M€$_{2008}$ by sector compared to the baseline, with flexible government consumption expenditures (2030 = period 2016–2045; 2050 = period 2036–2065)

It is a strong assumption in the macroeconomic model that government consumption expenditure is not allowed to change due to climate change (except for the effect originating in *cug* and *cam*) but that it is fixed. In the model this constraint is satisfied by a change of transfers to private households. Hence, whenever the government were confronted with higher (lower) consumption expenditures, transfers would be cut (raised).

In a separate model run, this constraint was lifted, to check whether the obtained results are robust when government consumption is flexible. It turns out that the effects on average annual GDP are stronger in that case (−0.12 % in 2030 and −0.19 % in 2050) as shown in Fig. 21.5. With flexible government consumption, the government is confronted with less revenue and is therefore forced to cut public consumption expenditures. As the typical government consumption goods and services are characterised by a relatively high labour intensity, the cut in government consumption leads to lower employment which in turn leads to lower labour tax income which feeds back to lower government consumption (a positive feedback loop emerges). Regarding sectoral activity effects, the sectors PUBL (Public administration and defence; compulsory social security) and HEAL (Health, social and residential care activities) are under the top losers in that case (compare Figs. 21.4 to 21.5). The detailed effects on government expenditures and revenues under flexible government consumption are given in the Supplementary Material Table 21.1.

21.6 Non-monetised Impact Chains and Model Limitations

The results presented reflect the damage to the Austrian economy triggered only by those impact chains which were quantified (and monetised) within the COIN project. Therefore it is important to be aware of the most important non-monetised impact chains and also of the limitations of the macroeconomic model (see also Chap. 7). First there are two impact fields where no monetization was carried out within the macroeconomic framework, namely Ecosystem Services as well as Human Health. Nevertheless those two impact fields are highly important for the agricultural sector (due to e.g. changes in pollination and pest control) and for the health sector (due to e.g. higher hospitalisation).

Second, within the remaining ten further impact fields a number of impact chains were not quantified. To give some examples: In Agriculture (sub-daily) heavy precipitation and hail events were not quantified. In the impact field Water, the decrease in precipitation in the vegetation period (water supply) as well as the increase in receiving water temperature (water sanitation) was not quantified. Regarding Buildings (heating and cooling), higher temperature levels in buildings and lower comfort of occupants due to higher temperature in summer could not be quantified. In the impact field Electricity, the change in supply and demand profiles and the change in reliability of electricity supply and change in probability of blackouts were excluded from the assessment (no natural hazards were included). In the impact field Transport, impacts of changes in precipitation were only considered for road infrastructure, but not for transport services nor for other transport modes nor for other climate change impact categories such as storm. In Manufacturing and Trade damages to infrastructure like office buildings or plants as well as delivery problems along the supply chain were not quantified. Regarding Cities and Urban Green loss of comfort in urban environments was excluded. In impact field Catastrophe Management, damages from storm events were not quantified. Finally, in impact field Tourism the change in the tourism sector's water and energy demand as well as business interruptions due to natural catastrophes were not quantified. For further information of neglected impact chains see Table 21.1.

Regarding extreme events, these are only poorly captured in both the sectoral models but also by the type of macroeconomic model (CGE) which is based on average annual numbers for climatic periods 2016–2045 and 2036–2065. Regarding average changes in extreme events, those are captured in the impact fields Catastrophe Management (floods, but no storms), Water and Transport. Thus, all effects need to be understood as higher costs which occur for an average year in the respective period in case of climate change in Austria compared to a (hypothetical) situation without climate change.

In addition to those impact chains mentioned as neglected in quantification, some important limitations emerge from the macroeconomic model itself. One crucial point is that autonomous adaptation of sectors, households and the government is implicitly allowed for, as agents adjust perfectly to price changes triggered by climate change. This may understate the results compared to a model environment with less flexibility. Another drawback of the CGE model is that even for goods with regulated or globally-given prices (such as water or agricultural goods), prices are adjusting endogenously subject to normal free market interactions, which does not depict reality very well.

Finally, it is important to stress that we are investigating climate change in Austria. All effects which might emerge from climate change elsewhere, but which might e.g. lead to changes in agricultural prices on international markets with significant repercussions for Austrian agriculture, or a shift in tourism destinations with eventual consequences for Austrian tourism, but also migration from other world regions due to more severe climate change impacts there, are neglected.

21.7 Discussion and Conclusions

The main finding of this chapter is that the modelled impact chains add up to a total macroeconomic effect on GDP of −0.1 %-points per year on average for the period 2016–2045 and −0.2 %-points in 2036–2065 when comparing the climate change scenario to the baseline scenario without climate change. The effect on welfare is stronger in both periods (−0.3 %-points and −0.5 %-points), as welfare is corrected for climate change induced forced consumption which is not welfare enhancing. When only looking at output quantities and hence neglecting (mostly positive) price effects, the effect is slightly stronger. For welfare, effects are similar to GDP effects in direction but stronger in total (when all impact chains are considered jointly), and additionally they differ in magnitude by impact field.

The negative GDP and welfare effects are the result of the net effect of negative effects from climate change impacts originating in Electricity, Forestry, Tourism and Catastrophe Management on the one hand and positive effects on the other from climate change impacts originating in Agriculture (due to higher productivity) and Buildings: Heating and Cooling (due to reduced heating which more than compensates for higher demand for cooling). The contribution of the remaining considered impact fields (Manufacturing and Trade, Cities and Urban Green, Water, Transport) to the total GDP and welfare effect are much smaller but also negative.

The modest negative effect of all modelled impact chains is in line with most of the findings of the European cost assessments such as the FP6/7 projects ADAM (Aaheim et al. 2012) and PESETA (Ciscar et al. 2011, 2012), which find negative costs of climate change for the coastal areas of Southern Europe and positive consequences for northern Europe, with central Europe falling in between and thus having weak effects. Higher damage costs are found by the the ClimateCost project (Watkiss 2011) which—contrary to our analysis here—also includes health effects. This estimate is also summarised in European Commission's climate impact assessment accompanying the adaptation strategy (EC 2013) and the EEA's assessment (EEA 2012).

Even though our model based assessment has a broader coverage of impact chains and a broader coverage of impact fields (sectors) compared to the international studies, many effects which emerge by region or sector are cancelled out at the national scale. While we cannot investigate the regional difference in our macroeconomic analysis due to our national scale CGE approach, we can look into the sectoral effects. Here strongest positive effects emerge for the construction sector (due to higher investments), with negative effects on output values and value added for most other sectors. Strongest negative effects (in terms of value added by sector) emerge for public health and other service sectors as well as for the sectors accommodation, electricity, trade and real estate activities.

Finally, effects on public budgets are confined on the one hand by direct public expenditures to compensate for direct damages and on the other hand by higher expenditures for unemployment benefits, which are partly offset by cuts in other transfers to households.

It is important to note that the modest effect of all modelled impact chains has to be viewed with caution as there are several major limitations. First, the type of model used for the assessment (a computable general equilibrium model) allows endogenously for autonomous adaptation, leading to lower costs compared to an assessment which does not allow for such an adjustment. Second, extreme events are captured only poorly in the model environment. Third, many qualitatively identified impact chains were not quantified and not monetised. Fourth, climate change is assumed to occur in Austria; potential climate change impacts on other world regions are ignored. But these effects will work via international markets and could be highly relevant for a small open economy like Austria's.

References

Aaheim A, Amundsen H, Dokken T, Wei T (2012) Impacts and adaptation to climate change in European economies. Glob Environ Change 22(4):959–968

Ciscar J-C, Iglesias A, Feyen L, Szabó L, Van Regemorter D, Amelung B, Nicholls R, Watkiss P, Christensen OB, Dankers R, Garrote L, Goodess CM, Hunt A, Moreno A, Richards J, Soria A (2011) Physical and economic consequences of climate change in Europe. Proc Natl Acad Sci U S A 108(7):2678–2683

Ciscar J-C, Szabó L, van Regemorter D, Soria A (2012) The integration of PESETA sectoral economic impacts into the GEM-E3 Europe model: methodology and results. Clim Change 112(1):127–142

EC (2013) Commission staff working document. Impact assessment: Part 2 accompanying document to the communication from the commission to the European parliament, the council, the European Economic and Social Committee and the Committee of the Regions, an EU strategy on adaptation to climate change, SWD(2013) 132 final

EEA (2012) Climate change, impacts and vulnerability in Europe 2012. An indicator-based report. EEA Report No 12/2012

IEA (2010) World energy outlook 2010. International Energy Agency, Paris

OECD-FAO (2011) World agricultural outlook 2011. OECD/FAO, Rome

Watkiss P (ed) (2011) The ClimateCost project. Final report, vol 1: Europe. Stockholm Environment Institute, Stockholm

Chapter 22
Climate Change Impacts at the National Level: Known Trends, Unknown Tails, and Unknowables

Karl W. Steininger, Gernot Wagner, Paul Watkiss, and Martin König

Abstract Economists attempting to evaluate the impacts of climate change are often caught between hard theory and exceedingly rocky empirics. Impact assessment models are necessarily based on highly aggregated—and sometimes highly simplified—damage functions. This study takes an alternative approach: a bottom-up, physical impact assessment and respective monetisation, attempting to cover *a much broader set of* impact fields, feeding directly into a macroeconomic and welfare analysis at the national level. To ensure consistency, our approach applies impact assessment at the sectoral impact chain level using shared socioeconomic pathways, consistent climate scenarios, computable general equilibrium evaluation, and non-market impact evaluation. The approach is applied to assess a broad scope of climate impacts in Austria. Results indicate significant impacts around 'known knowns' (such as changes in agricultural yield from climatic shifts), with uncertainty increased by 'known unknowns' (e.g. changes in water availability for irrigation, changes in pest and diseases) but also raises the question of unknowns and unknowables, which may possibly dominate future impacts (such as exceedance of critical ecosystem function for supporting agriculture). Climate change, ultimately, is a risk management problem, where insurance thinking warrants significant mitigation (and adaptation) action today.

K.W. Steininger (✉)
Wegener Center for Climate and Global Change, University of Graz, Graz, Austria

Institute of Economics, University of Graz, Graz, Austria
e-mail: karl.steininger@uni-graz.at

G. Wagner
Environmental Defense Fund, Cambridge, MA, USA
e-mail: gernot@gwagner.com

P. Watkiss
Paul Watkiss Associates, Oxford, UK
e-mail: paul_watkiss@btinternet.com

M. König
Environmental Impact Assessment and Climate Change, Environment Agency Austria, Vienna, Austria
e-mail: martin.koenig@umweltbundesamt.at

K.W. Steininger et al. (eds.), *Economic Evaluation of Climate Change Impacts*, Springer Climate, DOI 10.1007/978-3-319-12457-5_22

Analysis of the study results indicate that the current welfare damage of climate and weather induced extreme events in Austria is an annual average of 1 billion euros (large events only). This has the potential to rise to 4–5 billion euros by mid-century (annual average, known knowns of impact chains only), with an uncertainty range of 4–9 billion euros. When extreme events and the tails of their distribution are included, even for a partial analysis focused on extremes, damages are seen to rise significantly, e.g. with an estimated increase to 40 billion euros due to riverine flooding events alone by the end of the century. These highlight the need to consider the distribution of impacts, as well as the central values.

22.1 Introduction

What we know about climate change confirms it to be one of the major challenges facing humanity in the twenty-first century. However, what we don't (yet) know—and possibly won't know before it is too late to act—could drive up potential costs higher still. The Intergovernmental Panel on Climate Change (IPCC) in its Fifth Assessment Report confirms the bottom line reported in the peer-reviewed literature for decades: climate change is taking place with global mean temperature increase of almost 1 °C since 1880 and that it is predominantly caused by human activities (IPCC 2013). The IPCC also reports that left unabated, future emissions will lead to a temperature increase by the end of the century of 3.2–5.4 °C. The impacts of such a change are profound. Polar regions would warm by at least twice as much. Sea levels will rise for centuries [due to the slow process of heat uptake by the deep ocean and arising very long term (thermal expansion) commitment to sea level rise]. Furthermore, given that surface air temperature above oceans will warm by less than the global average, many regions, particular land-bound mountainous and continental climate zones, will face more substantial increases; e.g. a 4.5–6.6 °C increase by 2100 is projected for the Alpine region and thus for a country such as Austria (Jacob et al. 2013, APCC 2014).[1]

Even the most ambitious mitigation scenarios could potentially lead to dangerous climate change; i.e. even if global average warming is limited to 2 °C relative to pre-industrial levels [the current international goal agreed (UNFCCC 2010), noting that this is unlikely to be met]. Higher emissions pathways lead to increasingly costly impacts. The IPCC 5th Assessment Report (2014a, pp. 11–14) identifies which risks of climate change are "considerable at 1 or 2 °C above preindustrial levels" and all global risks to be "high to very high with global mean temperature increase of 4 °C or more above preindustrial levels in all of the reasons for concern"

[1] This range for Alpine regions refers to the "likely" range, i.e. the 17–83 percentile. To be fully comparable with the global temperature range given by IPCC, which refers to the 5–95 percentile, the range for the Alpine region would be larger.

(the latter being unique and threatened systems, extreme weather events, distribution of impacts, global aggregate impacts, large-scale singular events).

All of this points to the need to avoid high emission and warming scenarios with mitigation. It also indicates that adaptation to climate change will be needed, to the warming already observed and locked into the climate system over the next few decades (from past and near-term emissions), as well as to future emissions. This is likely to require complementary mixes of mitigation and adaptation (Warren et al. 2013), noting that the two address different risks, operate at different aggregation and temporal scales, and that there are limits to adaptation (IPCC 2014c). Nevertheless, both adaptation and mitigation require well informed decision making and thus knowledge and information on the type and magnitude of climate change impacts expected.

Over the last few years, a wide range of methodologies have emerged for assessing the costs of climate change. Global economic integrated assessment models assess the economic costs of climate change, combining the scientific and economic aspects of climate change within a single, iterative analytical framework. However, they use highly aggregated economic damage functions (usually based on global temperature increase as the sole aggregated climate parameter). They are applied to provide economic costs over time and thus for a specified rise in global mean temperature or for a specific future year, the net present values for future damages over time, and to estimate the marginal social costs of carbon (the damage cost of an extra tonne of greenhouse gas (GHG) emissions). While these provide valuable insights, these costs are extremely difficult to estimate and vary considerably, and are heavily influenced by the choice of discount rate and inclusion of equity weights as well as the coverage of impacts (Watkiss 2011a): their coverage is therefore recognised as partial and incomplete (IPCC 2014c). Their use has therefore been questioned. Pindyck (2013) emphasises the arbitrary choice of damage functions (especially for higher rates of warming) and neglect of many catastrophic outcomes; Weitzman (2009, 2012) and Wagner and Weitzman (2015) emphasise the deep-seated uncertainty around climate sensitivity that is not fully reflected in the models.[2]

There are two main approaches that differentiate impacts specifically. While these do not address all of the challenges above, they provide improved damage

---

[2] The three most often applied Integrated Assessment Models (IAMs) to date are DICE (Dynamic Integrated Climate and Economy), PAGE (Policy Analysis of the Greenhouse Effect), and FUND (Climate Framework for Uncertainty, Negotiation, and Distribution), with model descriptions given by Nordhaus (1991, 2011); Hope (2006)—on which the Stern review is based (Stern 2007)—and Tol (2002a, b), Anthoff and Tol (2010), respectively. The modelling aspects questioned most—for derivation of social costs of carbon by such means—include arbitrary parameter choice in social welfare functions, climate sensitivity (the temperature increase a GHG doubling implies), arbitrary and non-empirical based climate damage functions (usually a functional relationship between temperature increase and (regional) GDP loss, for FUND also distinguishing individual sectors), and neglect of consideration of possible catastrophic outcomes. For detailed discussions see Watkiss (2011a, b), Pindyck (2013), Stern (2013), and Wagner and Weitzman (2015).

functions and reduce aggregation errors. Their focus is primarily the regional, national, and/or sub-national scale. These approaches, as briefly presented in Chap. 2 of the present volume and discussed in more detail in Watkiss and Hunt (2010) are:

- *Scenario-Based Impact-Assessments* combine climate model outputs with sector impact models (or functional relationships) in order to estimate physical impacts, which are then valued so as to estimate welfare costs. However, these assessments are not able to capture cross-sectoral, economy-wide effects. There are a number of variations, including risk assessments, which focus on extreme (probabilistic) events such as flooding (using historical analogues or damage-loss relationships), and econometric assessments, which use historical relationships between economic production and climate and then apply these to future climate scenarios.
- *Computable General Equilibrium (CGE) models* provide multi-sectoral and macro-economic analysis of the economic costs of climate change. They have the advantage of capturing cross-sectoral linkages and economy-wide effects (and metrics), and they can also look at price and trade effects. However, they use aggregated representations of impacts and omit non-market impacts.

These approaches use different metrics, modelling approaches and assumptions. No one method is in principle right or wrong—their use depends on the given objectives. More recently, some studies have begun to combine these approaches in a single framework, to produce more complementary information. An example of such an analysis at the European level is presented in Chap. 2, which summarises results from the EU ClimateCost Project. There is also a clear demand for cost evaluations at the national level, as this is where climate change materialises and where the administration and governance of adaptation takes place.

The objective of this book was to provide a comprehensive impact assessment for a single country, spanning *as broad a* field of impact as possible. Methodologically, it draws from and combines the following:

- Scenario-Based Sector Impact-Assessment: to capture national impacts at the most detailed level available;
- Computable General Equilibrium (CGE) analysis: to capture cross-sectoral linkages and economy-wide effects;
- Qualitative analysis: to capture additional non-market effects.

To our knowledge this is one of only a small number of studies that have applied such a comprehensive approach at the national level (i.e. across many relevant impact fields). To date, national level studies have primarily focused on a few selected sectors (and impacts) covering for example agriculture, water, energy, human health, together with assessments of coastal impacts for non-landlocked countries [e.g. Ruth et al. (2007) for the US; Ciscar et al. (2011) for European countries; Ackerman and Stanton (2011) for an overview].

Increasing the sectoral comprehensiveness does, however, reveal more areas where we have insufficient knowledge. For some issues, incomplete knowledge is inherent in the nature of the problem. For others, it points to a to-do list for future work. Thus, our approach does not result in a 'final' figure with respect to total damages. In fact, the research lets us see that we remain some way from achieving such a figure. However, it does provide a more transparent picture of what we know, what we know we don't know, and to raise the potential to think more broadly and consider new aspects (current unknowns), which itself has profound implications for optimal policy.

Sectoral impacts, their economic costs and their macroeconomic feedback effects (taken in isolation for each impact field) are reported in Chaps. 8–19. The economic implications when all of these impacts are considered simultaneously are reported in Chap. 21. However, when testing for their aggregated effect, all such impacts are, first, assessed in terms of a single climate and socioeconomic scenario (considered to be a "medium" development) and, second, they do not focus on total weather and climate induced costs, but only on those costs triggered by additional climate change. The present chapter seeks to put these earlier aggregate results into perspective and to thus consider a more comprehensive evaluation.

To do this we move beyond simple aggregated results in three ways: First, total—and not only additional—weather and climate induced costs are considered. Second, we reconsider "known unknowns", such as biodiversity loss due to climate change. Third, we move beyond the expected value of damage alone, recognising this to be inadequate for risk management as it neglects analysis of the tails of the distribution of possible events and impacts. We thus take a closer look at more extreme or catastrophic events—the "fat tails" of climate impacts. By considering these three extensions, more adequate policy conclusions become possible.

Finally, we sketch the implications of potential "unknown unknowns" and "unknowables" in climate change—and the potential societal response.

The structure of this chapter is as follows. Section 22.2 covers the impacts that we know of, the "known knowns". Section 22.3 presents the picture of what we already know that we don't know, including an evaluation of ranges of developments. It also discusses fat tails in the context of tipping elements at a continental scale and illustrates the relevance of tail risks at the national level by considering three types of extreme events and their damage ranges. The final section concludes.

22.2 Known Knowns

22.2.1 Weather and Climate Induced Extreme Event Damage: Taking Stock of the Past

The first area of analysis is to estimate the current welfare damage of climate and weather induced extreme events in Austria.[3] MunichRe (2014) supplies the most

[3] In the literature this is also known as 'adaptation deficit'.

comprehensive database on weather and climate related damage for Austria. The data covers all large damage events and most medium ones,[4] and has comprehensive coverage since 2002. In 2002—an extreme year—the weather and climate related large damages in Austria amounted to $€_{2010}$ 3.67 billion (1.5 % of GDP),[5] mainly driven by flooding damage. Across the past decade (2001–2010), the annual average damage related to large and medium events in Austria was $€_{2010}$ 705 million (m), equivalent to slightly above 0.25 % of GDP. There is some (less detailed) data over a longer time-period that provides some basis for comparison. This indicates that damages have increased considerably over the last few decades, starting from an annual $€_{2010}$ 97 m in the 1980s, and rising to $€_{2010}$ 129 m in the 1990s (data coverage in the earlier decades is not complete, however, even on large events; König et al. (2014), p. 665), though this is most likely due to socio-economic change.

According to MunichRe NatCatService data, the climate and weather related premature death toll in Austria over the last decade (2001–2010) was 411, of which 334 were due to heat (330 alone in 2003) and the remaining 77 primarily due to avalanches (38), floods and storms.

The aggregated 2001–2010 total damages recorded in the database for Austria, totally $€_{2010}$ 7 billion, are attributed to the following types of events: 61 % are related to precipitation-triggered events (primarily large and small scale floods including flash floods and landslides/mudflows) while 23 % are storm-related. 7 % of the damage is related to cold spells and winter damage, 6 % to heat waves, droughts and forest fires and the remaining 3 % to hail damages.

However, these monetary estimates only include direct damage observed. Thus, neither the indirect disruption or follow-up costs, nor the non-market impacts (such as biodiversity losses, health inconveniences, morbidity and mortality, etc.) are included.

In Table 22.1 we provide monetary estimates for one of the non-market impacts, premature heat related deaths, annual average 2003–2012, using death tolls for this period from Table 11.6 (Chap. 11) and two different monetary evaluation methods as described in Watkiss (2011b). Adding these, we identify a stock of climate and weather induced damages in Austria (the 'adaptation deficit') at an annual average of $€_{2010}$ 1 billion for the first decade in the twenty-first century. Including the additional effects from indirect and non-monetary areas, as well as macroeconomic costs, would increase these estimates further, possibly by 25–100 % (e.g. see Hallegatte et al. 2007).

[4] Damage events covered by MunichRe (2014) concern catastrophes of UN classification level 3–6. With medium catastrophes (levels 3 and 4) characterized by damages larger than US$25 m (40 m/50 m/60 m) when occurring in the 1980s (1990s/2000s/2010s) and large and significant catastrophes (levels 5 and 6) characterized by damages beyond US$275 m (400 m/500 m/650 m) or respective death tolls (more than 100 and more than 500, respectively).

[5] This number relates to total damages (i.e. beyond those insured) of large and significant catastrophes (of levels 5 and 6), but covers direct damages only, i.e. it does not include indirect damages, macroeconomic consequences or non-market damages.

Table 22.1 Climate and weather induced damage, across sectors, quantified "known knowns" impact chains only, average annual totals for Austria (for periods 2016–2045 and 2036–2065)

Damage in m€ p.a. (2010 prices, undiscounted)		
(A) Stock of current damages (extremes)		
Damage observed to date (market & non-market)	**850–1,090**	
Annual average of extreme weather event damage (MunichRe, only larger damage, Ø for period 2001–2010)	705	
Non-market damage:		
Heat induced premature deaths (monetary value)	145–385	
Evaluation using value of statistical life (VSL)	385	
Evaluation using value of life years lost (LYL)	145	
	Ø2016–2045	Ø2036–2065
(B) Additional future damages		
Damage induced by future climate change	**995**	**1,955**
Welfare loss (reference socioeconomic development, mid-range climate change, see Chap. 21, Table 21.2)		
Additional damage induced by future socioeconomic change	**270**	**825**
Energy additional investment	99	298
Road infrastructure additional investment	8	20
Riverine flooding additional damage	163	507
Non-market damage:		
Heat induced premature deaths (monetary value)	**95–255**	**570–1,300**
Evaluation using Value of Statistical Life (1.6 million euros/SL)	255	1,300
Evaluation using value of life years lost (63,000 €/LYL)	95	570
(C) Total annual average (comprising current level plus future additional damages)	**2,210–2,610**	**4,201–5,170**

Note: Values for VSL and LYL from Watkiss (2011b), toll of heat induced premature deaths under reference socioeconomic scenario and mid-range climate scenario as given in Table 11.6 (first period 400, second 1060, including current damages), equivalent for LYL

It is highlighted that this captures the effects of extremes only: the current climate—and the variability between years—also affects many other areas, affecting crop productivity, winter heating and summer cooling, water flows and availability, etc.

22.2.2 *Future Additional Weather and Climate Change Induced Damages*

Considering a reference level of socioeconomic development and a mid-range climate change scenario, the additional economic net damage that climate change causes (i.e. relative to a baseline scenario of equivalent socioeconomic change, but no further climate change), measured in welfare terms amounts to 1 billion euros (annual average) for the period 2016–2045 and almost 2 billion euros for the period 2036–2065 (see Table 22.1). These numbers are based on sectoral impact models and the integration of their results in a national CGE model. Only those impact chain subsets are covered for which robust results can be derived. Chapter 21 is devoted to analysing and interpreting the above welfare results in detail.

Nations, however, are not only confronted with climate and weather damage triggered by *climatic change* (i.e. by a climate change signal beyond the state of climate we observe today). They also have to deal with the total climate and weather induced damage. This includes two further elements: the climate and weather induced damage already observed under the current state of climate (as covered in Sect. 22.2.1 above); and the climate and weather induced damage due to factors other than climatic change (e.g. population or infrastructure growth, but also changed lifestyle such as increase in living space by person) which are due to "socioeconomic development". In the present study we undertook a closer analysis of the latter in three core areas of climate and weather induced cost, i.e. electricity supply, road transport infrastructure, and riverine flooding. In each of these categories the change in socioeconomic factors alone (i.e. without any further climate change) tends to increase weather and climate induced costs: electricity supply has to react by additional investment to a (peak load) demand that increases with a higher share of air conditioning, an expanded road infrastructure network drives up weather induced damage costs, and the expansion in real values (e.g. of houses) increases riverine flooding damages. The estimated figures (yearly average, period 2036–2065) for additional costs occurring in each of these sectors are as follows:

- energy supply: annual additional investment 298 million euros (see Chap. 14; Sect. 14.4.5.2) to match rising electricity demand (including a rising share of air conditioning)
- road infrastructure: annual additional investment for road damage reconstruction 20 million euros (see Chap. 15; Sect. 15.4.4.2) due to a larger road network
- riverine flooding: annual additional riverine flooding cost of 507 million euros [see Chap. 18, Climate Cost approach, baseline damage as given in Sect. 18.4.5. 1 (820 million euros) subtracting damages observed to date already (Table 18.2; 313 million euros)] due to increased real values in flood prone areas.

For the period 2016–2045 the respective damage figures are 99 million euros (energy), 8 million euros (roads) and 165 million euros (floods) (sources as above).

Finally, one of the most significant non-market effects, future premature heat related deaths (as reported in Chap. 11), can be monetised. Based on Watkiss (2011a, b) we employ both the value of a statistical life (VSL) and the value of life years lost (VLY), see Table 22.1 for rates used.

22.2.3 Weather and Climate Related Damage: Known Totals

The structure of climate and weather induced damage is reported in monetary terms in Table 22.1. This comprises a sum of the components as described in Sects. 22.2.1 and 22.2.2 above. The structure of components may be different than in other studies. Note that CGE models are calibrated on historic data that already incorporate (and do not isolate) the current stock of damages. We identify the current stock of weather and climate induced damages separately (Sect. 22.2.1 above). Our CGE analysis identifies additional (net) damages, which will be triggered by future climate change, while the sectoral analysis allows us to isolate additional climate and weather induced costs that will be triggered by just the future change in socioeconomic factors (both as covered in Sect. 22.2.2 above). For a reference socioeconomic scenario and mid-range climate change we find climate and weather induced damage in Austria increases from a current annual average of approximately 1 billion euros, to between 2.2 and 2.6 billion euros in the 2030s, and to between 4.2 and 5.2 billion euros in the 2050s (all in 2010 €, undiscounted, to allow for direct evaluation for each of the future periods and comparison across them). These numbers don't include any of the "known unknowns" yet, i.e. unquantified impacts identified in Chaps. 8–19, such as increased irrigation or increased pest control costs in agriculture, increased soil erosion, increased impact of storm events in forestry, biodiversity losses etc. For a list of most relevant non-quantified impact chains see Table 22.3, for a full account of non-quantified impact chains see the Tables "Impact chains" in each of the Chaps. 8–19 and Table 21.1.

22.3 What We Know We Don't Know: Known Unknowns

22.3.1 Which Climate Change and Socioeconomic Scenario Will Materialise?

The climate change impacts quantified in monetary terms in Table 22.1 refer to one scenario, which we call "intermediate". More specifically, it is intermediate in two senses of that word: we use the "reference" socioeconomic scenario and the "mid-range" climate scenario to derive one intermediate cost estimate of climate and weather induced net costs. It is consistent across sectors, and thus also allows for a macroeconomic evaluation of impacts across sectors simultaneously.

However, focusing on intermediate scenarios also misses an important dimension of this book's overall analysis, which also covers climate model and socioeconomic uncertainty: what if parameter combinations are such that they lead to lower (higher) damages? How low (high) might the figures (which were reported in Table 22.1 for the intermediate case) for weather and climate induced damage become?

For each category of impact, Chaps. 8–19 identify the respective dimensions that determine the damage most significantly (e.g. for heat induced premature deaths one central parameter is driven by socioeconomic development, i.e. what percentage of the (old) population can reduce their risk by air conditioning, that is in addition to which climate scenario materialises). Each chapter's analysis, where the data basis allows to do so, varies these central parameters (if possible in both domains, socioeconomic and climate) in order to stress-test the intermediate values presented. This results in an additional low range and high range damage value for each category of impact.

There's a word of caution in order for the aggregated consideration of these, however. We cannot simply add low (high) range values across all impact fields, as there are also impacts that (at least partly) counterbalance across impact fields. For example, a warmer climate scenario tends to *in*crease premature deaths (and thus to increase damage), but at the same time tends to decrease (winter) heating expenses, creating an additional benefit that is *de*creasing overall damage. Acknowledging these interactions across impact fields gives a narrower range for damage, than just simply adding low (high) impact values across all sectors and impacts. Damage values for consistent low and high damage scenarios are presented in Table 22.2. These values do *not* cover the lowest and highest possible (for a further discussion on these see the section on fat tails below), they simply represent damage values originating from consistently varying central damage relevant parameters within a plausible range. Table 22.2 thus gives lower (higher) range damage values.

We find that, considering also lower and higher ranges of damages, figures for estimated damage in Austria range 2.1–4.2 billion euros/year on average in the period 2016–2045, and 3.8–8.8 billion euros in the period 2036–2065.

A significant share of damage is accounted for by heat induced premature deaths. While the damage number here also depends on which monetary unit is chosen for the valuation (VSL or LYL, see Table 22.2), we find that a far larger fraction of the range is determined by climate uncertainty, and the largest fraction by which socioeconomic scenario we choose. The latter is varied from "10 % of the population aged 65+ reduce their risk by 50 % due to air conditioning" (the "intermediate" case), to "20 % to do so" (the low damage case) and to "no additional air conditioning" (the high damage case) (Chap. 11, Table 11.6).

For future socio-economic uncertainty governing market damages, the variation in the damage value is most strongly driven by the share of future construction within/without flooding prone zones, and the thus flooding damage variability. The lower of the damage values is connected to all future buildings being only located in areas associated with a flooding recurrence period of less frequent than every 200 years, while the higher value is connected with new buildings being built in equal shares across flooding zones as they have been to date. Additional relevant driving factors arise in "heating and cooling of buildings", due to the thermal quality of buildings, the energy efficiency of heating and cooling systems and their energy characteristics in summer (all of these governed by building codes), and also required comfort levels and behaviour, technology, energy carrier mix, and energy price levels.

Table 22.2 Climate and weather induced damage, across sectors, quantified "known knowns" impact chains only, lower and higher range for average annual totals for Austria (for periods 2016–2045 and 2036–2065)

Damage in m€ p.a. (2010 prices)		
Stock of damages		
Damage observed to date (market & non-market)	**850–1,090**	
Annual average of extreme weather event damage (MunichRe, only larger damage, Ø for period 2001–2010)	705	
Non-market damage:		
Heat induced premature deaths	145–385	
Evaluation using Value of Statistical Life (VSL)	385	
Evaluation using Value of Life Years Lost (LYL)	145	
	Ø2016–2045	Ø2036–2065
(B) Additional future damages		
Damage induced by future climate change	**[890–1,211]**	**[1,825–2,280]**
Welfare loss (resulting from consistent low and high climate change mpact scenarios across impact fields)		
Additional damage induced by future socioeconomic change	**[268–314]**	**[800–1,080]**
Non-market damage:		
Heat induced premature deaths	**82–1,535**	**285–4,350**
Evaluation using Value of Statistical Life (1.6 million euros/SL)	[210–1,535]	[640–4,350]
Evaluation using Value of Life Years Lost (63,000 €/LYL)	[82–580]	[285–1,840]
(C) Total annual average (comprising current level plus future additional damages)	**2,090–4,150**	**3,760–8,800**

Note: Values for VSL and LYL from Watkiss (2011b), toll of heat induced premature deaths and life years lost across socioeconomic and climate scenarios from Table 11.6 in Chap. 11

Overall, the results show, that the effect of climate change model and socio-economic uncertainty (low or high damage scenarios) that we could quantify has most impact on variability in flooding damages and building cooling expenses.

22.3.2 Climate Change Impacts not Quantified

The determination of the costs of climate change impacts requires detailed and substantial research effort with respect to each of the fields and impact chains identified. The present project depended on substantial input from earlier research projects and sectoral impact models that had been developed and that could be employed within a consistent overall framework. Much of the available data remains, however, incomplete. This meant that a number of impact chains, of high potential relevance, could not be quantified here. The most important are named in Table 22.3. This clearly points to a need for future research.

Table 22.3 lists the most important climate change impact chains *not* quantified within the present project. These are thus *not* covered within the figures for damage given in this and previous chapters and thus likely to further extend the range of costs given here (for a fully detailed list of impact chains *not* quantified see Chap. 21, Table 21.1 and the Tables "Impact Chains" in each of the Chaps. 8–19).

22.3.3 Extreme Events and Tail Risks

At the global—and continental scale—there is the potential risk of catastrophic events from climate change, so called tipping points or tipping elements. These are defined (by Lenton et al. 2008) as '*subsystems of the Earth system that are at least sub-continental in scale and can be switched—under certain circumstances—into a qualitatively different state by small perturbations*'. These include major discontinuities such as abrupt solid ice discharge from the West Antarctic Ice Sheet—though the likelihood of these events, and the temperatures that might trigger their onset, is highly uncertain.

These risks—and the influence they have on policy—have also been recognised in the economic literature. From his seminal work on the 'dismal theorem' and the implications of catastrophic climate change, Weitzman (2009) concludes:

> In situations of potentially unlimited damage exposure like climate change, it might be appropriate to emphasize a slightly better treatment of the worst-case fat tail extremes—and what might be done about them, at what cost—relative to defining the calibration of most likely outcomes [. . .]. A clear implication [. . .] is that greater research effort is relatively ineffectual when targeted at estimating central tendencies of what we already know [. . .]. A much more fruitful goal of research effort might be to aim at understanding even slightly better the deep uncertainty (which potentially permeates the economic analysis) concerning less plausible scenarios located in the bad fat tail. (Weitzman 2009, p. 17).

To understand this, we consider private individual behaviour in the face of risk. When engaging in risk aversion, such as protecting ourselves against fire damage, we not only consider the expected damage, we also seek to avoid extreme damage (however small the likelihood). Similar reasoning may be applied to a society in the case of climate change. The only difference is that extreme (so-called 'tail') events are much more likely under climate change than is the likelihood of one's home burning down, thus raising the importance of the 'tail approach' even further. For instance, using the IPCC's own calibration of the equilibrium climate sensitivity parameter, Wagner and Weitzman (2015) calculate a 10 % chance of eventual global average surface temperatures exceeding 6 °C in a world with 700 ppm of CO_2-equivalent concentrations, a level expected to be reached by 2100 under the International Energy Information's baseline scenario.

Tab. 22.3 Most relevant climate change impact chains not quantified within the analysis that gives damage figures of Tables 22.1 and 22.2

Field of impact	Impact chains not quantified
Agriculture	Costs of irrigation
	Infestation pressure of pest, diseases, and weeds
	Heat induced labour productivity changes
	Heavy precipitation and hail events
	Flood damage
Forestry	Storm events
	Tree species change due to temperature rise
	Heat induced labour productivity changes
Ecosystem services	No impact in this field has been monetised, thus none of the impacts (loss of pest control and pollination, loss of fertilisation, loss of species, erosion control, water purification, soil functions....) has been considered in the cost of damage here
Human health	Temperature-related morbidity
	Health impacts of extreme precipitation events
	Air-pollution related mortality and morbidity
	Water- and food-borne diseases
	Vector-borne and rodent-borne diseases
	Effects of population displacement
Water supply and sanitation	Restoration cost due to increases in flood events
	Droughts and resulting investment profiles
	Increased need for water treatment due to lower surface water recharge
	Increased pollution due to increased floods
	Lower oxygen solubility in surface waters
Buildings	Lower comfort due to higher summer temperatures
	Increased storm frequency
Electricity	Change in supply and demand profiles
	Natural hazards (storm, floods, and other extremes) and their implications
Transport	Transport service interruption
	Storm events
	Temperature induced deformation of road surfaces
	Railways
	Air transportation
	Passenger discomfort in vehicles
Manufacturing and trade	Temperature and extreme event induced changes in Production processes Cooling and heating Infrastructure damages Shifts in consumption
Cities and urban green	Loss of climate comfort
	City tourism
	Heat related damage for pavements, tram rails etc.

(continued)

Tab. 22.3 (continued)

Field of impact	Impact chains not quantified
Catastrophe management	Disaster relief forces
	Volunteer relief labour
	Storm events
	Droughts
Tourism	Change in water and energy demand
	Change in environmental resources important for tourism extreme events (and business interruption)

Note: For a full list of impact chains not quantified see Chap. 21, Table 21.1

At the European scale, Levermann et al. (2012) assessed the potential transitions of six climatic subsystems (tipping elements) which would have large-scale impacts on Europe. Two of these relate to major ice sheets (West Antarctic and Greenland) and thus are of low relevance for Austria (at least directly—the indirect effects could still be important). The others—Atlantic Ocean circulation, Arctic stratospheric ozone, Arctic sea ice and Alpine glaciers—are more relevant, especially the latter two due to the impacts on Austria and them being potentially triggered already at relatively low warming levels.

These provide some examples of major discontinuities that strengthen the policy argument for mitigation, but they do not provide quantitative evaluations. To address this, the study has undertaken some analysis on extreme event distribution for Austria. These events are lower in scale than the major tipping points or fat tails above, but provide examples of the importance of capturing the distribution as well as the average impacts, when considering economic impacts. Three of the most relevant fields of impact are considered: drought-induced harvest damage in agriculture, premature deaths due to prolonged heat waves, and riverine flooding damage to buildings. See Chaps. 8, 11 and 18, respectively, for more details.

In agriculture, meteorological and agricultural droughts have been identified as major driver behind inter-annual yield variability in Central Europe (Hlavinka et al. 2009) and global food insecurity (IPCC 2014b, p. 37). For instance, the European drought and heat wave in 2003 affected a third of the EU and caused economic damage valued at around 13 billion euros (Tubiello et al. 2007). In 2013, Central Europe was hit by a severe summer drought and heat wave with negative impacts on crop harvests. In Austria, corn yields were 19 % below the previous year's production and 18 % below the 10 year average, as reported by Statistics Austria (2014). Due to climate change, drought conditions could potentially become more important in the future, which in some cases (and in the absence of adaptation) could lead to significant crop production losses (Olesen et al. 2011; IPCC 2014b, p. 30), though there is high uncertainty over these projections. A reference scenario (S1) and two drought scenarios (S2 and S3) for the period up to 2040 (Strauss et al. 2013; combining a dry day index with block-bootstrapping based on historical daily weather data for the period 1975–2007) were applied in order to assess the harvest implications for four crops, namely grain maize, winter wheat, winter rapeseed, and soybean considering three fertilisation intensities. Together, these crops represent 81 % of Austrian cropland. Drought anomalies during the growing seasons of grain maize and soybean as indicated by S3 occur every 10 years in the reference scenario S1. Compared to the ensemble of 31 climate models, these anomalies are expected to occur every 3 years in 2050. For the economic analysis, we assume an equal weight of the fertilisation intensity levels across Austria and similar crop shares as in the past. Real prices are based on the OECD-FAO (2013) projections whereas variable production costs are not taken into account. The results show a decreasing mean annual agricultural production value of about $€_{2010}$ 56 million in S2 and of about $€_{2010}$ 137 million in S3.

For human health, heat waves are of considerable importance. Following Kysely (2004) heat waves are defined here (for the analysis in Austria) as consecutive periods of at least 3 days during which the daily maximum/minimum temperature is ≥30 °C/20 °C ("Kyselý days"). The heat wave is said to persist as long as the maximum temperature of each following day does not fall below 25 °C and the mean temperature maximum during the whole period does not fall below 30 °C (Auer and Korus 2005). To estimate the effect of extremely hot years, those with a return period of 20 years (95th percentile) for a mid-range climate change scenario were selected. In such hot years the number of Kyselý days increases to 77 for the period 2036–2065 (the figure of the expected (medium) materialisation was 8–27). Evaluation using value of a statistical life, reveals that the economic cost—connected with a doubling of heat related premature deaths in these extreme years—increases to at least $€_{2010}$ 10.6 billion. Should climate change turn out even stronger (i.e. using a hotter than the mid-range climate scenarios) this number rises to $€_{2010}$ 14 billion in the high-range climate scenario. It should be stressed that these estimates account only for a higher number of heat days, not however for their intensity nor other stress increases by them being prolonged. Each of these changes are expected to lead to even more severe effects (D'Ippoliti et al. 2010; WHO and WMO 2012) beyond those quantified here.

Finally, the study has analysed riverine flooding with respect to an event with an average recurrence interval of 100 years (the 99 ‰). The associated damage cost of this extreme year was found to amount to $€_{2010}$ 6.9 billion for residential homes (only HORA method, see Chap. 18 for further details). An even broader sensitivity analysis (low and strong climate change, socioeconomic developments that diminish as well as enhance damage) was also analysed with respect to the end of the century (2071–2100). This found that at the end of the century, there is a 5 % likelihood that the annual cost of damage will be in the range of $€_{2010}$ 2.8–15 billion.

For a likelihood of 1 %, riverine flooding damage costs are in the range of $€_{2010}$ 8 to over 40 billion.[6] (To give some reference level: 1 % is still at least 10 times

[6] Chapter 18 supplies further details, see in particular Table 18.1.

more likely than an individual Austrian home burning down, for which the scale of potential damage level also is substantially lower).

These three case studies provide succinct examples that the tails of the distribution for climate change are important, and should be considered alongside any central estimates. The example of river floods also highlights that a strong increase in damages can arise when uncertainties are combined, e.g. the range of socio-economic and climate uncertainty leads to a range from 3 to 15 billion (a factor of five)—driven broadly equally from the socio-economic and climate elements.

These case studies—and their implications—also have high relevance for policy. Climate change has a high potential to increase the frequency and intensity of these types of extreme events (e.g. the three case studies above), thus a focus only on central trends is likely to miss the importance of larger and more frequent extremes. A policy maker is likely to be highly interested in the extreme event tails, not least because events of this scale have high political as well as social/economic consequences.

Perhaps one of the most poignant critiques of the standard approach comes from Weitzman (2009, p. 18):

> Perhaps in the end the climate-change economist can help most by *not* presenting a cost-benefit estimate for what is inherently a fat-tailed situation with potentially unlimited downside exposure as if it is accurate and objective [. . .] but instead by stressing somewhat more openly the fact that such an estimate might conceivably be arbitrarily inaccurate depending on what is subjectively assumed about the high-temperature damages function along with assumptions about the fatness of the tails and/or where they have been cut off. Even just acknowledging more openly the incredible magnitude of the deep structural uncertainties that are involved in climate-change analysis [. . .] might go a long way toward elevating the level of public discourse concerning what to do about global warming.

22.4 Conclusions

Climate change is a global, long-term challenge, with an enormous degree of uncertainty. In the present book, and in summary form in this chapter, we have identified what we know about the implications of climate change at the national level, exploring the impacts for one country, Austria, in detail. While the book as a whole offers a useful set of tools for devising a comprehensive and consistent approach to deriving the costs of climate change at the national level, we now focus here on the type of results to be expected from such an undertaking.

There is, first, the climate and weather induced damage currently observed. Insurance companies and national relief funds are key suppliers of some of this information, at least in terms of direct damage costs of extreme events (for Austria, related costs have exceeded 1 % of GDP in some years recently, the annual average figure to date amounts to around 0.25 % of GDP, or about 700 million euros, rising to 1 billion euros if average heat related mortality is added). Second, we employ a rich array of sectoral climate impact models to determine future weather and climate induced damage triggered by both additional climate change and socioeconomic development. We then merge the results in a cross-sectoral macroeconomic analysis (we use the CGE approach here), and non-market damage such as the costs related to future premature heat-related deaths are also added. Again, using Austria as an example, the analysis reveals that the cost of damage with respect to a 'medium climate and reference socioeconomic development' scenario will more than double by the 2030s and grow four to fivefold by the 2050s. And these figures only include the impact chains that can be quantified by available model chains. However, the range of uncertainty around these numbers is large—as an indication—typically a factor of two for each of the socioeconomic and climate dimensions.

Moreover, these estimates are the result of a standard economic analysis framework, which tends to focus on central estimates. This chapter highlights that it is as important to consider the extreme values, especially given the increase in frequency and intensity of many climate extremes with climate change. The analysis highlights the non-linear increase that can potentially arise, even in current '1 in 20 year events', and how these could lead to extremely large economic costs which have far-reaching consequences. It is therefore considered important to present this information alongside the central estimates.

Acknowledgement The authors thank Hermine Mitter for a revision of the agricultural drought information and Birgit Bednar-Friedl for extremely helpful comments on an earlier manuscript of this chapter.

References

Ackerman F, Stanton EA (2011) Climate economics: the state of the art. Stockholm Environment Institute, Somerville, MA

Anthoff D, Tol RSJ (2010) The climate framework for uncertainty, negotiation and distribution (FUND), technical description. Version 3.5. http://www.fnu.zmaw.de/

APCC (Austrian Panel on Climate Change) (2014) Austrian Assessment Report 2014, Summary for policy makers [Österreichischer Sachstandsbericht Klimawandel]. Austrian Academy of Sciences, Vienna

Auer I, Korus E (2005) The variability of heat waves and dry spells in the flat and mountainous regions of Austria, Final report. StartClim 2005: 604–607. StartClim, Vienna

Ciscar J-C, Iglesias A, Feyen L, Szabó L, Van Regemorter D, Amelung B, Nicholls R, Watkiss P, Christensen OB, Dankers R, Garrote L, Goodess CM, Hunt A, Moreno A, Richards J, Soria A (2011) Physical and economic consequences of climate change in Europe. Proc Natl Acad Sci USA 108(7):2678–2683

D'Ippoliti D et al (2010) The impact of heat waves on mortality in 9 European cities: results from the EuroHEAT project. Environ Health 9:37

Hallegatte S, Hourcade J-C, Dumas P (2007) Why economic dynamics matter in assessing climate change damages: illustration on extreme events. Ecol Econ 62(2):330–340

Hlavinka P, Trnka M, Semerádová D, Dubrovský M, Zalud Z, Mozný M (2009) Effect of drought on yield variability of key crops in Czech Republic. Agr Forest Meteorol 149:431–442

Hope C (2006) The marginal impact of CO_2 from PAGE2002: an integrated assessment model incorporating IPCC's five reasons for concern. Integr Assess 6(1):19–56

IPCC (2013) Climate change 2013: the physical science basis, working group I contribution to the fifth assessment report of the intergovernmental panel on climate change. Oxford University Press, Oxford

IPCC (2014a) Climate change 2014: impacts, adaptation, and vulnerability, summary for policy makers. IPCC and Oxford University Press

IPCC (2014b) Climate change 2014: impacts, adaptation and vulnerability, fifth assessment report, working group II, Chap. 7: food security and food production systems. IPCC

IPCC (2014c) Climate change 2014: impacts, adaptation and vulnerability, fifth assessment report, working group II, Chap. 16: adaptation opportunities, constraints, and limits. IPCC

Jacob D et al (2013) EURO-CORDEX: new high-resolution climate change projections for European impact research. Reg Environ Chang 14(2):563–578. doi:10.1007/s10113-013-0499-2

König M, Loibl W, Steiger R, Aspöck H, Bednar-Friedl B, Brunner KM, Haas W, Höferl KM, Huttenlau M, Walochnik J, Weisz U (2014) Climate change impacts on the anthroposphere [Österreichischer Sachstandsbericht Klimawandel 2014 (AAR14)]. Austrian Panel on Climate Change (APCC), Austrian Academy of Sciences, Vienna, pp 641–704

Kysely J (2004) Mortality and displaced mortality during heat waves in the Czech Republic. Int J Biometeorol 49:91–97

Lenton TM, Held H, Kriegler E, Hall JW, Lucht W, Rahmstorf S, Schellnhuber HJ (2008) Tipping elements in the Earth's climate system. Proc Natl Acad Sci USA 105(6):1786–1793

Levermann A, Bamber JL, Drijfhout S, Ganopolski A, Haeberli W, Harris NRP, Huss M, Krüger K, Lenton V, Lindsay RW, Notz D, Wadhams P, Weber S (2012) Potential climatic transitions with profound impact on Europe: review of the current state of six 'tipping elements of the climate system'. Clim Change 110:845–878. doi:10.1007/s10584-011-0126-5

Munich Re (2014) NatCatSERVICE [WWW Document]. http://www.munichre.com/de/reinsurance/business/non-life/nat-catservice/index.html. Accessed 10 Jun 2014

Nordhaus WD (1991) To slow or not to slow: the economics of the greenhouse effect. Econ J 101(407):920–937

Nordhaus WD (2011) Estimates of the social cost of carbon: background and results from the RICE-2011 model. National bureau of economic research working paper 17540

OECD-FAO Organisation for Economic Co-operation and Development-Food and Agriculture Organization of the United Nations (2013) OECD-FAO Agricultural Outlook 2013–2022. OECD Publishing, Paris

Olesen JE, Trnka M, Kersebaum KC et al (2011) Impacts and adaptation to European crop production systems to climate change. Eur J Agron 34:96–112

Pindyck RS (2013) Climate change policy: what do the models tell us? J Econ Lit 51(3):860–872

Ruth M, Coelho D, Karetnikov D (2007) The US economic impacts of climate change and the costs of inaction. University of Maryland, College Park, MD

Statistics Austria (2014) Crop production and production of permanent grasslands 2013. Statistics Austria, Vienna

Stern N (2007) The economics of climate change: the Stern review. Cambridge University Press, Cambridge

Stern N (2013) The structure of economic modeling of the potential impacts of climate change: grafting gross underestimation of risk onto already narrow science models. J Econ Lit 51(3): 838–859

Strauss F, Moltchanova E, Schmid E (2013) Spatially explicit modeling of long-term drought impacts on Austrian crop production. Am J Clim Change 2:1–11

Tol R (2002a) Estimates of the damage costs of climate change, Part I: benchmark estimates. Environ Resour Econ 21(1):47–73

Tol R (2002b) Estimates of the damage costs of climate change, Part II: dynamic estimates. Environ Resour Econ 21(2):135–160

Tubiello FN, Amthor JS, Boote KJ et al (2007) Crop response to elevated CO2 and world food supply: a comment on "Food for Thought..." by Long et al. Science 312:1918–1921, 2006.

UNFCCC (2010) The Cancun agreements. United Nations framework convention on climate change. http://unfccc.int/meetings/cancun_nov_2010/meeting/6266.php

Wagner G, Weitzman ML (2015) Climate shock: the economics consequences of a hotter planet. Princeton University Press, New Jersey, NJ

Warren R, Benzie M, Arnell N, Nicholls R, Hope C, Klein R, Watkiss P (2013) Scoping study: modelling the interaction between mitigation and adaptation for decision making. Working paper for the AVOID project: avoiding dangerous climate change. http://www.metoffice.gov.uk/media/pdf/t/9/AVOID_WS2_D1_39.pdf

Watkiss P (2011a) Aggregate economic measures of climate change damages: explaining the differences and implications. Wiley interdisciplinary reviews—climate change 2(3):356. Published online, 2 May 2011

Watkiss P (ed) (2011b) The ClimateCost project. Final report, vol 1: Europe. Stockholm Environment Institute, Stockholm

Watkiss P, Hunt A (2010) Review of adaptation costs and benefits estimates in Europe for the European Environment State and Outlook report 2010. Technical report prepared for the European Environmental Agency for the European Environment State and Outlook report 2010, Copenhagen, Denmark. March 2010. Published online by the European Environment Agency, Copenhagen

Weitzman ML (2009) On modeling and interpreting the economics of catastrophic climate change. Rev Econ Stat 91(1):1–19

Weitzman ML (2012) GHG targets as insurance against catastrophic climate damages. J Public Econ Theory 14(2):221–244

WHO and WMO (2012) Atlas of health and climate change. World Health Organization and World Meteorological Organization: WMO-No 1-98. WHO Press, Geneva

List of Authors

Ivonne Anders Scientist at ZAMG, the Central Institute for Meteorology and Geodynamics in Vienna. She studied Cartography and Software Technology at the Dresden University of Technology and is going to finish her doctorate in Meteorology at the University of Hamburg in 2014. Key activities: regional climate modelling, statistical analysis, climate change in mountainous regions.

Gabriel Bachner Researcher at the Wegener Center for Climate and Global Change at the University of Graz. He holds a Master in "Environmental System Sciences" with major in Economics from the University of Graz. His fields of research are the economics of climate change, impacts and adaptation, computable general equilibrium modelling (CGE) as well as greenhouse gas mitigation in the transport sector.

Birgit Bednar-Friedl Associate Professor at Department of Economics, University of Graz, and Wegener Center for Climate and Global Change. Her current research interests span the economics of climate change, resource and energy economics, and international economics. Her most recent research covers several aspects of unilateral and multilateral climate policy, international burden sharing in climate policy, the macroeconomic costs of climate change impacts on various fields such as agriculture, electricity, tourism and transport, and the role of harvest costs for resource use in intertemporal general equilibrium.

Herbert Formayer Assistant Professor and head of the climate working group at the Institute of Meteorology, BOKU-University, Vienna. He studied meteorology at the University Vienna and holds a Ph.D. from BOKU. The main research topics are regional climate modelling, applied climate analyses and integrated climate impact assessments. He is experienced in a number of inter- and trans-disciplinary projects on the national and international level.

Martin Götzl Employed at the Environment Agency Austria. He studied biology at the University of Vienna. Focus of his work: the assessment, quantification and mapping of ecosystem services and ecosystem service trade-offs. He is author of the

K.W. Steininger et al. (eds.), *Economic Evaluation of Climate Change Impacts*,
Springer Climate, DOI 10.1007/978-3-319-12457-5

Austrian inventory on agricultural ecosystem services. Additionally, he is national coordinator of the Global Biodiversity Information Facility in Austria and head of the Austrian GBIF Participant Node.

Willi Haas Senior researcher and lecturer at the Institute of Social Ecology, part of the Faculty of Interdisciplinary Studies/Alpen-Adria University. He is interested in the complex relations between societal dynamics and environmental change. Further, he explores how analytical knowledge on transitions can be used to guide societal developments onto more sustainable pathways. In this context he focuses on material, energy, time use and health at various levels of scale.

Stefan Hochrainer-Stigler Senior Research Scholar with the Risk, Policy and Vulnerability Program at the International Institute for Applied Systems Analysis (IIASA). He is a member of the Board of Directors of the Integrated Disaster Risk Management (IDRiM) Society and a lecturer at the University of Vienna, the Karlsruhe Institute of Technology and the UME Graduate School at the Institute for Advanced Study in Pavia. His main research interests include risk management of extreme events, stochastic modelling of rare events, extreme value theory, dependency of risks, econometrics and multivariate analysis.

Marcus Hummel Research assistant at Vienna University of Technology, Energy Economics Group. He is working on national and international research projects and takes care of project acquisition and coordination. His research includes efficient and renewable energy systems, energy economics and policies in industry as well as for space heating and cooling. He focuses on developing sound political measures for the transition to low-carbon energy systems.

Robert Jandl Forest ecologist and soil scientist, coordinator of climate change research activities at the Austrian Forest Research Center. Special interest in forest ecosystem services.

Claudia Kettner Researcher at the Austrian Institute of Economic Research (WIFO). She holds a Master in Economics from University of Graz and a Master in Renewable Energies from Vienna University of Technology. Key areas of research: EU and Austrian climate and energy policy, transformation of the energy system and sustainable development.

Judith Köberl Research scientist in the research group Regional Science, Risk and Resource Economics at the Institute for Economic and Innovation Research of JOANNEUM RESEARCH. She studied Environmental System Sciences (Subject Focus Economics) at the University of Graz. Main research topics: the economics of weather and climate risks, climate change impact assessment, predictive analytics, and tourism and recreation research.

Katharina Köberl Research assistant at the Austrian Institute of Economic Research (WIFO). She holds a Master in Political Science from University of Vienna and a Master in Environmental Science from University Hagen and Fraunhofer Institute.

Martin König Senior expert at the Environment Agency Austria. He has coordinated activities on setting up national CCIVA research programs, initialised and coordinated EU projects for better cooperation of national climate research programs throughout Europe (CIRCLE) and is currently involved in adaptation projects providing background, instruments and DSS for different sectors and scales—from regional via national to European. For him, his involvement in various climate impact studies is one important base for meaningful policy consulting on response strategies.

Angela Köppl Researcher at the Austrian Institute of Economic Research (WIFO) since 1992. She holds a Doctor's degree in Economics from the University of Vienna. From 1985 to 1987 post graduate student in economics at the Institute for Advanced Studies. In 2002 she was visiting scholar at the MIT. Key areas of research are economics of climate change and climate policy, transformation of the energy system and incentive based instruments in environmental economics.

Dominik Kortschak Research scientist in the research group Regional Science, Risk and Resource Economics at the Institute for Economic and Innovation Research of JOANNEUM RESEARCH. He holds a doctoral degree in technical mathematics of the Graz University of Technology. During his doctoral thesis work he was a research scientist at the Johann Radon Institute for Computational and Applied Mathematics (RICAM) of the Austrian Academy of Science (ÖAW). Thereafter he worked as a research scientists at the Department of Actuarial Science (DAS), Faculty of Business and Economics (HEC), University of Lausanne and the Institut de Science Financière et d'Assurances (ISFA) of the University Lyon 1 Claude Bernard. His research interests are risk modelling and the economics of weather and climate risks.

Lukas Kranzl Senior researcher at the Institute of Energy Systems and Electrical Drives at Vienna University of Technology. He studied mechanical engineering and wrote his Ph.D. thesis about the macro-economic impact of bioenergy systems. His research activities focus on future perspectives of sustainable energy systems with a focus on building related energy demand and supply, scenario development and analysis of policy instruments for promoting renewable and efficient energy systems.

Markus Leitner Scientist at the Environment Agency Austria. He holds an MSc in Landscape Ecology and Landscape Planning and is specialist for Environmental Impact Assessment (EIA), Strategic Environmental Assessment (SEA) and climate change adaptation. His main fields of expertise range from project coordination and management to research management and development like the FP6 project CIRCLE and the FP7 project CIRCLE-2 in the field of Climate change impacts, vulnerability and adaptation in Europe. He contributed with his climate change adaptation expertise to diverse project like CcTalK! Envisage-CC, adapt2to4, COIN, ARISE. C3-Alps, ETC/CCA and the EU Adaptation Strategy. Since 2012, he is lecturer at BOKU on Applied EIA.

Manfred J Lexer Associate Professor (Silviculture and Vegetation Modelling) at the Institute of Silviculture, BOKU. Main research interests are understanding and modelling forest vegetation dynamics, adaptive forest management, ecosystem services and planning and decision making.

Wolfgang Loibl Senior scientist and deputy head of the business unit Sustainable Buildings and Cities of Energy Department at the AIT—Austrian Institute of Technology. He is coordinating the climate change impact assessment and adaptation research activities. Since a decade he was and still is coordinating projects and work packages at the national level (future.scapes, reclip:century, UFT-ADI) and the EU level (SENSOR, PLUREL, SPARD, MANFRED, EU-Lakes, UrbanAPI, TRANSFORM) dealing with spatial modelling tasks for urban and regional development, for climate impact assessment and climate change adaptation with special interest in urban climate topics.

Philipp Ernst Maier Scientific project assistant at the Institute of Social Ecology at the University of Klagenfurt. In 2014 he is finishing his master's degree at the Institute of Social Ecology and studied Social and Cultural Anthropology (BA) at the University of Vienna.

Reinhard Mechler Deputy director of the 'Risk, Policy, Vulnerability' research program at the International Institute for Applied Systems Analysis (IIASA) and a senior lecturer at the University for Economics and Business in Vienna. He has more than 15 years of experience working on the economics of disaster risk, risk modelling and climate change. Specific interests of his include catastrophe risk modelling, the impacts of extreme events and climate change on development, the use of novel risk financing mechanisms for sharing disaster risks, and the assessment of the efficiency and equity of climate change response measures.

Klemens Mechtler Responsible for plant variety and crop production issues in oil crops at AGES Institute for Sustainable Plant Production. He studied agriculture at the University of Agricultural Sciences at Vienna. Current working areas: Austrian national list trials for variety registration in oil crops, project work related to variety aspects for improvement of quality of agricultural products, food security and crop production potentials, also in context with climatic change.

Ina Meyer Research associate at the Austrian Institute of Economic Research (WIFO) in the area environment, agriculture and energy. Her research interests lie in climate change and energy economics with a focus on sustainability in the transportation and agricultural sectors. She studied economics at the Freie Universität Berlin and received her Ph.D. from the University of Potsdam in collaboration with the Potsdam Institute of Climate Impact Research.

Hermine Mitter Doctoral researcher at the Institute for Sustainable Economic Development and in the Doctoral School of Sustainable Development at the University of Natural Resources and Life Sciences. She holds a master in landscape planning and a master in agricultural sciences. Her main research areas are the analysis of climate change adaptation measures with statistical techniques. Furthermore, she is interested in transdisciplinary research processes.

Andreas Müller Senior researcher at Energy Economics Group at the Vienna University of Technology. He holds a degree in industrial engineering at the Vienna University of Technology. His research areas are modelling of energy systems with focus on the built environment, analyses of low temperature energy demand, energy efficiency, and analysis of medium to long term competiveness and potentials of renewable energy carriers.

Stefan Nabernegg Junior Scientist at the Wegener Center for Climate and Global Change at the University of Graz. He studied Economics as well as Environmental System Sciences with major in Business Administration at the University of Graz. His field of research is the economics of climate change and computable general equilibrium modelling (CGE).

Imran Nadeem Ph.D. student at Institute of Meteorology, BOKU-University, Vienna. He holds a Master degree in Physics from Quaid-i-Azam University, Islamabad, Pakistan. He has worked on several national and international projects including EU funded FP6 project CECILIA. He is an expert in high resolution climate modelling using sub-grid scale approach to achieve high resolution of up to 1 km. He has also experience with running regional climate models with different initial and lateral boundary conditions.

Roman Neunteufel Senior scientist, lecturer, and project manager with the University of Natural Resources and Life Sciences. Since 2003 Dr. Neunteufel is active within the expert committee of the Austrian Association for Gas and Water (sector: economy; working group: benchmarking) and since 2010 member of the Academic Senate of the University. Dr. Neunteufel holds a M.Sc. degree in Land and Water Management and Engineering and a Ph.D. related to benchmarking in the water supply sector from University of Natural Resources and Life Sciences, Vienna.

Ivo Offenthaler A forest ecologist and plant physiologist by training, Ivo Offenthaler has worked several years in environmental monitoring including human biomonitoring and health before moving professional focus to climate change adaptation. Across topics, his approach relies heavily on computer-assisted data analysis and visualisation.

Reinhard Perfler Senior Lecturer and Senior Scientist at the University of Natural Resources and Life Sciences in Vienna. He is Deputy Head of the Institute of Sanitary Engineering and Water Pollution Control and coordinates the "Freshwater/ Drinking Water Working Group". His actual working focus is on water supply including water treatment, sustainable water management, risk and quality management and water safety planning.

Franz Prettenthaler Head of the research group Regional Science, Risk and Resource Economics at the Institute for Economic and Innovation Research of JOANNEUM RESEARCH. He lectures insurance economics (Graz University of

Technology, A) and holds degrees in economics and environmental systems sciences (Graz), public economics (Paris X and Cergy) and philosophy (St Andrews). He pioneered the analysis of the national risk transfer mechanism in Austria and is engaged in risk quantification and mechanism design issues in the context of natural hazards and climate change with a focus on Europe and a high spatial resolution.

Irene Schicker Scientist at the Devision Data, Methods, Models at Central Institute for Meteorology and Geodynamics. Before, Institute of Meteorology at the University of Natural Resources and Life Sciences, Vienna. Study of Meteorology at the University Innsbruck. Research area: high resolution and very high resolution meteorological and climate model in mountainous regions, land-atmosphere interactions, boundary layer meteorology, environmental and micro-meteorology, climate impact research, glaciology.

Thomas Schinko Researcher at the Wegener Center for Climate and Global Change. Research assistant at the International Institute for Applied Systems Analysis (IIASA). He studied environmental system sciences with a major in economics at the University of Graz and Uppsala University. His main fields of scientific interest include the economics, ethics and (perceived) risks of energy and climate change mitigation and adaptation strategies at the national and international level.

Erwin Schmid Professor for Sustainable Land Use and Global Change at BOKU University of Natural Resources and Life Sciences, Vienna. He studied agricultural economics at BOKU and achieved a Venia Docendi in Agricultural and Resource Economics. Main research interest is the methodological development of integrated models in the fields of agricultural economics and agronomy.

Fabian Scholz Scientific project assistant at the Institute of Social Ecology at the University of Klagenfurt. He is currently finishing his master's degree at the Institute of Social Ecology and studied Social and Cultural Anthropology (BA) at the University of Vienna and Social Work (MA) at the University of Applied Science Vienna (Fh-Campus Wien).

Martin Schönhart Scientist at the Institute for Sustainable Economic Development, BOKU University of Natural Resources and Life Sciences, Vienna. He studied agricultural sciences at BOKU. Main research interests are the development of integrated land use models for evaluation of agricultural policies and integrated assessments of land use strategies particularly in the field of climate change adaptation.

Reimund Schwarze Head of the research group on climate change and senior researcher at the Helmholtz Center for Environmental Research in Leipzig and Professor for Environmental Economics at the University of Frankfurt/Oder. He graduated in the field of economics at the University of Berlin and worked for the Environmental Protection Agency of Germany in Berlin. He has served as an environmental expert for the German Parliament and the German Ministry of Environmental Affairs and is acting as scientific advisor to the German Committee for Disaster Prevention of UN-ISDR. He has coordinated the EU FP7 project ConHaz on Costing of Natural Hazards.

Franz Sinabell Senior researcher at WIFO, the Austrian Institute of Economic Research and lecturer on economics and agricultural policy at BOKU—University of Natural Resources and Life Sciences. His work includes analyses of the European agricultural policy, climate change, market risks and natural hazards.

Karl W. Steininger Professor at the Department of Economics and head of the socioeconomic research group at the Wegener Center for Climate and Global Change, both at the University of Graz, Austria. He studied Economics and Computer Science at the University of Vienna and UC Berkeley, and specialised in environmental and climate economics, and in international trade. He is lecturer at the Vienna based University of Life Sciences, and Research Professor at the German Institute of Economic Research (DIW), Berlin. Previously he held positions in the World Bank (Environment Department) and at the University of Triest, Italy.

Agne Toleikyte Research associate at the Institute of Energy Systems and Electrical Drives at Vienna University of Technology. She holds a degree in Management of Environment and Bio Resources at the University of Natural Resources and Life Sciences in Vienna. Her research fields include energy efficiency in buildings, evaluation of policy measures and instrument impacts as well as assessment of economic feasibility of renovation actions.

Gerhard Totschnig Senior researcher at the Institute of Energy Systems and Electrical Drives at Vienna University of Technology. He received his Ph.D. in chemical engineering, fuel, and environmental technology in 2002 from TUWIEN and studied theoretical physics at the University of Vienna. His main research interests are the modelling of the optimal integration of renewables into the power system, optimal investment planning strategies, uncertainty management with optimization models and the interplay between demand side, supply side and transmission options for RES-E integration.

Herwig Urban Researcher at the Wegener Centre for Climate and Global Change at the University of Graz. He studied Economics as well as Environmental System Sciences with a major in Sustainability-oriented Management at the University of Graz. His fields of research are the economics of climate change and computable general equilibrium modelling (CGE) as well as the evaluation of methods for assessing the feasibility of landfill mining projects.

Gernot Wagner He serves as lead senior economist at the Environmental Defense Fund, teaches as adjunct associate professor at Columbia University's School of International and Public Affairs, and is the co-author, with Harvard's Martin L. Weitzman, of *Climate Shock: the economic consequences of a hotter planet* (Princeton University Press, 2015). For more see: www.gwagner.com

Paul Watkiss Independent researcher (Paul Watkiss Associates) specialising in multi-disciplinary research and policy support in the area of climate change impacts and adaptation, as well as a Senior Visiting Research Associate at the ECI, University of Oxford.

Ulli Weisz Senior staff member, researcher and lecturer at the Institute of Social Ecology Vienna. She is a qualified nurse, studied ecology/socio-economy at the University of Vienna and worked at the Ludwig Boltzmann Institute for the Sociology of Health and Medicine, Vienna. Her research foci are: sustainable development and health, health co-benefits of climate mitigation strategies, and transdisciplinary research.

Brigitte Wolkinger Researcher at the Wegener Center for Climate and Global Change at the University of Graz. She studied in the Master Program 'Environmental and System Sciences' with major Economics at the University of Graz. Her fields of research cover topics of environmental economics and econometrics like greenhouse gas mitigation policies on the regional level and adaptation to climate change.

Klaus Peter Zulka Senior expert at Environment Agency Austria (EEA) in Vienna. He studied zoology in Stuttgart-Hohenheim and Vienna. He has worked on species conservation at EEA since 2000 and as a lecturer for conservation biology from 1997 to 2003 and since 2013. He was responsible for the concept, the coordination and the editing of the Austrian Red Lists of threatened animals. Besides, he has worked on Environmental Impact Assessment, biodiversity effects of renewable energies and species richness in agricultural landscapes.

MIX
Papier aus verantwortungsvollen Quellen
Paper from responsible sources
FSC® C105338

If you have any concerns about our products,
you can contact us on
ProductSafety@springernature.com

In case Publisher is established outside the EU,
the EU authorized representative is:
Springer Nature Customer Service Center GmbH
Europaplatz 3, 69115 Heidelberg, Germany

Printed by Libri Plureos GmbH
in Hamburg, Germany